Biology of the Arctic Charr

Biology of the Arctic Charr: Proceedings of the International Symposium on Arctic Charr

Edited by

LIONEL JOHNSON AND BONNIE BURNS

UNIVERSITY OF MANITOBA PRESS

Printed in Canada

Photographs following page xiv courtesy of Laurie Tait.

All other photographs courtesy of the authors.

Cover design by Ken Armistead.

Canadian Cataloguing in Publication Data

International Symposium on Arctic Charr (1981 : Winnipeg, Man.)
Biology of the Arctic charr

ISBN 0-88755-137-8

1. Arctic charr — Congresses. I. Johnson, Lionel, 1924- II. Burns, Bonnie, 1959- III. Title.
QL638.S2I58 1981 597′.55 C85-091126-5

Contents

Preface

The International Symposium on Arctic Charr was held in Winnipeg between the 4 - 8 May 1981, on the campus of the University of Manitoba. In this setting representatives from most of the circumpolar countries graced by Arctic charr and its close relatives, that together form the *Salvelinus alpinus* complex, gathered together to probe the intellectual boundaries of its niche. The truly wide-ranging distribution of the charr may be judged from the countries of origin of the delegates: Austria, Great Britain, Iceland, Japan, Norway, Sweden, Switzerland, the United States of America, West Germany and Canada.

Arctic charr is the freshwater and anadromous species that lives closest to the North Pole, occupying the fresh waters of the most northerly lands far beyond the limits attained by other freshwater fish species. In the most northerly latitudes some populations of Arctic charr live, feed and breed solely in fresh water; other populations spend part of their life cycle at sea. These anadromous charr spend between four and eight years in fresh water before making their first trip to the sea where they feed on the relatively abundant fauna of the Arctic littoral and sub-littoral regions that develops for a short time during the brief openwater season. They then visit the sea each year for about six weeks except in the year of spawning when they either stay all summer in the fresh water or return early from the sea. Spawning in these latitudes takes place at a maximum of once every two or three years. The charr live to a considerable age and attain great size. Charr with 36 rings on their otoliths have been recorded but this may under-represent the real age. In these northern conditions, charr have few predators and may be one species that regularly lives out its physiological life-span.

In more southerly latitudes, generally south of approximately 60°N, anadromy no longer takes place, but many non-migratory, so-called ''land-locked'' stocks, are found. These stocks exist throughout Sweden, Northern Russia, Finland,

Great Britain, and Japan, with outlying populations in the lakes of the European Alps, and a small concentration in the headwaters of the Lena and Angara Rivers in the region around Lake Baikal in central Asia. In all these regions the charr appears to occupy the 'leaner' environments where only a physiological and ecological generalist can survive through utilization of all or any resources. In North America, Arctic charr has not penetrated to the deeper waters of lakes in the central land mass as has occurred in Europe; in these waters its place has been taken by its relative the lake trout or lake charr, *Salvelinus namaycush*.

It is only in comparatively recent times that the intensive investigation of the northern anadromous stocks has been undertaken. This is not surprising in view of the difficulties of travel in the Arctic prior to the development of air transportation. Before World War II much of the information on charr was confined to the more southerly regions where it formed an intriguing but rather odd species without general interest because of its apparent local distribution.

As knowledge from all regions began to accumulate it became apparent that it would be beneficial for all those interested in the species to gather and exchange information. This was stimulated further by the publication of "Charrs: salmonid fishes of the genus *Salvelinus*" edited by E.K. Balon (Dr. W. Junk, The Hague). It was recognized that the genus *Salvelinus*, especially *Salvelinus alpinus*, had a great deal to contribute to biology. Accordingly events were put in train and the Symposium at Winnipeg resulted.

The concept for this Symposium was to bring together all those interested: research workers, managers, processors and those dependent on the stocks for part of their livelihood. For this reason there were no concurrent sessions and no restrictions on subject matter beyond the fact that it must be centered on Arctic charr or its close relatives, Dolly Varden, *Salvelinus malma*, and white-spotted charr, *Salvelinus leucomaenis*. The conference thus maintained a very generalist approach, in line with the characteristics of the species concerned. The content of the Symposium was therefore what reviewers of proceedings usually describe as "uneven". Nevertheless the mixture worked at the practical level and the exchange and interaction was enthusiastic, and I believe that the resulting Proceedings provide a unique compendium on a single species.

Arctic charr emerges as a species of great theoretical significance as well as being of great practical importance. Few species can contribute more to an understanding of sympatric and allopatric speciation. The variability of the stocks that occur together in many lakes immediately raises the question of their origin; did they originate in the lakes in which they are found today or did successive waves of different species migrate at different times from different centres of origin?

These different stocks make the taxonomy of the species exceedingly difficult, and boldness and imagination are needed to devise formal relationships that are likely to withstand the assaults of future workers. Different views must be resolved; the rewards for their resolution are scientifically great.

The physiology of the Arctic charr is also of great signficance: how do they make a living in the lean environments they inhabit? What physiological adaptations do they have with respect to sense organs or metabolism, that allows them to survive in what appears to be some of the most depauperate environments in the world? How do they manage to escape predation in the sea where large mammalian predators, seals, white whales and narwhal, are relatively abundant?

Arctic charr also provide unique opportunities for the study of populations since they are the only fish species in many Arctic lakes and these lakes have not been touched by man since their formation in the Late Pleistocene. These lakes represent the closest approach to a laboratory culture that is likely to be found in nature.

The parasites of charr, relatively unknown twenty years ago, have proved exceedingly interesting. The effect of anadromy on the parasitic fauna can be compared with sympatric stocks that do not go to sea, and the various parasites in different stocks have contributed significantly to understanding post-Pleistocene distribution patterns.

Finally there is the question of the utilization of the species by man. In this respect there was total accord within the symposium participants; there is no area of research where the subject of the investigations can be utilized with greater enjoyment, whether smoked or fresh cooked or frozen, preferably accompanied with a dry white wine. Although recognized as a gourmet food item in many regions, particularly in 18th and 19th century England, where the limited stocks were severely depleted at this time, it is a staple in the diet of the peoples of the far north.

The Laplanders of Fenno-Scandia and the inhabitants of the Alpine regions of Europe are known to have extended the range of charr by transporting it to lakes that previously held no fish. This is an extremely intriguing area as differing views have been advanced as to how certain lakes became populated; a matter of great importance to understanding the present distribution and how the species arrived in its present locations. The discussion continues, but new and exciting findings are meanwhile emerging on both sides of the controversy.

Perhaps the people most dependent on Arctic charr for their livelihood, in former times even more than at present, are the Canadian Inuit, whose representative at the Symposium, Mr. Billy Lyall of Cambridge Bay, Victoria Island gave a graphic account of the relationships, both the past and present, between the Inuit and the charr and the methods in use today as well as those formerly employed.

What more can a species offer? Perhaps some of the most beautiful and peaceful places left on earth in which to carry out investigations, even if the waters are chilly and the snow occasionally flies in early August. It is no wonder that workers on Arctic charr tend to be fanatical; the excitement of the controversy and the flavour of the remoter regions of the earth combine to imbue this species with very special properties. These properties were explored in great depth and much enthusiasm in Winnipeg between the 4-8 May, 1981.

A Symposium of this sort does not emerge without the support and effort of a large number of people. First I thank Dr. G.H. Lawler, Director General, Western Region, Department of Fisheries and Oceans, and Dr. H.E. Welch, Chairman, Department of Zoology, University of Manitoba, for the sponsorship of their respective organizations, their moral support and financial assistance. I also thank most sincerely, all those who so nobly served on committees and in more practical veins: Bob Moshenko, Al Kristofferson, Red Clarke, Eric Marshall, Cathy Austin, Dave Iredale, Eric Gyselman, Lothar Dahlke, Gary Carder, Judy Fredette, Dale McGowan, Mary Smith and our conference technological advisor, Sharon Leonhard, as well as the many more who contributed so generously of their time and effort. I particularly thank Red Clarke and Eric Marshall for their support on the Editorial Committee.

Lionel Johnson
Convenor
Winnipeg, 1983

61

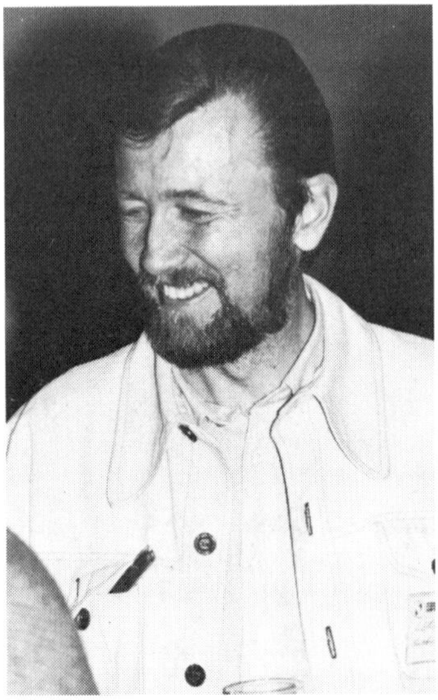

ANADROMOUS MIGRATION

ANADROMOUS MIGRATION

PECHLANER

Biology of the Arctic Charr

Organizing the Diversity of the Arctic Charr Complex

ROBERT J. BEHNKE

Department of Fishery and Wildlife Biology, Colorado State University,
Fort Collins, Colorado, U.S.A.

BEHNKE, R.J. 1984. Organizing the diversity of the Arctic charr complex. p. 3–21. *In* L. Johnson and B.L. Burns [eds.] Biology of the Arctic charr, Proceedings of the International Symposium on Arctic Charr, Winnipeg, Manitoba, May 1981. Univ. Manitoba Press, Winnipeg.

The subgenus *Salvelinus* is divided into the Dolly Varden complex and the Arctic charr complex. The Arctic charr, *S. alpinus*, is viewed to be composed of three major groups with origins associated with Europe, Siberia, and the Arctic Ocean in the Chukotsk Sea-Bering Sea region. *S. leucomaenis*, *S. confluentis*, and *S. albus* are assigned to the Dolly Varden complex. Except for sympatry between *S. malma* and *S. alpinus*, examples of sympatric charr populations almost always involve populations derived from within the geographical area of occurrence and they are very closely related to each other. The origins of most sympatric charr are attributed to divergences during and since the last glacial epoch. Recommendations are given for the preservation and utilization of the diversity present in charrs.

INTRODUCTION

There is something about the beautiful fish known as charr or char that excites the imagination and emotions. The question might be asked why I should write another paper on *Salvelinus* systematics in view of the fact that I so recently gave opinions in a systematic review of the genus (Behnke 1980) published in a large volume of works devoted to charr (Balon 1980). An examination of the works of the various authors in this volume on charrs reveals a range of opinions quite divergent from my position on the evolutionary relationships and classification of the genus. Immediately following my paper, Savvaitova (1980a) discusses how her opinions on charr systematics differ from mine.

A paper by Leggett (1980) discusses the reproductive behaviour of the Dolly Varden charr, *S. malma*, of Kootenay Lake, British Columbia; but, according to the work of Cavender (1980) in the same volume, the Kootenay Lake charr is the "bull trout," *S. confluentus*. Cavender (op. cit.) recognized *S. imbrius* and *S. pluvius* of Japan as valid species whereas in my paper I considered *imbrius* as a synonym of subspecies of *S. leucomaenis* and *pluvius* as a subspecies of *S.*

malma. In this paper I align *pluvius* with *leucomaenis*. Savvaitova (1980b) repeated her stout belief that the Dolly Varden charr, *S. malma*, should be treated as a synonym of Arctic charr, *S. alpinus*, but she discussed ''alpinoid,'' ''high Arctic'', and ''malmoid'' types of charr. Most authors in the charr volume followed the classification given in McPhail and Lindsey (1970) which regarded the common charr in North American waters westward from the Mackenzie River (tributaries to the Beaufort Sea), as *S. alpinus*, whereas Morrow (1980) and I (Behnke 1980) pointed out that this group of charr is identical to *S. malma* of the type locality (Kamchatka). Johnson (1980) presented an excellent and comprehensive review on *S. alpinus* but had to admit that he could only arbitrarily draw taxonomic boundaries for the species because of the problems in separating *S. alpinus* from *S. malma*.

Viktorovsky et al. (1981) described a new species of charr from a head-water lake in the Anadyr River basin of central Chukokst Peninsula. The new species has 39 to 51 gillrakers, considerably more than any other species in the subfamily Salmoninae, and is sympatric with a large charr (to 1 m) with 23 to 31 gillrakers.

Thus, on the basis of the divergent opinions found in the most recent and comprehensive compendium on the genus, and new information, it is obvious that the status of systematic study and opinions on *Salvelinus* is one of active ferment. It is now appropriate that I again try to organize the diversity into a taxonomic scheme and attempt a better definition of the limits of the subgenus *Salvelinus*, the Arctic charr species complex and the Arctic charr, *S. alpinus*. Recent literature, colour slides, and specimens of Far Eastern charr, kindly sent to me by I.A. Chereshnev and associates, have contributed to my latest ''revision.'' I also borrow heavily from my studies on a group of salmonid fishes, the Pacific basin trouts of the subgenus *Parasalmo* (Behnke 1979 MS), which are similar in their degree of evolutionary divergence (within a comparable era of geological time), relatedness, and evolution of diverse morphological, behavioural and ecological forms, giving rise to comparable taxonomic problems.

THE SUBGENUS *Salvelinus*

The first step toward a more precise definition of ''Arctic charr'' is to divide the genus into smaller phylogenetic units. In my last two papers on charr, I recognized three subgenera; one for the North American lake charr, *S. namaycush*, one for the North American brook charr, *S. fontinalis*, and one for all other members of the genus, but pointed out that the correct position of the Far Eastern ''kundscha'' charr, *S. leucomaenis*, is in doubt (Behnke 1972, 1980). *S. leucomaenis* has some similarities with *S. namaycush*, but these may be the result of convergent evolution as a predacious species. Until convincing evidence is forthcoming regarding the phylogenetic branching sequence in the genus leading to *S. leucomaenis*, I include, provisionally, the kundscha charr in the subgenus *Salvelinus*.

Thus, based on zoogeographic evidence correlated with degrees of divergence, it appears that an early (Pliocene?) branch of the genus became established in North America, eventually giving rise to the two widely distributed living species, *S. namaycush* and *S. fontinalis*. The former species specialized as a large lacustrine predator whereas the latter species, in the general absence of other salmonine fishes, evolved a more generalized way of life, utilizing a wide range of environments, more typical of the trouts in the genus *Salmo*. Conveniently ignoring *S. leucomaenis*, the evolutionary divergence in the rest of the genus resulted in two major groups or species complexes — the Arctic charr group and the Dolly Varden group. This split may have occurred during an early Pleistocene glacial period when the common ancestral charr was divided into northern (precursor of the Arctic charr) and southern (precursor of the Dolly Varden group) groups in the Pacific and Arctic ocean drainages.

In view of the contentious nature of my recognition of Arctic charr and Dolly Varden charr not only as separate species but as separate species groups or complexes, the evidence supporting my position must be presented.

Reality of Arctic Charr and Dolly Varden Species Groups

DeLacy and Morton (1943) presented the first conclusive evidence that Dolly Varden charr and Arctic charr are valid species by finding representative populations of the two species occurring sympatrically in Karluk Lake, Alaska where they exhibited sharp differences in life history and taxonomic characters. McPhail (1961) found other Alaskan lakes where the two species were sympatric with no evidence of hybridization, further strengthening the case for full species recognition between Dolly Varden and Arctic charr. The problem was that all of these examples involved the southern form of North American Dolly Varden which formerly was generally assumed to characterize the taxon *S. malma*.

Consideration of the taxonomic data on Dolly Varden throughout its range reveals a major divergence occurring between northern and southern Dolly Varden that was probably initiated by isolation during a glacial period subsequent to the separation of Dolly Varden and Arctic charr. A relatively homogeneous northern Dolly Varden occurs in both Asia and North America. In the Far East, in the northern Okhotsk Sea drainages, Kamchatka, and northward around the Chukotsk Peninsula, the Dolly Varden is typically characterized by 21 to 23 gillrakers, 66 to 68 vertebrae, and 25 to 30 pyloric caeca. This same form of Dolly Varden (which should be recognized as *S. malma malma* because Kamchatka is the type locality for the name *malma*) occurs in North America from the Alaska Peninsula (north side) to the Mackenzie River. McPhail (1961) and McPhail and

Lindsey (1970) were not aware of the similarities between the Alaskan and Far Eastern northern forms of Dolly Varden and they considered the northern North American form of Dolly Varden as the western Arctic form of *S. alpinus*.

South of the Alaska Peninsula to Washington, the southern form of North American Dolly Varden occurs and it is differentiated from the northern form by fewer gillrakers and vertebrae (typically 17-18 and 62-64, respectively). In the Far East, the "southern" subspecies of Dolly Varden occurs on Sakhalin, Hokkaido, and mainland drainages south of the Amur River mouth. The Far Eastern southern Dolly Varden is typically characterized by 19-23 gillrakers and 61-63 vertebrae. Viktorovsky (1978) gave the karyotype of the Far Eastern southern Dolly Varden (*S. malma krascheninnikovi*) as 2N = 82-84 with 98 chromosomal arms. Cavender (pers. comm.) found 2N = 82 with 98 arms in the karyotype of North American southern Dolly Varden, indicating a recent common ancestry. The karyotype of the northern Dolly Varden, *S. malma malma*, according to Viktorovsky (op. cit.) is 2N = 76-78 with 96 arms.

Formerly (Behnke 1972, 1980), I considered the charr of Honshu Island, "*S. pluvius*" as an extreme form of Dolly Varden (*S. m. pluvius*). The low meristic values of the *pluvius* charr, 16-19 gillrakers, 58-61 vertebrae, and 20-25 pyloric caeca, are typical of *S. leucomaenis*. The *pluvius* charr, however, has red spots on the body, which are lacking in all other populations of *S. leucomaenis*, and this character led me to assign *pluvius* to *S. malma*. Based on the structure of the gillrakers (Cavender 1980), greater familiarity with Japanese studies (fostered by this symposium), and the karyotype values given by Viktorovsky (1978), I now consider *pluvius* as a subspecies of *S. leucomaenis*. As I previously pointed out (Behnke 1980), a charr with meristic characters identical to *pluvius* of Honshu occurs on the Kuril Islands, and the name *curilus* has priority over *pluvius*. Viktorovsky (op. cit.) found "*curilus*" and *leucomaenis* specimens to have identical karyotypes, 2N = 84-86 with 100 arms.

All of the above forms that I have identified as Dolly Varden are differentiated from all North Pacific forms of Arctic charr by their general phenotypic appearance (smaller spots on the body, a more blunt snout), fewer gillrakers and fewer pyloric caeca. There is no evidence of sympatric occurrence between northern and southern forms of Dolly Varden. The diversity within the species can be arranged into three broadly distributed subspecies — *S. malma malma*, the northern form in both Asia and North America; *S. malma krascheninnikovi*, the southern Asiatic form of Hokkaido, the Sakhalin Islands and the Asiatic mainland south of the Amur River; and the southern North American form, which I have tentatively recognized as *S. malma lordi* (Behnke 1980).

The validity of full species recognition of *S. malma* depends on its degree of reproductive isolation where it has come into contact with members of the Arctic charr complex. If Dolly Varden occur sympatrically with Arctic charr over a broad area in Asia and North America, with sympatric populations maintaining their integrity through reproductive isolation, then it strongly suggests that *S. malma* is indeed a "real" species of relatively long (perhaps early Pleistocene) separation from any member of the Arctic charr complex, and not as Savvaitova (1980) believed, simply local populations of Arctic charr that happen to be so similar because of convergent evolution since the last glacial period.

In my recent paper (Behnke 1980) I reviewed the records of sympatric occurrence of both the southern and northern forms of North American Dolly Varden with Arctic charr in several lakes in Alaska. Typically, in Alaska, the Dolly Varden is anadromous, overwintering in the lakes and spawning in tributaries (or resident in tributaries) to the lake whereas the Arctic charr is strictly lacustrine (and the more piscivorous species). *S. malma* most commonly exists with the form of Arctic charr called the Bristol Bay-Gulf of Alaska form of *S. alpinus* by McPhail (1961), but sympatry also occurs between the northern Dolly Varden and the "eastern Arctic" form of *S. alpinus* in some lakes west of the Mackenzie River mouth. Thus, in North America, there is sympatric occurrence over a broad geographical area between populations of Dolly Varden and Arctic charr; two subspecies of Dolly Varden (northern and southern) and two subspecies of Arctic charr (Bristol Bay-Gulf of Alaska and eastern Arctic) are involved. There is no real evidence of intermediacy of hybridization between Dolly Varden and Arctic charr.

In Asia, there is widespread sympatry between the northern Dolly Varden and Arctic charr. The Asiatic Arctic charr of the Chukotsk Peninsula and Kamchatka is very similar to the Bristol Bay — Gulf of Alaska form of Arctic charr of Alaska. It is characterized by large spots, typically 23-26 gillrakers, 40-55 pyloric caeca, and 66-68 vertebrae. Two names have been given for this charr from the Chukotsk area—*S. andraishevi* and *S. taranetzi*.

In Kamchatka, the Arctic charr occurs as lacustrine relict populations in Lake Dalnye and Lake Nachinskoye where it is sympatric with populations of Dolly Varden. In lakes and rivers draining the Chukotsk Peninsula, Arctic charr and Dolly Varden are sympatric, often with both species being anadromous (Chereshnev 1978a, 1978b, 1979). A paper by Volobuyev et al. (1979) casts doubt on Chereshnev's conclusions that the Arctic charr and Dolly Varden are two distinct species. The authors examined a series of charr specimens from the Kukekkuyum River, in southern Chukatsk (previously studied by Chereshnev), emphasizing that intermediate (hybrid) specimens between the large-spotted Arctic charr and small-spotted Dolly Varden occurred in their sample. They did this by emphasizing the overlap in the ranges of numbers of gillrakers and pyloric caeca that occur between the two species and attributing the overlap to hybridization.

The publication by Volobuyev et al. (1979) is an inappropriate attempt to support a preconceived position that *malma* and *alpinus* are not valid species. It is obvious that their paper did not receive critical review before publication. Chereshnev (1982) criticized the work of Volobuyev et al. (op. cit.) in great detail, convincingly demolishing the supposition of one polymorphic species for Chukotsk charrs; hopefully, the notion that *S. malma* is part of the polymorphic Arctic charr species and a synonym of *S. alpinus*, has been put to rest for all time. Chereshnev (op. cit.) presented data from sympatric populations of Dolly Varden and "Taranetz" charr from 7 Chukotsk rivers. A total of 585 Dolly Varden and 756 "Taranetz" charr specimens were examined. Clear-cut differences between the species were found in many characters such as spotting, colouration, body shape, presence of kype in mature Dolly Varden and its absence or feeble development in the "Taranetz" charr, and gillraker structure — long, attenuate in Dolly Varden; short, thick in "Taranetz" charr. Pyloric caeca counts of specimens identified as Dolly Varden and "Taranetz" charr revealed little overlap; 578 of 585 Dolly Varden had less than 35 caeca, and 744 of 756 "Taranetz" charr had 35 or more caeca. Critical comparison of the 1341 specimens of 7 sympatric populations found no evidence that hybridization occurs.

A problem, I must point out, concerns the relationships of the north Pacific form of Arctic charr (*S. anadraishevi* or *S. taranetzi*). Is it a subspecies of *S. alpinus*? Where would its branching point be in the phylogeny of *Salvelinus*? Its large spots are somewhat similar to *S. leucomaenis*. Is this type of spotting an example of convergent evolution?

Glubokovsky and Chereshnev (1981) compared the cranial bones of "Taranetz" charr with *S. alpinus* (specimens from northwest USSR near type locality, and assumed to be *S. alpinus alpinus*), *S. malma malma, S. leucomaenis,* and "stone" charr, *S. albus*. Both a numerical taxonomic arrangement (overall phenotypic similarity) and a cladistic interpretation were worked out. The "Taranetz" charr was grouped with *S. alpinus* and, more distantly, *S. leucomaenis*, *S. malma*, and *S. albus* were grouped together. I would tentatively agree with this phylogenetic interpretation but it should be emphasized that cladistic ordering of cranial characters of *Salvelinus* does not rest on hard evidence. The divergences in size and shape of the various bones are subject to convergent and independent evolution and primitive and derived states can not be determined with great confidence. Glubokovsky and Chereshnev (1981) claimed that the "Taranetz" charr occurs sympatrically with *S. alpinus* and *S. malma* in the northwestern part of its range in the Amguema and Erguveem river basins, but gave no further details. In any event, considering the significance of the "Taranetz" charr as an integral component for resolving the classification problems of the Arctic charr complex and its uncertain relationships to *S. alpinus* and *S. leucomaenis,* I propose to recognize the "Taranetz" charr as a full species, at least until its relationships are better understood.

As I have previously discussed (Behnke 1980), the correct species name for the "Taranetz" charr is in doubt. The type locality of *S. andriashevi*, Lake Estikhet, and the type locality of *S. taranetzi*, Lake Achchen, have both "Taranetz" charr and Dolly Varden occurring in sympatry (Chereshnev 1982). Barsukov (1960) noted that the holotype of *S. andriashevi* (Zool. Inst. 31107) has 19 gillrakers on the right and 21 on the left side. These gillraker counts strongly indicate that *S. andriashevi* is a synonym of *S. malma malma*. Kaganovsky (1955) published a short description of *S. taranetzi* that was obviously based on specimens of both Dolly Varden and "Taranetz" charr, as can be noted from pyloric caeca counts of 32-42. The specimen illustrated for the description has small spots and a relatively long jaw, and is almost certainly *S. malma*. Kaganovsky (op. cit.) did not designate a type specimen and the number of specimens examined was not given except for a table of morphometric characters stating that data from 10 specimens are given.

Added confusion is introduced to this muddled nomenclatorial problem by notes I made from examination of 3 specimens designated as *S. taranetzi* in the Zoological Institute's collection in Leningrad (32104). The collection date listed for collection 32104 is July 12, 1949 and the collection site is Lake Achchen. The specimens used by Kaganovsky (op. cit.) for his description of *S. taranetzi* were collected August 12, 1948. Thus, collection 32104 are not types. The 3 specimens of collection 32104 are from 152 to 193 mm standard length and are very thin, almost emaciated in appearance. My gillraker counts are 24, 26, 28, quite typical of the values expected of "Taranetz" charr according to Chereshnev (1982). I noted, however, when I examined these specimens in 1960, that the gillrakers are "extremely long and fine". All "Taranetz" charr discussed by Chereshnev (op. cit.) and specimens I have examined from Alaska, have short, thick gillrakers. Perhaps a resident lacustrine charr exists in Lake Achchen, differentiated from both the anadromous Dolly Varden and "Taranetz" charr.

Although *S. taranetzi* has been completely characterized by Glubokovsky and Chereshnev (1981) and by Chereshnev (1982), the original description of *taranetzi* is based on a composite of 2 (possibly 3) species, and the name is a "nomen dubium" until someone obtains the collection of specimens used by Kaganovsky (1955) (syntypes) and designates a lectotype that conforms to the later characterizations of *taranetzi*. If the original collection is lost, topotype specimens of "Taranetz" charr could be used for a redescription of the species.

The phylogenetic affinities of two other Pacific drainage species — the stone charr, *S. albus*, of Kamchatka, and the bull trout, *S. confluentus*, of North

America are not clear. I discussed the stone charr in my 1980 paper but I did not realize that the description of *Salvelinus albus* by Glubokovsky (1977), for a charr from the Kamchatkan River basin, was actually a description of the stone charr. There is some confusion surrounding the stone charr as I discussed in my 1980 paper. Viktorovsky (1975) discussed various forms of charr from Lake Kronotskoye, Kamachatka, using common names — the white (bely) charr, the longheaded (dlinnogolovy) charr, and the nose or snout (nosaty) charr. Glubokovsky (1977), in naming the ''white'' charr, *S. albus*, from the Kamchatka River, mentioned that it is identical to the white charr of Lake Kronotskoye. Viktorovsky (1978) revised the Lake Kronotskoye charr. He decided that the white charr is actually the typical *S. malma*; the nose charr was described as a new subspecies of *malma, S. m. schmidti*, and the longheaded charr was described as a new species, *Salvelinus kronocius*. From the descriptions given by Glubokovsky and Viktorovsky and from previous data on the stone charr supplemented by two specimens of *S. albus*, sent to me by I.A. Chereshnev, there is little doubt that *S. albus* and *S. kronocius* both represent the stone charr of Kamchatka. Thus *S. kronocius* is a synonym of *S. albus*. Unfortunately, the confusion with the ''white'' charr of L. Kronotskoye led to the inappropriate name *S. albus*: the stone charr could be called the black charr as the colouration in freshwater specimens is decidedly dark.

As mentioned in my previous paper, I believe the closest relationship of the stone charr is with the North American bull trout, *S. confluentus*. There are some real differences in meristic characters (about what might be expected from two branches of monophyletic origin that have been separated for more than 100,000 years), but there are basic similarities in the shape of the head and jaws and overall appearance. I believe the bull trout and stone charr represent an early divergence in the Dolly Varden complex.

I suggest that the relationship of *S. leucomaenis* is also with the Dolly Varden complex: the low values of meristic characters and the karyotype of *S. leucomaenis* (2N = 84-86) (Viktorovsky 1978) indicate close affinity of *leucomaenis* with the Dolly Varden complex.

The Arctic Charr Complex

By a process of elimination at the subgeneric and 'complex' levels the rest of the diversity in the genus can be ascribed to the Arctic charr complex. That is, the phylogenetic branching points leading to all living forms in the Arctic charr complex should connect to a common ancestor after the separation of the Dolly Varden.

The key to defining the limits of the species *S. alpinus* is provided by the charr of the type locality area: ''Swedish Lappland'' as given by Linnaeus. All forms designated *S. alpinus* or ''Arctic charr complex'' should be more closely related (more recent phylogenetic separation) to Scandinavian *S. alpinus* than they are to *S. malma* of Kamchatka.

The typical Scandinavian form of *S. alpinus* has about 63-64 vertebrae, 30-50 pyloric caeca and 21-28 gillrakers. This form is distributed in the Kola and Karelian region of the USSR, Finland, northern Sweden, Norway, Great Britain and Ireland. A closely related group, *S. a. salvelinus*, occurs in the Alpine lakes of central Europe. Typically, the central European ''normal-saibling'' or ''wildfangsaibling'' has slightly more gillrakers (typically 25-30).

There is an interesting group of Arctic charr concentrated in the Taimyr Peninsula region of Siberia and in the Arctic Ocean drainages of North America east of the Mackenzie River (eastern Arctic form of *S. alpinus*), possibly monophyletic in origin but now having a broad and disjunct distribution. These charr have the highest numbers of gillrakers, from 25 to 35; they are also characterized by about 65 to 70 vertebrae and about 40 to 50 pyloric caeca. They may attain a weight of 12 kg or more and are typically piscivorous. They occur as anadromous and resident lacustrine populations (resident lacustrine stocks are much more common in Siberia where *S. namaycush* is absent); disjunct relict populations may occur far inland in Siberia in lakes of the Yenesei, Lena, and Kolyma river basins and in the Baikal region (*S. erythrinus*). Headwater transfer from the Kolyma basin has established this form of charr (*S. neiva*) in lakes of the Okhota River basin, draining to the Okhotsk Sea. This charr (the eastern Arctic charr of McPhail and Lindsey 1970) occurs as resident lacustrine populations in lakes west of the Mackenzie basin where they are sympatric with anadromous *S. malma*. I also believe this same form of charr occurs in Lake Coomarsharn, Ireland (*S. fimbriatus*). The distribution of the Arctic cisco, *Coregonus autumnalis*, exhibits many similarities with this particular group of charr. The Arctic cisco also occurs as relict lacustrine populations in Ireland where it is known as the pollan (Ferguson et al. 1978).

The Taimyr Peninsula is of special interest in regards to sympatry between populations of Arctic charr. I have previously discussed a charr from L. Taimyr (Behnke 1972, 1980) named *S. taimyricus* that has a distinctive morphology. Savvaitova et al. (1980) recorded *S. taimyricus* from Lake Keta and described four sympatric *Salvelinus* populations from Lake Lama. All have meristic characters similar to the generalized anadromous Taimyr charr and most likely all represent minor divergences within this particular ''major'' divergence of the Arctic charr complex. The life history and ecological differences between the various Taimyr charrs are pronounced. There are anadromous and lacustrine forms; the various lacustrine forms differ in feeding specialization, depth distribution, growth rate and maximum life span.

For a classification of *S. alpinus* best denoting evolutionary relationships among the major divergences making up the species, I suggest that all of these charr (Taimyr region, eastern Arctic North America, interior Siberia lakes, and Lake Coomarsharn) be considered as one subspecies (*S. alpinus erythrinus*). The great ecological diversity encompassed in such classification (or any taxonomic arrangement of *S. alpinus* conforming to the international rules of zoological nomenclature), readily reveals the severe limitations of our binomial or trinomial system of nomenclature to describe and classify accurately the enormous diversity manifested in the Arctic charr complex. For proper management of this diversity, the various ecological forms, particularly in sympatry, should be regarded as if they were different species.

One other form of *S. alpinus* that probably deserves subspecific recognition is the group of disjunct lacustrine relicts of northern New England and southern Quebec known as Sunapee, blueback, and red ''trout'', *S. alpinus oquassa*. This group of charr appear to be an early derivative of the evolutionary line leading to the northern European *S. a. alpinus*. They differ mainly in having fewer gillrakers (18 to 24 vs. 21 to 28). Their distribution appears to be largely the result of exclusion or extermination by *S. namaycush*. Native populations occurred only where *namaycush* was absent and they soon disappeared in those lakes where *namaycush* was introduced.

From a preliminary synthesis of all information, I provisionally recognize three major centers of evolution in the Arctic charr complex, or *S. alpinus* in the broadest sense: a European origin of the north European *S. alpinus alpinus*; the central European *S. a. salvelinus* (in eastern North America *S. a. oquassa* is probably most closely associated with this group); a Siberian origin for the vast and diverse group that, for purely practical taxonomic purposes, can be recognized as *S. a. erythrinus* having a multitude of derivative forms such as *taimyricus, neiva, drjaqini*, etc; and thirdly, I provisionally include the peculiar form of charr with very large spots, from Asiatic and North American drainages at the Chukotsk and Bering seas (where it is widely sympatric with *S. malma*). This form, the ''Taranetz'' charr, is without a valid scientific name, as discussed previously. The name *S. taranetzi* is a nomen dubium that requires redescription. The extremely large spots of this form of Arctic charr may be an example of character displacement that reinforces reproductive isolation with *S. malma*. It is also possible that the ''Taranetz'' charr is derived from a common ancestor with *S. leucomaenis*.

From what I have been able to determine, examples of sympatry in the Arctic charr complex are always between members of the same evolutionary group, and are the result of recent divergences initiated during or since the last glacial period. That is, the three major divergences that, for practical purposes of communication, I have classified as *S. a. alpinus, S. a. erythrinus*, and the "Taranetz" charr, do not occur in sympatry with each other. The sympatric populations in Lake Windermere, England, and in many Scandinavian and Karelian lakes are more closely related to other members of the broad and inclusive *S. a. alpinus* than they are to any member of the *S. a. erythrinus* or "Taranetz" charr evolutionary lines. This also appears to be true for the numerous sympatric populations in lakes of Taimyr; they are all probably derived from within the *S. a. erythrinus* evolutionary lineage, after separation from the *S. a. alpinus* branch. From a practical point of view and in relation of principles of animal taxonomy, it seems absurd to classify all four of the sympatric populations in Lake Lama as a single subspecies. The most useful taxonomic arrangement of the subgenus *Salvelinus* (for other than evolutionary or systematic purposes) will most probably stray considerably from a strict cladistic interpretation.

The "tiefseesaibling," *S. profundus*, is a striking exception to the rule that sympatric populations (except for *malma* and *alpinus*) have their origins in relatively recent geological time. The degree of differentiation of the tiefseesaibling is extreme in the genus. The uniform, dull colouration, the broad and overhanging snout, the peculiar structure of the gillrakers, the low numbers of gillrakers, vertebrae and pyloric caeca (20-25, 61-63, and about 25-35, respectively) set off *S. profundus* from any other known form in the genus. I have examined specimens of tiefseesaibling only from the Bodensee but it has been reported from the following Alpine lakes of Central Europe: Neuchatel, Walchensee, Ammersee, Tegernsee, Achensee, Plansee, Attersee, and Traunsee (Berg 1932, Nerseheimer 1937). Brenner (1980) described the tiefseesaibling from the Attersee. Brenner concluded that tiefseesaibling are not a distinct taxon but are only "local forms created in response to different environmental and feeding conditions." Such a statement is difficult to accept for anyone familiar with the range of variability in charr specimens throughout the distribution of the genus. If it were true that sympatric populations of tiefseesaibling evolved independently in each lake inhabited by *Salvelinus* where suitable conditions were present, then there would be tiefseesaibling throughout the Holarctic region and not only in central European lakes. The rich whitefish (subgenus *Coregonus*) fauna of the European Alpine lakes, has probably severely restricted the distribution and abundance of tiefseesaibling by competitive exclusion. This would suggest to me that the evolution of the tiefseesaibling occurred before *Coregonus* invaded the Alpine lakes or it occurred in a contiguous area where *Coregonus* was absent. In any event, the tiefseesaibling is morphologically very different from any other form of charr. I include it in the Arctic charr complex only on zoogeographical grounds.

The Malorotaya Charr

A most significant new event in *Salvelinus* systematics was the discovery of two extremely differentiated charrs occurring in Lake Elgygytgyn, a headwater lake in the Anadyr River basin, central Chukotsk. The extreme differentiation of the new species, *S. elgyticus*, with 39-51 gillrakers, denotes a divergence, perhaps comparable in magnitude to the tiefseesaibling of Europe. Viktorovsky et al. (1981) described *S. elgyticus*, the malorotaya (smallmouth) charr based on 67 specimens from Lake Elgygytgyn. They also obtained 38 specimens of a large charr (to about 1 m) with 23-31 gillrakers from the same lake, which they identified as *S. boganidae* (*S. boganidae* was described from the Taimyr region and I consider it as part of the *S. a. erythrinus* complex discussed above).

Through the kind cooperation of Soviet ichthyologists I obtained a copy of the publication containing the description of the new species and also a specimen of *S. elgyticus*. The new species is characterized by an extreme number of long, fine gillrakers (my specimen has 45 gillrakers), a large eye, and a small mouth, the upper jaw barely reaching to beneath the pupil of the eye. There are large spots, comparable to the spots of the ''Taranetz'' charr, sparsely distributed on the body. The small mouth does not seem to be in agreement with the great development of the gillrakers which indicate a specialization for pelagic feeding on zooplankton, and, for such feeding, a larger mouth would be assumed to be advantageous (as is found in the cisco group of planktonic feeding whitefishes). The large eye indicates specialization for life in deep water and the small mouth may be an adaptation to feeding on small benthic organisms that are retained by the numerous gillrakers.

The basic questions concerning the origin and phylogenetic affinities of *S. elgyticus* can not, at present, be answered with any authority. A large part of the Chukotsk Peninsula was a glacier refuge area (Viktorovsky 1978) comparable to the Yukon refugium in North America. Lake Elgygytgyn lies near the drainage divide of the Anadyr River basin and the Chaunsk Bay drainage of the East Siberian Sea. The lake is at an elevation of 489 m and was not submerged by the late-Pleistocene marine transgression. The origin of the lake, which is 12 km in diameter and 169 m deep, is reputedly from a meteorite crater of latePliocene age. The only other fish species found in the lake besides charr are the sculpin, *Cottus cognatus* and grayling, *Thymallus arcticus*, both of which occur in Chaunsk Bay and Anadyr drainages. It would appear that *S. elgyticus* represents a relict of an ancient evolutionary line. Perhaps by the time of the next charr symposium we will have a better understanding of the phylogenetic position of this interesting new discovery.

Recent Genetic Studies

The hope for a significant contribution to the resolution of taxonomic problems of *Salvelinus* from electrophoretic studies is yet to be fulfilled. A major reason is that most studies have been made on the northern European charr (*S. a. alpinus*) which exhibits extremely low levels of enzyme polymorphism (Ferguson 1981) while almost nothing is known of genetic variation in the many divergent forms from diverse geographical areas. The limitations that this ''provincialism'' of genetic characterization has on *Salvelinus* systematics is exemplified by the work of Nyman et al. (1981) who identified three ''species'' of Scandanavian charr on the frequency of a serum esterase allele (the ''F'' or ''fast'' allele or EST-1 (115) allele according to Ferguson (1981)). *Salvelinus alpinus* was identified by an EST (115) allelic frequency of .9-1.0; *S. salvelinus* was assigned a frequency of this allele of .4-.7; and *S. stagnalis* .0-.2. The taxonomic significance of type localities and topotype specimens for redefining taxa was acknowledged but then ignored. It should be obvious that all charr and every living species must have EST (115) frequencies that range from .0 to 1.0. I would agree that a certain allelic frequency characterizing a group of populations of a defined geographical area clearly differentiating them from other populations, is an indication of different ancestral origins; however, the esterase enzyme is labile, readily changed in evolution and thus subjected to convergence and misinterpretation. As such, it must be used with caution as an indicator of phylogenetic affinity. Nyman et al. (1981) presented no corroborative evidence such as additional genetic or morphological data to support their taxonomic interpretation of the EST allele. It was claimed that correlation exists between EST allelic frequencies and size and ''competitive ability'', but no such clear-cut relationship emerges from the discussion in the text. Rather a hodge-podge of allelic frequencies and ecological forms are apparent which are attributed to ''introgression'' between the ''three species''. I do agree, however, with the final words of this publication that their work '. . . adds a new dimension to the view of treating all Mendelian polymorphisms as simple population markers.'' It is indeed a ''new dimension'' to treat Mendelian polymorphism as distinct species whereby all *Salvelinus* can be assigned to three species based on the EST (115) allelic frequency and all intermediate values are the result of introgression.

Kornfield et al. (1981) presented data on esterase of New England and Quebec charr (*S. a. oquassa*). The frequenices of two alleles varied greatly as in Scandanavian charr. Does this mean the same ''three species'' and introgression between them is involved despite the fact of the consistently uniform differences in taxonomic characters between *S. a. alpinus* and *S. a. oquassa*?

McCart (1980) obtained esterase data from Canadian eastern Arctic charr. A total of 78 specimens from 5 lakes were examined; all had the ''fast'' (EST (115)) allele. According to Nyman et al. (1981), the Canadian eastern Arctic charr must be the same taxon as the Scandanavian charr with the fixed fast allele ("*S. alpinus*") which, according to their interpretation, is the smallest of charrs in the Arctic charr complex. The eastern Arctic charr, along with the Taimyr charr of Siberia with similar taxonomic characters, is actually the largest of the Arctic charrs (to 12 kg and more), and it is differentiated from Scandanavian charrs by about 4 more vertebrae and gillrakers. McCart (op. cit.) also presented esterase data on 212 specimens of *S. malma malma* from 13 Canadian sites, in which he found 209 of the 212 specimens (98.6 percent) to possess the "slow" allele. Thus, the northern Dolly Varden would become a synonym of "*S. stagnalis*" according to the interpretation of Nyman et al.

Kornfield et al. (1981) included a sample of Eastern Arctic charr from Cambridge Bay in their analysis of charr enzymes. The allele PGI-3 (105) occurred with 80 percent frequency in the Eastern Arctic charr sample but did not occur at all in the five samples of New England and Quebec charr. Also the LDH-5 (or LDH-E) (92) allele occurred with 50 percent frequency in eastern Arctic charr but not at all in the five samples of *S. a. oquassa*. Thus, there is some indication that electrophoresis can detect genetic divergences between the major subspecies in the Arctic charr complex, but the data must be critically analyzed and weighted to detect unique events as is done for other taxonomic characters. Typically electrophoretic data are mechanically analyzed by means of a quantitative comparison of overall phenotypic similarity. As such, the great number of monomorphic loci obscure and homogenize the truly significant differences.

It is evident from the publication by Stoneking et al. (1981) that the North American brook charr *S. fontinalis*, exhibits considerable genetic polymorphism. A manuscript submitted for publication by Leary and Allendorf on natural hybrids between *S. fontinalis* and bull trout, *S. confluentus*, in Montana, reveals considerable polymorphism in both species and considerable genetic divergence between the species (genetic distance .373).

May we look forward to a major contribution from electrophoresis, or other genetic research, toward a better understanding and taxonomic interpretation of *Salvelinus* phylogeny by the time of the next charr symposium?

DISCUSSION

I believe the charrs in the subgenus *Salvelinus* like the trouts of the subgenus *Parasalmo* have undergone rapid morphological and ecological divergence: that is, the morphological part of the genome has speciated much more rapidly than the metabolic part of the genome. If this is true then it can be expected that information obtained from the metabolic genes, typically from electrophoretic studies of proteins, will not be in full accord with the morphological divergence. This matter was lucidly assessed by Clayton (1981). It was also made clear by Todd (1981) that electrophoretic data may not always be effective in identifying closely related species of salmonid fishes. Phelps and Allendorf (1983) found complete identity between two species of sturgeon; 34 loci were monomorphic and there was no consistent allelic frequency difference at the 3 polymorphic loci. Thus, until techniques are refined and new biochemical technologies developed, an authentic delineation of phylogenetic branching sequences in *Salvelinus* is not likely to come from electrophoretic analysis alone.

The data on *Salvelinus* karyotypes indicate the general trend of Robertsonian fusion — going from higher chromosome numbers in primitive form to lower numbers in derived forms. The karyotype of *S. leucomaenis* is given as 2N = 84-86 with 100 arms. *S. malma malma* is listed as 78-80 with 96 arms; the Far Eastern form, intermediate in morphological characters (*S. m. krascheninnikovi*), also has an intermediate karyotype with a diploid number according to Viktorovsky (1978) of 82-84 and 98 arms. *S. fontinalis* and *S. namaycush* have diploid numbers of 84 with 100 arms. *S. alpinus* (Sweden) has 80 chromosones with 100 arms. The stone charr of Kamchatka and the bull trout of North America evidently possess a common karyotype of 2N = 78 with 100 arms (Viktorovsky 1978 and T. Cavender pers. comm.).

I would not necessarily interpret a smooth progression in karyotype speciation from this data. That is, I don't believe the list of karyotypes can be used as strong evidence that *S. leucomaenis* and all subspecies of *S. malma* were already evolved before *S. malma malma* gave rise to the Arctic charr complex.

Extrapolating the karyotype information from *Parasalmo* (Behnke 1979 MS; Thorgaard 1983) to *Salvelinus* suggests that when all of the diverse geographical forms of charr have been karyotyped, the comparative data may not be a decisive factor for determining the position of the phylogenetic branching points in *Salvelinus*.

Mednikov et al. (1980) published an interesting attempt to resolve taxonomic problems in *Salvelinus* by the use of DNA hybridization. I pointed out (Behnke 1980) that DNA hybridization data cannot be trusted because of the noise of experimental error when dealing with groups such as the Arctic charr complex and the Dolly Varden complex. Their evolutionary separation has not had sufficient time to accumulate the amount of genetic differentiation needed to overcome the noise of experimental error.

Thus, despite modern methodology, I doubt that a taxonomic arrangement of charrs will ever be constructed that will be universally accepted and endorsed. This is particularly true because of the conflict between the goal of a phylogenetic classification (in the strict cladistic sense) and that of a practical classification, useful in fisheries management, enabling degrees of morphological and ecological divergence to be recognized where sympatric forms occur.

The management, preservation, and utilization of this great range of diversity found in the various groups of charrs must be mentioned. Compared with other salmonid fishes, most of the distribution of charrs occurs in areas of the world with low human population density that are, as yet, relatively lightly influenced by man; by contrast, I estimate that the cutthroat trout species in western North America has lost about 98 percent of its original distribution (with concommitant loss of genetic diversity) in interior waters during the past 100 years; there can be great ecological and life history differences between local populations of the same subspecies that may be of major significance for management purposes (Hickman and Behnke 1979, Behnke 1981). A long range goal should be the maintenance of geographical and ecological diversity still present in *Salvelinus*, in order that the base of the natural resource (its genetic diversity) be available for future utilization.

For many years I have emphasized the values of the conservation of genetic resources, particularly of intraspecific units of a species. I am glad to note that this theme has gained considerable popular appeal in recent years. Volume 38, number 12 (1981) of the Canadian Journal of Fisheries and Aquatic Sciences was devoted to the proceedings of the Stock Concept Symposium. Ryman (1980) edited a publication promoting the conservation of fish gene pools and the FAO (1982) of the United Nations published a paper on the conservation of the genetic resources of fish. Chereshnev (1982) correctly pointed out that a classification that considers all diversity in Arctic charr and Dolly Varden as a single polymorphic species is a negative influence for the proper management and rational exploitation of populations with hereditarily based differences in life history. If valid species difference are not recognized, how can proper management of stocks at the subspecies level be implemented?

The natural distribution of the Arctic charr complex, with the exception of the European Alpine lakes and some lakes in the interior of Siberia, is largely restricted to within 100 km of the Arctic coast and associated islands. The Arctic charr is an effective agent to channel the food resources in waters of very low primary productivity into a highly desirable product. There is a great potential for the introduction of various forms of Arctic charr into suitable new waters for the enhancement of the fishery value of these waters. In 1978, I examined specimens of an unknown charr sent to me from a lake in the Rocky Mountains of Idaho. I identified the specimens as *S. alpinus oquassa*. Later, it was discovered that charr from Sunapee Lake, New Hampshire, were stocked into a few Idaho lakes in 1925 (the native charr in Sunapee Lake later became extinct after *S. namaycush* was introduced). The Sunapee charr is now known to persist in two lakes in

Idaho. These lakes are ultra oligotrophic but the Sunapee charr attain weights of about 2 kg whereas *S. fontinalis* in these same lakes seldom exceed 100 g. In many such lakes in the Rocky Mountains of North America, having no suitable spawning inlets to maintain trouts of the genus *Salmo, S. fontinalis* is often the only fish species; typically they are represented by stunted populations of little fishery value. There must be thousands of similar lacustrine situations in the world where the introduction of members of the Arctic charr complex would greatly improve the fisheries by maximizing salmonid biomass production, making efficient use of the pelagic and benthic zones, and because of their range of food utilization and long life span, producing specimens of large size. Innovative ways to maximize growth and production could be tried such as hybrid combinations and attempts to create sympatric stocks through introduction from diverse parental sources. Consideration of the diversity of the potential niches of the diverse forms of Arctic charr and how these potential niches could be converted into realized niches through interaction with other species (or other sympatric populations of Arctic charr), offers a wealth of opportunity for theoretical and practical management studies. I believe the Arctic charr is a fish that will become more widely appreciated, despite its confused taxonomy.

REFERENCES

Balon, E.K., [ed.] 1980. Charrs: salmonid fishes of the genus *Salvelinus*. Dr. W. Junk Publishers, The Hague. 928 p.

Barsukov, V.V. 1960. The systematics of Chukotsk char of the genus *Salvelinus*. Voprosy Ikhtiologii 14:3-17 (In Russian).

Behnke, R.J. 1972. The systematics of salmonid fishes of recently glaciated lakes. J. Fish. Res. Board Can. 29:639-671.

Behnke, R.J. 1979 MS. Monograph of the native trouts of the genus *Salmo* of western North America. U.S. Fish Wildlife Service, Denver, Colo. 215p.

Behnke, R.J. 1980. A systematic review of the genus *Salvelinus*, p. 441-480. *In* E.K. Balon [ed.] Charrs: salmonid fishes of the genus *Salvelinus*. Dr. W. Junk Publishers, The Hague.

Behnke, R.J. 1981. Systematic and zoogeographical interpretation of Great Basin trouts. *In* R.J. Naiman and D.L. Soltz [ed.] Fishes in North American deserts. John Wiley Pub. Co. (In press).

Berg, L.S. 1932. Ubersicht der verbreitung der susswasserfische Europas. Zoogeographica 50:107-208.

Berg, L.S. 1948. Freshwater fishes of the U.S.S.R. and adjacent countries. Zool. Inst., Acad. Sci. USSR 1(27) Vol. 1:466 p. (English Translation 1962).

Brenner, T. 1980. The Arctic charr, *Salvelinus alpinus salvelinus*, in the prealpine Attersee, Austria, p. 765-772. *In* E.K. Balon [ed.] Charrs: salmonid fishes of the genus *Salvelinus*. Dr. W. Junk Publishers, The Hague.

Cavender, T.M. 1980. Systematics of *Salvelinus* from the North Pacific Basin, p. 295-322. *In* E.K. Balon [ed.] Charrs: salmonid fishes of the genus *Salvelinus*. Dr. W. Junk Publishers, The Hague.

Chereshnev, I.A. 1978a. The systematics of charrs, genus *Salvelinus*, of the Bering Sea region of the Chukotsk Peninsula. Biol. Morya 1:36-46. (In Russian).

Chereshnev, I.A. 1978b. The systematics of charrs, genus *Salvelinus*, of the Amguema River (Chukotsk), p. 276-289. *In* V. Ya. Levanidov [ed.] Systematics and biology of freshwater organisms of northeast Asia. Acad. Sci. USSR Far East Sci. Center, Vladyvostok (In Russian).

Chereshnev, I.A. 1979. Systematics of anadromous charr (genus *Salvelinus*) from the Ionivyeem River basin (Chukotsk Peninsula). Pages 114-125. *In* V. Ya. Levanidov, editor. Systematics and ecology of fishes of the continental waters of the Far East. Acad. Sci. USSR Far East Sci. Center, Vladyvostok (In Russian).

Chereshnev, I.A. 1982. The taxonomic status of sympatric diadromous charrs of the genus *Salvelinus* (Salmonidae) from eastern Chukota. J. Ichthyology 22(6):22-38 (English Translation of Voprosy Ikhtiologii).

Clayton, J.W. 1981. The stock concept and the uncoupling of organismal and molecular evolution. Can. J. Fish. Aquat. Sci. 38(12):1515-1522.

De Lacy, A.C. and W.M. Morton. 1943. Taxonomy and habits of the charrs, *Salvelinus malma* and *Salvelinus alpinus* of the Karluk drainage system. Trans. Am. Fish. Soc. 72:79-91.

FAO. 1982. Conservation of the genetic resources of fish: problems and recommendations. FAO Fish. Tech. Pap. 217: 43p.

Ferguson, A. 1981. Systematics of Irish charr as indicated by electrophoretic analysis of tissue proteins. Biochem. System. Ecol. 9(2/3):225-232.

Ferguson, A., K-J.M. Himberg, and G. Svardson. 1978. Systematics of the Irish pollan (*Coregonus pollan*): an electrophoretic comparison with other Holarctic Coregoninae. J. Fish. Biol. 12:221-233.

Glubokovsky, M.K. 1977. Taxonomic relationships of charr of the genus *Salvelinus* of the Kamchatka River basin. Biol. Morya 3:24-35.

Glubokovsky, M.K., and I.A. Chereshnev. 1981. Unresolved problems concerning the phylogeny of chars (*Salvelinus*) of the Holarctic: I. Migratory chars of the East Siberian Sea basin. J. Ichthyology 21(5): 1-15 (English translation of Voprosy Ikhtiologii).

Hickman, T.J. and R.J. Behnke. 1979. Probable discovery of the original Pyramid Lake cutthroat trout. Prog. Fish Cult. 41:135-137.

Johnson, L. 1980. The Arctic charr, *Salvelinus alpinus*, p. 15-98. *In* E.K. Balon [ed.] Charrs: salmonid fishes of the genus *Salvelinus*. Dr. W. Junk Publishers, The Hague.

Kaganovsky, A.G. 1955. Char of the Bering Sea basin. Voprosy Ikhtiologii 3:54-56.

Kornfield, I., K.F. Beland, J.R. Moring, and F.W. Kircheis. 1981. Genetic similarity among endemic char (*Salvelinus alpinus*) and implications for their management. Can. J. Fish. Aquat. Sci. 38(1):32-39.

Leggett, J.W. 1980. Reproductive ecology and behavior of Dolly Varden charr in British Columbia, p. 721-738. *In* E.K. Balon [ed.] Charrs: salmonid fishes of the genus *Salvelinus*. Dr. W. Junk Publishers, The Hague.

Loudenslager, E.J. and R.M. Kitchin. 1979. Genetic similarity of two forms of cutthroat trout, *Salmo clarki*, in Wyoming. Copeia (4):673-678.

McCart, P.J. 1980. A review of the systematics and ecology of Arctic char, *Salvelinus alpinus*, in the western Arctic. Can. Dept. Fish. Oceans Tech. Rep. 935:89p.

McPhail, J.D. 1961. Study of the *Salvelinus alpinus* complex in North America. J. Fish. Res. Board Can. 18:793-816.

McPhail, J.D. and C.C. Lindsey. 1970. Freshwater fishes of northwest Canada and Alaska. Fish. Res. Board Can. Bull. 173:381 p.

Mednikov, B.M., V.A. Maksimov, and K.A. Savvaitova. 1980. Genetic divergence of Eurasian charrs, p. 357-364. *In* E.K. Balon [ed.] Charrs: salmonid fishes of the genus *Salvelinus*. Dr. W. Junk Publishers, The Hague.

Morrow, J.E. 1980. Analysis of the Dolly Varden char, *Salvelinus malma*, of northwestern North America and northeastern Siberia, p. 323-338. *In* E.K. Balon [ed.] Charrs: salmonid fishes of the genus *Salvelinus*. Dr. W. Junk Publishers, The Hague.

Neresheimer, E. 1937. Die Lachsartigen (Salmonidae). Handb. Binnenfischerei Mitteleuropas 3:219-370.

Nyman, L., J. Hammar, and R. Gydemo. 1981. The systematics and biology of landlocked populations of Arctic char from northern Europe. Inst. Freshw. Res. Drottningholm Rep. 59:128-141.

Phelps, S.R. and F.W. Allendorf. 1983. Genetic identity of pallid and shovel-nose sturgeon (*Scaphirhynchus albus* and *S. platorynchus*). Copeia 3:696-700.

Ryman, N. [ed.] 1980. Fish gene pools. Ecol. Bull. (Stockholm) 34:111p.

Savvaitova, K.A. 1980a. Comments to the systematic review of the genus *Salvelinus*, p. 480-481. *In* E.K. Balon [ed.] Charrs: salmonid fishes of the genus *Salvelinus*. Dr. W. Junk Publishers, The Hague.

Savvaitova, K.A. 1980b. Taxonomy and biogeography of charrs in the *Palearctic*, p. 281-481. *In* E.K. Balon [ed.] Charrs: salmonid fishes of the genus *Salvelinus*. Dr. W. Junk Publishers, The Hague.

Savvaitova, K.A., V.A. Maksimov, and E.L. Nesterov. 1980. Systematics and ecology of charrs of Taimyr Peninsula. Voprosy Ikhtiologii 20:195-210 (In Russian).

Stoneking, M., D.J. Wagner, and A.C. Hildebrand. 1981. Genetic evidence suggesting subspecies differences between northern and southern populations of brook trout (*Salvelinus fontinalis*). Copeia 4:810-819.

Thorgaard, G.H. 1983. Chromosomal differences among rainbow trout populations. Copeia 3:650-661.

Todd, T.N. 1981. Allelic variability in species and stocks of Lake Superior ciscoes (Coregoninae). Can. J. Fish. Aquat. Sci. 38(12):1808-1813.

Viktorovsky, R.M. 1975. Karyotypes of the endemic charrs of Lake Kronotskyoye. Tsitologiya 17:464-466 (In Russian).

Viktorovsky, R.M. 1978. Mechanisms of speciation in the charrs of Lake Kronotskoye. Acad. Scil USSR Far East Sci. Center, Vladyvostok. 110 p. (In Russian).

Viktorovsky, R.M., M.K. Glubokovsky, L.N. Yermolenko, and M.B. Skopets. 1981. Charrs of the genus *Salvelinus* from Lake Elgygytgyn (central Chukota), p. 67-78. *In* V.Ya. Levanidov, I.M. Levanidova, and I.A. Chereshnev (eds.) Fishes and ecosystems of salmonid rivers of the Far East. Acad. Sci. USSR. Far Eastern Sci. Cent., Vladyvostok (In Russian).

Volobuyev, V.V., E.D. Vasilyeva, and K.A. Savvaitova. 1979. Systematic status of the anadromous Chukotsk charr of the genus *Salvelinus*. Voprosy Ikhtiologii 19:408-418 (In Russian).

Management of Allopatric and Sympatric Populations of Landlocked Arctic Charr in Sweden

LENNART NYMAN

Institute of Freshwater Research, S-170 11 Drottningholm, Sweden

NYMAN, L. 1984. Management of allopatric and sympatric populations of landlocked Arctic charr in Sweden, p. 23–34. *In* L. Johnson and B.L. Burns (eds.) Biology of the Arctic charr, Proceedings of the International Symposium on Arctic Charr, Winnipeg, Manitoba, May 1981. Univ. Manitoba Press, Winnipeg.

The native charrs of Sweden may be placed in three taxa having overlapping meristic and morphological characteristics but also recognizable on genetic and ecological grounds; this necessitates a multiple approach to taxon recognition.On this complicated genetic background must be superimposed the manifold ways in which man has interacted with the charrs. The resulting picture was then examined to discern rational principles of charr management. Recommendations for charr population improvement involve six general principles and six more dependent on local conditions. Both allopatric and sympatric populations are considered.

INTRODUCTION

Sweden has a long tradition of research on Arctic charr. Petrus Artedi (1738) and Carl von Linné (1758) — Linnaeus — provided a firm beginning in the first half of the 18th century. The landlocked populations of charr had been little affected by man in those days, in great contrast to today's situation. In Linné's time, and maybe for a couple of thousand years before, Laplanders (Sameh) were the only exploiters of the charr stocks in the Swedish north; their only means of catching fish was by spear, or hook and line. Needless to say the sparse nomadic Sameh population never posed a threat to the lacustrine stocks. The last two hundred years has seen the arrival of settlers from the south, people of different culture settling permanently in log cabins. The same period has also seen the coming of the net, and no matter how crude the first gillnets may seem today, they initiated a new era in fishery techniques. Unlike the Sameh, the settlers were physically bound to a limited number of accessible lakes and with the growth of the settled population, overexploitation of charr stocks in the small low-yielding lakes soon became a reality. In the larger lakes, however, the home-made large-meshed nets rarely caused overexploitation; surprisingly, many instances of improved charr quality were recorded following the "netting period". The crucial point is, however, to differentiate between size increase and improved fish quality; as pointed out by Filipsson and Svärdson (1976) the modest fishing pressure of those days probably caused the latter condition. The twentieth century brought a number of very significant changes, among which may be mentioned the declining use of beach seines, the use of mass-produced standardized gillnets of rapidly improving quality, the increasing popularity of ice fishing with jiggers to set nets, and the fundamental ecological change produced by the conversion of lakes and rivers into hydroelectric reservoirs.

Another aspect of the charr problem in Sweden was also realized as far back as the 18th century, viz. the existence of several forms of charr, differences not only occurring from lake to lake but with individual lakes having up to three different forms. The typological taxonomy of those days, exclusively based on morphological and meristic data, failed to provide simple taxonomic criteria for the recognition of species. Even today, with refined supplementary techniques such as chromosome karyology and the various electrophoretic methods, much controversy remains.

In summary, the landlocked charr populations of northern Sweden have been subjected to dramatic changes in the last century; these changes, superimposed on the different taxa of charr inhabiting the major lakes have given rise to many practical problems. Is it possible to suggest satisfactory management procedures for a given lake, and if so, can generalized management programs be employed to help reverse the trend from large numbers of stunted charr back to a more pristine condition reflected either by improved fish quality or increased average size? This paper describes some of the difficulties inherent in attempting to answer these questions; it also provides a summary of the knowledge gathered during the implementation of various measures employed in the improvement of charr stocks for human utilization.

ANALYTICAL PROCEDURES

The strategy developed in the present paper, and in other contributions to this symposium (for example, those by Fürst, Gydemo and Hammar) originated in the realization that all taxonomic arrangements are the result of human reasoning and as such are bound to be only approximate translations of biological truth. Consequently, we have tried to combine a number of more or less unrelated techniques used for discriminating charr stocks, thus attempting an evalution of the taxonomic position from all the data available on each stock. This multiple approach method for charr taxon recognition has evolved over the past 16 years. The basic elements are ecological and electrophoretic; this pool of knowledge has resulted in numerous articles (Nyman 1965, 1972; Nyman and Filipsson

1972; Henricson and Nyman 1976; Fürst, Boström and Hammar 1978; Olsen 1980; Nyman 1980; Gydemo 1980; Hammar 1980; Nyman and Gydemo 1981). Some aspects of the method have been employed by non-Swedish authors (Klemetsen and Grotnes 1980; Nilsen 1980; Holthe 1981; Child 1977, 1980; Ferguson 1980). The data now analyzed by the Swedish authors mentioned above involves more than 25,000 charr samples derived from hundreds of bodies of water, collected over a large geographical area encompassing Sweden, Finland, Norway (including Spitsbergen), Iceland and Greenland. Scattered samples have also been analyzed from Arctic Canada and Newfoundland and a single sample from Lake Windermere, England.

Since some of the following papers will deal with this multiple approach model I will attempt to summarize the basic principles of charr taxon recognition. Figure 1A schematically shows the simultaneous representation of a number of fish from a certain lake. Each fish is plotted on the length at age diagram as its serum esterase phenotype - F (for the "fast" allele), S (for the "slow" allele) or F/S (for the heterozygote). The resulting frequency of the F allele is roughly 0.50. Theoretically, a large number of fish will be needed to minimize deviation from the actual "population frequency", but practical limitations often preclude the analysis of more than 100 fish from each stock; this figure has been used as the minimum sample size allowed. The population samples, however, vary within much broader limits, from 3 to some 1,000 fish per stock, but the vast majority fall close to the 100 specimens per stock average.

Figure 1B shows the idealized picture obtained in a sympatric population of charr, viz. a different growth rate and a significantly different allele frequency. The last example (Fig. 1C) shows another common feature, interpreted as a case of introgression between two taxa, resulting in a varied growth rate and different allele frequencies among slow and fast growing fish respectively. A clue as to which taxa have contributed in the introgressed stock may be obtained by analyzing samples from adjacent lakes, inhabited by seemingly allopatric populations, according to the Figure 1A criterion. With known frequencies of allopatric populations of the contributing taxa and the pooled gene frequency of the total sample in the introgressed stock, the relative proportion of the contributing taxa may be roughly calculated.

Carrying the analysis one step further additional ecological data are included. These parameters may be differences in time and place at spawning, horizontal and vertical distribution in the lake, food preference as evidenced by stomach analysis and incidence of parasites. Two frequent sympatric combinations are summarized schematically in Figure 1D and 1E.

The use of allele frequencies as a basic element in taxon recognition requires extremely detailed knowledge of the polymorphism used *and* the geographical

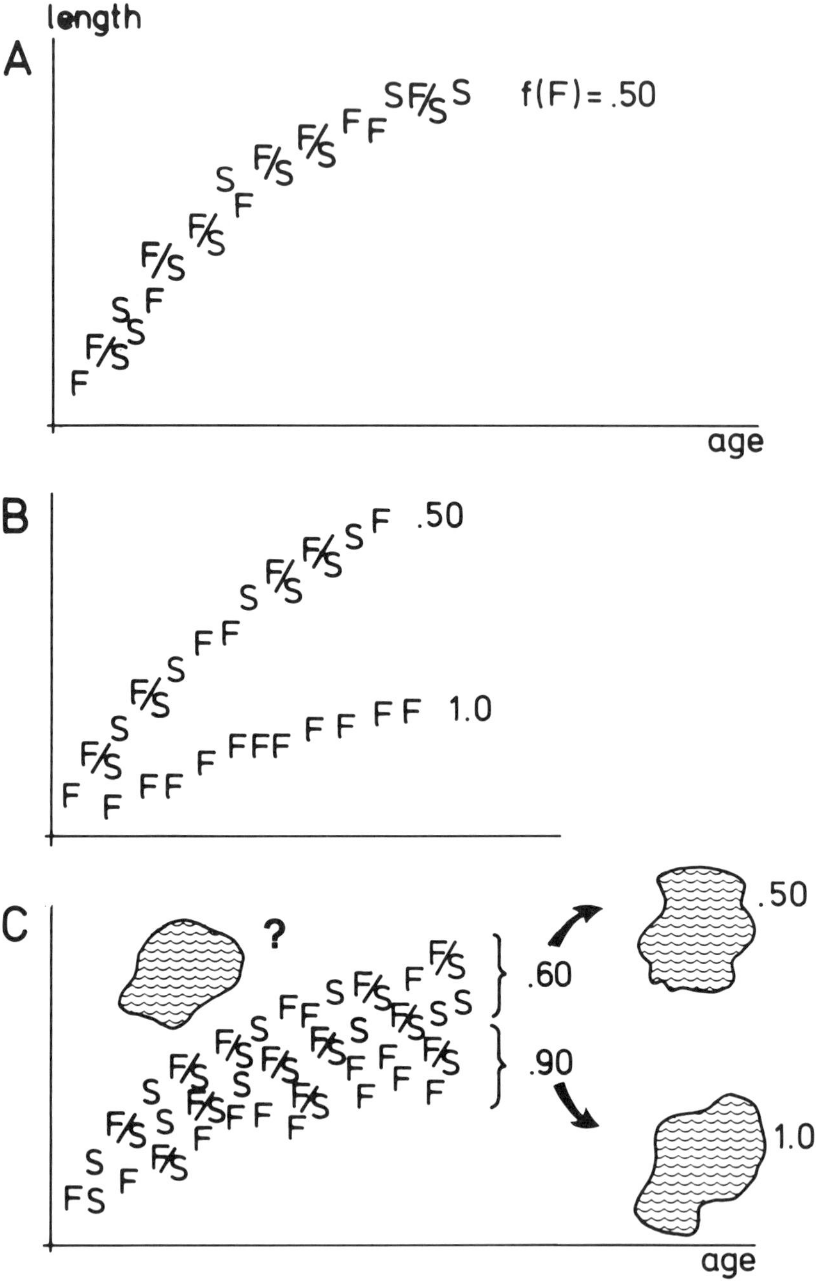
length
A
f(F) = .50
age
B
.50
1.0
C
?
.60
.90
.50
1.0
age

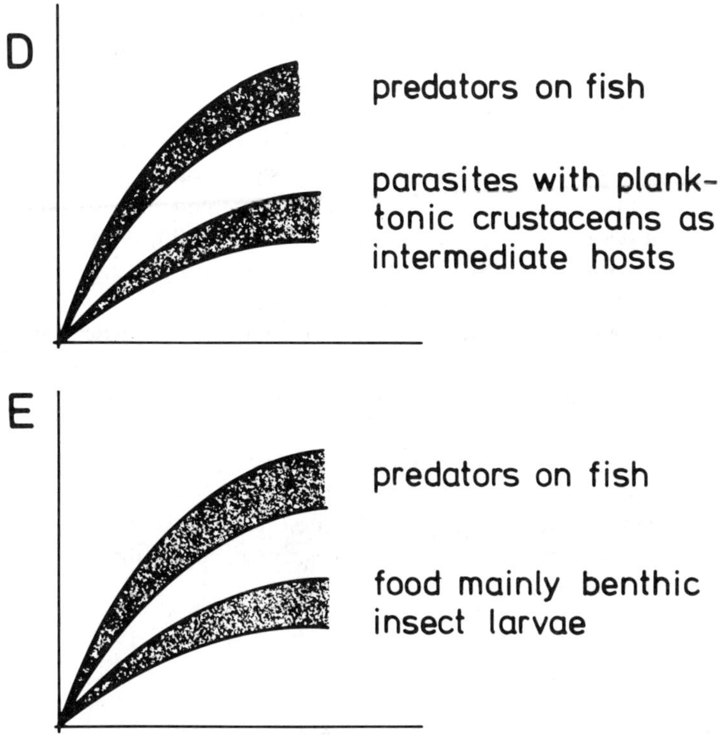

FIG. 1 Schematic presentation of basic criteria for charr taxon recognition.

A) allopatric population

B) sympatric population

C) backtracking of taxon origin in introgressed populations — adjacent allopatric populations of different taxa are used for comparison

D) one additional ecological parameter added to basic evaluation — fast growing N charr above and slow growing S charr below

E) same as D, but the slow growing taxon is the littoral/benthic F charr

In all figures the age and length of each fish is given on the horizontal and vertical axis respectively. Instead of usual plot techniques each fish is indicated by its serum esterase 'phenotype'. Allele frequencies in all figures depict the frequency of the F (fast) allele. Further information is given in the text.

and temporal stability of the frequency range found within each taxon. Since 1966 controlled breeding experiments have been performed on the recognized stocks and their offspring followed for several generations in natural waters (Nyman 1972). These fish retained the genetic and ecological characteristics of the parent taxa (Nyman 1980 and in press).

A number of cases, where fish of known origin were transplanted into lakes where the species was previously absent, were reported by the present author in 1972; those data confirmed the temporal stability of the allele frequency over several decades, no matter how different the selective forces of the old and new lakes. Numerous other cases have been found since: in stocks from the majority of waters analyzed, the criteria of age, length and esterase data have proved sufficient for recognition of three taxa of charr in allopatric populations having allele frequency ranges of roughly .40-.70, .90-1.0 and 0.0-.20, respectively (Nyman, Hammar and Gydemo 1981). The corresponding ecological characteristics of the three taxa are also summarized in the same article.

So far only four polymorphic loci have been found in tissues of Arctic charr, viz. esterases (in many tissues), transferrins, eye LDH and muscle MDH. Because so little is known about the resolving power for taxon recognition of the three last loci, only the esterases have been routinely employed. Other merits of the esterases are the strict genetic control of the polymorphism, and the fact that age, size and sex of the fish has no influence on the esterase phenotype — and neither have environmental factors. Last but not least it should be emphasized that, usually, esterases are accessible in many tissues, are tolerant to long storage and inexpensive to expose electrophoretically.

TAXONOMY

Taxonomy is not a matter of whether "osteology holds the only key to sound systematics" or "who finds the largest number of polymorphic loci", but rather the drawing together of conclusions from whatever methods provide convergent results. To give an example: no taxonomist would give equal taxonomic weight to the *number of rays* found in a particular fin compared with the *total number of fins* when separating two species of fish. The same reasoning goes for enzyme polymorphisms. By what divine power are we endowed with the ability to say that *all* polymorphisms exposed by electrophoresis have equal resolving power as population and species discriminants? If, for example, we look at many intensely studied species of fish, such as cod or herring, some polymorphisms from wide geographic areas yield gene frequency data with little variation, whereas others display considerable and significant differences at a local level. Which of these polymorphisms tells most about the local population? The charr example shows another dimension of this problem: if we were not aware of the three taxa of charr and their fairly stable intertaxa esterase frequencies, gene frequency collections would prove meaningless for stock analysis or any other genetic interpretation. A true situation is depicted in Figure 2; Figure 2A shows the pooled frequency of all charr sampled in each lake, and Figure 2B the resulting picture after the samples have been subdivided according to all supplementary data. The capital letters N, F and S stand for the three taxa of charr found, and the frequency data may be given the following interpretation. One lake is inhabited by an allopatric population of taxon N (compare with the frequency ranges given in the latter part of the Analytical Procedures section above). The other two lakes are inhabited by all three taxa of charr found; one lake shows no obvious introgression between the three taxa, the other shows introgression between taxa F and S. Evaluation of frequency data alone (Fig. 2A) would, in all probability, have been interpreted as showing each lake with a homogeneous stock of *one* taxon of charr.

It may seem that too much attention has been focused on the taxonomic side of the Arctic charr complex, but I feel that one of the basic prerequisites of sound management programs designed for sympatric taxa of fish must be detailed knowledge of each taxon, i.e. its genetic and ecological characteristics. The second step, which is of more academic interest, is the organization of this biological information into a taxonomic arrangement and, if warranted, ascribing specific or subspecific recognition to the taxa concerned. It should once more be stated that this presentation deals exclusively with landlocked populations of charr, even though all the three major taxa occur as diadromous populations.

MANAGEMENT

The usual complaint by sports fishermen and commercial fishermen in Sweden is the size reduction of charr stocks following exploitation by gillnetting. This development is of course shared with most other stocks of fish where man superimposes an unnatural selection pressure whereby the largest and fastest growing fish are caught; this is in great contrast to the natural predator-dependent situation where small fish are favoured as prey. It is important to develop management measures that will at least halt the negative impact, but this can only be done if it is recognized how this situation has developed and how different fishing methods may indeed affect different charr stocks.

The introduction of monofilament nylon gillnets in the 1950s, coincident with a drastic reduction in mesh size necessitated by the drop in average fish size, had a further negative impact. Increased netting efficiency also increased the fishing pressure on the predatory brown trout, known to exert a positive influence on the quality of charr stocks (Filipsson and Svärdson 1976). This combination of negative events led to today's situation where most charr populations consist of small-sized fish of little commercial value although still quite attractive to tolerant sports fishermen. The cultural history of the Sameh people provides one example of how a generalized management procedure may revert the trend in

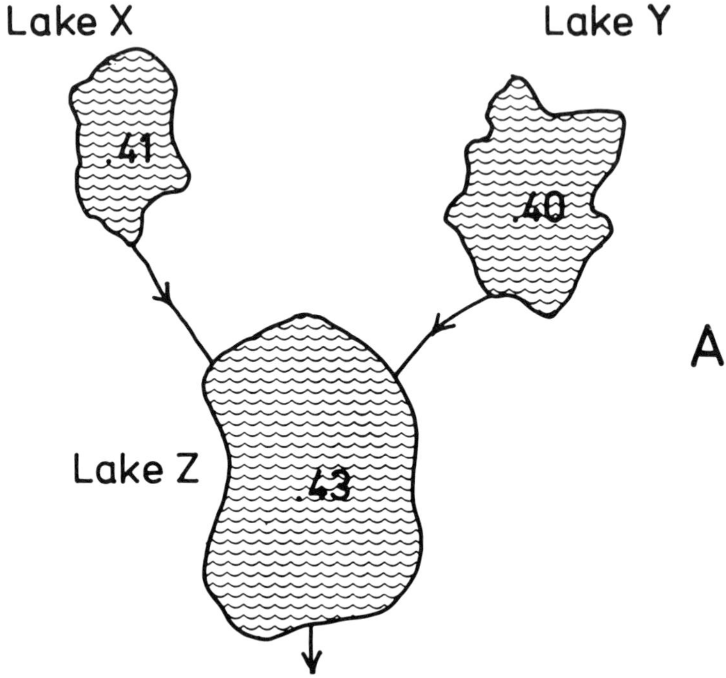

FIG. 2 a) Pooled frequency data of esterase polymorphism mentioned in the text in three inter-connected lakes.

b) The resulting figure when ecological data are used to supplement the protein marker. Lake X has allopatric populations of taxon N, Lake Y harbours all three taxa with no signs of introgression between them, and Lake Z — which also has all three taxa of charr — shows distinct signs of introgression between taxa S and F.

certain bodies of water. The Sameh were basically nomadic, tending migrating reindeer herds. To this people it seemed practical to fish the usually small lakes in a rotational way which developed into a system of letting lakes lie fallow for periods of several years. This practice has continued into present time so we have first hand information of the effects of such a system of exploitation. The results are usually positive in so far as the development of stunted charr populations rarely takes place (Filipsson and Svärdson 1976).

A thorough account of most aspects of charr management is found in Filipsson and Svärdson (1976) which draws most of its material from a contemporaneous article by Svärdson (1976). Most of the basic recommendations in the present paper are based on these publications but I have tried to stress even more the importance of recognizing the three taxa of charr before management policies are adopted. All principles developed recognize that all three taxa of charr are

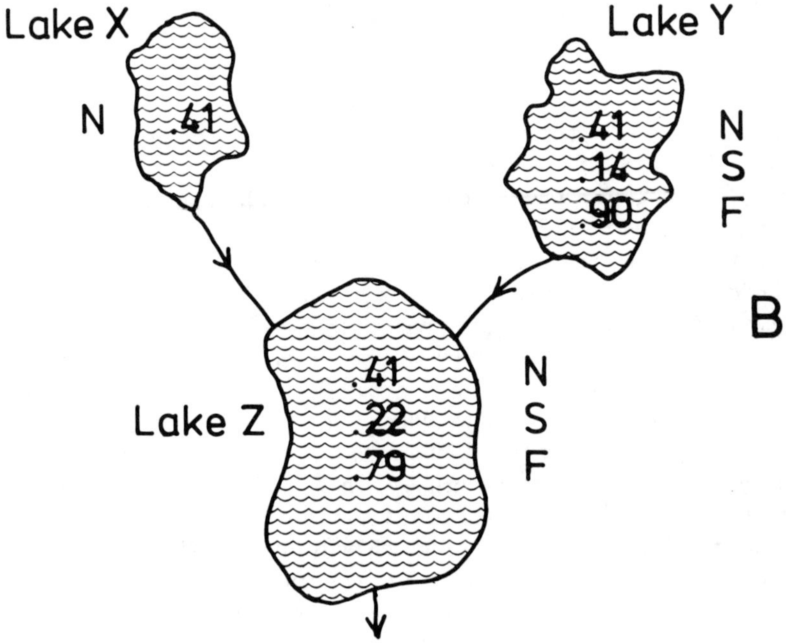

sensitive to interaction with other species of fish, and almost all species — with the exception of the brown trout — exert such a powerful influence on the charrs that sympatric occurrence often leads to extinction of the charr. This leads to the first basic principle in charr management: 1) never introduce whitefish, cisco (vendace), pike, perch, roach (*Rutilus*) or any other cyprinid fish into a charr habitat if charr are considered important by the local fishermen. The second principle allows predation on small charr by brown trout: 2) protect brown trout habitat, and as a consequence: 3) minimize netting in the littoral zone. Two other basic principles are also linked: 4) increase mesh sizes, so as to: 5) select for the largest charr possible. The sixth basic principle maintains stock purity: 6) never stock lakes with species of charr other than those already present. This recommendation stems from the negative effects of charr introgression, i.e. the depletion of locally adapted gene pools and the negative outcome of inter-taxon introgression/interbreeding.

In addition to these general principles local conditions must be considered: these may be grouped into six categories: 1) let small lakes with allopatric charr or charr/brown trout lie fallow (the merits of the rotational system practiced by the Sameh is outlined above); 2) in impounded lakes or rivers non-littoral crustaceans may be introduced to improve flesh quality and possibly average size (there is a potential danger to be borne in mind that introduced food items may cause a better growth rate even for ''suppressed'' F and S charr, thereby initiating introgression with N charr — if present); 3) in impounded lakes or rivers with a depleted littoral zone do not introduce F charr unless spawning streams are available (the F charr has an ecology similar to the brown trout, often spawning in running water; it is tied to the littoral zone and feeds on benthic crustaceans or even fish); 4) introduce N charr when forage fish are present and S charr when little but planktonic food is available (the different ecological characteristics of the three taxa are outlined in more detail in Nyman, Hammar and Gydemo 1981); 5) in sympatric charr populations use bottom-set gillnets, particularly in deep lakes where there is vertical segregation of the sympatric taxa (the ecologically suppressed dwarfed charr populations may then be selectively removed); 6) in sympatric populations try partial treatment with rotenone or antimycin on the spawning grounds of undesired taxa of charr (F or S) living sympatrially with N charr (this procedure is likely to reduce interactive segregation and/or the potential risk of introgression from ''inferior'' taxa — from a human standpoint).

These recommendations for the management of landlocked charr populations necessitate the development of standardized stock-taking and the development of individual management programmes for each body of water. The transplantation of fish has been common in Scandinavia for many centuries (Ekman 1910), but in recent times, with rapid air and ground transportation, there has been a great extension of the practice; fortunately most transplantation has been accompanied with detailed records of the date and origin of the fish employed. This has provided a valuable tool for evaluation of the practice and allowed explanation of discrepancies and ''aberrations'' otherwise difficult to include within a single formulation.

REFERENCES

Artedi, P. 1738. Ichtyologia. (Posthuma). C. Linnaeus [ed.] (Reprint 1967 by J. Cromer.) Wheldom and Wesley, Ltd. and Hafner Publishing Co., Codicote, Herts. and New York, N.Y. 554 p.

Balon, E.K. [ed.] 1980. Charrs: salmonid fishes of the genus *Salvelinus*. Dr. W. Junk Publishers, The Hague. 928 p.

Child, A.R. 1977. Biochemical polymorphism in charr (*S. alvelinus alpinus* L.) from Llynnau Peris, Padarn, Cwellyn and Bodlyn. Heredity 38(3):359-365.

Child, A.R. 1980. Electrophoretic analysis of the Arctic charr of Cumbrian lakes. *In* Proceedings of the first ISACF Workshop on Arctic charr. Inst. Freshw. Res., Drottningholm.

Ekman, S. 1910. Om människans andel i fiskfaunans spridning till det inre Norrlands vatten. Ymer 30:133-140. (In Swedish).

Ferguson, A. 1980. Electrophoretic analyses of tissue proteins of Irish charr. *In* Proceedings of the first ISACF workshop on Arctic charr. Inst. Freshw. Res., Drottningholm.

Filipsson, O. and G. Svärdson. 1976. Principer för fiskevården i rödingsjöar.Inform. Inst. Freshw. Res. Drottningholm (2). 79 p. (In Swedish).

Fürst, M., U. Boström and J. Hammar. 1978. Effects of new fish food organisms in Lake Blåsjön. Inform. Inst. Freshw. Res. Drottningholm (15). 94 p. (In Swedish with English summary).

Gydemo, R. 1980. Electrophoretic studies on Arctic charr in the county of Västerbotten, northern Sweden and Iceland: a survey of the distribution of the Arctic charr species. *In* Proceedings of the first ISACF workshop on Arctic charr. Inst. Freshw. Res., Drottningholm.

Hammar, J. 1980. Ecology and taxonomy of Arctic charr in lake reservoirs in Sweden. *In* Proceedings of the first ISACF workshop on Arctic charr. Inst. Freshw. Res., Drottningholm.

Henricson, J. and L. Nyman. 1976. The ecological and genetical segregation of two sympatric species of dwarfed char (*Salvelinus alpinus* (L.) species complex). Inst. Freshw. Res. Drottningholm Rep. 55:15-37.

Holthe, H. 1981. To sympatriske former av røye, *Salvelinus alpinus* (L.) i Ellasjøen, Bjørnøya. En sammenligning av genotypiske og fenotypiske trekk. Hovedfagsoppgave i ferskvannsbiologi, Univ., Tromsø. 76 p. (In Norwegian).

Klemetsen, A. and P.E. Grotnes. 1980. Coexistence and immigration of two sympatric Arctic charr, p. 757-763. *In* E.K. Balon [ed.] Charrs: salmonid fishes of the genus *Salvelinus*. Dr. W. Junk Publishers, The Hague.

Linné, C. von. 1758. Systema Naturae. Regnum Animale. Editio Decima. 824 p.

Nilsen, P. 1980. Stasjonaer- og anadrom røye, *Salvelinus alpinus* (L.). En sammenligning av økologiske og genetiske trekk hos bestander med sannsynlig felles opprinnelse. Hovedfagsoppgave i akvatisk biologi, Univ., Tromsø. 84 p. (In Norwegian).

Nyman, L. 1965. Variation of proteins in hybrids and parental species of fishes. Rep. Swed. Salm. Res. Inst. 13:11 p.

Nyman, L. 1972. A new approach to the taxonomy of the "*Salvelinus alpinus* species complex". Inst. Freshw. Res. Drottningholm Rep. 52:103-131.

Nyman, L. 1980. The population genetics of Arctic charr: a state of the art. *In* Proceedings of the first ISACF workshop on Arctic charr. Inst. Freshw. Res., Drottningholm.

Nyman, L. and O. Filipsson. 1972. The chars of Lake Yraf. Inform. Inst. Freshw. Res., Drottningholm (10). 20 p. (In Swedish with English summary).

Nyman, L., J. Hammar and R. Gydemo. 1981. The systematics and biology of landlocked populations of Arctic charr from northern Europe. Inst. Freshw. Res. Drottningholm Rep. 59:128-141.

Olsen, P. 1980. Population development of introduced *Mysis relicta* and impact on char and brown trout. Doctor's Thesis. Scripta Limnol. Upsaliensia 504: 41 p.

Svärdson, G. 1976. Interspecific population dominance in fish communities of Scandinavian lakes. Inst. Freshw. Res. Drottningholm Rep. 55:144-171.

NOTE ON THE SPELLING OF CHARR:

I have been rather disturbed ever since the publication of "Charrs" (Balon 1980); the effects of which we can all observe in the original contributions to this symposium. I am referring to the "char versus charr controversy". William Markham Morton wrote a very significant contribution in that book where he reviewed the history of the spelling of charr with one or two "r's". He did so thoroughly and wittily, but to my mind forgot one important aspect of the issue: everybody working with the taxonomic aspects of charr soon realizes that charr taxa like all other organisms are subject to change, usually slowly and gradually, by a process we term evolution. Man — being part of the same dynamic system

— is also subject to change, not only physically but also mentally and linguistically. We do not use the language of Shakespeare today but an evolved version of it (at best). This process is natural and neither can nor should be avoided. With this type of reasoning I fail to see the significance of reverting to the 18th century spelling acknowledged by Artedi and Linné without performing a complete revision of the spelling and usage of English words back to bygone days. Why not stick to the long accepted spelling of charr with one ''r'', which at least is an invention of the 20th century like so many other words in current use. There is no turning back in evolution! Or better still, why not let natural selection decide which spelling is ''right'' by allowing the simultaneous use of both.

Ecological Characters of Different Combinations of Sympatric Populations of Arctic Charr in Sweden

JOHAN HAMMAR

Institute of Freshwater Research, S-170 11 Drottningholm, Sweden

HAMMAR, J. 1984. Ecological characters of different combinations of sympatric populations of Arctic charr in Sweden, p. 35–63. *In* L. Johnson and B.L. Burns (eds.) Biology of the Arctic charr, Proceedings of the International Symposium on Arctic Charr, Winnipeg, Manitoba, May 1981. Univ. Manitoba Press, Winnipeg.

Long term experimental introductions of new fish food organisms as a means of managing salmonid stocks in impounded lakes have contributed to the knowledge of the taxonomy and ecology of the Arctic charr complex in Sweden. Sympatric populations of landlocked charr have been analyzed in a number of reservoirs. Ecological characters such as distribution patterns within river systems and in lakes, growth, feeding habits, spawning behaviour and interactive competition support the theory of three sibling species. Frequencies of serum esterases were used as gene markers to identify the lowland "normal" charr and two high altitude charrs, types F and S. The benthic, and sometimes stream spawning, mountain charr, type F, is the most common species in high altitude lakes in Sweden; it is sensitive to competition from the smaller, pelagic, lake spawning "S charr". The last immigrant, the big "normal" charr, is the most competitive species and is dominant in all sympatric population combinations. The order of interactive competition supports the theory of the order of immigration, F, S and then "normal" charr. All combinations of the three sibling species are found in Sweden, and sympatric populations of big "normal" charr and either dwarfed "F" or "S" charr are most common in the reservoirs studied.

INTRODUCTION

The family Salmonidae inhabit the northern hemisphere in areas influenced by recent glaciations. Glacial and interglacial periods created a series of events which led to successive periods of isolation interspersed with access provided by migration routes either by way of ice-dammed lakes or the incursion of marine or brackish waters. allowed ecological or genetic variation that led to the creation of species, sibling species and species complexes. To these processes is attributed the great variation within the Salmonid fishes as a group and their attainment of a completely circumpolar distribution.

Many examples of the results of these changes are to be found in the mountain lakes of Scandinavia where the salmonids "Arctic charr" and brown trout, and "whitefish" show many instances in which populations of closely allied forms live together in the same lake or stream. The present study attempts to develop a pattern of occurrence with respect to ecological, morphological and genetic characteristics within the Arctic charr complex.

Central to this discussion is the mechanism of origin of the various groupings, or taxa, within the genus: have they developed or evolved sympatrically as a result of ecological conditions or through local isolation and recombination or by isolation in some area distant from their present distribution which they have reached by a number of waves of immigration? According to Svärdson (1961, 1979) speciation in the whitefish and Arctic charr complexes of Scandinavia was preglacial, sibling species having evolved outside the area; in post-glacial times their immigration history has been similar. Behnke (1972) regards some sympatric, reproductively isolated populations of Arctic charr as the product of "micro-geographical" speciation from a common ancestor since the last glaciation, recognizable in minor differences in the number of gillrakers.

Both genetic and ecological characters are used to distinguish sibling species within each complex; numerical differences in the arrangement of parts in ecologically isolated populations are regarded as evidence of introgression of originally well-separated species (Svärdson 1979). A complex pattern of biotopes with slight differences in requirements has allowed many interactive confrontations to develop between originally separated species; this has created a number of sympatric populations with great variation in the degree to which introgression has occurred between them: distinct species A and B have changed to highly modified populations of A or B. Svärdson's (1979) opinion is that the process can be characterized as "speciation in regress", from an initial complement of three sibling species within the Arctic charr complex. The reality of these various forms of Arctic charr is confirmed by an extensive series of local names designated by the fishermen of the area; not only do these names refer to distinct morphological types but they are also validated by the recognition of particular spawning sites and times (Fig. 1).

The first attempt to separate these groups on a scientific basis was carried out by Määr (1949, 1950) on the Arctic charr of lakes of the Faxälven system; he separated different biological groups by means of number of eggs, spawning time and place. Subsequent analyses have compared growth and food habits (Nilsson 1955; Nilsson and Filipsson 1971; Klemetsen and Grotnes 1975, 1980; Fürst, Boström and Hammar 1978), habitat choice (Nilsson and Filipsson 1971; Klemetsen and Grotnes 1975, 1980; Fürst, Boström and Hammar 1981), spawning times (Frost 1965, Gönczi 1971), quality and quantity of parasite infestation (Henricson and Nyman 1976) and finally genetically with the aid of electrophoretic methods (Nyman 1965, 1972; Nyman and Filipsson 1972; Child 1980;

FIG.1 The morphological characters of two Arctic charr species in Lake Blåsjön, April 1977. Above: 'Normal charr', 211 mm, 84 g, age 4 years. Below: Tita or 'S-charr', 194 mm, 71 g, age 6 years. Photo: J. Hammar

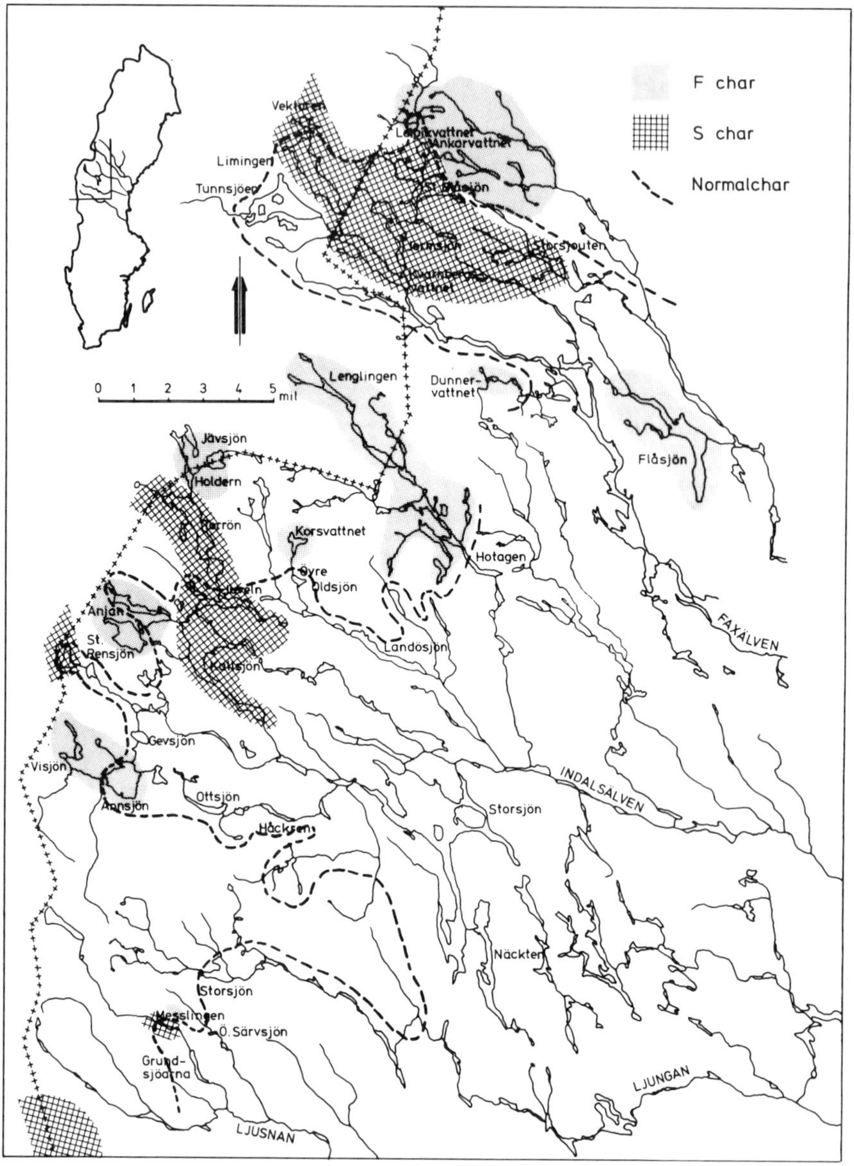

FIG. 2 Distribution of three sibling species within the Arctic charr complex in the Province of Jämtland, northern Sweden. The heavy line shows the upper limit of 'Normal charr' populations.

Hammar 1980; Nyman 1980). Nyman (1972) found that the dwarfs in some sympatric populations in Sweden belong to two genetically different species.

Many of the lakes with Arctic charr populations have been converted to hydroelectric reservoirs with consequent changes in their level regimes and, in addition, many of these reservoirs have had new fish food organisms introduced. Careful study of the effect of these changes (Fürst et al. 1978, 1981; Hammar unpubl.) has provided what is, in effect, a large scale experiment. As Arctic charr is the most important species in these reservoir lakes it is possible to investigate in detail the sympatric populations, their life histories and ecological requirements under the influence of change, co-ordinated with the taxonomy of the three postulated sibling species within the Arctic charr complex (Hammar 1980; Nyman et al. 1981).

MATERIALS AND METHODS

Long term experiments with new fish food organisms have been made in lake reservoirs and evaluated partly by means of recurrent standardized test-fishing (Fürst et al. 1978, 1980, 1981). Benthic and pelagic stations of different characteristics and depths were studied to get a picture of the main habitats and ecology of different fish species. Biological gillnets with mesh sizes ranging from 10 to 50 mm were used and fish length, weight, sex, maturation, flesh colour, parasites and special physical characters were noted as well as station and depth for each fish specimen. Otoliths, gillrakers, stomachs and blood samples were taken for further analysis in the laboratory.

Ages of Arctic charr were determined from otoliths according to Nordeng (1961) and Filipsson (1967); unfortunately, for those fish caught before 1966 (Runnström 1951), only scale ages are available. Documentation of the different food items in the stomachs was made in percent of volume. The polymorphism in serum esterases discovered by Nyman (1965, 1972) was studied in horizontal starch gel electrophoresis of blood samples from the heart. The buffers and enzyme substrates were the same as described by Nyman (1967).

In some lakes Arctic charr were bought from local fishermen in months of special interest in order to study critical periods in the food habits. Interviews and discussions with local people were held, particularly with respect to spawning times and spawning grounds.

Lake Torrön, in the Indalsälven system (Fig. 8) and Lake Blåsjön, in the Faxälven System (Fig. 3), are both classical examples of lake reservoir studies; they have been the main study areas for the effects of impoundment and water level regulation on fish (Runnström 1951, Nilsson 1955, 1961) and the effect of introductions of the two relict crustaceans, *Mysis relicta* Lovén and *Pallasea*

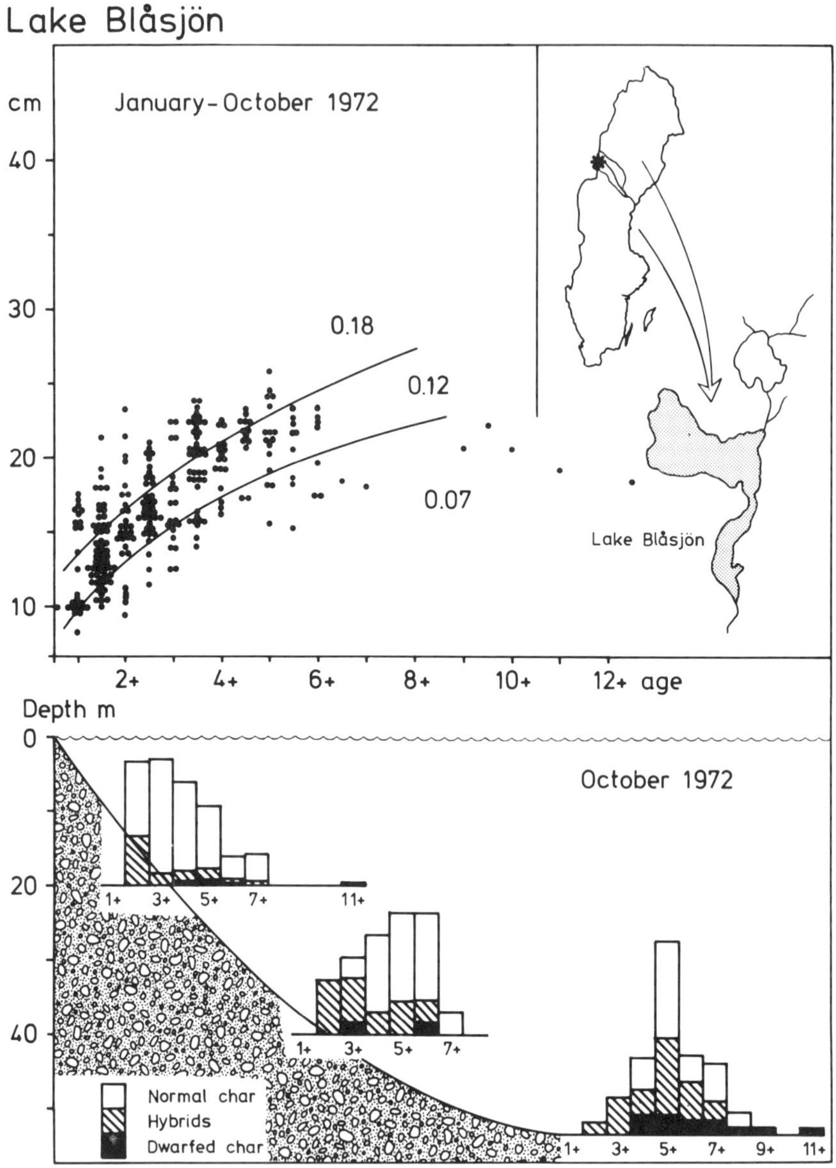

FIG. 3 Growth, gene frequencies of different growth fractions (above left), age structures and depth distribution (below) of two sympatric populations of Arctic charr in Lake Blåsjön. Each graph represents 100 percent of the specimens caught at each special depth interval. The location of the lake reservoir is shown (above right).

quadrispinosa Sars (Fürst et al. 1978, 1981). Kvarnbergsvattnet, Faxälven (Fig. 9), Lake Ajaure, Umeälven (Fig. 5) and Lake Suorva, Stora Luleälven (Fig. 7) are also reservoirs where new fish food organisms have been introduced. Lake Yraf, Laisälven, Övre Björkvattnet, Umeälven and Lakes Upper Fättjaure and Lower Fättjaure and Borkajaure, River Vojman, have been described by Nyman and Filipsson (1972), Nilsson and Filipsson (1971) and Henricson and Nyman (1976).

The otolith examinations were performed by O. Filipsson (Lake Blåsjön, Lake Suorva), O. Lindh (Kvarnbergsvattnet) and the author (Lake Ajaure, Lake Torrön); electrophoretic analyses of charr in Lake Blåsjön 1972 and Lake Suorva 1976 were made by L. Nyman and the remainder by myself.

Combined with the investigation of ecological characteristics was an investigation of gene markers to assist in the identification of the three sibling species within the Arctic charr complex. These sibling species are identified by frequencies of the F-allele, or fast allele, in serum esterases: the so-called ''F-charr'' (frequencies (F) = 0.90-1.0), ''S-charr'' (f(F) = 0.00-0.20) and ''Normal charr'' (f(F) = 0.40-0.70, 1.00) (Hammar 1980, Nyman 1980, Nyman et al. 1981, Nyman this volume). Poor fit with Castle-Hardy-Weinberg equilibrium in samples with excess homozygotes is regarded as an indication of possibly more than one spawning population. With the knowledge of distribution and gene frequencies of surrounding allopatric and sympatric populations the degree of introgression is measurable. Fixation of the F-allele in allopatric populations in small headwaters is common and in some sparse populations of lowland deep-water lakes also due to genetic drift (Nyman 1972). All three sibling species occur allopatrically and all four combinations of sympatry are found in Sweden. Three species have also been identified in Norway (Nyman et al. 1981), Iceland (Gydemo this volume), Svalbard (Hammar unpubl. data) and probably also in Canada (Nyman 1972) and Greenland (Hammar unpubl. data). The populations in Sweden are landlocked, but all three species are recorded as anadromous in Arctic areas (Nyman et al. 1981).

In almost all river systems studied the ''F-charr'' is found allopatrically in high altitude headwater lakes, small creeks and springs. The ''S-charr'' inhabits some large mountain lakes close to the Norwegian border to the west, and the ''Normal charr'' is found allopatrically in low altitude deep clear-water lakes; it is often seen sympatrically with either of the two mountain charr species as well as other fish genera. As an example, the distribution of these three sibling species in the Province of Jämtland, latitude 63°-65°N, is shown in Figure 2. The map is based on 70 lakes analyzed (Nyman 1972; Hammar 1980, 1981; Hammar and Nyman in prep.). A general survey of the systematics and distribution of the

Table 1. Distribution of phenotypes and agreement with CHW equilibrium in the five lakes with sympatric populations of Arctic charr. (Total = Total sample, Norm = 'Normal charr', X = intermediate charr, S = 'S-charr', F = 'F-charr', Benth = Benthic catch, Pelag = Pelagic catch, Stream = Stream spawning population)

Locality	Date	Pop.	Total	fF	Est obs. F/F	Est obs. F/S	Est obs. S/S	Est exp F/F	Est exp F/S	Est exp S/S	x^2(1d.f.)	Probability of worse fit	95% c.i.
Lake Blåsjön	Jan-Dec 1971	Total	311	0.12	4	67	240	4.5	65.7	240.8	0.08	0.80-0.70	0.09-0.15
		Norm.	80	0.18	1	26	53	2.5	23.1	54.5	1.31	0.30-0.20	0.12-0.24
		X	150	0.12	3	29	118	2.2	31.7	116.2	0.55	0.50-0.30	0.08-0.16
		S	81	0.07	0	12	69	0.4	10.5	70.1	0.63	0.50-0.30	0.03-0.11
Lake Ajaure	July 1979	Pelag.	19	0.74	10	8	1	10.4	7.3	1.3	0.15	0.70-0.50	0.60-0.88
		Benth.	196	0.71	100	78	18	98.5	80.9	16.6	0.24	0.70-0.50	0.66-0.76
		Norm.	49	0.54	11	31	7	14.3	24.3	10.4	3.72	0.10-0.05	0.44-0.64
		X	64	0.70	33	24	7	31.4	26.9	5.8	0.64	0.50-0.30	0.62-0.78
		F	83	0.81	56	23	4	54.5	25.5	3.0	1.22	0.30-0.20	0.75-0.87
Lake Suorva	Aug 1976	Total	206	0.44	46	89	71	39.9	101.5	64.6	3.11	0.10-0.05	0.39-0.49
		Norm.	147	0.41	27	68	52	24.7	71.1	51.1	0.37	0.70-0.50	0.35-0.47
		Dwarf	59	0.50	19	21	19	14.8	29.5	14.8	4.83	0.05-0.025	0.41-0.59
		S	30	0.22	1	11	18	1.4	10.3	18.3	0.17	0.70-0.50	0.11-0.33
		F	29	0.79	18	10	1	18.1	9.6	1.3	0.09	0.80-0.70	0.68-0.90
Lake Torrön	Aug 1979	Total	112	0.14	3	25	84	2.2	27.0	82.8	0.46	0.50	0.09-0.19
		Benth.	68	0.14	2	15	51	1.3	16.4	50.3	0.51	0.50-0.30	0.08-0.20
		Pelag.	44	0.14	1	10	33	0.9	10.6	32.5	0.05	0.90-0.80	0.07-0.21
R. Holderströmmen	Oct 1976	Stream.	10	0.30	1	4	5	0.9	4.2	4.9	0.02	0.90-0.80	0.10-0.50
R. Gånälven	Sep 1979	Stream.	25	0.24	3	6	16	1.4	9.1	14.4	3.06	0.10-0.05	0.12-0.36
Lake Kvarnbergsv.	Aug-Sep 79	Total	129	0.40	24	54	51	20.6	61.9	46.4	2.03	0.20-0.10	0.34-0.46
		Norm.	32	0.58	13	11	8	10.8	15.6	5.6	2.83	0.10-0.05	0.46-0.70
		X	58	0.38	8	28	22	8.4	27.3	22.3	0.04	0.90-0.80	0.29-0.47
		S	39	0.27	3	15	21	2.8	15.4	20.8	0.03	0.90-0.80	0.17-0.37
R.Brännälven	Sep-Oct 1979	Stream	20	0.33	2	9	9	2.2	8.8	9.0	0.02	0.90-0.80	0.18-0.48

Salvelinus alpinus complex was published in the Report series of the Drottningholm Institute (Nyman et al. 1981).

Theory suggests that the Arctic charr complex in Scandinavian waters can be explained on the basis of three sibling species of preglacial origin. In their extreme forms they can be described as follows:

1) an early immigrating, littoral, benthos- and fish-eating, stream spawning high mountain charr, sensitive to other fish species and ecologically very similar to the brown trout (*Salmo trutta* L.) in Europe or the Dolly Varden (*Salvelinus malma* Walbaum) in North America.
2) a smaller, pelagic, plankton-eating, lake spawning mountain charr from a later immigration period and dominating the earlier immigrant; and
3) a large, ecologically more plastic, fish-eating, lake spawning lowland charr dominating the two earlier immigrants.

The objective of the present paper is to test this theory in the light of the known ecological features of the various groups or taxa within the Arctic charr complex in Swedish waters.

RESULTS

Three different combinations of sympatric populations of lake spawning Arctic charr and two combinations of lake spawning and stream spawning populations are studied for patterns of ecological characteristics. These patterns are compared with the pattern of serum esterase distribution (Table 1).

Sympatric Lake Spawning Populations

1) ''Normal Charr'' + ''S-Charr''

This combination dominates the upper parts of Indalsälven, Faxälven, Fjällsjöälven and Laisälven.

Growth: Figure 3 shows the growth pattern in Lake Blåsjön; ''Normal charr'' are large, the ''S-charr'' dwarfed. Populations are highly introgressed and gene frequencies of 0.18 and 0.07 suggest a ''genetically'' dominating ''S-charr''. This picture also characterizes the relationships of sympatric populations in Lake Ankarvattnet (0.18, 0.07) upstream and Kvarnbergsvattnet (0.58, 0.27) (Fig. 9) downstream of Lake Blåsjön. The same situation is found in Lake Yraf (0.49, 0.28) (Nyman and Filipsson 1972) and downstream in Lake Gautosjön (0.44, 0.29) (Hammar 1978), Laisälven where the cline in introgression is obvious.

Habitat: In Lake Blåsjön dwarfed ''S-charr'' are found in deep water with ''Normal charr'' inhabiting shallower regions; a third ''Intermediate'' group can also be recognized. The depth distribution of the three parts of the population is shown in Figure 3; in October 1967 and October 1974 the following mean depths were recorded:

	1967	1974
"Normal charr"	16.9 m (n = 208)	24.0 m (n = 39)
"Intermediate charr"	27.3 m (n = 85)	25.8 m (n=39)
"S-charr"	36.0 m (n = 19)	47.3 m (n = 39)

In all growth fractions there are correlations between age and depth: the younger fishes being closer to the shore, the older ones in deeper water. Pelagic fishing has never been successful in this lake since impoundment (Filipsson pers. comm.).

Reproduction: In Lake Blåsjön "Normal charr" spawn in shallow water around October 10, the dwarfed charr three weeks later on spawning grounds located in deeper water. In recent years "Normal charr" have spawned later; the overlap in spawning time may allow increased introgression. Males of both species are known to spawn as four-summer-old (3+) and females as five-summer-old (4+) individuals. In 1967 the smallest spawners among the dwarfs were 158 (♂) and 165 (♀) mm. Both males and females of the "Normal charr" were between 210 and 220 mm as smallest spawners.

There are no records of stream spawning charr in Lakes Blåsjön and Ankarvattnet.

Food Habits: Nilsson (1955, 1961) studied the effects of the food habits of the Arctic charr in Lake Blåsjön before and after the start of water level regulations. In Fürst et al. (1978) the effects of introductions of the new fish food organisms *Mysis relicta* Lovén and *Pallasea quadrispinosa* Sars are intensely analyzed. Approximately 9,000 stomachs from Arctic charr in Lake Blåsjön have been analyzed.

In Figure 4 the food habits and the feeding activity, measured as degree of stomach fullness, are shown. The deep-living "S-charr" feed to a greater extent on benthic invertebrates and terrestrial insects showing a constant activity in feeding during the whole year, as measured by the degree of stomach fullness. The plankton part of the food habits of the "Normal charr", the consequence of regulation and interspecific segregation with brown trout, was still of striking importance in 1971. Today the amount of *Mysis* in the stomachs of "Normal charr" is significant and several specimens pass the critical size and start feeding on smaller fish. In March-April very few "Normal charr" had anything in their stomachs. For further information see Fürst et al. (1978).

2) "Normal Charr" + "F-Charr"

This combination is very common and dominates river systems such as River Ume (Gydemo 1980), River Skellefte and River Torne (Hammar 1979).

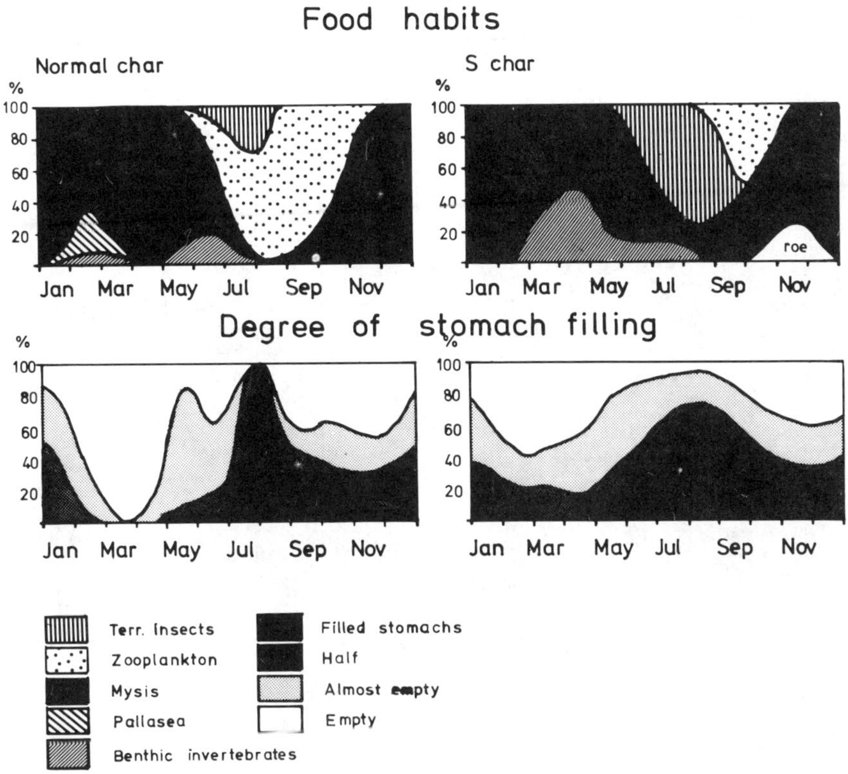

FIG. 4 The food habits and the degree of stomach fullness of 'Normal charr' and dwarfed 'S-charr' in Lake Blåsjön, 1971. The different food items and the classifications of the amount of food are shown below.

Growth: In all cases "Normal charr" dominate the size relationships; "F-charr" are dwarfed. In Lake Ajaure, River Ume, two distinct species with the gene frequencies 0.54 and 0.81 are identified in a growth diagram (Fig. 5). The dwarf or "blattjen" is an extremely deep-living charr with a blunt nose and pale green colouration with stirr patches not unlike a perch. In Lake Övre Björkvattnet, River Ume, a similar situation is described by Nilsson and Filipsson (1971). In this lake the gene frequency of the "F-charr" is fixed to 1.00 and the frequency of the "Normal charr" is 0.73 (Nyman 1972, Hammar unpubl. data). The morphological appearance of the two species in Lake Övre Björkvattnet is shown in a drawing (Fig. 6). The "blattjen" charr in Lake Övre Björkvattnet was known by elderly fishermen to be much larger: specimens up to 8 kg are mentioned (O. Filipsson pers. comm.). This growth pattern is further found in Lake Stora

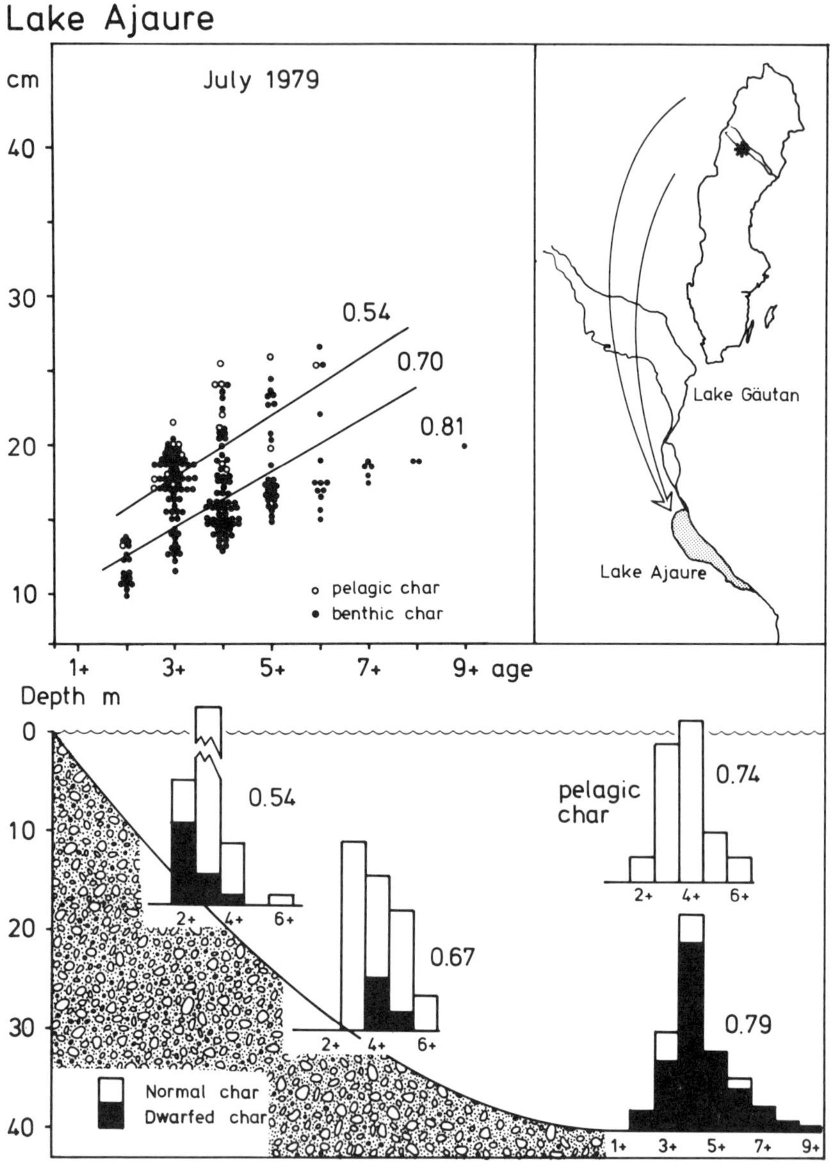

FIG. 5 Growth, gene frequencies of different growth fractions of catch (above left), age structures and depth distribution (below) of two sympatric populations of Arctic charr in Lake Ajaure. Each graph represents 100 percent of the specimens caught at each special depth interval. The location of the lake is shown (above right).

Rösjön, (0.75, 0.89), River Dalälven (Andersson et al. 1971, Nyman 1972, Lindström and Hammar unpubl. data).

Habitat: The cline in gene frequencies found in the growth pattern in Lake Ajaure is found also in the depth distribution. The majority of the dwarfs are found at depths between 30 and 40 metres, although some are close to the shore. The fast growing "Normal charr" are caught mainly at depths less than 15 metres. The proportions of the two species and the age structures at different depths are shown in Figure 5. The same picture is found in Lake Övre Björkvattnet. A pelagic population of very fast growing, high gene frequency, charr is also found in Lake Ajaure, similar to the fast growing "F-charr" in Lake Gäutan, upstream (Gydemo 1980).

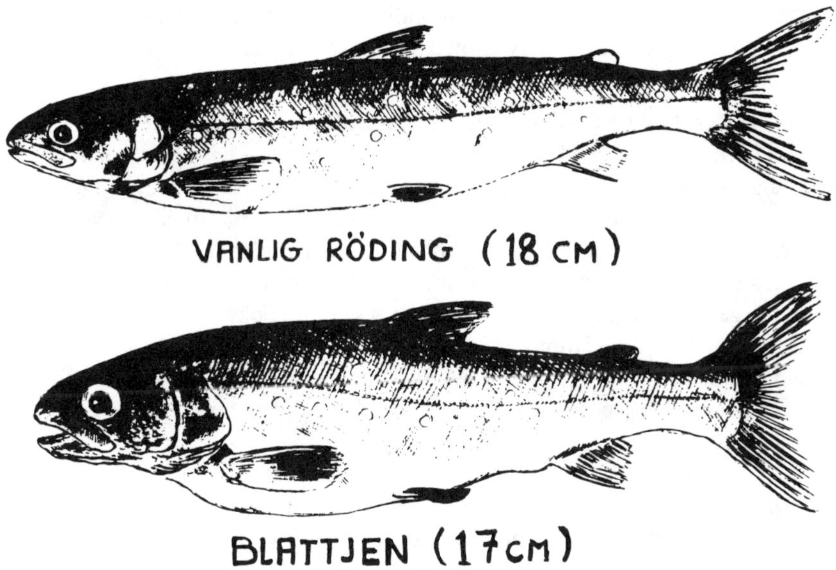

FIG. 6 The morphological characters of two Arctic charr species in Lake Övre Björkvattnet. Above: 'Normal charr'. Below: Blattjen or 'F-charr'. Drawing: L. Sörensen

Reproduction: At present no data are available about reproduction in Lake Ajaure. According to local fishermen no reproduction has been possible since the impoundment of 18 metres in 1967, a statement which I doubt. However, in Lake Övre Björkvattnet the "Normal charr" was found to spawn from mid-September to early October with the "blattjen" spawning later (Nilsson and Filipsson 1971). In Juksbäcken, a small tributary upstream of Lake Ajaure and in some other nearby parts of the river system, stream spawning charr is known to occur in the first half of September (Lindström 1954, 1958, Pettersson 1961). In

Lake Övre Björkvattnet no such reproduction is recorded. In Lake St. Rösjön the dwarfs are known to inhabit running water in summer.

Food Habits: Both species in Lake Ajaure utilize the introduced amphipod *Pallasea quadrispinosa* Sars but sufficient material has not yet been analyzed to compare the food habits. In Lake Övre Björkvattnet, Nilsson and Filipsson (1971) studied the food habits; they found "blattjen" feeding on amphipods, benthic cladocerans, molluscs and fish in late summer and autumn while the "Normal charr" were feeding on pelagic cladocerans such as *Daphnia* sp. and *Bosmina* sp. and terrestrial insects.

3) "Normal charr" + "S-charr" + "F-charr"

All three species are found sympatrically in at least three headwater lakes of River Stora Lule, so far the only known system in Sweden where it is possible to study the entire complex together in the same lake.

Growth: In August 1976 and 1979 test nettings were made and the specimens were classified in the field as "Normals" and dwarfs. The growth patterns in 1976 and 1979 of fish from the same station, were not distinct, suggesting accelerated introgression (Fig. 7). The fast growing "Normal charr" have a gene frequency of 0.41, identical with the allopatric populations upstream in the Padjelanta area (Nyman in press, Hammar and Nyman unpubl. data). The dwarfs showed a gene frequency of 0.50 with an excess of homozygotes which could be split up into two populations 0.79 and 0.22, "F-charr" and "S-charr", in equilibrium with Castle-Hardy-Weinberg (Nyman in Press).

It is however not possible to distinguish the dwarfed species in the growth diagram (Fig. 7).

Habitat: The "Normal charr" is found in shallow water, closer to the surface than whitefish (Hanson and Lindström 1979). The dwarfs are found at greater depths; in some parts of the huge reservoir the "F-charr" dominates and in others the "S-charr". The evaluation of the different ecological "borders" of the three sibling species in Lake Suorva has just started.

Reproduction: In Lake Sitasjaure, with the same sympatric combination, just northwest of Lake Suorva, the charr was found to spawn very late in December-January (S. Andreasson pers. comm.). In Lake Suorva where only Lapps have been fishing during the open water season there is, so far, no information of spawning grounds and spawning times. Vallin (1938) describes two types of Arctic charr: the dwarf is blunt nosed with brown and yellow patches almost like a burbot; even during spawning time there was no red colouration and the smallest mature dwarfs caught were only 110 to 120 mm. The normal charr was caught spawning at lengths around 300-350 mm, weighing 0.3-0.4 kg. Specimens classified "between" these two groups were also known. In River Vuojatätno

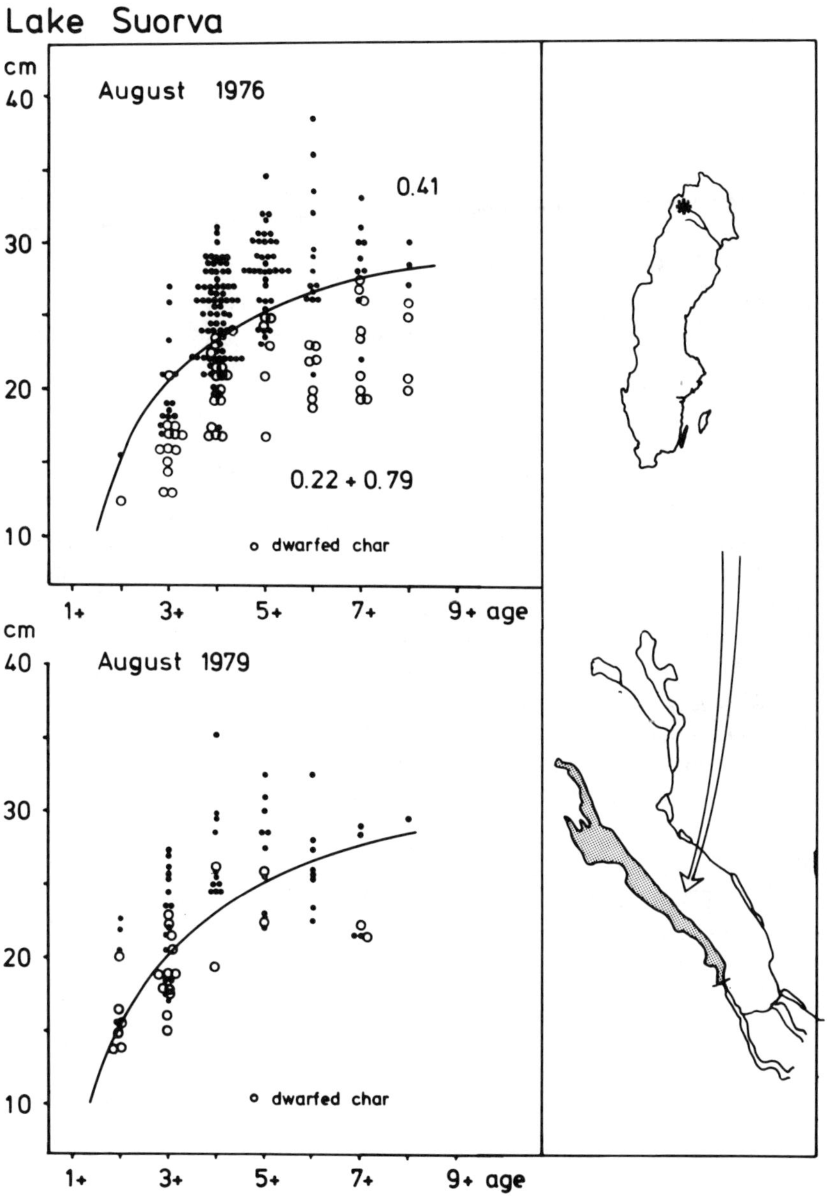

FIG. 7 Growth, gene frequencies and field 'dwarf' classifications in August 1976 (above) and August 1979 (below) in Lake Suorva. The location of the lake reservoir is shown to the right.

(B. Fredrikson pers. comm.) and some small tributaries, Svaltajokk, Ratojokk (Å. Pettersson pers. comm.), large stream-living Arctic charr are recorded.

Food habits: The correlation of food items with charr taxonomy was first made after the observation of a dense population of the introduced *Mysis relicta*. In 1976 taxonomic analyses were made at one station only (Vaisaluokta-Änonjalme). The dwarfs were feeding on benthic invertebrates and larvae and imagos of Trichoptera and Ephemeroptera, while the larger "Normal charr" were eating plankton. There were no differences reported between the "F-charr" and the "S-charr" in the area studied (Hanson and Lindström 1979). The excess of food probably blurred niche specialization and feeding habits in 1979. Today *Mysis relicta* dominates the stomach contents of both normal and dwarfed Arctic charr, brown trout and whitefish (Hanson in prep.).

Lake Spawning and Stream Spawning

1) Lake Spawning "S-charr" + Stream Spawning "F-charr"

The Arctic charr of Lake Torrön, River Indalsälven, have long been considered as an allopatric "S-charr" population, f(f) = 0.14 (Nyman 1972, Fürst et al. 1981, Hammar in prep.). However, it was early known that rather large charr were caught each autumn, spawning in the small tributary, River Hobergsån (C. Löfgren pers. comm.). After the impoundment in 1937 this population disappeared but in the two major inlets, River Gånälven and River Holderströmmen, now available when the water level is high in early autumn, two distinct stream spawning populations (f(F) = 0.24) have become established. According to A. Gönczi (pers. comm.) another stream spawning charr from Blåsjöälven (Runnstrom 1964) was introduced to lake Torrön after the impoundment.

Growth: The lake spawning population in the northern part of the lake shows a rather dense growth curve with a horizontal level after the age of 5+. The specimens caught in the streams are found to have a better growth rate (Fig. 8).

Habitat: The pelagic part of the lake spawning population is very conspicuous. Young specimens dominate the age structure of the catch in deep waters and the older fishes are mainly found closer to the shore. The year classes 3+, 4+ and 5+ seem to be missing in the benthic gillnets but very pronounced in floating nets (Fig. 8). The larger stream spawners have been seen by the local fisherman P. Anderson in schools close to the surface where they can be caught with a so-called otter.

Reproduction: The spawning grounds are very well known and the spawning takes places from September 10 to 25; in Rivers Holderströmmen and Gånälven it occurs in the first days of September. In Lake Holdern and Lake Jävsjön, two lakes upstream of Lake Torrön, stream spawning is dominant; in River

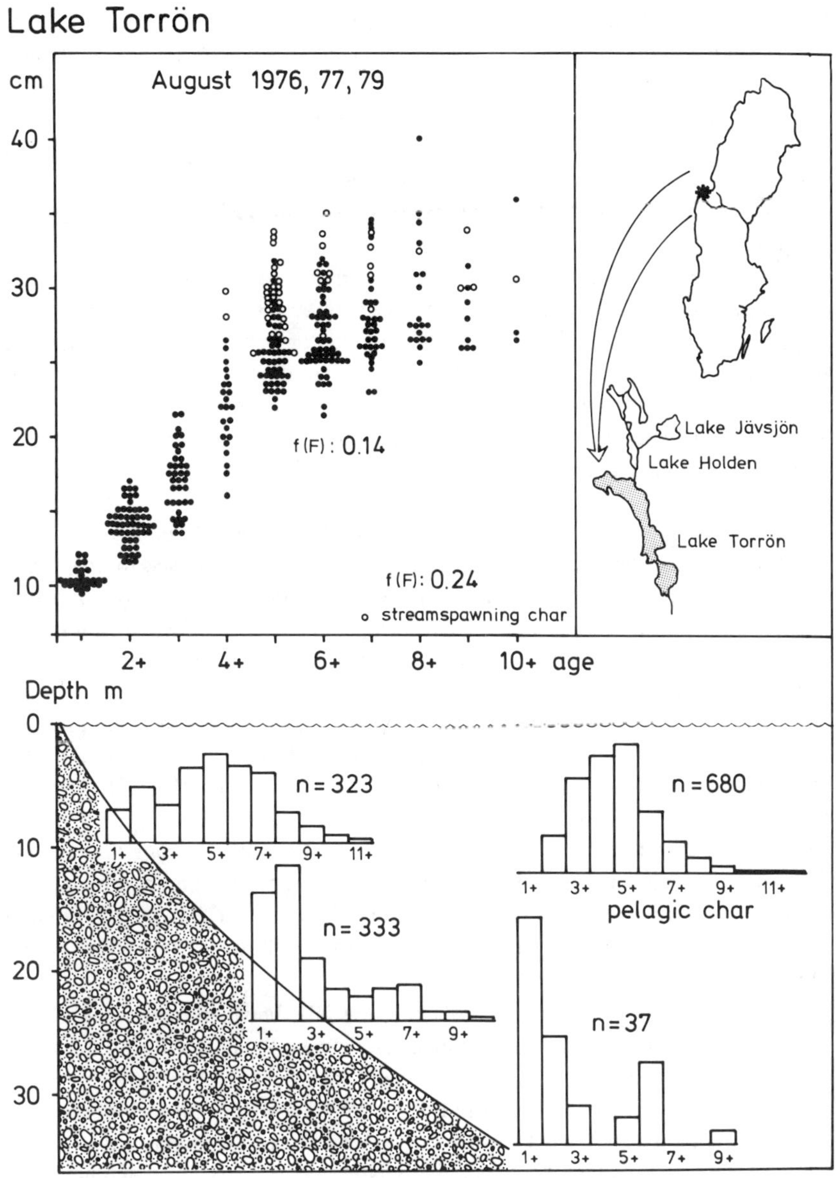

FIG. 8 Growth, gene frequencies (above left), age structures and depth distribution (below) of benthic and pelagic fractions of a lake spawning population of Arctic charr in Lake Torrön. Each graph represents 100 percent of the specimens caught at each special depth interval. The location of the lake reservoir is shown (above right).

Jävsjöströmmen it starts about September 10 (A. Gönczi pers. comm.) and in River Björkvattenån, a mountain brook, as late as the last days of September (Hammar and Öhman unpubl. data). Both lakes are inhabited by allopatric "F-charr" populations. Their locations are shown in Figure 8.

The lake spawning population of L. Torrön has furthermore been found to be very sensitive to the lowering of the water level in spring resulting in severe damage to the recruitment (Hammar 1980, Fürst et al. 1981, Fürst and Hammar this volume).

Food habits: *Mysis relicta* was introduced in 1957 but planktonic food in summer is still predominant in the lake spawning "S-charr". The stream spawning population with a higher gene frequency can only be recognized for sure in the spawning streams when stomachs, usually, are empty. The higher degree of infestation of *Diphyllobothrium* plerocercoids on the intestines and the larger sizes of the stream spawning charr suggest fish predation. There are as yet no records of fish-eating lake spawning charr in Lake Torrön.

2) Two Lake Spawning Populations + a Stream Spawning Population

In Kvarnbergsvattnet, Faxälven (Fig. 9), local fishermen used to distinguish three different types of Arctic charr: "Skiva", "ålrör" and "smårör" meaning "disc-charr", "eel-charr" and "small charr" (Nyström 1862). It is not certain whether these three populations correspond to the sympatric lake-spawning populations of "Normal charr" and "S-charr" and the stream spawning "Lindnasröra" (so-called after the geographical name of the spawning place).

Growth: As previously mentioned, in combinations of "Normal charr" and "S-charr", "Normal charr" dominates the size relationships (Fig. 9). The gene frequencies 0.58 and 0.27 suggest genetic relationships more with the Lake Limingen branch than with Lake Jorm and Lake Blåsjön. The big charr, called "Skiva", was also known in Lake Limingen where, before the impoundment, it might reach weights close to 2 kg (Aass 1963). The stream spawning population from Brännälven, with the F-frequency 0.33 ($n = 20$) shows a much better growth rate.

Reproduction: Spawning takes place in the lake between September 15 and October 15 but the deep-living specimens are known to be ripe during the whole winter (J. Nordström pers. comm.). Stream spawning specimens were caught twice close to the former outlet from Lake Limingen, in the spillway of the dam, between September 25 and October 5, 1979. In Lake Limingen, Aass (1963) describes a "rare behaviour in southern Norway", a stream spawning charr population in Röyrvikelva.

Habitat: The results of the test fishing in August-September 1979, showed no distinct cline in depth distribution (Fig. 9). The variability of gene frequencies

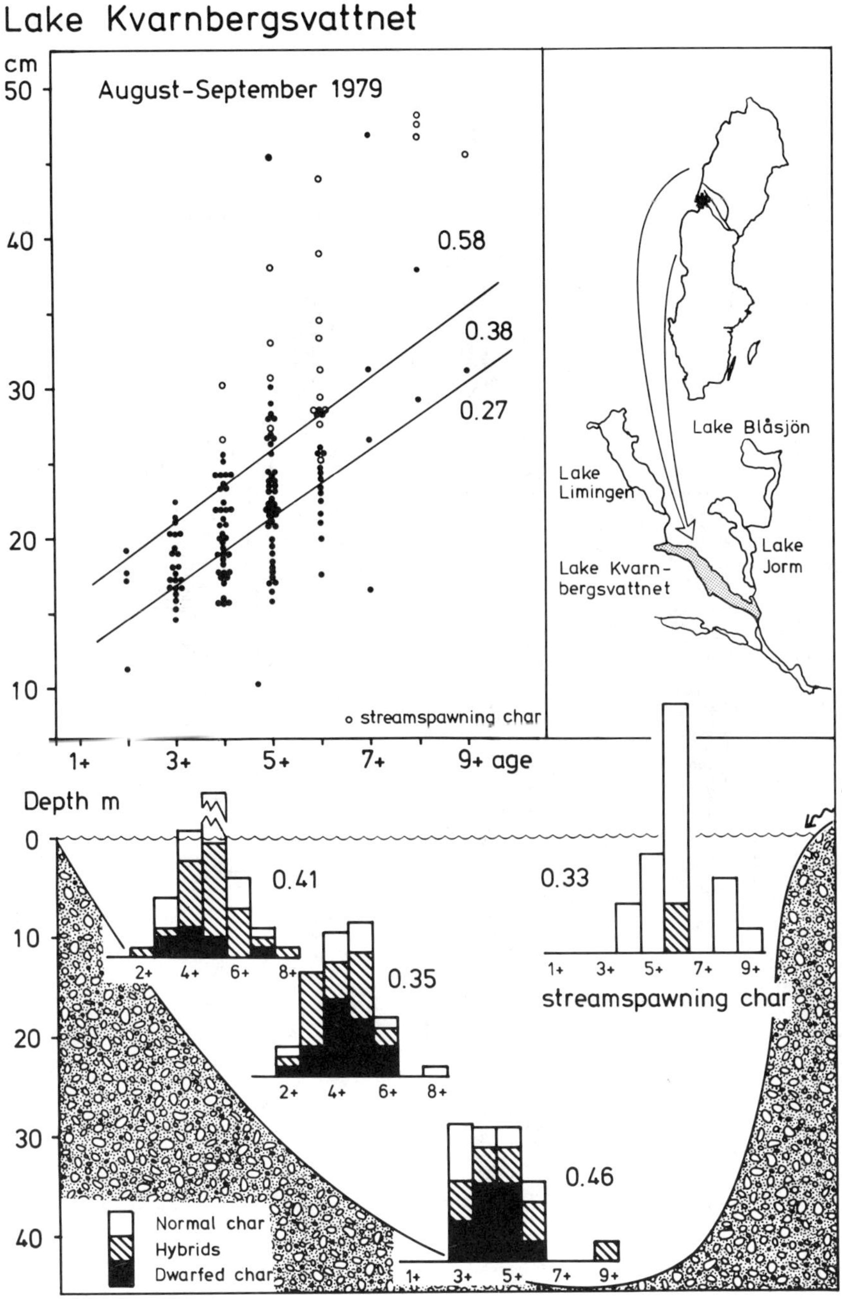

FIG. 9 Growth, gene frequencies of different growth fractions of benthic catch (above left), age structures and depth distribution (below) of two sympatric lake spawning populations and a stream spawning population in Lake Kvarnbergsvattnet. Each graph represents 100 percent of the specimens caught at each special depth interval. The location of the lake reservoir is shown (above right).

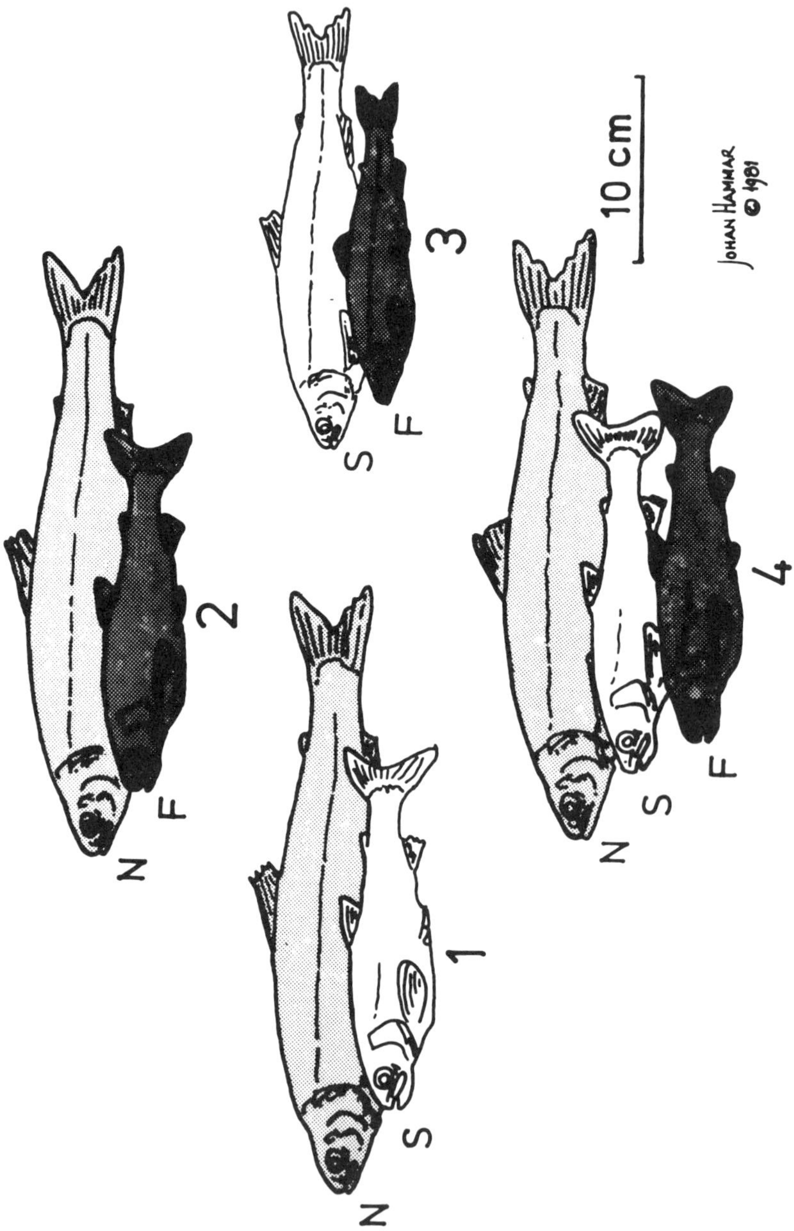

FIG. 10 The size relationships in four possible combinations of sympatric populations of three Arctic charr sibling species. N = 'Normal charr', S = 'S-charr', F = 'F-charr'. Drawing: J. Hammar

between the different stations ranged from 0.00 to 0.62 suggesting a very complex horizontal habitat distribution of the three genetically "isolated" populations mentioned. The two fast growing, probably stream spawning specimens, caught in the lake were found opposite the spawning stream close to the shore, at a depth of 25 m, near the main inlet from Lake Jorm. At greater depths, the dwarfs dominate.

Food habits: No stomach analysis has so far been done. According to J. Nordström (pers. comm.) the stomachs contain the introduced *Mysis relicta* but in winter the specimens caught with hook and line through the ice were often deep-living forms with empty stomachs.

DISCUSSION

If successive migrations of different sibling species have occurred in post-glacial periods and these populations have remained genetically isolated, then ecological or ethological specializations must have been in existence to maintain this segregation.

Growth, Competition and Immigration

The growth pattern of the three sympatric combinations of lake spawning charr (Fig. 3, 5, 7) mentioned show the dominance in size by the "Normal charr" over both "F-charr" and "S-charr". In Lake Fättjaure and two down-stream lakes where Henricson and Nyman (1976) found sympatric "F-charr" and "S-charr", the latter was larger.

Figure 10 shows the general size relationships in the four possible sympatric combinations. Since allopatric "F-charr" populations, sensitive to the other sibling species within the complex, as well as other fish species, Inhabit all headwaters in the rivers studied, it is tempting to suggest that the "F-charr" was the first immigrant to arrive in the Scandinavian mountain lakes. The distribution map (Fig. 2) also suggests the "S-charr" to be immigrant number two. The "Normal charr" which certainly inhabits the southern Swedish lakes as the only sibling species today, was probably the last species to arrive in the northern Swedish rivers after the glacial period. The order of immigration is then opposite to the order of dominance in competition — which is very likely.

Lundberg (1899) suggests immigration from the Atlantic of the northern type (*S. alpinus*) through the outlets from large ice lakes dammed between the land ice and the Scandinavian Mountain Ridge. Immigration of the southern type (*S. salvelinus*) was explained as having a more southwest direction and was believed to be the oldest one. Immigration from the east was thought to be excluded due to absence of populations along the Baltic Coast. Ekman (1910), however, meant that the early western outlets from the Scandinavian ice-cap were too steep to allow a western immigration and suggested an eastern origin of postglacial populations. He emphasized the numerous examples of transplantations of Arctic charr to mountain lakes by man; he also thought that the sparse populations in Swedish lowlands were caused by high postglacial temperatures. The first ''zoo-geographer'' to realize and understand the physical and biological restrictions affecting the distribution of Arctic charr in Scandinavia was Huitfeld-Kaas (1918). He considered the Arctic charr to be the first freshwater fish to immigrate to Norway and describes in detail several possible pathways across the mountain ridge for Arctic charr of eastern origin from Finland and Sweden, as well as the routes of coastal populations into south-western Norway inhabited by former western anadromous stocks. He was aware of the interspecific segregation with whitefish and its effects on the distribution of charr and he knew numerous cases of man-made introductions of brown trout and Arctic charr or both, to high altitude lakes in Norway and Sweden. He was also open to the idea of a series of dammings and drainings of ice lakes elevating charr populations of both western and eastern origin into mountain lakes, theories used in several more recent studies. Ekman (1953) listed six lakes on the mountain border between Norway and Sweden, having outlets in both directions, suggesting possible immigration from both west and east.

The pattern recognized in the present paper shows an eastern immigration of ''Normal charr'' and possibly also of the ''F-charr'' corresponding with the theories of Huitfeld-Kaas (1918) and the whitefish studies summarized by Svärdson (1979). The immigration of Arctic charr into northern Norway is also discussed by Klemetsen (this volume) and Nilsen and Klemetsen (this volume).

The scattered patches of ''S-charr'' populations oriented to the west with presence in several Norwegian coastal and mountain lakes disturb the picture; they are probably examples of immigration from the west.

According to Nilsen (1980) no ''S-charr'' have been found in the northern and eastern part of Northern Norway and Bear Island, but during the Arctic expedition Ymer 80 I managed to collect three landlocked specimens of SS-homozygotes at the northern coast of Nordaustlandet, Svalbard (Hammar unpubl. data) suggesting a presence north of Norway.

With the aid of genetic markers like electrophoretic techniques and useful specific enzyme systems, the immigration pathways will probably be solved convincingly when a sufficient number of Arctic charr populations surrounding Sweden or Scandinavia have been analyzed.

Food habits: The food habits of Arctic charr treated as a group have been intensively studied and found to be very variable in relation to effects of impoundment (Nilsson 1961), interactive segregation with other sibling species within the complex and with brown trout (Nilsson 1955, 1960, 1965; Nilsson and Filipsson 1971; Klemetsen and Grotnes 1975, 1980) and introductions of new fish food organisms (Fürst et al. 1978, 1981, Hammar 1980). It is my opinion that the food habits of the different sibling species of Arctic charr are most obvious and specialized under the stressed conditions of sympatry.

Both the "F-charr" and "Normal charr" have been observed to be predators on benthos and fish in Lake Ajaure (Hammar unpubl. data), Övre Björkvattnet (Nilsson and Filipsson 1971), Korsvattnet (Nilsson 1955, Hammar unpubl. data), Lake Stora Rösjön (Hanson 1976) and Lake Blåsjön (Fürst et al. 1978). The "Normal charr" has also been found to feed on planktonic crustaceans in lakes with dominant populations of brown trout (Nilsson 1960, 1965) and as an effect of water level regulations (Nilsson 1961, Fürst et al. 1978). The "Normal charr" can be regarded as a very plastic species while the "F-charr" seems to be more or less benthic in its way of feeding.

The "S-charr" population in Lake Torrön is merely feeding on plankton in summer (Fürst et al. 1981) and in Lake Fättjaure species "A", ("S-charr") was feeding on "pelagic food" while species "B" ("F-charr") was feeding on "bottom food" in September-October (Henricson and Nyman 1976). These differences in food habits also affected their parasitic faunas: plankton feeding "S-charr" were more heavily infected with *Diphyllobothrium* spp. and *Proteocephalus* spp. (parasites that use planktonic copepods as intermediate hosts). In combinations with "Normal charr", benthic "S-charr" seem to feed on benthic and surface food (Klemetsen and Grotnes 1975, 1980; Fürst et al. 1978). In sympatric populations with extreme size and habitat segregation, vertical migrations to the surface by deep-living dwarfs during the night have been observed in Lake Blåsjön (Fürst et al. 1978) and Lake Suorva (Hanson and Lindström 1979), verified by the amounts of terrestrial food objects in their stomachs.

In allopatric populations both the "Normal charr" and the "F-charr" reach weights close to 10 kg, while no "S-charr" so far have been found larger than 1 kg. This reflects the main food habits of the three species, but it should also be stressed that stunted allopatric populations of the three species are even more common in Sweden. The special physical appearance of deep-living dwarfs — blunt nose, large eyes, large belly and greenish colour (Fig. 1, 6) — is thought to be an effect of age and the deep-living way of life rather than a relation to the systematic position.

Reproduction: The time of spawning is probably dependent more on altitude and latitude, i.e. the effect of temperature, rather than genetics; deep spawning populations seem to spawn later than shallow spawners. This is also well documented for Windermere, in England, where the deep, spring spawning, probably "F-charr" population, hatches in May when the temperature in shallow water may be lethal to Arctic charr (Frost 1965, Child 1980).

Lake spawning is recorded for all three sibling species within the complex; in sympatric populations the dominant species spawns close to the shore, probably on what is regarded as the best spawning grounds. In Lake Storsjouten, Fjällsjöälven, a sympatric but introgressed population of "Normal charr" and "S-charr", similar to Lake Blåsjön (Fig. 3), shows a distinct genetic cline on a spawning ground with the highest gene frequencies at the top and lower frequencies in the deeper parts (Hammar and Lindh unpubl. data).

The best spawning place in many lakes is definitely in the tributaries. These are most commonly used by brown trout but in some cases when the brook gets dry in summer, Arctic charr is favoured by the short flowing water season during autumn, winter and early spring, sufficient for the charr eggs to hatch and for the juveniles to "migrate" down to the lake (Gönczi 1972).

In many of the headwaters inhabited by "F-charr", stream spawning in larger brooks is very common as mentioned before; I am convinced that many of the reported stream spawning populations with higher gene frequencies mentioned in the text are remnants of stream spawning "F-charr" in headwater lakes. It is very interesting to see that these populations also show better growth rates which may be a sign of their greater competitive ability.

In the management of Arctic charr fisheries in lake reservoirs, stream spawning populations unaffected by recruitment damage from water level regulation (Fürst and Hammar this volume), must be protected and regarded as valuable genetic resources for future stocking in ruined reservoirs (Hammar 1980).

Habitat: The recurrent pattern in the sympatric populations studied is the occurrence of large, pelagic fish and benthic dwarfs. This is the result of some unknown part of complex interactive segregation between sibling species. It is also possible that an allopatric population can split into groups with different growth rates caused by differences in food habits during benthic and pelagic stages in the lifecycle. If furthermore, we add the presence of brown trout (*Salmo trutta* L.) a more graduated picture of three habitats occur.

In Båtsvatn (Fig. 11A) the existence of a littoral "S-charr" and a pelagic "Normal charr" (in the absence of brown trout) is described by Klemetsen and Grotnes (1975). In Lake Blåsjön, and several other lakes (Fig. 11B), the "S-charr" is deep-living in the presence of a littoral brown trout. In lakes with only "N" and "F", e.g. Lake Stora Rösjön, the same picture as in Figure 11A occurs, only with the "F-charr" littoral (Fig. 11C). In lakes with the same combination (and in the presence of brown trout e.g. Övre Björkvattnet and Lake Ajaure) the "F-charr" just moves a little deeper suggesting a stronger competitive character than the "S-charr" to the brown trout (Fig. 11D). In Lake Fättjaure the "S-charr"

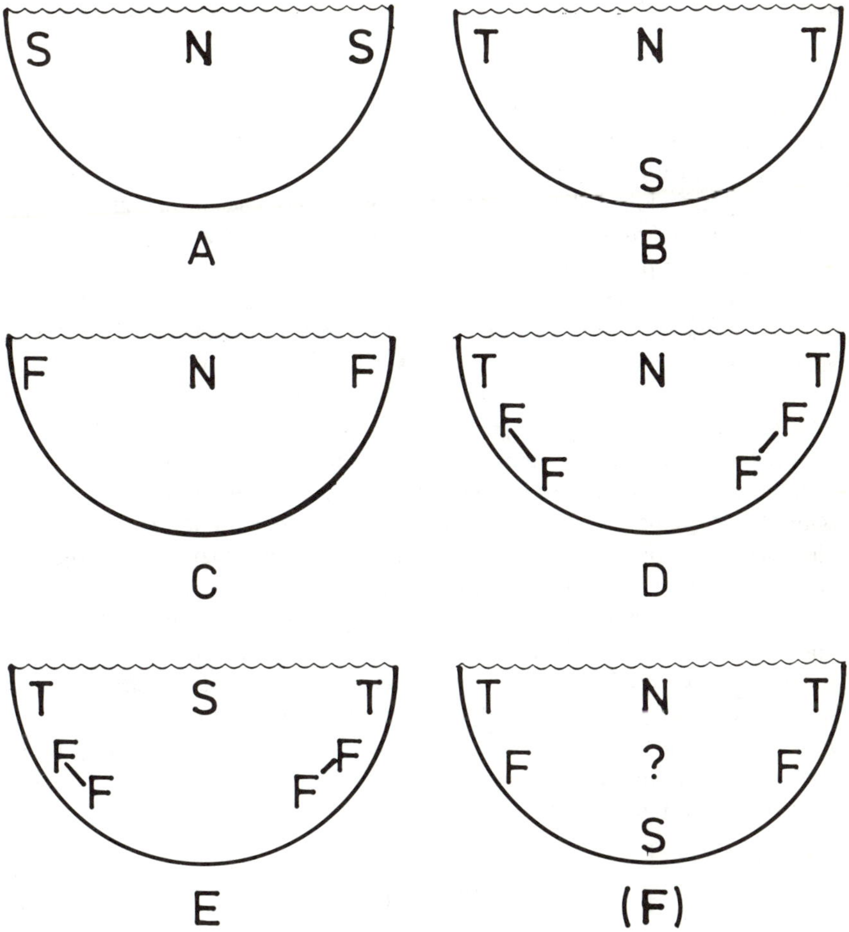

FIG. 11 Habitat choice of different combinations of three sympatric Arctic charr sibling species and brown trout (*Salmo trutta* L.). N='Normal charr', S='S-charr', F='F-charr', T=brown trout. The pattern shown in Figure 11F is a suggestion based on the ecology mentioned in the text.

is pelagic and the same benthic co-existence of brown trout and "F-charr" occurs in the littoral zone (Henricson and Nyman 1976 (Fig. 11E)). At present no lakes with "F-charr" and "S-charr" in the absence of brown trout have been studied; in the headwaters of Stora Luleälven where all three sibling species occur, the evaluation of the test fishing result has just started. With the present knowledge of ecological characteristics of the three sibling species within the Arctic charr complex, it is tempting to suggest that in Lake Suorva, where the full "combination" is found, an arrangement as described in Figure 11F will be encountered.

The population of "S-charr" in Lake Torrön seems to have a life-cycle involving changes of habitat, similar to those of anadromous stocks (Fig. 8). This was observed by Klemetsen and Grotnes (1980) for "Normal charr" in Båtsvatn.

The presence of burbot, *Lota lota*, a bottom dwelling predator, will, in all probability, affect the habitat of "dwarf" populations, causing alteration of their growth rates, and turning the sympatric population into an apparently allopatric or monomorphic population.

In general, the concept of three sibling species having invaded Scandinavia in successive waves of immigration, is compatible with the results of ecological investigations carried out over many years. Charr taxonomy does not let itself be revealed by "single-method" techniques. A recent example of this (Andersson et al. 1983) is based entirely on standard electrophoretic procedures. Andersson et al. conclude that the low amount of genetic divergence found precludes any sub-division of charr species in northern Europe. They can therefore be grouped with other "lumpers" who have also relied on single criteria. The *absence* of electrophoretically discernable characteristics, or any other method showing similarity, can be no proof of genetic identity, if the species/populations studied have life-history parameters clearly displaying all signs of reproductively isolated gene pools. The final criterion of taxonomic rank must be the proof provided by the population itself, the biological species status, not the data given by techniques of purely human design. It is therefore, extremely important to have full knowledge of both the ecology and taxonomy of the existing Arctic charr stocks before introducing new management techniques, or new introductions of charr and/or food organisms.

REFERENCES

Aass, P. 1963. Limingenreguleringens virkninger på fisket. 40 p. (Mimeographed in Norwegian).

Andersson, G., K.J. Gustafson and T. Lindström. 1971. Rödingen i Rösjöarna på Fulufjäll. Inform. Inst. Freshw. Res. Drottningholm (8). 19 p. (In Swedish).

Andersson, L., N. Ryman and G. Ståhl. 1983. Protein loci in the Arctic charr, *Salvelinus alpinus* L.: electrophoretic expression and genetic variability patterns. J. Fish Biol. 23: 75-94.

Behnke, R.J. 1972. The systematics of salmonid fishes of recently glaciated lakes. J. Fish. Res. Board Can. 29 (6): 639-671.

Child, A. 1980. Electrophoretic analysis of the Arctic char of Cumbrian lakes. *In* Proceedings of the first ISACF workshop on Arctic char. Inst. Freshw. Res. Drottningholm.

Ekman, S. 1910. Om människans andel i fiskfaunans spridning till det inre Norrlands vatten. Ymer 30:133-140. (In Swedish).

Ekman, S. 1953. Über Fishpasse in der skandinavischen Hochgebirgskette, Arch. f. Hydrobiol. 48:92-96.

Filipsson, O. 1967. Åldersbestämning av röding med hjälp av otoliter. Inform. Inst. Freshw. Res. Drottningholm (5). 10 p. (In Swedish).

Frost, W. 1965. Breeding habits of Windermere char, *Salvelinus willugbi* (Gunther) and their bearing on speciation in these fish. Proc. R. Soc. Edinb. B. 163:232-284.

Fürst, M., U. Boström and J. Hammar. 1978. Effects of new fish-food organisms in Lake Blåsjön. Inform. Inst. Freshw. Res. Drottningholm (15). 94 p. (In Swedish with English summary and legends).

Fürst, M., U. Boström and J. Hammar. 1980. Effects of introduced *Mysis relicta* on Fish in Lake Vojmsjön. Inform. Inst. Freshw. Res. Drottningholm (3). 42 p. (In Swedish with English summary and legends).

Fürst, M., U. Boström and J. Hammar. 1981. Effects of introduced *Mysis relicta* on fish in Lake Torrön. Inform. Inst. Freshw. Res. Drottningholm (1). 48 p. (In Swedish with English summary and legends).

Fürst, M. and J. Hammar. 1984. Effects of water level fluctuations on the recruitment of Arctic charr. Paper presented at Int. Symp. on Arctic Charr, Winnipeg, 1981 (This volume).

Gydemo, R. 1980. Populationsgenetisk undersökning av röding i Västerbottens län. Del 3. Fiskenämnden, Lantbruksnämnden, Västerbottens län, Umeå. 71 p. (Mimeographed in Swedish).

Gydemo, R. 1984. Preliminary survey results on the distribution of the Arctic charr species complex in Iceland. Paper presented at Int. Symp. on Arctic Charr, Winnipeg, 1981 (This volume).

Gönczi, A. 1971. Fält-och akvarieförsök med lekröding. Inform. Inst. Freshw. Res. Drottningholm (3). 13 p. (In Swedish with English summary).

Gönczi, A. 1972. Studies on spawning char. Inform. Inst. Freshw. Res. Drottningholm (2). 12 p. (In Swedish with English summary).

Hammar, J. 1978. Resultat av populationsgenetisk analys av röding från Falehaure, Piteälven och Gautosjön, Laisälven, September 1978. Inst. Freshw. Res. Drottningholm 8 p. (Mimeographed in Swedish).

Hammar, J. 1979. Resultat av populationsgenetisk analys av röding från Norrbotten. Inst. Freshw. Res., Drottningholm 16 p. (Mimeographed in Swedish).

Hammar, J. 1980. Ecology and taxonomy of Arctic char in lake reservoirs in Sweden. *In* Proceedings of the first ISACF workshop on Arctic char. Inst. Freshw. Res., Drottningholm.

Hammar, J. 1981. P.M. ang taxonomiska analyser av rödingpopulationer, Krokoms kommun, Jämtland. Inst. Freshw. Res., Drottningholm. 3 p. (Mimeographed in Swedish).

Hanson, M. 1976. The biology of an acid mountain lake as illustrated by the food of Arctic char. Inform. Inst. Freshw. Res. Drottningholm (5): 13 p. (In Swedish with English summary).

Hanson, M. and T. Lindström. 1979. Suorva: a lake with char and whitefish of good size. Inform. Inst. Freshw. Res. Drottningholm (4): 33 p. (In Swedish with English summary).

Henricson, J. and L. Nyman. 1976. The ecological and genetical segregation of two sympatric species of dwarfed char (*Salvelinus alpinus* (L.) species complex). Inst. Freshw. Res. Drottningholm Rep. 55:15-37

Huitfeldt-Kaas, H. 1918. Ferskvandsfiskenes utbredelse og invandring i Norge. Fiskeriinspektørens inberetning om ferskvandsfiskerierne for aaret 1915. Anhang II. Lantbruks-departementet. 139 p. (In Norwegian).

Klemetsen, A. 1984. The Arctic charr speciation problem as seen from northern Norway. Paper presented at Int. Symp. on Arctic Charr, Winnipeg, 1981 (This volume).

Klemetsen, A. and P.E. Grotnes. 1975. Food and habitat segregation by two sympatric Arctic char populations. Verh. int. Ver. Limnol. 19:2521-2528.

Klemetsen, A. and P.E. Grotnes. 1980. Coexistence and immigration of two sympatric Arctic char, p. 757-763. *In* E.K. Balon [ed.] Charrs: salmonid fishes of the genus *Salvelinus*. Dr. W. Junk Publishers, The Hague.

Lindström, T. 1954. Non-reproductive migrations in the charr, *Salmo alpinus*, L. Inst. Freshw. Res. Drottningholm Rep. 35:118-132.

Lindström, T. 1958. Ungdomsstadier av röding i rinnade vatten. Svensk Fisk. Tidskr. 67(4,5):54-58, 71-80. (In Swedish).

Lundberg, R. 1899. On the Distribution of Swedish Freshwater Fishes. Medd. K. Lantbr. Styr. 10(58). 87 p. (In Swedish with English Summary).

Määr, A. 1949. Fertility of char (*Salmo alpinus* L.) in the Faxälven water system, Sweden. Inst. Freshw. Res. Drottningholm Rep. 29:57-70.

Määr, A. 1950. A supplement to the fertility of char (*Salmo alpinus* L.) in Faxälven water system, Sweden. Rep. Inst. Freshw. Drottningholm Rep. 31: 127-136.

Nilsen, P. 1980. Stasjonaer- og Anadrom røye, *Salvelinus alpinus* (L.). En sammenligning av økologiske og genetiske trekk hos bestander med sannsynlig felles opprinnelse. Hovedfagoppgave i akvatisk biologi. Univ. Tromsø. 84 p. (In Norwegian).

Nilsen, P. and A. Klemetsen. 1984. Anadromous, resident and landlocked Arctic charr of the coast of Finnmark, Northern Norway. Paper presented at Int. Symp. on Arctic Charr, Winnipeg, 1981 (This volume).

Nilsson, N.-A. 1955. Studies on the feeding habits of trout and char in North Swedish lakes. Inst. Freshw. Res. Drottningholm Rep. 36:163-225.

Nilsson, N.-A. 1960. Seasonal fluctuations in the food segregation of trout, char and whitefish in 14 North-Swedish lakes. Inst. Freshw. Res. Drottningholm Rep. 41:185-205.

Nilsson, N.-A. 1961. The effect of water-level fluctuations on the feeding @habits of trout and char, in the lakes Blåsjön and Jormsjön, North Sweden. Inst. Freshw. Res. Drottningholm Rep. 42:238-261.

Nilsson, N.-A. 1965. Food segregation between salmonid species in North Sweden. Inst. Freshw. Res. Drottningholm Rep. 46:58-78.

Nilsson, N.A. and O. Filipsson. 1971. Characteristics of two discrete populations of Arctic char (*Salvelinus alpinus* L.) in a north Swedish lake. Inst. Freshw. Res. Drottningholm Rep. 51:90-108.

Nordeng, H. 1961. On the biology of char (*Salmo alpinus* L.) in Salangen, North Norway. I. Age and spawning frequency determined from scales and otoliths. Nytt Mag. Zool. 10:67-123.

Nyman, L. 1965. Variation of proteins in hybrids and parental species of fishes. Rep. Swed. Salm. Res. Inst. 13. 11 p.

Nyman, L. 1967. Protein variation in *Salmonidae*. Inst. Freshw. Res. Drottningholm Rep. 47:5-38.

Nyman, L. 1972. A new approach to the taxonomy of the "*Salvelinus alpinus* species complex". Inst. Freshw. Res. Drottningholm Rep. 52:103-131.

Nyman, L. 1980. The population genetics of Arctic char: a state of the art. *In* Proceedings of the first ISACF workshop on Arctic char. Inst. Freshw. Res., Drottningholm.

Nyman, L. 1984. Management of allopatric and sympatric populations of land-locked Arctic charr of Scandinavia. Paper presented at Int. Symp. on Arctic Charr, Winnipeg, 1981 (This volume).

Nyman, L. and O. Filipsson. 1972. The chars of Lake Yraf. Inform. Inst. Freshw. Res. Drottningholm (10). 20 p. (In Swedish with English summary).

Nyman, L., J. Hammar and R. Gydemo. 1981. The systematics and biology of land-locked populations of Arctic char from northern Europe. Inst. Freshw. Res. Drottningholm Rep. 59: 128-141.

Nyström, C.L. 1862. Om fiskfaunan och fiskerierna i Jämtlands lan. Reseberättelse, Kungl. Vetensk. Akad. 44 p. (Mimeographed in Swedish).

Pettersson, B. 1961. P.M. över rödinglek i Ume älv 1960. Inst. Freshw. Res., Drottningholm. 2 p. (Mimeographed in Swedish).

Runnström, S. 1951. Population of char, *Salmo alpinus*, Linne, in a regulated lake. Inst. Freshw. Res. Drottningholm Rep. 32:60-78.

Runnström, S. 1964. Kontroll av Jormsjöns lekrödingbestånd i Blåsjöälven genom en fiskspärr samt studier över dess vandringar genom märkningsförsök. Inform. Inst. Freshw. Res. Drottningholm (4). 35 p. (In Swedish).

Svärdson, G. 1961. Rödingen (The Arctic char). Fiske 1961:25-37 (In Swedish).

Svärdson, G. 1979. Speciation of Scandinavian *Coregonus*. Inst. Freshw. Res. Drottningholm Rep. 57. 95 p.

Vallin, S. 1938. Redögörelse för fiskeribiologiska undersökningar i Luleälvens övre sjösystem speciellt med hänsyn till verkan av regleringen genom Suorva-dammen. Inst. Freshw. Res., Drottningholm, 69 p. (Mimeographed in Swedish).

The Arctic Charr Speciation Problem as Seen from Northern Norway

ANDERS KLEMETSEN

Institute of Biology and Geology, University of Tromsø, N-9001 Tromsø, Norway

KLEMETSEN, A. 1984. The Arctic charr speciation problem as seen from northern Norway, p. 65–77. *In* L. Johnson and B.L. Burns [eds.] Biology of the Arctic charr, Proceedings of the International Symposium on Arctic Charr, Winnipeg, Manitoba, May 1981. Univ. of Manitoba Press, Winnipeg.

Arctic charr samples from N. Norway had blood serum esterase frequencies which gave indication of three different population groups. The low Est^F group was found inland and in the southern part of the sampling area. Landlocked, high f(F) charr occurred along the coast of the northern part. Moderate value f(F) populations were found as anadromous and stationary populations in the whole area. In this group an increase of f(F) values was indicated from south to north. Gillraker and pyloric caeca numbers were not diagnostic. The three groups may be taken to support the suggestions by Nyman (1972) of three sibling charr species. These may have been allopatrically derived in glacial refugia on the Siberian coast (anadromous), on the SW European coast (anadromous) and to the E/SE of the Weichselian ice shield (freshwater). N. Norway takes a central position in relation to immigration from these possible refugia. Two distinct spawning groups of charr from Ellasjøen, Bear Island could not be separated by serum esterase analysis. Incipient sympatric speciation can not be ruled out for this situation. The Arctic charr complex may be caused by a mixture of sympatric and allopatric speciation process.

INTRODUCTION

The charr monograph (Balon 1980 a), notably the review by Johnson (1980) and the contributions by Balon (1980 b,c), Behnke (1980) and Savvaitova (1980 a,b) showed that several controversial items have to be clarified before a theory explaining the systematics of the *Salvelinus alpinus* complex can emerge. Much of the debate centers around two sets of contradictory hypotheses. The first concerns the question of one polytypic or two or more closely allied species of Arctic charr (McPhail 1961, Svärdson 1961, Frost 1965, Behnke 1972, 1980, Nyman 1972, McCart 1980, Savvaitova 1980 a,b). The second concerns the question of allopatric or sympatric speciation modes for these fish (Svärdson 1961, Behnke 1972, Balon 1980 b,c, Savvaitova 1980 a,b).

The present contribution discusses the speciation problem of the north Fennoscandian Arctic charrs in relation to these sets of hypotheses. The discussion is based on studies of *S. alpinus* in N. Norway undertaken by the University of Tromsø during the 1970 decade, some of which have already been treated by Klemetsen et al. (1972), Grotnes (1975), Klemetsen and Grotnes (1975, 1980), Nilsen (1980), Holthe (1981) and Nilsen and Klemetsen (this volume).

North Norway is well suited to the study of Arctic charr. A wide variety of lentic and lotic habitats abound. Arctic charr is probably the most common freshwater fish occurring in most areas. It is found both as anadromous and stationary (resident) populations, and in all sorts of ecological situations, from just above sea level to the high mountains.

METHODS

First we set out to look for sympatric morphs of charr. If found, we wanted to determine whether these represented real, discrete biological units. The following parameters were used to compare sympatric pairs: growth, age distributions, size and age of maturation, food and habitat choice, gillraker and pyloric caeca counts, and frequencies of blood serum esterase isozymes.

A second approach was to study several populations spread over a large portion of N. Norway. The aim was to determine whether genetic clines existed along geographical gradients, or if sharp changes in genetic characters occurred at any points. For this purpose the well-established character of the serum esterase allele F frequency, f(F) (Nyman 1972), was chosen as a genetic tag. This was assessed for all populations studied. Some of the other parameters mentioned above were also compared for several allopatric populations.

The techniques employed were standard and are described by Klemetsen et al. (1972) and Nilsen and Klemetsen (this volume). Starch gel electorphoresis was performed according to Nyman (1967).

Throughout the text the term *resident* refers to stationary charr living sympatrically with the freshwater phases of anadromous charr. The term *landlocked* is reserved for stationary charr with no present contact with anadromous charr.

RESULTS

The localities sampled covered a large part of N. Norway (Table 1, Fig. 1). Two samples from Bear Island in the Barents Sea (18-19) (figures in parentheses refer to localities in Table 1) were also included. The mainland range included both samples from water courses originating close to the main Fennoscandian water divide (1-4, 9-10), and samples from the outer coast (12-17) and archipelago (5-8, 11). The main water course group comprised both upstream (1-3, 9)

Table 1. List of North Norwegian Arctic charr localities. All are lakes except Grønvika (5) and Pollen (16) which are marine sampling sites close to estuaries.

No.	Name	Water Course	Elevation m.a.s.l.	Appro imate area, ha	Sampling year	Other fish species
1	Båtsvatn reservoir	Skjoma	859	1550	1970-78	None
2	Cuorvejavre	Skjoma	923	335	1978	None
3	Lake 1025	Skjoma	1025	5	1978	None
4	Øvervatn	Salangselv	8	250	1975	Salmo salar S.trutta, Gasterosteus aculeatus, A.anguilla
5	Grønvika	Åelv(estuary)	-	-	1980	-
6	Åvatn	Åelv	7	30	1976	As no. 4
7	Rødsandvatn	Åelv	19	100	1976	As no. 4
8	Storvatn	Tennelv	173	230	1976	S.trutta
9	Gamasjavre	Barduelv - Målselv	647	410	1979	S.trutta, Esox lucius, L.lota, Phoxinus aphya
10	Takvatn	Takelv - Målselv	214	1525	1979	S.trutta G.aculeatus
11	Skogsfjordvatn	Skogsfjordelv	19	1330	1976	As no. 4
12	Oksfjordvatn	Oksfjordelv	5	500	1979	S.salar S.trutta
13	Neverfjordvatn	Kvalsundelv	128	80	1979	S.trutta
14	Brensvikvatn	Brensvikelv	28	60	1979	S.trutta
15	Eidevatn	Smørfjordelv	81	70	1978	S.trutta
16	Pollen	Smørfjordelv (estuary)	-	-	1978	-
17	Brennelvvatn	Brennelv	47	60	1978	S.trutta
18	Ellasjøen	Ellasjøelv, B.I.	19	73	1978	None
19	Stevatn	Lakselv, B.I.	40	30	1977	None

downstream (4) and tributary samples (10). Four samples consisted of anadromous charr (4, 5, 12, 16). Lake elevations varied from close to sea level to over 1000 m (Table 1). This coincides well with the elevation range of fish-bearing lakes of northernmost Fennoscandia. All lakes were oligohumic and oligotrophic, some tending to the ultraoligotrophic side (the Skjomen lakes) and some to the mesotrophic side (Åvatn and Takvatn). Ellasjøen was a potentially eutrophic lake due to heavy loading with droppings from large numbers of bathing kittiwakes, *Rissa tridactyla*. Lake sizes varied from 5 to 1550 hectares (Table 1). Most were deep, dimictic lakes. Exceptions were Åvatn, Eidevatn and Stevatn which were too shallow to allow the development of stable stratification; three were regular (Ellasjøen) or facultative (Båtsvatn reservoir, Cuorvejarve) cold monomictic lakes.

FIG. 1 Geographic position of the Arctic charr localities of the present study. County and national borders are stippled. The main Fennoscandian water divide is marked by a solid line with hatching on one side. The main water courses of N. Norway which start at the water divide are outlined.

One clearcut case of stationary, sympatric populations, that of the Båtsvatn reservoir, occurred in this material. Båtsvatn hydroelectric impoundment united three lakes in 1973. All three had initially double populations of Arctic charr. The case has been under study for several years (Klemetsen et al. 1972, Grotnes

Table 2. Gillraker and pyloric caeca counts for Arctic charr populations from North Norway. Nos. 15-17 after Nilsen and Klemetsen (this volume), no. 18 after Holthe (1981).

		Gillrakers				Pyloric caeca			
No.	Population	N	Mean	St.dev.	Range	N	Mean	St.dev.	Range
1	Båtsvatn res., dwarf charr	40	24.50	0.20	22-27	42	33.95	4.30	26-42
1	Båtsvatn res., normal charr	87	25.31	1.17	21-27	89	38.53	4.43	28-52
2	Cuorvejavre	122	25.00	1.28	22-28	129	34.26	4.09	25-44
3	Lake 1025	26	26.20	1.39	24-29	24	32.92	2.92	29-38
6	Åvatn	17	25.65	1.06	24-28	18	41.52	6.60	30-55
7	Rødsandvatn	48	25.15	1.29	23-30	47	38.30	4.31	31-50
8	Størvatn	39	25.21	1.26	22-29	51	41.45	5.11	31-55
11	Skogsfjordvatn	89	26.13	1.14	23-29	88	48.41	4.70	39-60
15	Eidevatn	160	22.0	0.80	20-24	204	37.2	3.50	35-59
16	Pollen	37	24.0	1.39	22-27	50	39.9	3.67	33-48
17	Brennelvvatn	221	23.7	1.07	19-25	227	46.3	4.68	30-47
18	Ellasjøen	558	25.1	1.19	17-30	502	35.2	2.46	20-52

1975, Klemetsen and Grotnes 1975, 1980). In summary the two morphs differed in outer appearance (morphology, colour), size, flesh colour, habitat and food choice, maturation age and maximum age. There were also indications of different times and places of spawning. Presorted material agreed well with the expected values for two different Hardy-Wienberg equilibria of serum esterase distributions. The charr of Cuorvejarve on the Swedish-Norwegian border upstream from the Båtsvatn reservoir had exactly the same F-frequency as the dwarf charr of the reservoir (Fig. 2, Table 4). Growth pattern also closely matched that of the dwarf charr (Table 3). The development of maturation (ages for first maturation, half the year class ripe and all fish ripe) was 3, 5 and 8 years respectively in the Cuorvejarve sample. This was similar to the dwarf charr pattern in the reservoir (Table 2 of Klemetsen and Grotnes 1980).

Other cases of two charr morphs were found on Bear Island, both in Ellasjøen (Holthe 1981) and in Stevatn (Klemetsen, unpublished). Ellasjøen is the best studied, and some of Holthe's (1981) results are, briefly: two morphs were readily sorted by external characters. Size distributions of mature fish fell into two discrete groups, with no overlap in fork lengths between them. Differences in growth, habitat, food and relative fecundity were documented. In spite of this, no segregation in serum esterase frequencies was found between the morphs (Table 4).

Table 3. Age and growth of some North Norwegian Arctic charr populations. See also Nilsen and Klemetsen (this volume). Values given are mean fork length (mm) for each age group (otolith readings). Standard deviations are not calculated when N < 4.

Populations	N	Age groups 2+	3+	4+	5+	6+	7+	8+	9+	10+	11+	12+	13+	14+
1 Båtsvatn res.	38	-	119	-	-	168	169	175	174	195	170	132	203	-
dwarf charr			-			-	±15.5	±14.5	±21.7	-	-	-	-	
1 Båtsvatn res.	33	-	-	122	227	248	251	294	300	308	340	-	-	-
normal charr				-	-	±24.2	±41.7	±25.7	±19.1	-	-			
2 Cuorvejavre	119	-	111	119	157	161	165	169	162	177	187	-	-	172
			±7.9	±21.8	±31.1	±20.6	±18.3	±18.1	±11.2	±23.3	-			-
3 Lake 1025	28	92	104	107	113	120	-	-	-	-	-	-	-	-
		-	-	±5.5	±14.5	±18.5								
6 Åvatn	16	-	-	-	272	279	274	-	-	313	-	-	-	-
					±29.3	±6.5	±13.2			-				
7 Rødsandvatn	33	-	200	201	205	215	251	-	-	-	-	-	-	-
			-	±32.5	±34.7	±20.2	±81.7							
8 Storvatn	35	140	173	195	209	256	245	258	220	310	-	320	-	-
		-	±35.3	±12.5	±29.1	±34.9	-	-	-	-		-		
11 Skogsfjordvatn	67	-	189	196	206	204	213	205	216	225	-	210		
			±17.0	±18.4	±14.9	±7.8	±9.2	±9.0	±22.9	-		-		

Growth rates varied considerably among N. Norwegian charr (Table 3). No systematic variations along geographic gradients could be demonstrated. The mean values of pyloric caeca and gillraker counts varied little between samples, and ranges overlapped extensively (Table 2). Again, no systematic trends were discernable. This applied to these characters when treated separately, in combinations and combined with esterase frequencies.

The frequencies of serum esterases varied greatly among samples, from 0.19 to homozygosity for the Est^F allele. Comparisons of the F frequency and lake

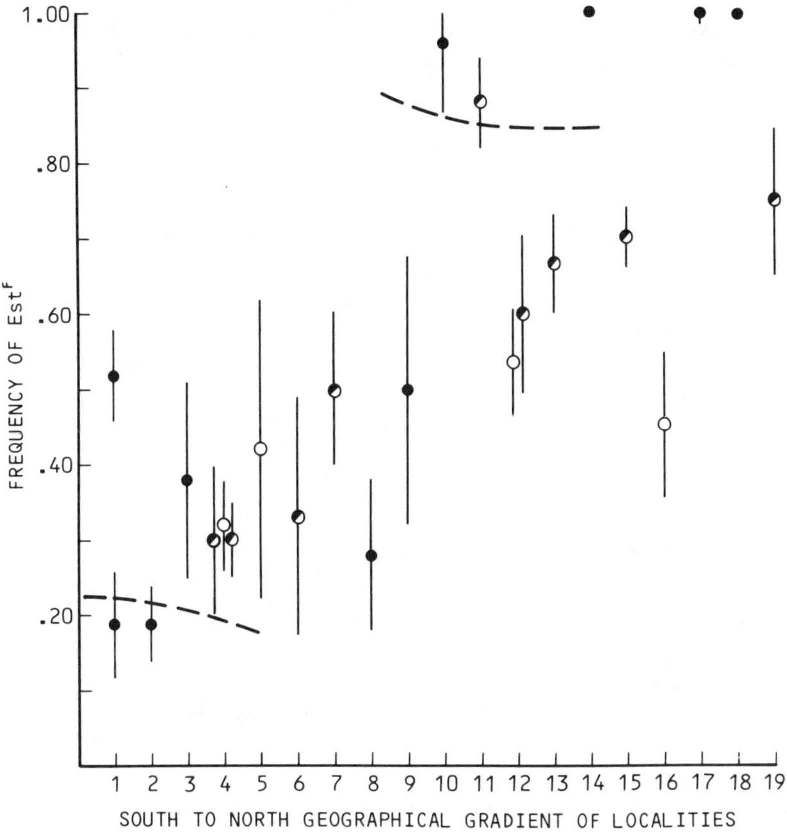

FIG. 2 Frequencies of the blood esterase isozyme Est^F of N. Norwegian Arctic charr populations ordered along a S-N gradient. Locality numbers refer to Table 1 and Figure 1. Open circles denote anadromous charr, half filled circles resident charr and filled circles landlocked charr. 95 percent confidence limits are given. The possible segregation of three population groups indicated by stippled lines is discussed in the text.

size, elevation or the gradient inland - coast - archipelago - remote island did not give any meaningful correlations. However, by simply ordering the localities along the coast several patterns appeared (Fig. 2). An abrupt increase in the F-frequency occurred in Takvatn (population 10) and all localities northward.

With the exception of Skogsfjordvatn these high f(F) samples were from landlocked populations of charr. In recent years sea-run charr have been very scarce in Skogsfjordvatn (H. Raste, pers. comm.). Low f(F) charr was not found at all north of the two Skjomen lakes (Fig. 2). A cline of several populations with moderate but steadily increasing f(F) appeared along the whole span of the geographic gradient. The anadromous charr samples all belonged here. All samples except Storvatn (8) and the anadromous samples from Øvervatn (4) agreed with expected values under Hardy-Weinberg equilibria (χ^2; 95% significance level, see Table 4).

DISCUSSION

The two charr populations of the Båtsvatn reservoir are discrete biological units. Ecological (habitat, food, schooling behaviour) and life history (growth, size, maturation pattern, time and place of spawning) parameters were distinctly different. In spite of spatial mixing in the lake their different genetic equilibria for serum esterases show that they are reproductively isolated. Their external differences are sufficient to separate them in this locality; however, the characters used (colouration, parr marks, fin form, eye diameter, snout length) are highly variable in charr and may not be diagnostic in other situations. Klemetsen and Grotnes (1980) proposed that these populations represented sibling species of charr. Mayr's (1977) concept of sibling species is not contradicted by the available information. It was also proposed (op. cit.) that Fennoscandian charrs may be descendants of one freshwater ancestor invading from the east, and one anadromous ancestor invading from the south-west. The present results support this hypothesis. No low f(F) charr were found along the coast of N. Norway. The only new S-charr was found *upstream* from Båtsvatn reservoir; this lake is closer to the E/W water divide, having a river outlet not ascendable by fish. Apparently it lives alone in this lake. A mixed sample of normal and dwarf charr would not have demonstrated genetic equilibrium as the present one did (Table 4). The other characters studied also matched those of the dwarf charr of the reservoir. Furthermore the many populations along the coast with median F-frequencies, including anadromous charr, support the hypothesis that the normal charr is a descendant of a sea-run ancestor. Upstream dispersal by man has most probably been important. Ekman (1910) stated that transport of fish between lakes has been common practice in N. Sweden. This probably applies also to Norway. The Sameh people with their long indigenous history and their reliance upon freshwa-

Table 4. Blood serum esterase electrophoresis of Arctic charr samples from N. Norway. No. 4 populations recalculated after H. Nordeng (in lit.); No. 1 populations after Klemetsen and Grotnes (1980) and No. 18 after Holthe (1981). See also Nilsen and Klemetsen (this volume).

Population	Observed phenotypes F/F	F/S	S/S	Frequency of allele F	$\underline{X}^2$ deviation from expected values
1 Båtsvatn res., dwarf charr	3	17	40	0.19	0.18
1 Båtsvatn res., normal charr	42	72	35	0.52	0.17
2 Cuorvejavre	4	49	94	0.19	0.65
3 Lake 1025	5	11	12	0.38	0.76
4 Øvervatn, small stationary charr	14	63	74	0.30	0.02
4 Øvervatn, large stationary charr	4	18	21	0.30	< 0.01
4 Øvervatn, anadromous charr	18	40	62	0.32	6.28
5 Grønvika	3	4	5	0.42	1.18
6 Åvatn	3	6	9	0.33	3.14
7 Rødsandvatn	10	30	10	0.50	2.00
8 Storvatn	0	25	19	0.29	6.43
9 Gamasjavre	5	4	5	0.50	2.57
10 Takvatn	37	3	0	0.96	0.06
11 Skogsfjordvatn	43	14	0	0.88	1.19
18 Ellasjøen	1408	5	0	1.00	0.01
19 Stevatn	20	17	1	0.75	1.27

ter fish resources has most certainly contributed significantly to the dispersal of charr.

Recently Rahman (1980) described a double charr situation in Kalandsvatn near Bergen, W. Norway. One of the sibling species was a low f(F) charr. This does not at first sight fit the hypothesis of the S-charr having freshwater origin. However, the presence of freshwater coregonids at Jaeren further south on the west coast has been explained by dispersal along the brackish Norwegian coastal current (Huitfeldt-Kaas 1918). This current originates partly in the Baltic. During heavier meltwater discharge the salinity of the coastal current was probably lower than today. It may then have offered a possible immigration route for some eastern freshwater charr to the west coast of Norway.

The latitude/temperature hypothesis of Nyman and Shaw (1971) and Nyman (1972) seems to be contradicted by the present results. Figure 2 indicates an increasing Est^F frequency from south to north for moderate f(F) charr. This is contrary to the cline presented by Nyman (1972).

This disagreement calls for more research into the physiological biochemistry of charr esterase isozymes. Very little is actually known about them. A better understanding could probably lead to more insight into charr evolution under different selection pressures.

The Takvatn charr was introduced from the nearby Fjellfrøskvatn in the late 1920's. A sample from this lake had a similar high value: $f(F)=0.95$ ($N=40$, χ^2-deviation 0.11). Further south no high f(F) populations were recorded. This applied both inland in a major tributary to Målselv (9) and on the island of Senja (5-8). These results may be taken to support Nyman's (1972) hypothesis that Scandinavian charrs actually belong to three species, one marked by very high, one by moderate and one by very low F-values. The F-charr appears to be the most cold-adapted of the three (Nyman 1972 and pers. comm., Hammar this volume). This fits with the northern distribution of the high F populations in the present material (Fig. 2).

Nilsen and Klemetsen (this volume) suggest that present F-charr populations may be descendants of an ancestor which survived on the Siberian coast to the east of the Weichselian ice sheet. Recent geological literature reveals the possibility of a refugium in the Laptev Sea region (Boulton 1979, Vorren and Rommevik 1979). The distance to N. Norway from this region is roughly the same as from SW Europe. Via the White Sea connection the Baltic may also have been reached by a NE ancestor. The geographic overlap of high and moderate f(F) populations of charr in N. Norway (Fig. 2) is readily explained by this postulate because it implies two immigration routes in opposing directions. By reason of its commonness the normal charr seems to be the most successful invader. It may even have reached much further east along the Siberian coast, and it is possibly the only existent anadromous form of N. Norway. Preliminary results from inland Finnmark indicate that the normal charr also dominates there (Holthe et al. 1979). However, samples from two lakes (Mathisvatn and Fidnajavre) had high F values (0.75 and 0.87) which were not in Hardy-Weinberg equilibria (χ^2-deviations 29.85 and 29.53, N=220 and 206). Further investigations are necessary to reveal whether or not these lakes hold sympatric F-charr and normal charr. Other preliminary results from Raisajavre in inland N. Troms and Akkajavre in inland E. Finnmark also indicate normal charr (Klemetsen unpubl.).

Serious objections to this proposition may be raised. It is based on only one genetic character, and classical morphological characters such as gillraker and pyloric caeca numbers are not diagnostic. However, the one character should not be taken for more than it actually is: a genetic tag which may indicate the need for an explanation. Whether these possible taxa should have species rank or not is still not settled. Svärdson (1961) and Nyman (1972) prefer to designate them as species. I tend to agree, but would also like to stress that northern salmonids have put the species concept in biology to a serious test. The discussion by Behnke (1980) and Savvaitova (1980b) illustrates the point.

The results from Bear Island complicate the picture further. Ellasjøen holds two distinct spawning groups of Arctic charr. Despite several other segregating parameters (growth, food, habitat, maturation age, fecundity) they could not be separated by serum esterase frequencies. This situation suggests that incipient sympatric splitting may be at work. This possibility has been raised repeatedly in biology and has been refuted many times, notably by Mayr (1977). Maynard-Smith (1966) claimed that such speciation is theoretically possible. Later papers, especially in entomology, have confirmed this (see for instance Tauber and Tauber 1977). Scudder (1974) argued for the possibility and lately Balon (1980b, c) and Savvaitova (1980a) have proposed it again for Arctic charr. Lindsey et al. (1970) also believe in sympatric speciation in northern salmonids (see also C.C. Lindsey in Klemetsen and Grotnes (1975), the verbal discussion).

Balon's (1980b, c) hypothesis that paedomorphosis by progenesis may represent one possible line of evolution is interesting. Gould's (1977) thorough treatment of heterochrony in evolution certainly supports this in a general way. A better analysis to explain the sympatric evolution of two optimal reproductive strategies for charr in some, but not in other situations is, however, needed.

Nordeng (1961) described three morphs of charr in the Salangen river system. These are included at location no. 4 (Fig. 2). All three had similar Est^F frequencies (Table 4, after H. Nordeng in lit.). Nordeng (1983) claims that they belong to the same population. The f(F) values seem to support this.

Two conclusions may be drawn from these studies. It is confirmed that North Norway has a central geographical position for Arctic charr. The distances to three possible glacial refugia are about equal, and there are indications that three species may occur in the region. Secondly, the possibility of sympatric speciation among these fish must be taken seriously. At present I can see no objection to accepting that both allopatric and sympatric speciation may have contributed to produce the confusion we term the *Salvelinus alpinus* complex.

ACKNOWLEDGEMENTS

Hans Nordeng generously permitted the inclusion of serum esterase analyses from Salangen charr in this paper. Odd Gulseth supplied blood samples from Oksfjordvatn and Martin Svenning from Takvatn charr. Per Grotnes gave critical comments to the manuscript and Jeff Wallace helped with the English. Their assistance is gratefully acknowledged.

REFERENCES

Balon, E.K. [ed.] 1980a. Charrs: salmonid fishes of the genus *Salvelinus*. Dr. W. Junk Publishers, The Hague. 928 p.

Balon, E.K. 1980b. Early ontogeny of the lake charr, *Salvelinus (Cristivomer) namaycush*, p. 485-562. *In* E.K. Balon [ed.] Charrs: salmonid fishes of the genus *Salvelinus*. Dr. W. Junk Publishers, The Hague.

Balon, E.K. 1980c. Early ontogeny of the European landlocked Arctic charr — altricial form, *Salvelinus (Salvelinus) alpinus alpinus*, p. 607-630. *In* E.K. Balon [ed.] Charrs: salmonid fishes of the genus *Salvelinus*. Dr. W. Junk Publishers, The Hague.

Behnke, R.J. 1972. The systematics of salmonid fishes of recently glaciated lakes. J. Fish. Res. Board Can. 29:639-671.

Behnke, R.J. 1980. A systematic review of the genus *Salvelinus*, p. 441-480. *In* E.K. Balon [ed.] Charrs: salmonid fishes of the genus *Salvelinus*. Dr. W. Junk Publishers, The Hague.

Boulton, G.S. 1979. A model of Weichselian glacier variation in the North Atlantic region. Boreas 8:373-395.

Ekman, S. 1910. Om människans andel i fiskfaunans spridning til det inre Norrlands vatten. Ymer:133-140. (In Swedish).

Frost, W.E. 1965. Breeding habits of Windermere char, *Salvelinus willughbii* (Gunter), and their bearing on speciation of these fish. Proc. Roy. Soc. (B) 163:232-283.

Gould, S.J. 1977. Ontogeny and phylogeny. Belknap of Harvard University Press. Cambr., Mass., 501 p.

Grotnes, P.E. 1975. The consequences of impoundment on an Arctic char lake system. An analysis by simulation modelling. Verh. int. Ver. Limnol. 19:2529-2537.

Huitfeldt-Kaas, H. 1918. Ferskvandsfiskenes utbredelse og indvandring i Norge. Kristiania, 106 p. (In Norwegian).

Johnson, L. 1980. The Arctic charr, *Salvelinus alpinus*, p. 15-98. *In* E.K. Balon [ed.] Charrs: salmonid fishes of the genus *Salvelinus*. Dr. W. Junk Publishers, The Hague.

Holthe, H. 1981. To sympatriske former av røye, *Salvelinus alpinus* (L) i Ellasjøen, Bjørnøya. En sammenligning av genotypiske og fenotypiske trekk. Cand. real. thesis, University of Tromsø, 71 p. (In Norwegian).

Holthe, H., P. Nilsen and T. Hesthagen. 1979. Foreløpig rapport om utbredelse og forekomst av dvergrøye (*Salvelinus alpinus*) i Finnmark. Directorate of wildlife and freshwater fish, Trondheim, 30 p. (In Norwegian, mimeographed).

Klemetsen, A. and P.E. Grotnes. 1975. Food and habitat segregation by two sympatric Arctic char populations. Verh. int. Ver. Limnol. 19:2521-2528.

Klemetsen, A. and P.E. Grotnes. 1980. Coexistence and immigration of two sympatric Arctic charr, p. 757-763. *In* E.K. Balon [ed.] Charrs: salmonid fishes of the genus *Salvelinus*. Dr. W. Junk Publishers, The Hague.

Klemetsen, A., P.E. Grotnes and C. Andersen. 1972. Preliminary studies of North Norwegian Arctic char lakes. Verh. int. Ver. Limnol. 18:1107-1113.

Lindsey, C.C., J.W. Clayton and W.A. Franzin. 1970. Zoogeographic problems and protein variation in the *Coregonus clupeaformis* whitefish species complex, p. 127-146. *In* C.C. Lindsey and C.S. Woods [eds.]. Biology of coregonid fishes. University of Manitoba Press, Winnipeg.

Maynard-Smith, J. 1966. Sympatric speciation. Amer. Nat. 100:637-650.

Mayr, E. 1977. Populations, species and evolution. Belknap of Harvard University Press. Cambr., Mass. 453 p.

McCart, P.J. 1980. A review of the systematics and ecology of Arctic char, *Salvelinus alpinus* in the western Arctic. Can. Tech. Rep. Fish. Aq. Sci. 935:1-89.

McPhail, J.D. 1961. A systematic study of the *Salvelinus alpinus* complex in North America. J. Fish. Res. Board Can. 18:793-816.

Nilsen, P. 1980. Stasjonaer og anadrom røye, *Salvelinus alpinus* (L). En sammenlikning av økologiske og genetiske trekk hos bestnader med sannsynlig felles opprinnelse. Cand. real. thesis, University of Tromsø. 84 p. (In Norwegian).

Nordeng, H. 1961. On the biology of charr (*Salmo alpinus* L) in Salangen, North Norway. I. Age and spawning frequency determined from scales and otoliths. Nytt Mag. Zool. 10:67-123.

Nordeng, H. 1983. Solution to the "charr problem" based on Arctic charr (*Salvelinus alpinus*) in Norway. Can. J. Fish. Aquat. Sci. 40: 1372-1387.

Nyman, L. 1967 Protein variations in Salmonidae. Inst. Freshw. Res. Drottningholm Rep. 47:6-38.

Nyman, L. 1972. A new approach to the taxonomy of the "*Salvelinus alpinus* species complex." Inst. Freshw. Res. Drottningholm Rep. 52:103-131.

Nyman, L. and D.H. Shaw. 1971. Molecular weight heterogeneity of serum esterases in four species of salmonid fish. Comp. Biochem. Physiol. 40B:563-566.

Rahman, S.B. 1980. The complex and the muddled complexity. A study of two sympatric species of the assumed *Salvelinus alpinus* complex. Cand. real. thesis, University of Bergen, 47 p.

Savvaitova, K.A. 1980a. Taxonomy and biogeography of charrs in the Palearctic, p. 281-294. *In* E.K. Balon [ed.] Charrs: salmonid fishes of the genus *Salvelinus*. Dr. W. Junk Publishers, The Hague.

Savvaitova, K.A. 1980b. Comments to the "Systematic review of the genus *Salvelinus*", p. 480-481. *In* E.K. Balon [ed.] Charrs: salmonid fishes of the genus *Salvelinus*. Dr. W. Junk Publishers, The Hague.

Scudder, G.E. 1974. Species concepts and speciation. Can. J. Zool. 52:1121-1134.

Svärdson, G. 1961. Young sibling fish species in northwestern Europe, p. 44-52. *In* K. Blair [ed.]. Vertebrate speciation. University of Texas Press, Austin.

Tauber, C.A. and M.J. Tauber. 1977. A genetic model for sympatric speciation through habitat diversification and seasonal isolation. Nature, Lond. 268:702-705.

Vorren, T. and H.C. Rønnevik. 1979. Sokkelens naturhistorie. Ottar 118:9-26 (In Norwegian).

Anadromous, Resident and Landlocked Arctic Charr on the Coast of Finnmark, N. Norway

PETTER NILSEN[1] and ANDERS KLEMETSEN

Institute of Biology and Geology, University of Tromsø N-9001, Tromsø, Norway

NILSEN, P., AND A. KLEMETSEN. 1984. Anadromous, resident and landlocked Arctic charr on the coast of Finnmark, N. Norway, p. 79–89. *In* L. Johnson and B.L. Burns [eds.] Biology of the Arctic charr, Proceedings of the International Symposium on Arctic Charr, Winnipeg, Manitoba, May 1981. Univ. Manitoba Press, Winnipeg.

Arctic charr from Brennelvvatn, Smørfjord, were almost homozygous for the blood serum esterase allele F. Postglacial elevation of the land has resulted in the isolation of this population for 4-5000 years. The anadromous population of the neighbouring river (Smørfjordelva) had a F-frequency of 0.45. A resident lake population in the same watercourse (Eidevatn) had 0.70 F. Until very recently, sea-run charr ascended into Eidevatn. The three populations differed in growth rate and also had slight differences in numbers of gillrakers and pyloric caeca. Their-spawning-age patterns were the same. In a comparable geographical situation further west (Kvalsund), nearly identical esterase frequencies were found for landlocked and resident charr. In Oksfjordvatn (N. Troms), resident and anadromous charr esterase patterns were similar to those for Smørfjord. Apart from resident and anadromous charr, sympatric populations were not found in any of the lakes studied.

[1] Present address: N-9620 Kvalsund, Norway.

INTRODUCTION

Klemetsen and Grotnes (1980) suggested that the two sympatric populations of Arctic charr found in the Båtsvatn reservoir in the Skjomen mountains of northern Nordland (Klemetsen et al. 1972, Klemetsen and Grotnes 1975, 1980) may be two allopatrically derived sibling species. They further proposed that Fennoscandian charrs may be descendants of charr from two different glacial refugia; one basically freshwater species from an inland refugium to the E/SE of the glacial sheet and one basically anadromous species from a coastal refugium to the SW of the glacial sheet. The species immigrated to Fennoscandia largely via the Ancylus Lake phase of the Baltic. The anadromous species immigrated along the coast and descendants exist today both as anadromous and stationary charr. This is an extension of the classic immigration theory of Huitfeldt-Kaas

(1918), but stresses the possible existence of sibling species, two widely separated glacial refugia and anthropogenic upstream dispersal.

An implication of these hypotheses is that anadromous and stationary populations on the outer coast of Finnmark should belong to the same species, provided that immigration via water courses originating close to the main Fennoscandian water divide is not possible. A demonstration of sympatric sibling species in this region would therefore contradict the immigration hypothesis. The same would be the case if a charr species taxonomically similar to the dwarf charr in Skjomen was found. Completely landlocked populations, resident populations in lakes with sea-run charr and sea charr of coastal Finnmark should all show the same characters.

METHODS

The Smørfjord area on the W. side of Porsangerfjord was chosen as the main study area (Fig. 1). This area is geographically well separated from the inner coast and central Finnmark where freshwater immigrants like whitefish (*Coregonus* spp.), perch (*Perca fluviatilis*) or pike (*Esox lucius*) occur. Only anadromous species or their descendants like brown trout (*Salmo trutta*) and salmon (*S. salar*) are found. Supplementary material was taken in the Kvalsund area and in Oksfjordvatn further west (Fig. 1). The same general characteristics apply to these areas; the immigration of pure freshwater fish is unlikely.The Smørfjord area had three types of charr populations (landlocked, resident and anadromous) closely situated to each other. The possibility of a common ancestor was likely on geographic grounds. Brennelvvatn is situated close to the sea (Fig. 1) at 47 m a.s.l. The upper marine transgression limit of this region is 52 m. Today several waterfalls effectively block upstream migration (Fig. 1). The Brennelvvatn charr, therefore, is a true landlocked population. The lake has an area of 0.6 km^2 and a maximum depth of 32 m. It is an oligotrophic, dimictic lake with a narrow littoral zone. The field work gave the impression of a poor fish food fauna. Brown trout occurred along with charr. About 20 years ago charr from Brennelvvatn were transferred to a small fishless lake further up the same watercourse, Røyevatn (Fig. 1). The transfer to Røyevatn founded a new population of charr. A limited number of samples were taken from Røyevatn. The Smørfjordelva river system empties into the sea about 4 km from Brennelva. Pollen, the almost enclosed estuary of this system, was the main site for sampling ascending anadromous charr (Fig. 1). Eidevatn is situated at 81 m a.s.l. and 5 km upstream from the estuary. The lake holds resident charr and brown trout. Up to about 10 years ago a substantial population of sea-run charr reached Eidevatn. Today anadromous charr are very few and possibly in the process of disappearing from Eidevatn. A possible reason for this is that the Spring flood loosened some

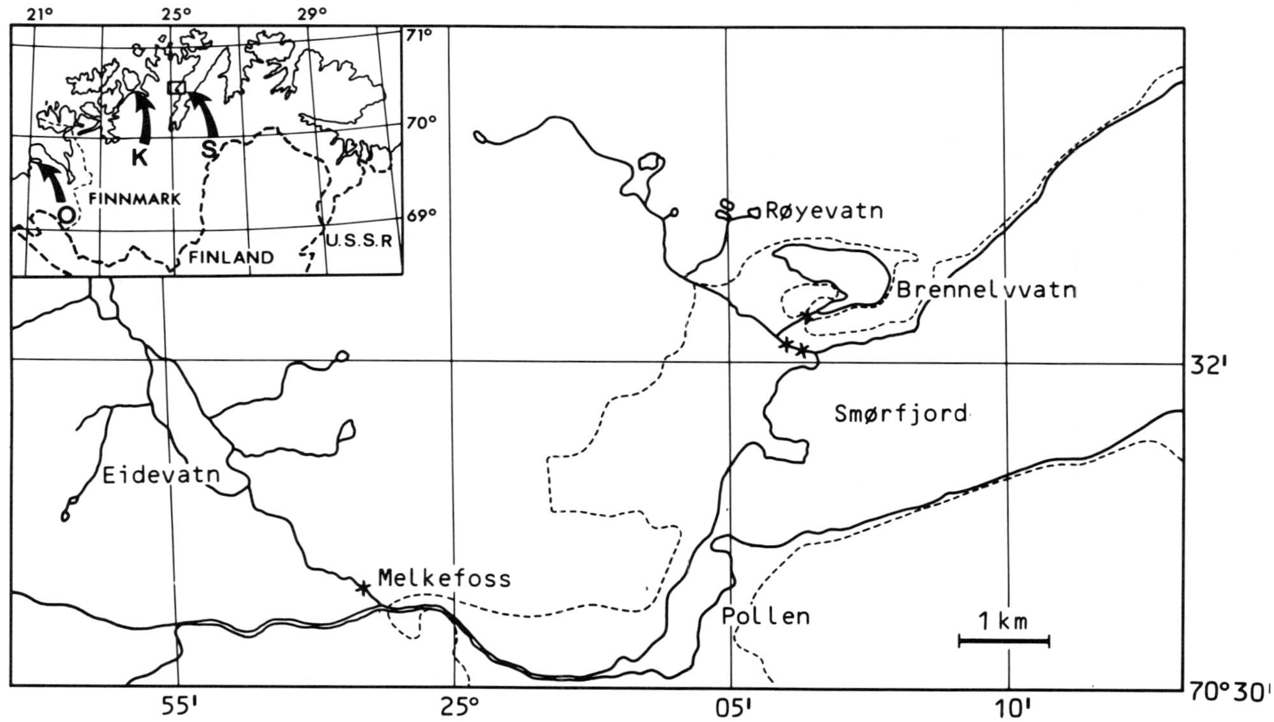

FIG. 1 Map of the study area at Smørfjord. The stippled line shows the maximal postglacial marine transgression. Crosses denote waterfalls which are impassable for ascending charr today. Inserted is a situation map showing the location of the sampling localities. Q: Oksfjordvatn, K; Kvalsund area, S: Smørfjord area.

boulders in the Melkefoss waterfall (Fig. 1). This created a much higher fall and has made Melkefoss almost impassable for ascending fish. The result is that the resident charr population of Eidevatn has almost lost contact with the sea-run charr. In the terminology of this paper the Eidevatn population is probably moving from a resident (stationary but sympatric with freshwater phases of anadromous charr) to a landlocked situation (allopatric from sea-run charr). Eidevatn has an area of 0.7 km^2 and is a shallow (4 m) lake with a well developed littoral zone. Foraging conditions for fish seem to be good. Both brown trout and Arctic charr are present in the lake. The lakes Brensvikvatn and Neverfjordvatn in Kvalsund duplicate the situation in Smørfjord. Both are small lakes at low elevation, and both have charr and trout. The charr of Brensvikvatn are landlocked while the stationary charr of Neverfjordvatn form a resident population. According to local fishermen the sea-run charr of Neverfjordvatn have been less common in recent years. In addition to these populations a sample from Oksfjordvatn (Fig. 1) was included in the material. This lake holds both resident and anadromous charr along with salmon and trout. The prime genetic character used to study these populations was blood serum esterases. Nyman (1972) demonstrated that serum esterases of Arctic charr occur in two co-dominant allelic forms governed by a single locus. Serum esterase analysis was performed on all populations. The Smørfjord populations were also analyzed for gillraker and pyloric caeca numbers. These are much used characters in *Salvelinus* taxonomy (see for example Behnke 1972, 1980, Johnson 1980). Spawning age characteristics were also studied for these populations. Recent reports by Naevdal et al. (1977) and Ritter and Newbold (1977) indicate that spawning age rather than spawning size may be inherited in salmonids.

The Smørfjord populations were sampled in 1978, Brennelvvatn in July, August and September, Eidevatn in July and September and Pollen in August and September. The supplementary material from Røyevatn was taken in July 1978, from the Kvalsund lakes in July 1979 and from Oksfjordvatn in August 1978.

Fish sampling was performed with standard 1.25 by 25 m sinking gillnets set from the shore outwards in series of 2 to 7 nets. Knot to knot mesh sizes varied from 10 to 39 mm. The anadromous charr in Pollen were fished with floating nets.

The field measurements for the Smørfjord populations included weight, fork length, sex, maturation stage and flesh colour of each fish. Otoliths, the front left gill arch, stomachs and pyloric caeca were removed and preserved in ethanol. Blood samples were drawn, usually from the sinus venosus, into heparinized capillaries and deep frozen. Scale samples were taken from the Eidevatn charr in order to identify possible anadromous specimens by sea-run checks in the scales

(Nordeng 1961). McPhail (1961) and Rechetnikov (1961) discuss the possibility of a positive correlation between fish size and number of pyloric caeca, at least for small fish. To avoid this possible source of error the smallest specimens were discarded before counting the caeca. No correlations between caecal numbers and fish lengths were found in the remaining samples. Neither were any correlations found between fish lengths and gill-raker numbers.

The ethanol conservation usually rendered otoliths clear enough for direct reading. Additional clearing was achieved by placing the otolith in glycerol. Gillrakers and pyloric caeca were counted under slight magnification, the caeca, by removing them one by one. All gill arches which bore signs of being incomplete were discarded. Starch gel electrophoresis of blood samples was performed according to the procedure described by Nyman (1967).

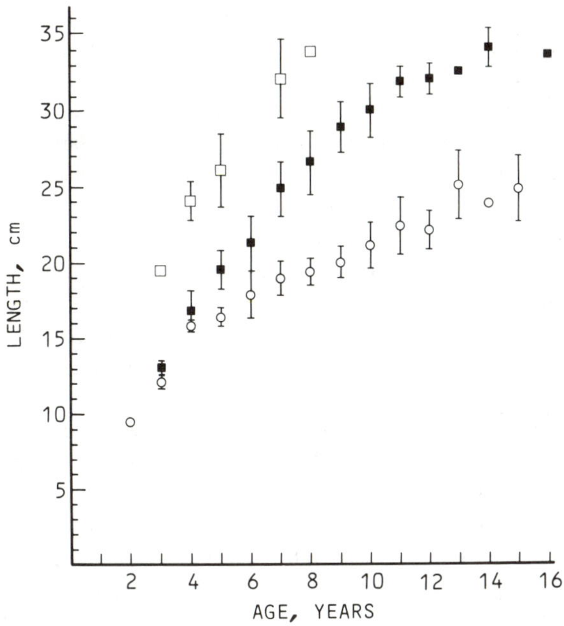

FIG. 2 Length for otolith ages in charr samples from Pollen (anadromous, open squares), Eidevatn (resident, filled squares) and Brennelvvatn (landlocked, circles). Ranges given are standard deviations.

RESULTS

The anadromous charr had the fastest growth rate of the three main populations investigated in the Smørfjord area (Fig. 2). From age 3-4 the sea-run charr

Table 1. The development of maturity with age for Eidevatn and Brennelvvatn charr.

	Age group for:			
	first ripe fish	half the year class ripe	all fish ripe	maximum age observed
Eidevatn	4+	7+	10+	16+
Brennelvvatn	5+	7+	10+	15+

showed a significantly greater length for age. The resident Eidevatn charr showed a significantly better growth than the landlocked Brennelvvatn charr from age 5 and upwards. The growth retardation of the Brennelvvatn charr was marked; no fish over 250 mm occurred in a total of 411 fish. The charr transplanted into Røyevatn (not included in Figure 2 for clarity) showed good growth, very similar to that of the resident charr in Eidevatn.

The development of maturity with age was almost the same for Brennelvvatn and Eidevatn charr (Table 1). At age 7, half the year class was ripe in both populations. This corresponds to about 190 mm fish length in the landlocked and 250 mm in the resident charr (Fig. 2).

Gillraker and pyloric caeca counts from the Smørfjord populations are shown in Table 2. Gillraker numbers overlapped extensively between populations. The slight differences recorded were statistically significant (χ-test, 95% level). Caecal numbers also overlapped extensively, but again, mean values were significantly different.

Serum esterase distributions from Smørfjord, Kvalsund and Oksfjord charr populations are shown in Table 3. The frequencies of Est^F for the same populations are illustrated in Figure 3. Samples from all populations agreed with Hardy-Weinberg equilibria (χ, 95% level), with the exception of the resident Oksfjordvatn population. The Eidevatn material also differed from the expected HW — values. Specimens with sea-run checks on their scales (8 fish) and all specimens smaller than 210 mm fork length were then discarded from the material in order to exclude anadromous charr, both fish which had been to sea and possible presmolts. This stripped sample did not deviate from the expected values under HW-equilibrium.

Among the Smørfjord populations, the sequence anadromous-resident-landlocked coincided with an increasing frequency of the Est^F allele. In the sea-run charr the F and S alleles were about equal ($f(F)=0.45$), whilst the population which had been landlocked over a long period of time was almost

Table 2. Gillraker and pyloric caeca counts for charr populations from the Smørfjord area.

	Gillrakers				Pyloric caeca			
Population	N	Mean	St.dev.	Range	N	Mean	St.dev.	Range
Eidevatn	160	22.0	0.80	20-24	204	37.2	3.50	35-59
Pollen	37	24.0	1.39	22-27	50	29.9	3.67	33-48
Brennelvvatn	221	23.7	1.07	19-25	227	46.3	4.68	30-47

homozygous for F. Only one heterozygote F/S was found in this sample (Table 3). The resident charr sample took an intermediate position (f(F) = 0.70).

The supplementary material from Kvalsund lacked an anadromous sample. For resident and landlocked charr the esterase frequencies were almost identical to the Smørfjord populations of similar ecological status (Fig. 3). The sample of resident charr had an F-frequency of 0.67 (Table 3), and the landlocked charr sample was homozygous for F. The F-frequency of the sample of anadromous charr from Oksfjordvatn was similar to the value found in sea charr from Pollen in Smørfjord. Resident charr from that lake showed a higher value (0.60), but this sample was not in HW-equilibrium.

DISCUSSION

Apart from anadromous and resident charr, no case of sympatric populations was demonstrated in the present study. The existence of sibling species pairs similar to Skjomen (Klemetsen and Grotnes 1980), was therefore not indicated in the areas investigated. Furthermore, the material gave no indication of a species with low F frequency, similar to the dwarf charr in Skjomen. The geographic position of the present charr localities are well outside the probable immigration range of pure freshwater fish. The present results, therefore, do not contradict the hypothesis that one strict freshwater species of charr invaded Fennoscandia, and that this is the low f(F) species.

Growth patterns of *S. alpinus* are known to vary greatly with the ecological situation (Johnson 1980). The differences observed among the Smørfjord populations are, therefore, most probably phenotypic. The establishment of the Røyevatn population, which grows much better than the parental population (Brennelvvatn), confirms this. Neither growth data nor the progress of maturation with age (Table 1) gave support to the existence of genetic differences between the populations (Table 1).

Behnke (1980), in his diagnosis of *S. a. alpinus*, lists 21-28 gillrakers as typical for charr from Northern Sweden. He argues that this is typical also for the

Table 3. Blood serum esterase electrophoresis of charr samples from Oksfjordvatn, the Kvalsund area and the Smørfjord area.

Populations	Observed phenotypes F/F	F/S	S/S	Frequency of allele F	χ^2 deviation from expected values
1 Oksfjordvatn (anadromous)	21	53	15	0.53	3.81
2 Oksfjordvatn (resident)	11	28	3	0.60	6.16
3 Neverfjordvatn (resident)	36	44	6	0.67	2.50
4 Brensvikvatn (landlocked)	73	0	0	1.00	-
5 Pollen (anadromous)	10	25	15	0.45	0.00
6 Eidevatn (resident)	69	68	9	0.70	2.13
7 Brennelvvatn (landlocked)	390	1	0	1.00	<0.01
8 Røyevatn (landlocked)	39	0	0	1.00	-

rest of north Fennoscandia and the Kola peninsula. The present observations fall within this range, and also agree with the values given by Savvaitova (1980) for European Arctic charr. The differences of the mean values between the populations are too small to have any taxonomic significance for allopatric populations. The same conclusion applies to the pyloric caeca counts. They fall within the ranges given by Behnke (1980) and Savvaitova (1980), and the differences between populations are not large enough to indicate any taxonomic divergence. The only possible exception may be the landlocked Brennelvvatn charr which had a markedly higher mean value than the two populations from the Smørfjordelv water course (Table 2). This may be taken to be caused by selection pressure during the several thousand years of probable isolation of the Brennelvvatn charr. However, some authors claim that food or other environmental factors may change pyloric caeca numbers phenotypically (Rechetnikov 1961, Lindsey 1964, Savvaitova 1980). We conclude that pyloric caeca numbers alone do not indicate that the present populations had different ancestors.

Several analyses have shown that the 'normal' or 'big' charr of Scandinavia has frequencies of the serum esterase allele F of around 0.5 (Nyman 1972, Klemetsen and Grotnes 1980, Nyman, this volume and Hammar, this volume).

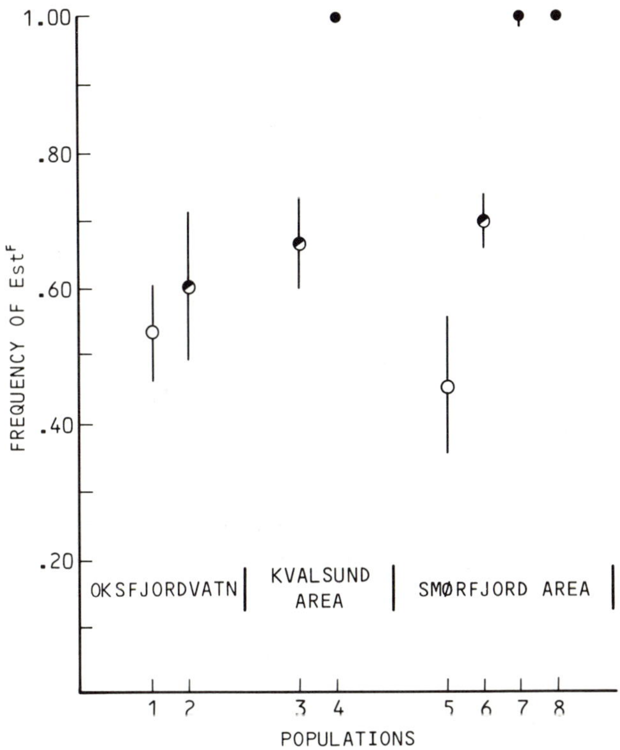

FIG. 3 Frequencies of the serum esterase allele F for the different charr populations studied. Population numbers refer to Table 3. Open circles denote anadromous charr, half filled circles resident charr and filled circles landlocked charr. 95 percent confidence limits are given.

Klemetsen and Grotnes (1980) put forward the hypothesis that the normal charr of Fennoscandia is conspecific with the anadromous charr of the North Norwegian coast. At that time no analyses of esterase frequencies of such anadromous charr had been published. The present results confirm the hypothesis. Both anadromous populations had F-values close to 0.5 (see also Klemetsen, this volume).

Perhaps the most interesting results of the serum esterase analysis from the Smørfjord area was the demonstration of a nearly homozygous F distribution in the Brennelvvatn population. We think it is highly probable that this lake, like most water courses along the Norwegian coast, was colonized by an anadromous ancestor after the retreat of the Weichselian ice cap (Huitfeldt-Kaas 1918). The contact with the sea was of relatively short duration because the lake's present elevation is slightly below the maximal marine transgression of the area (47 versus 52 m) and because impassable waterfalls were established only a short distance downstream from the outlet of the lake (Fig. 1). A landlocked population of charr was, therefore, probably established in Brennelvvatn as long as 4-5000 years ago. Genetic drift like the one proposed for Welsh monomorphic populations by Child (1977) was not very likely in Brennelvvatn. The demonstration of another F-homozygous, landlocked population in Brensikvatn, Kvalsund (Table 3, Fig. 3) confirmed that Brennelvvatn was not a special case. Recently, Holthe (1981) reported another landlocked charr (Ellasjøen, Bear Island) which was nearly homozygous for the allele F. The lack of Hardy-Weinberg equilibrium for the resident charr of Oksfjordvatn may have been caused by mixing with some presmolt sea-run charr, as was found in Eidevatn. Presumably the sorting out of possible anadromous specimens would have given a HW-equilibrium also for Oksfjordvatn resident charr, and increased the F-frequency of the sample. Accepting this, all resident populations had essentially higher f(F) values than the anadromous populations (Table 3, Fig. 3). This may suggest that the S isozyme is selected against during transition from anadromous to resident life cycles, for instance due to different fat types in marine and freshwater food animals. Support for this is, however, not indicated in other N. Norwegian charr samples (Klemetsen, this volume). At least these results highlight the need for more understanding of the physiological function of charr serum esterase isozymes. The only analysis of this known to us is the one by Nyman and Shaw (1971). The matter is further discussed by Klemetsen (this volume).

Nyman (1972) proposed that three species of Arctic charr are encountered in Scandinavia. The present results may be taken to support this hypothesis by claiming that the landlocked F-populations on the Finnmark coast are conspecific with Nyman's F-charr. If so, there is need for a plausible theory to explain survival throughout the glaciation and the subsequent migrations to present localities. Recent quarternary geological discussion indicates that a possible refugium existed on the Siberian coast to the east of the ice cap (Vorren and Ronneberg 1979). The question is further discussed by Klemetsen (this volume) on the basis of a more extensive North Norwegian material.

ACKNOWLEDGEMENTS

We gratefully acknowledge the technical assistance with the electrophoresis by Jan Evjen, the critical reading of the manuscript by Per Grotnes and the correction of the English by Malcolm Jobling.

REFERENCES

Behnke, R.J. 1972. The systematics of salmonid fishes of recently glaciated lakes. J. Fish. Res. Board Can. 29:639-671.

Behnke, R.J. 1980. A systematic review of the genus *Salvelinus*, p. 441-480. *In* E.K. Balon [ed.] Charrs: salmonid fishes of the genus *Salvelinus*. Dr. W. Junk Publishers, The Hague.

Child, A.R. 1977. Biochemical polymorphism in char (*Salvelinus alpinus* L.) from Llynnau Peris, Padarn, Cwellyn and Bodlyn. Heredity 38:359-365.

Holthe, H. 1981. To sympatriske former av røye, *Salvelinus alpinus* (L.), i Ellasjøen, Bjørnøya. En sammenligning av genotypiske og fenotypiske trekk. Cand. real. thesis, University of Tromsø. 71 p. (In Norwegian).

Huitfeldt-Kaas, H. 1918. Ferskvandsfiskenes utbredelse og innvandring i Norge. Kristiania, 167 p. (In Norwegian).

Johnson, L. 1980. The Arctic charr, *Salvelinus alpinus*, p. 15-98. *In* E.K. Balon [ed.] Charrs: salmonid fishes of the genus *Salvelinus*. Dr. W. Junk Publishers, The Hague.

Klemetsen, A., and P.E. Grotnes. 1975. Food and habitat segregation by two sympatric Arctic char populations. Verh. int. Ver. Limnol. 19:2521-2528.

Klemetsen, A., and P.E. Grotnes. 1980. Coexistence and immigration of two sympatric Arctic charrs, p. 757-763. *In* E.K. Balon [ed.] Charrs: salmonid fishes of the genus *Salvelinus*. Dr. W. Junk Publishers, The Hague.

Klemetsen, A., P.E. Grotnes, and C. Andersen. 1972. Preliminary studies of North Norwegian Arctic char lakes. Verh. int. Ver. Limnol. 18:1107-1113.

Lindsey, C.C. 1964. Problems in zoogeography of the lake trout, *Salvelinus namaycush*. J. Fish. Res. Board Can. 21:977-994.

McPhail, J.D. 1961. A systematic study of the *Salvelinus alpinus* complex in North America. J. Fish. Res. Board Can. 18:793-816.

Naevdal, G., M. Holm, O. Ingebrigtsen, and D. Møller. 1977. Variation in age at first spawning in Atlantic salmon (*Salmo salar*). J. Fish. Res. Board Can. 35:145-147.

Nordeng, H. 1961. On the biology of char *Salmo alpinus* L. in Salangen, North Norway. I. Age and spawning frequency determined from scales and otoliths. Nytt Mag. Zool. 10:67-123.

Nyman, L. 1967. Protein variations in Salmonidae. Inst. Freshw. Res. Drottningholm Rep. 47:5-38.

Nyman, L. 1972. A new approach to the taxonomy of the "*Salvelinus alpinus* species complex". Inst. Freshw. Res. Drottningholm Rep. 52:103-131.

Nyman, L., and D.H. Shaw. 1971. Molecular weight heterogeneity of serum esterases in four species of salmonid fish. Comp. Biochem. Physiol. 40B:563-566.

Rechetnikov, J.S. 1961. On the connection between the amount of gillrakers and the character of feeding in chars of the genus *Salvelinus*. Zool. Zh. 10:1574-1577. (In Russian, English summary).

Ritter, J.A., and K. Newbold. 1977. Relationship of parental and smolt age to age at first maturity of Atlantic salmon (*Salmo salar*). Int. Council Explor. Sea, C.M. 1977/M:32 Anacat. Comm., 4 p. (Mimeogr.).

Savvaitova, K.A. 1980. Taxonomy and biogeography of charrs in the Palearctic, p. 281-294. *In* E.K. Balon [ed.] Charrs: salmonid fishes of the genus *Salvelinus*. Dr. W. Junk Publishers, The Hague.

Vorren, T.O., and H.C. Ronnevik. 1979. Sokkelens naturhistorie. Ottar. 118:9-26. (In Norwegian).

Preliminary Survey Results of the Distribution of the Arctic Charr Species Complex in Iceland

ROLF GYDEMO

Institute of Freshwater Research, S-170 11 Drottningholm, Sweden

GYDEMO, R. 1984. Preliminary survey results of the distribution of the Arctic charr species complex in Iceland, p. 91–107. *In* L. Johnson and B.L. Burns [eds.] Biology of the Arctic charr, Proceedings of the International Symposium on Arctic Charr, Winnipeg, Manitoba, May 1981. Univ. Manitoba Press, Winnipeg.

Arctic charr is the most abundant freshwater salmonid species in Iceland. Serum esterase frequencies from 38 populations were determined electrophoretically. All three Scandinavian species of the Arctic charr complex occur, probably, as both landlocked and anadromous populations. Distribution and ecology of the landlocked populations appear to be similar to the better studied populations of Scandinavia, i.e. charr with high frequency of the F-allele has a higher altitudinal distribution thus suggesting it to be the first immigrant. Charr with an intermediate frequency of the F-allele dominates at lower altitudes, while charr with the lowest frequency of the F-allele so far has been found only in one instance in a landlocked population. The anadromous populations have gene frequencies deviating somewhat from the landlocked populations in Scandinavia. One population with a probable fixation of the F-allele and one with fixation of the S-allele were found, indicating that there could be distinct anadromous populations. The most common gene frequencies lie within the higher part of the range of those found in the ''intermediate'' species. The slight deviation from the gene frequencies found in landlocked populations could be the result of gene-flow due to straying. The distribution of gene frequencies is not correlated with geographical distribution, thus no immigration direction could be determined from the data available.

INTRODUCTION

In any discussion of the taxonomy, systematics and distribution of Arctic charr (*Salvelinus alpinus* L. species complex), Iceland is an area of great interest. Situated in the north Atlantic ocean, half-way between Europe and North America on the Brito-Arctic tertiary formation, in a volcanically active area, it has been open to postglacial invasion by Arctic charr for something less than 10,000 years.

Iceland is situated between 63°24′ and 66°32′N and 13°30′ and 24°32′W. The climate is maritime with cold summers and mild winters. Vegetation is sparse, trees growing only in sheltered areas. This leaves the lakes exposed to winds.

Geologically it is a young country, its oldest rocks dating back to early Tertiary (Thorarinsson et al. 1959). Situated on the mid-Atlantic ridge, it is still in the process of formation by volcanic activity. During the last glacial period, which ended about 10,000 years ago, the ice cap covered the whole country out to the present 100 m depth contour (Einarsson 1968). There are still large glaciers present, the largest one being Vatnajokull in the southeast. The limit of postglacial submergence, as marked by the altitude of the highest strandline, varies in different parts of the country. In the Lake Thingvallavatn area the former sea level is 110 m above that of today, although, in general, the marine limit is about 50 m a.s.l. (Einarsson 1968). The area of Reykjavik has probably risen about 5 m in the last 3,000 years (Einarsson 1968), which means that the lakes in that area are almost all of older date. Further north in the area of Borgarfjordur near the Hvita river, the highest shoreline is about 57 m a.s.l. indicating that Skorradalsvatn, for example, has been cut off from the sea for about 2,800 years (Einarsson 1968).

Thingvallavatn, Mývatn, Hlidarvatn (Hnappadalur) and Oddastadsvatn are all situated in volcanically active areas and have all been influenced by vulcanism in postglacial times. Because of the well-known history of human colonization which began about 1,100 years ago, the years of great eruptions, earthquakes and volcanic activity are known. Mývatn was formed about 3,800 years ago by an eruption and concomitant lava flow which dammed the River Laxá. During the years 1724-29 the lake was exposed to a series of earthquakes which raised and lowered the water level and a lava flow that reached the north shore of the lake in July 1729 (Thorarinsson 1979). This probably had great influence on the fauna of the lake by heating, release of sulphur into the water and desiccation of the northeast part (Lamby 1941). In the last few years geothermal activity has again increased in the Mývatn area.

FISH SPECIES

Iceland has only two non-salmonid freshwater fish species: the three-spined stickleback, *Gasterosteus aculeatus* L., and the European eel, *Anguilla anguilla* L.; of the salmonids the most abundant and widespread is Arctic charr which occurs as both landlocked and anadromous populations. The other species are Atlantic salmon, *Salmo salar* L., and brown trout, *S. trutta* L..

In spite of the wide occurrence and great economic importance few investigations have been made on the taxonomy and systematics of Arctic charr in Iceland. Most investigations concerned management problems of the often overpopulated charr lakes (Gudjonsson 1954, Kristjansson 1973, 1978a and b). Saemundsson (1909) and Fridriksson (1939) studied the sympatric charr in Thingvallavatn, and they concluded, respectively, that there were four and three sympatric varieties or races of Arctic charr within the lake. Lamby (1941) investigated the feeding of two forms of Arctic charr in Mývatn but made no taxonomic studies. He suggested only that the dwarf form was equivalent to the "murta" of Thingvallavatn. Oliva (1951) examined three specimens of charr from Reydarvatn in western Iceland but did not discuss their systematic position. Behnke (1980) discussing Icelandic charrs on the basis of results obtained by Fridriksson (1939) and Lamby (1941), and his own examination of a limited number of charr from Thingvallavatn and Mývatn, left the question of their systematic position open, stressing, however, that Iceland should be a high-priority area for the study of Arctic charr systematics.

The opportunity to work at the Institute for Freshwater Fisheries in Reykjavik for short periods in 1978 and 1979, opened the possibility of starting a survey programme on the occurrence and distribution of the Arctic charr complex in Iceland. Although I was not primarily concerned with Arctic charr, samples were collected when the opportunity arose. This resulted in samples of varying size; in many cases they were almost too small to permit definite conclusions, but they covered a large number of localities around the country. The results from the 1978 and 1979 investigations were briefly presented at the ISACF workshop in 1980 (Gydemo 1980b). At that time there was still hope of collecting more material from populations of dubious status. The bulk of material obtained in 1980 was from the northeastern part of Iceland, kindly collected by Mr. Jon Kristjansson. However, in spite of the doubts with regard to some populations, I consider it useful to present the results gathered so far as a background for further studies. One such programme underway is the inter-nordic study of Thingvallavatn and its ecology, of which charr investigations form an important part.

MATERIALS AND METHODS

Svärdson (1961) suggested that three sibling species of Arctic charr invaded Scandinavia post-glacially from independent refugia. Nyman (1965, 1967, 1972) presented a method that made possible extensive surveys of the distribution of the Arctic charr with a reasonable amount of effort. The method, a combination of ecological data and gene frequency data of serum esterases, has been successfully employed in an extensive survey program on the distribution of the three Arctic charr species in Sweden (Gydemo 1978, 1979, 1980a, Nyman et al. 1981). The results have given a more thorough knowledge of the ecology of the species which has been of great value in charr management.

It is the purpose of this paper to show the occurrence and indicate the distribution of the Arctic charr complex in Iceland by the use of serum esterase analyses in combination with ecological criteria.

During 1978, 1979 and 1980, 1248 charr were collected from 36 localities around the country. Of these, 10 populations were fully anadromous or mixed with resident charr; the rest were landlocked.

Most of the material was collected with gillnets; only a minor number being taken by electrofishing or midwater trawling.

Fish length was measured as total length to the nearest 5 mm. Otoliths were extracted for age determination. Sex and maturity were noted in most cases as well as flesh colouration and differences in external morphology. Blood, sampled for electrophoretic studies of serum esterases, was kept frozen until analyzed in horizontal starch gel electrophoresis as described by Nyman (1972). The gene frequency of the fast (F) allele was determined for each population and individual genotypes were plotted in a length-age relationship diagram. In cases where gene frequencies were not fixed in equilibrium, (i.e. inconsistent with the Hardy-Weinberg formulation) the material was subdivided to investigate the possible existence of sympatric populations (Nyman and Filipsson 1972).

The gene frequencies described for allopatric populations of the three sibling species of Arctic charr in Scandinavia (Nyman et al. 1981) are: ''F-charr'' (F-dwarf, F-tita): f(F) = 0.90-1.00, ''Normal charr'' (N-charr): f(F) = 0.40-0.70 and ''S-charr'' (S-dwarf, S-tita): f(F) = 0.00-0.20. The gene frequencies of the Icelandic material are compared with these values.

The approximate location of each of the waters sampled is given in Figure 1 and Table 1. The different localities are presented for each of the four parts of Iceland in clockwise order starting on the west coast.

For each locality sampled, data are given in Table 1 on its approximate position, altitude above sea level, area in km^2, maximum and mean depths, other fish species present, year of sampling, and type of population (anadromous, resident or landlocked).

RESULTS

In Figure 2a-d and 3a-c growth curves are presented for most populations studied. Table 2 gives data on esterase allele frequencies, total number of fish analyzed, frequency of the F-allele, chi-squared value, probability of worse fit, 95 percent confidence interval and whether the population is of allopatric or sympatric type. The general findings from the four areas of Iceland are given. Numbers in the text and figures correspond to the numbering of lakes and rivers given in Tables 1 and 2.

Western Iceland (Localities 1-8)

Of the eight bodies of water sampled, six were lakes with landlocked populations of charr; the two remaining were rivers with anadromous charr. Four

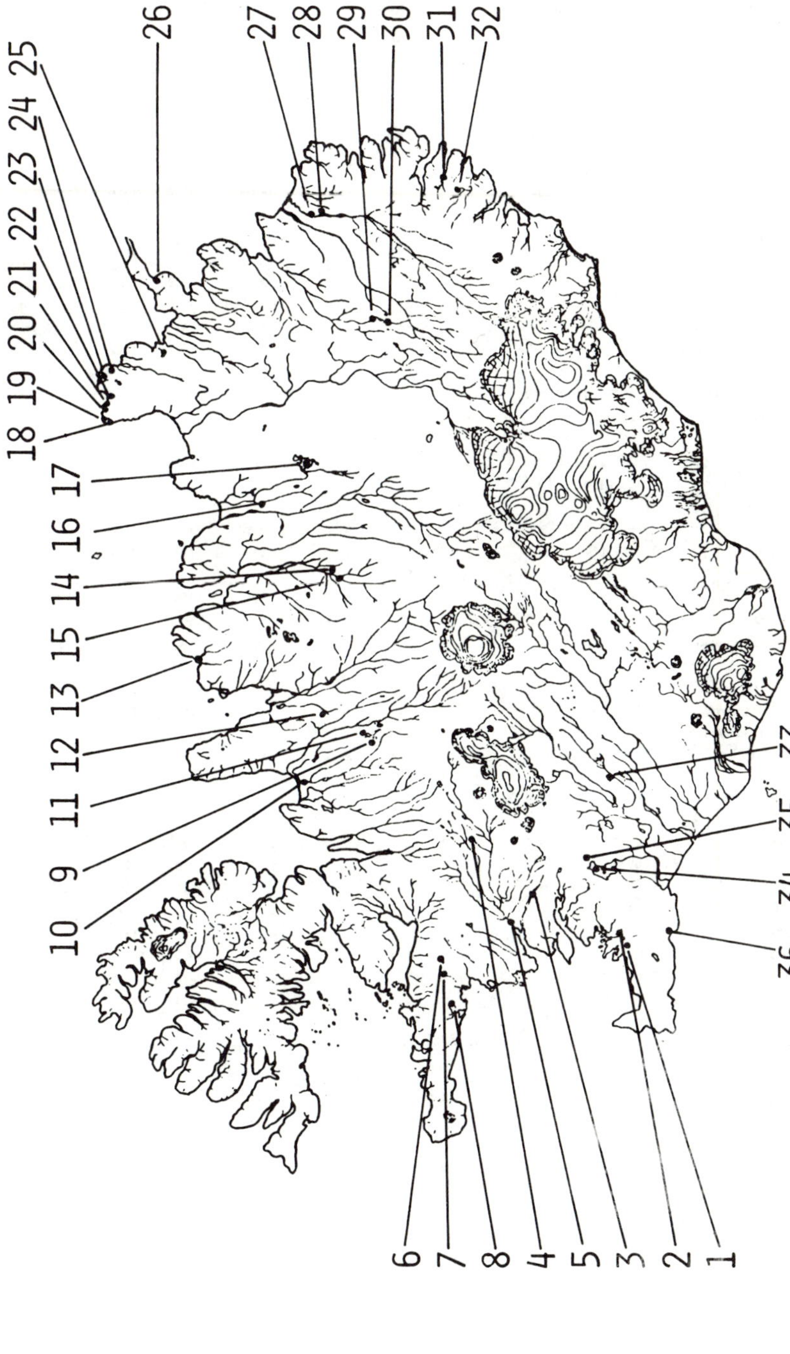

FIG. 1 Map of Iceland showing the localities sampled for Arctic charr. The numbers refer to Table 1.

Table 1. General data on the waters sampled. Abbreviations: B = brown trout, S = Atlantic salmon, E = European eel, TS = threespine stickleback. Where no figures are given, data is lacking.

No.	Locality	Approximate location		Altitude m a.s.l.	Area km²	Depth max.	Depth mean	Other species	Sampled year	Type of water	Population type
1	Vifilsstadavatn	64°03'N	21°50'W	20	0.6		1.5	B E TS	1978	lake	landlocked
2	Ellidavatn	64°05'N	21°40'W	20	1.8		1.5	B TS E?	1978	lake	landlocked
3	Skorradalsvatn	64°30'N	21°30'W	57	14.3	55	25	B TS	1978	lake	landlocked
4	Hólmavatn	64°47'N	20°50'W	363	2.4	14		B TS	1978	lake	landlocked
5	Hvitá	64°40'N	21°40'W					B S E TS	1978,80	glacial river	anadromous
6	Hlidarvatn, Hnappadalur	64°13'N	22°10'W	78	4.4	21	4.7	B S	1978	lake	landlocked
7	Oddastadsvatn	64°53'N	22°15'W	65	3.0	15	5.4	B S TS	1978	lake	landlocked
8	Laxá i Miklholtshreppur	64°50'N	22°30'W					B S	1978	river	anadromous
9	Vestre Fridmundarvatn	65°17'N	19°50'W	441	6.0	2.3	1.2	TS	1978	lake	landlocked
10	Húnavatn	65°35'N	20°20'W	0.3	1.0		1.5	B S E TS	1978	estuary	anadromous
11	Austre Fridmundarvatn	65°19 N	19°50'W	435	2.4	1.1	0.8	B TS	1978	lake	landlocked
12	Vatnshlidarvatn	65°30'N	19°40'W	280	0.75		1.0		1980	lake	landlocked
13	Miklavatn i Fljótum	66°05'N	19°10'W	0	5	23		B S TS E	1980	estuary	resident
14	Hólavatn	65°22'N	18°15'W	100	0.15			?	1979	lake	landlocked
15	Eyjafjardará	65°23'N	18°15'W					B	1979	river	anadromous
16	Vestmannsvatn	65°50'N	17°25'W	25	2.5	2-4		B S	1979	lake	landlocked
17	Mývatn	65°36'N	17°00'W	277	38	14		B TS	1979	lake	landlocked
18	Kötluvatn	66°30'N	16°30'W	0				TS	1980	lake	landlocked
19	Skerjalón-	66°31'N	16°27'W	5	0.8	1-4		B TS	1980	lake	resident
20	Örfærutjörn	66°31'N	16°25'W	0	0.2			B TS	1980	estuary	anadromous
21	Sudurvatn	66°30'N	16°18'W	1.8	5		1-2	TS	1980	lake	resident
22	Sigurdarstadavatn	66°30'N	16°03'W	1.7	1		1-1.5	TS	1980	lake	resident +anadr.
23	Hraunhafnarvatn	66°31'N	15°55'W	3.8	2		4	B	1980	lake	resident
24	Höskuldarnesvatn	66°30'N	15°55'W	1	0.5			B	1980	lake	resident
25	Stora Vidarvatn	66°14'N	15°50'W	2.6	151		60	-	1980	lake	landlocked
26	Eidisvatn	66°15'N	15°00'W	2.3	0.5	shallow		TS	1979	lake	resident + anadr.
27	Saenautavatn	65°17'N	15°30'W	1.8	525			-	1979	lake	landlocked
28	Ánavatn	65°12'N	15°30'W	5	522			-	1979	lake	landlocked
29	Eidavatn	65°22'N	14°20'W	1.0	32			B	1979	lake	landlocked
30	Lagarfljót	65°30'N	14°25'W					B	1979	glacial river	resident
31	Dalsá	64°55'N	14°10'W					B	1979	river	anadromous
32	Gilsá	64°50'N	14°20'W					B	1979	river	anadromous
33	Núpsvatn	64°12'N	20°10'W	0.7	300				1979	lake	landlocked
34	Thingvallavatn	64°11'N	21°10'W	83.7	100.5	34.1	114	B TS	1978,79,80	lake	landlocked
35	Gjá at Thingvellir	64°15'N	21°08'W		appr.105				1978	lava cleft	landlocked
36	Hlidarvatn, Selvogi	63°52'N	21°53'W	3.3	1	2.9	5		1979	lake	resident + anadr.

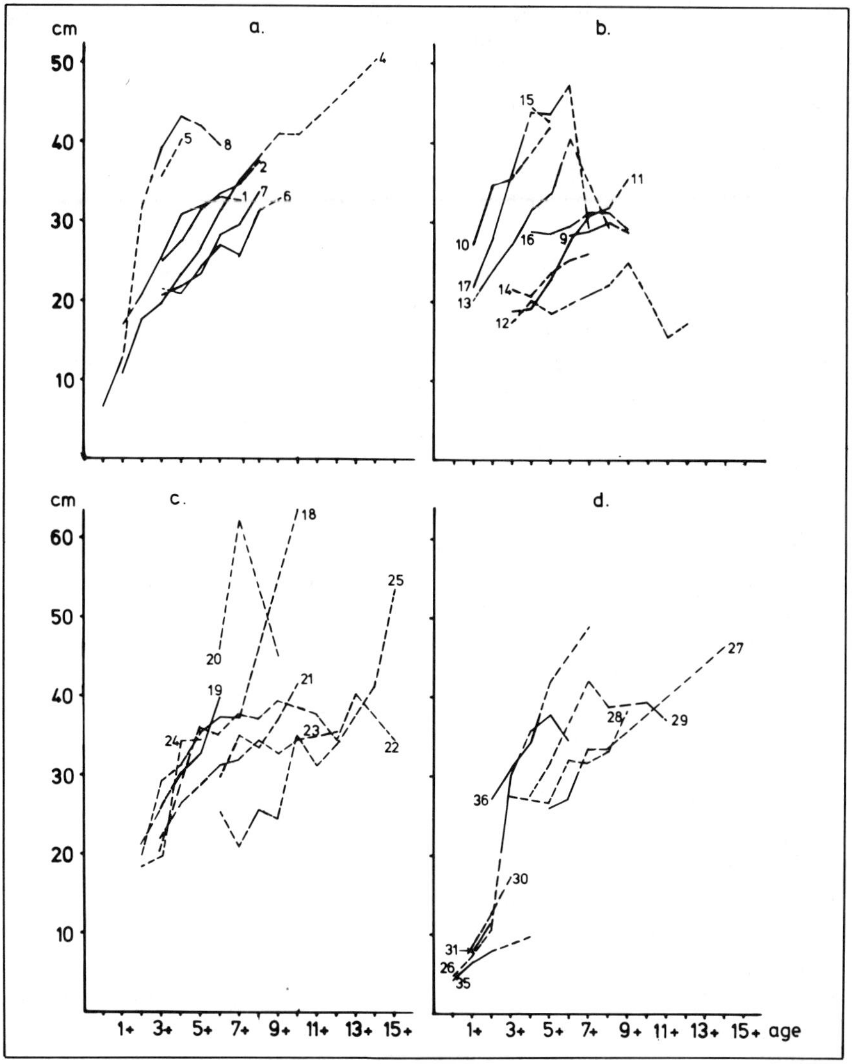

FIG. 2 Growth curves (length/age) for the sampled populations from the different geographical areas.

a) Western Iceland

b) Northern Iceland, western part

c) Northern Iceland, eastern part

d) Eastern and Southern Iceland

Numbers refer to Table 1.

of the landlocked populations (2, 4, 6 and 7) had high frequencies of the F-allele, close to 1.00 (Table 2). Two populations had lower gene frequencies. The charr in Vifilstadavatn (1) had a gene frequency of the F-allele of 0.76 which is intermediate between "N-" and "F-charr", so the possibility of introgression cannot be excluded. The second case is Skorradalsvatn (3) where the total gene frequency of the charr population was 0.60 which is close to that of "normal" charr. However, two morphologically different forms were observed: one dwarfed, dark coloured with a blunt nose, some having large eyes; the other had more silvery colouration, a normal head and apparently also a better growth rate. When the genotypes are plotted in a length-age relationship diagram (Fig. 3a), the possible existence of two sympatric stocks of charr is further supported. The length range within single year-classes increases from 100 mm for 3+ charr to 285 mm for 8+ charr. When dealt with separately the 14 fastest growing and the 16 slowest growing fish differ in mean age of maturity, the fastest growing being mature at a mean age of 6.3 years while the slowest growing charr show a mean age at maturity of 4.6 years. The gene frequencies for the two groups were 0.64 for the fastest growing compared with 0.44 for the slowest growers. The two groups had overlapping confidence intervals. In some specimens the maxillae extended behind the posterior rim of the eye; this appeared to be correlated to the fish length since it only occurred in fish over 240 mm in length. The gene frequency of these was 0.93. The slowest growing charr with large eyes and blunt nose had a gene frequency of 0.36. This indicates the presence of sympatric species of Arctic charr in Skorradalsvatn. The charr with good growth rate and long maxillae had a frequency of the F-allele that suggests it to be "F-charr", while the dwarf has a gene frequency between those of "S-" and "normal" charr. It is, however, more likely an S-charr with an elevated gene frequency due to introgression than a "normal" charr with a lowered gene frequency. The presence of F-charr in surrounding lakes and anadromous S-charr in the River Hvitá (5) also supports the view of introgressed F- and S- charr in Skorradalsvatn. The last population studied from this area is the charr from the River Laxá in Miklholtshreppur which showed a gene frequency of 0.79. This gene frequency is intermediate between "F-charr" and "normal" charr and could indicate sympatry where one species is "F-charr" and the second species "S-" or possibly "normal" charr. The status of the populations is still unclear and a more thorough study is needed. Thus the results so far indicate that in western Iceland "F-" and "S-charr" are the dominating species.

Northern Iceland (Localities 9-25)

With possibly two exceptions, which will be presented below, "F-charr" and "normal" charr dominated the samples from this area. Of the nine F-charr

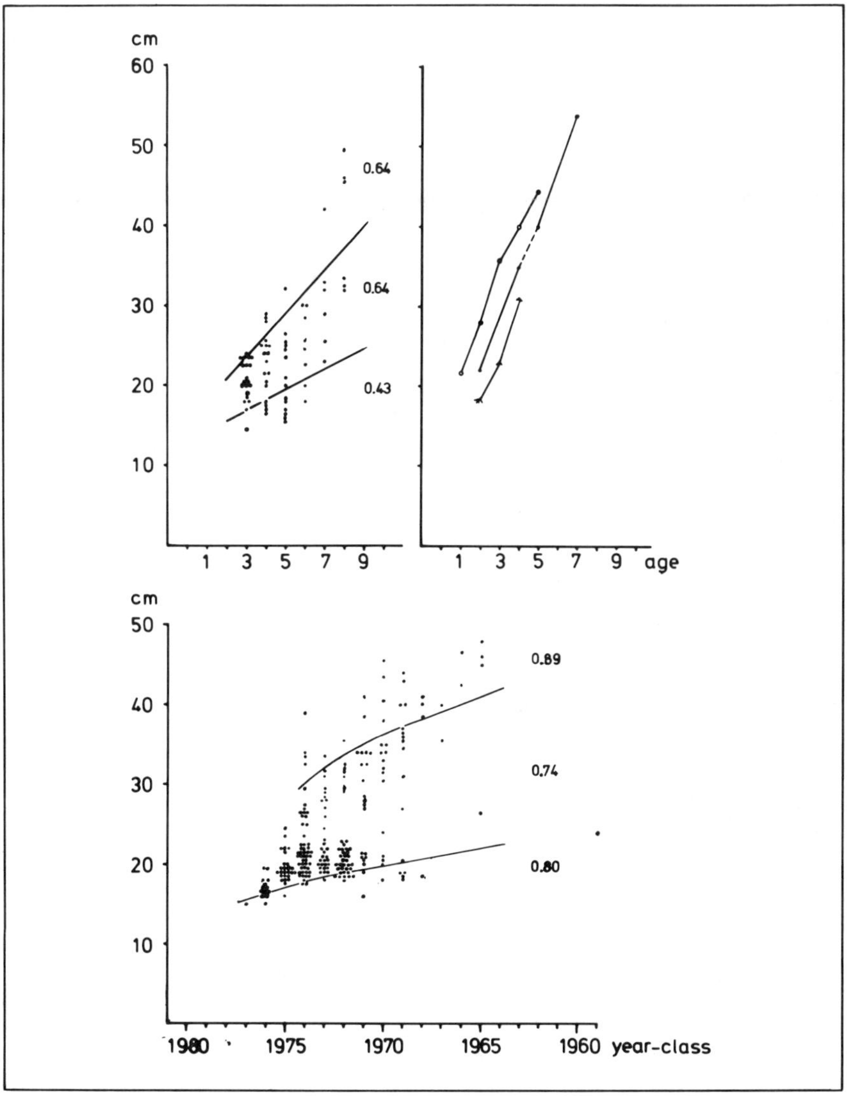

FIG. 3 a) Lake Skorradalsvatn: plot diagram of length/age with gene frequencies f(F) for different growth fractions

b) Lake Mývatn: growth curves for Arctic charr; ∘ this paper; ● Lamby 1941; △ Adalsteinsson 1979

c) Lake Thingvallavatn: plot diagram of length/age with gene frequencies f(F) for different growth fractions.

Table 2. Gene frequency data of the populations studied. Agreement of observed and expected values.

No.	Locality	Est	F/F	F/S	S/S	Total	fF	X^2(1 d.f.)	Probability of worse fit	95% confidence interval	Type of population
1	Vifilsstadavatn		19	18	0	37	0.76	3.87	0.025-0.05	± 0.10	sympatric ?
			21.4	13.5	2.1						
2	Ellidavatn		42	0	0	42	1.00	-	-		allopatric
			42	0	0						
3	Skorradalsvatn		25	44	9	78	0.60	2.41	0.10-0.20	± 0.08	sympatric
			28.1	37.4	12.5						
4	Hólmavatn		90	12	0	102	0.94	0.42	0.50-0.70	± 0.03	allopatric
			90.1	11.5	0.4						
5	Hvitá		0	1	12	13	0.04	0.00003	>0.99	± 0.08	allopatric
			0.02	1.00	11.98						
6	Hlidarvatn, Hnappadalur		57	0	0	57	1.00	-	-		allopatric
			57	0	0						
7	Oddastadsvatn		32	7	0	39	0.91	0.36	0.50-0.70	± 0.06	allopatric
			32.3	6.4	0.3						
8	Laxá i Miklholtshreppur		37	27	0	64	0.79	4.54	0.025-0.05	± 0.07	sympatric ?
			39.9	21.3	2.8						
9	Vestre Fridmundarvatn		25	0	0	25	1.00	-	-		allopatric
			25	0	0						
10	Húnavatn		15	9	6	30	0.65	2.85	0.05-0.10	± 0.12	allopatric
			12.7	13.6	3.7						
11	Austre Fridmundarvatn		70	0	0	70	1.00	-	-		allopatric
			70	0	0						
12	Vatnshlidarvatn		9	7	2	18	0.69	0.127	0.70-0.80	± 0.15	allopatric
			8.6	7.7	1.7						
13	Miklavatn i Fljótum		2	15	7	24	0.40	2.26	0.10-0.20	±0.14	sympatric (?)
			3.8	11.5	8.7						
14	Hólavatn		7	0	0	7	1.00	-	-		allopatric
			7	0	0						
15	Eyjafjardará		1	2	3	6	0.33	0.358	0.50-0.70	± 0.27	sympatric ?
			0.7	2.6	2.7						
16	Vestmannsvatn		9	10	7	26	0.54	1.34	0.20-0.30	± 0.14	allopatric
			7.6	12.9	5.5						
17	Mývatn		75	0	0	75	1.00	-	-		allopatric
			75	0	0						
18	Kötluvatn		11	10	3	27	0.67	0.095	0.70-0.80	± 0.14	allopatric
			10.8	10.6	2.6						

Table 2. continued.

No.	Locality	Est	F/F	F/S	S/S	Total	fF	X^2(1 d.f.)	Probability of worse fit	95% confidence interval	Type of population
19	Skerjalón		14	11	1	26	0.75	0.427	0.50-0.70	± 0.12	sympatric
			14.6	9.8	1.6						
20	Örfaerutjörn		4	0	0	4	1.00	-	-		allopatric
			4	0	0						
21	Sudurvatn		11	2	0	13	0.92	0.083	0.70-0.80	± 0.11	allopatric
			11.0	1.9	0.1						
22	Sigurdarstadavatn		20	7	0	27	0.87	0.599	0,30-0,50	± 0.09	allopatric
			20.4	6.1	0.5						
23	Hraunhafnarvatn		7	12	1	20	0.65	2.031	0.10-0.20	± 0.15	sympatric ?
			8.5	9.1	2.4						
24	Höskuldarnesvatn		11	1	0	12	0.96	0.027	0.80-0.90	± 0.08	allopatric
			11.1	0.9	0						
25	Stora Vidarvatn		21	0	0	21	1.00	-	-		allopatric
			21	0	0						
26	Eidisvatn		11	14	2	27	0.67	0.75	0.30-0.50	± 0.13	sympatric ?
			12.1	11.9	2.9						
27	Saenautavatn		14	0	0	14	1.00	-	-		allopatric
			14	0	0						
28	Ánavatn		14	0	0	14	1.00	-	-		allopatric
			14	0	0						
29	Eidavatn		3	4	1	8	0.63	0.023	0.80-0.90	± 0.24	allopatric
			3.1	3.8	1.1						
30	Lagarfljót		10	1	0	11	0.95	0.001	0.95-0.99	± 0.09	allopatric
			9.9	1.0	0.0						
31	Dalsá		3	6	0	9	0.67	2.250	0.10-0.20	± 0.22	allopatric
			4.0	4.0	1.0						
32	Gilsá		2	0	1	3	0.67	3.190	0.05-0.10	± 0.38	allopatric ?
			1.4	1.3	0.3						
33	Núpsvatn		3	0	0	3	1.00	-	-		allopatric
			3	0	0						
34	Thingvallavatn		129	86	12	227	0.76	0.250	0.50-0.70	± 0.04	sympatric
			131.1	82.8	13.1						
35	Gjá at Thingvellir		0	5	7	12	0.21	0.783	0.50-0.70	± 0.17	allopatric
			0.5	4.0	7.5						
36	Hlidarvatn, Selvogi		29	21	0	50	0.79	3.521	0.20-0.30	± 0.08	sympatric
			31.2	16.6	2.2						

populations found, five were landlocked (9, 11, 14, 17 and 25), three were resident (21, 22 and 24) and one was anadromous (20). Five of the populations had gene frequencies between 0.54 and 0.69 which correspond to the frequency of ''normal'' charr. Of these, three were landlocked (12, 16 and 18), one resident (23) and one anadromous (10). The resident charr of Lake Skerjalón (19) has a gene frequency of 0.75. This could be a case of sympatry between ''F-charr'' which are anadromous in Lake Örfaerutjörn (20) through which Lake Skerjalón drains to the sea, and resident ''normal'' charr as in the nearby Kötluvatn (18).

In two cases the gene frequencies of the sampled populations are intermediate between ''S-'' and ''normal'' charr. The estuarine Miklavatn i Fljótum (13) has stocks of anadromous as well as resident charr. The sample of resident charr showed a gene frequency of 0.40. A separation of the fastest growing and the slowest growing charr in a length-age relationship diagram results in 0.4 for the fastest growing and 0.2 for the slowest growing charr. The River Eyjafjardará (15) has a stock of anadromous charr. The six charr examined had a gene frequency of 0.33. The sample sizes, however, are too small to allow the results to be anything but indicative of the possible presence of ''normal'' and ''S-charr''.

In short, ''F-'' and ''normal'' charr are the most common species among the studied populations of northern Iceland. Both occur as landlocked, resident and anadromous populations.

Eastern Iceland (Localities 26-32)

Of the seven populations sampled from eastern Iceland, three were high-frequency ''F-charr''; two of these were landlocked, one in the high altitude lakes, Saenautavatn (27) and Ánavatn (28). The third ''F-charr'' population (30) sampled in the lower part of the glacial River Lagarfljót just above the power plant at Lagarfoss, was probably resident. The remaining four populations studied all had gene frequencies ranging from 0.63 to 0.67 which indicates that they are ''normal'' charr. The charr in Eidisvatn (26) are resident although anadromous charr are also present. Eidavatn (29) has a landlocked population of charr while the charr in the rivers Dalsá (31) and Gilsá (32) are anadromous.

The problem of small sample sizes affects the last three populations mentioned and the gene frequencies are therefore only indicative. There are, however, no traces, to date, of low frequency ''S-charr'' from this part of the country and probably only ''F-'' and ''normal'' charr are present in the material.

Southern Iceland (Localities 33-36)

Only four localities were sampled in southern Iceland, all being situated toward the west where charr is more common than further east. Three populations were landlocked. The charr of Nupsvatn (33) may be ''F-charr'' since the three individuals analyzed were all homozygous for the F-allele which is at least an indication.

Thingvallavatn (34) with its complex of morphologically different forms had a total gene frequency of 0.76. Further analysis of the material showed that at least two species of Arctic charr were probably present; possibly ''F-charr'' and ''normal'' charr. Dwarf charr having a gene frequency of 0.21 inhabit lava clefts at Thingvellir close to the lake.

Hlidarvatn in Selvogi (36) has a population of resident charr mixed with a small portion of anadromous charr. The total gene frequency was 0.79. From a plot diagram two fractions, each with differences in growth rate and gene frequency, can be distinguished. The faster growing charr had a gene frequency of 0.71, the slower growers: 0.90. This indicates sympatry between introgressed ''F-'' and ''normal'' charr.

DISCUSSION

The material upon which this study is based is, in some cases, inadequate to permit a definite statement regarding the gene frequencies in some populations. The gene frequencies do, however, even in these cases, indicate their possible status. With or without these dubious populations, it is clear that all of the species from Scandinavia as defined by Nyman et al. (1981), are present in Iceland also. Charr with a high frequency of the F-allele occur as anadromous and landlocked populations both allopatrically and sympatrically; it is also the most common charr in high altitude lakes, thereby showing the same pattern of distribution as in Scandinavia and suggesting it to be the first immigrant after the last glaciation.

The charr with an intermediate gene frequency of the F-allele (the ''normal'' charr) seems to have a slightly higher gene frequency in the anadromous, and possibly also in landlocked, populations than the corresponding ones from Scandinavia. Nyman (1972) demonstrated a cline in the frequency of the F-allele in Sweden, the more southerly populations having a higher gene frequency; this he suggested may be temperature (climate) dependent. Iceland, in spite of its high latitude, has a mild climate; this could explain a higher gene frequency among ''normal'' charr.

The two most well-known lakes in Iceland are Mývatn and Thingvallavatan. From both of these, sympatric stocks of charr are reported, in Mývatn by Lamby (1941) among others, and in Thingvallavatn by several authors including Saemundsson (1909).

Lamby (1941) separated two forms of Arctic charr in Mývatn by their feeding habits (Fig. 3b). The smaller form (age 2+ - 4+) with length 220-350 mm, was caught in large numbers in the southeastern part of the lake in the autumn. It was called "krus" by the local fishermen. The large form (age 5+ - 7+) with length 400-540 mm, was caught all over the lake. There was some difference in choice of food. Adalsteinsson (1979) cites Lamby's opinion of two species and compares growth data from his own studies with those given by Lamby (Fig. 3b). There is a good agreement between the two curves. Kristjansson (pers. comm.), on the other hand, considers only one species of Arctic charr to be present in the lake. The results of this investigation support the latter view. The growth curve of Kristjansson is similar in appearance to that of Lamby and Adalsteinsson, although corresponding to fish one year younger (Fig. 3b). The gene frequency of 1.00 shows no influence from another species. Hence the conclusion that only "F-charr" is present in the lake. This also leads to the conclusion that what Lamby reported as two forms were probably young and old fish with differences in feeding; Filipsson and Svärdson (1976) provide similar examples of dietary shifts in charr during their life span.

The largest lake in Iceland is Thingvallavatn; it is also the site of the world's oldest known parliament, dating back to the Ninth Century, hence there has been considerable sophisticated human activity in the area for over 1,100 years. The occurrence of sympatric stocks of Arctic charr in the lake was first described by Saemundssen (1909) and later by Fridriksson (1939). Saemundsson described two forms, the large "bleikja" and the dwarf "murta". Fridriksson (1939) studied the "murta" to determine whether it was a young form of bleikja or a separate charr. He found significant differences in vertebrae numbers between the large "bleikja", the "murta" and a third form he named "svartmurta" (black "murta"). There were also differences in age at maturity.

It is apparent from Figure 3c that there are two groups present that are identifiable by their growth rates; the gene frequency for the fastest growing group had a value of 0.89 and for the slower group 0.80; the intermediate individuals had a value of 0.74. It is therefore impossible to separate the groups on these grounds. However it must be recognized that the samples were obtained from the commercial catch and this resulted in the virtual absence of young fish. These results indicate the presence of "F-" and "normal" charr, but they cannot be separated.

The volcanic clefts close to Thingvallavatn contain isolated populations of dwarf charr with gene frequencies of 0.21, almost within the range of values for "S-charr". In view of the long period of occupancy of this region it is by no means impossible that some Viking or his descendent transplanted charr from Thingvallavatn. In the clefts, charr are exposed to a harsh environment with a constant water temperature between 3-4°C. Saemundsson (1909) describes these charrs as a variety, *Salmo alpinus* var. Thingvallensis. It may be that they are an original population that may or may not have relatives in Thingvallavatn although no low frequency forms have, to date been found there. However, since up to four morphological forms have been described from the lake and only two analyzed so far, there is still the possibility that one may be present.

Thus there is only one clear-cut case of ''S-charr'', the anadromous population of the River Hvitá, but four more dubious cases: the anadromous charr of the River Eyjafjardará, the extreme dwarfs of the Thingvellir clefts, the dwarfs in Skorradalsvatn and the slowest growing charr in Miklavatn. Skorradalsvatn has sympatric stocks of charr, but from the material available it is not possible to state which species have contributed to the total stock. The overall picture, however, is very similar to the F-S relationship of charr in Lake Fättjaure, Sweden (Henricson and Nyman 1976). The presence of ''S-charr'' in the River Hvitá, draining into the same estuary as the outflow of Skorradalsvatn, and the presence of ''F-charr'' in other western Icelandic lakes, support the assumption that landlocked ''S-charr'' are also present in Skorradalsvatn. The dwarfs in the Thingvellir clefts may have arisen due to genetic drift, or possibly the result of a founder effect.

In his investigation of landlocked resident and anadromous populations from northern Norway, Nilsen (1980) found that landlocked charr had gene frequencies close to 1.00 while anadromous charr had gene frequencies around 0.50. Resident populations had intermediate gene frequencies between 0.60-0.70. Of the landlocked populations, those that had been landlocked for a considerable time, probably for about 4,000-5,000 years, all had frequencies close to 1.00. One recently landlocked population was, until 1970, to some extent in contact with anadromous charr with a gene frequency of 0.68, which is close to the gene frequency of resident populations still in contact with anadromous charr. The two populations of anadromous charr had gene frequencies of 0.45 and 0.60. Klemetsen, in Nilsen (1980), gives data on a resident population on Bear Island that had a gene frequency of 0.75. These data show a close resemblance to the findings in Iceland.

The occurence of all three charr in anadromous form supports the theory that the landlocked charrs were already separated into ''species'' when the immigration occurred. It has been suggested that in Sweden the ''F-charr'' and ''normal'' charr immigrated from the east and ''S-charr'' from the west. When examining the distribution of the charrs in Iceland, no definite solution to the problem of origin is obvious but there is a tendency for ''S-charr'' to occur mainly in the western and northwestern part (with the possible exception of the Thingvellir dwarfs). ''Normal'' charr has possibly a more easterly distribution while ''F-charr'' is present all over the country. It is apparent that much more information on the distribution of the various gene frequencies is necessary before firm conclusions can be reached.

REFERENCES

Adalsteinsson, H. 1979. Size and food of Arctic char *Salvelinus alpinus* and stickleback *Gasterosteus aculeatus* in Lake Mývatn. Oikos 32(1/2):228-231.

Behnke, R.J. 1980. A systematic review of the genus *Salvelinus*, p. 441-480. *In* Charrs: salmonid fishes of the genus *Salvelinus*. Dr. W. Junk Publishers, The Hague.

Einarsson, Th. 1968. Jardfraedi (Geology). Mál og Menning, Reykjavik. 335 p. (In Icelandic).

Filipsson, O. and G. Svärdson. 1976. Principles for the management of char populations. Inform. Inst. Freshw. Res. Drottningholm (2). 79 p. (In Swedish with English summary).

Fridriksson, A. 1939. On the murta in Thingvellir Lake in relation to the char and trout in the lake. Nátturúfraedingnum (1):1-36. (In Icelandic with English summary).

Gudjonsson, T. 1954. Veidin i Medalfellsvatni. 1953. Veidimadurinn 27:24-31. (In Icelandic).

Gydemo, R. 1978. Populationsgenetisk undersökning av röding i Västerbottens län. Del 1. Fiskenämnden och Lantbruksnämnden i Västerbottens län. 72 p. (Mimeographed in Swedish).

Gydemo, R. 1979. Populationsgenetisk undersökning av röding i Västerbottens län. Del 2. Fiskenämnden och Lantbruksnämnden i Västerbottens län. 41 p. (Mimeographed in Swedish).

Gydemo, R. 1980a. Populationsgenetisk undersökning av röding i Västerbottens län. Del 3. Fiskenämnden och Lantbruksnämnden i Västerbottens län. 20 p. (Mimeographed in Swedish).

Gydemo, R. 1980b. Electrophoretic studies on Arctic char in the county of Västerbotten, northern Sweden and Iceland: a survey of the distribution of the Arctic char species. *In* Proceedings of the first ISACF workshop on Arctic charr. Inst. Freshw. Res., Drottningholm.

Henricson, J. and L. Nyman. 1976. The ecological and genetical segregation of two sympatric species of dwarfed char (*Salvelinus alpinus* (L.) species complex). Inst. Freshw. Res. Drottningholm Rep. 55:15-37.

Kristjansson, J. 1973. Fiskirannsoknir i Medalfellsvatni. Veidimálastofnunin. Fjolrit 8. 8 p. (Mimeographed in Icelandic).

Kristjansson, J. 1978a. Growth rates of brown trout and Arctic char in Iceland. Islenskar Landbunadarrannsoknir (J. Agric. Res.) 2:125-134. (In English with Icelandic summary).

Kristjansson, J. 1978b. Silungsrannsoknir i Mývatni 1977. Veidimálastofnunin Fjölrit 21. 10 p. (In Icelandic).

Lamby, K. 1941. Zur Fischereibiologie des Mývatn, Nord-Island. Z. Fisch. 39(5):749-805.

Nilsen, P. 1980. Stasjonaer- og anadrom røye, *Salvelinus alpinus* (L.). En sammenligning av økologiske og genetiske trekk hos bestander med sannsynlig felles opprinnelse. Hovedfagsoppgave i akvatisk biologi, Univ. Tromsø. 84 p. (In Norwegian).

Nyman, L. 1965. Variation of proteins in hybrids and parental species of fishes. Rep. Swed. Salm. Res. Inst. 13. 11 p.

Nyman, L. 1967. Protein variations in Salmonidae. Inst. Freshw. Res. Drottningholm Rep. 47:5-38.

Nyman, L. 1972. A new approach to the taxonomy of the "*Salvelinus alpinus* species complex". Inst. Freshw. Res. Drottningholm Rep. 52:103-131.

Nyman, L. and O. Filipsson. 1972. The chars of Lake Yraf. Inform. Inst. Freshw. Res., Drottningholm (10). 20 p. (In Swedish with English summary).

Nyman, L., J. Hammar and R. Gydemo. 1981. The systematics and biology of landlocked populations of Arctic charr from Northern Europe. Inst. Freshw. Res. Drottningholm Rep. 59: 128-141.

Oliva, O. 1951. Description of *Salvelinus alpinus* (Linneaus) from Iceland. Copeia (1):91-92.

Saemundsson, B. 1909. Oversigt over Islands fiske med oplysninger om deres forekomst, vigtigste biologiske forhold og økonomiske betydning. Skr. Komm. Havunders., København 5. 140 p. (In Danish).

Svärdson, G. 1961. Rödingen (The Char). Fiske 61:25-38. (In Swedish).

Thorarinsson, S., T. Einarsson and G. Kjartansson. 1959. On the geology and geomorphology of Iceland. Geogr. Ann. Stockh. 41:135-169.

Thorarinsson, S. 1979. The postglacial history of the Mývatn area. Oikos 32 (1/2):17-28.

Life Histories of Arctic Charrs: An Epigenetic Explanation of Their Invading Ability and Evolution

EUGENE K. BALON

Department of Zoology, College of Biological Science, University of Guelph, Guelph, Ontario, N1G 2W1, Canada

BALON, E.K. 1984. Life histories of Arctic charrs: an epigenetic explanation of their invading ability and evolution, p. 109–141. *In* L. Johnson and B.L. Burns [eds.] Biology of the Arctic charr, Proceedings of the International Symposium on Arctic Charr, Winnipeg, Manitoba, May 1981. Univ. Manitoba Press, Winnipeg.

Within the salmonids, the prolongation of generalized juvenile characters and adaptability into later ontogeny is most noticeable in Arctic charrs, *Salvelinus alpinus*. They constitute one of the best examples of paedomorphosis in fishes. Forms of this species complex include numerous landlocked relics and the northernmost freshwater and anadromous invaders throughout the holarctic, as well as large and dwarf or deep-water glacial relics in the alpine or subalpine lakes of Eurasia and North America. By comparing the ontogenies of a generalist (altricial) and a specialist (precocial) form of Arctic charrs, models of life history styles are presented. For example, changes in timing of epiboly, germ ring closure, cardiac pulse, hatching, calcification of skeletal tissues, appearance of parr marks, onset of exogenous feeding, endogenous and exogenous oxydative metabolism, filling of the swimbladder, maturation, variations in fecundity, growth rate and in selection of spawning substrates are explained as exaptations or adaptations to environmental perturbations.

Altricial and precocial dynamic states seem to operate continuously in ontogeny, creating life history styles of a generalist nature in unpredictably perturbed and of a specialist nature in less perturbed and competitive environments. Life history states which represent this dichotomy can accumulate into a speciation event which, in time, combines into a skewed evolutionary pattern of low frequency speciation within altricial and successively higher frequency within the precocial traits. In spite of a constant epigenetic tendency to specialize, the process of juvenilization often reverses specialization, leading to extinction, reinstating some of the variation lost. The sympatric evolution of life history states, e.g. stocks or species, is a reflection of the "altricial $\rightleftharpoons$ precocial dynamic states" and the juvenilization process constantly at work, whereas the vulnerability of precocial relics are examples of specialization overdone. Consequently, the forms or stocks of charrs like "dwarf" and "normal", sunapee and anadromous *Salvelinus alpinus*, lake charr and siscowet, and brook and aurora *S. fontinalis*, recognized by ecologists but a mystery to taxonomists, are altricial and precocial pairs, the homeostatic epiphenotypes of the evolution in making.

INTRODUCTION

> *Epigenesis* is the word preferred by C.H. Waddington for his central field of interest, whose old name was *embryology*. It stresses the fact that every embryological step is an act of *becoming* (Greek *genesis*) which must be built upon (Greek *epi*) the immediate status quo ante.
>
> Gregory Bateson (1979)
> in Mind and Nature

The processes responsible for producing life history traits — epigenesis and juvenilization — seem to be also responsible for speciation, i.e. evolution. Epigenetically induced variations in ontogeny may be responsible for the appearance of new phenotypes as well as for the formation of evolutionary units. I have presented earlier, in greater detail, most of the arguments in favour of saltation, heterochrony and paedomorphosis in development (Balon 1979a,b, 1981a,b), including their relevance to the formation of life history traits and sympatric speciation. I have frequently used charrs as an example. Much of what I say here will be repetitive although I hope that the arguments may become clearer if developed around a particular organism — the Arctic charrs. It worked well when we similarly reviewed the life histories of tilapine cichlids (Noakes and Balon 1982). The theoretical justification for the interpretation of charr case histories presented here are given elsewhere (Balon 1983).

In ''Tools for Thought'' the late Conrad Hal Waddington (1977) asked: ''What is common sense except the philosophy which your parents or your peers have soaked you in when you weren't realizing what you were absorbing? Philosophy of one kind or another cannot be avoided or evaded or given up like sin in Lent''. And one paragraph earlier he reasoned: ''. . . it is just as well to know something of the main alternative philosophies, if only to see why the other chap is talking such nonsense, and why he simply does not seem able to get the drift of what you are saying yourself. It is likely to be not his intelligence, but his philosophy, which is responsible''.

Much of the interpretation of charr taxonomy, evolution and life history is the result of such consciously or subconsciously held philosophical beliefs rather than scientific truth (if such exists). Charr taxonomy will differ according to beliefs in Linnean, biological, evolutionary, multispecies or terminal species concepts (Van Valen 1976, Løvtrup 1979, Wiley 1980). To continue in the spirit of my opening quotation, many have ''soaked in'' the bold claims that sympatric speciation is impossible to such a degree that strong evidence to the contrary is either ignored or laboriously bent to fit the ''soaked in belief''. And this, even after they have been deserted by their main apostle . . . for how else would you interpret the words of Ernst Mayr (1976, p. 144) in the Preface to the reprint of his earlier essay on sympatric speciation which started the allopatric orthodoxy?: "It seemed to me that the best strategy was to state the case against sympatric speciation as uncompromisingly as possible in order to provoke refutation if such was possible. (. . .) My strategy was successful. The vague claims of sympatric speciation were henceforth replaced by well worked-out, fully documented cases". Our charr specialists of one or other faith may feel more comfortable knowing that both gangs still remain in opposite camps (e.g. White 1978a, Futuyma and Mayer 1980).

The belief in genetic mutations as the main source of variation on which natural selection acts seems badly eroded (Van Valen 1976, Rosen and Buth 1980). Alternative philosophies may therefore be useful and are in fact available provided that our beliefs have not yet become fanatical faith (Eigen and Schuster 1979, Steele 1979). Settle (1979, p. 522) expressed this concern as follows: "Many thinkers, seeing that the search for truth is an unending quest, abandon it (in despair perhaps), and settle for agreement with their fellows. If they are right that it is concensus rather than truth that ought to be aimed for in science, then the picture that emerges (and this picture is pretty explicit in some writings — for example in Kuhn's) is gloomy. It is not just that there will be pockets of scientists who agree among themselves and disagree with the gangs in other pockets — that often happens — nor just that such rival groups may not understand each other — that too happens. What's particularly gloomy, is that the agreement itself becomes vacuous if it is not constrained by external matters".

THE THEORY OF SALTATORY ONTOGENY AND MODELS OF LIFE HISTORY

That the growth and formation of cells, tissues and organs in ontogeny proceeds at various rates is a truism, but saltatory processes in ontogeny are more than variable developmental rates (Vorobyeva et al. 1981). Structures which together form a system (organ) align their rates of development to become functional all at the same time and initiate a new vital function (integrative action sensu Adolph 1982 or steady state sensu Balon 1983) at highly accelerated rates (Fig. 1). This "high speed" switch is what I have called a threshold, a rapid transition from one steady state of ontogeny or life style to another (Balon 1971, 1979a, 1980, 1981a,b). Depending on the number of coinciding structures and functions and their significance, the "strengths" of thresholds will differ and so will the functional quality of the preceding and following life history states (intervals). This is the basis of my hierarchical life history model (Balon 1975a), mistaken by some as nothing but an exercise in semantics (Richards 1976, Balon 1976).

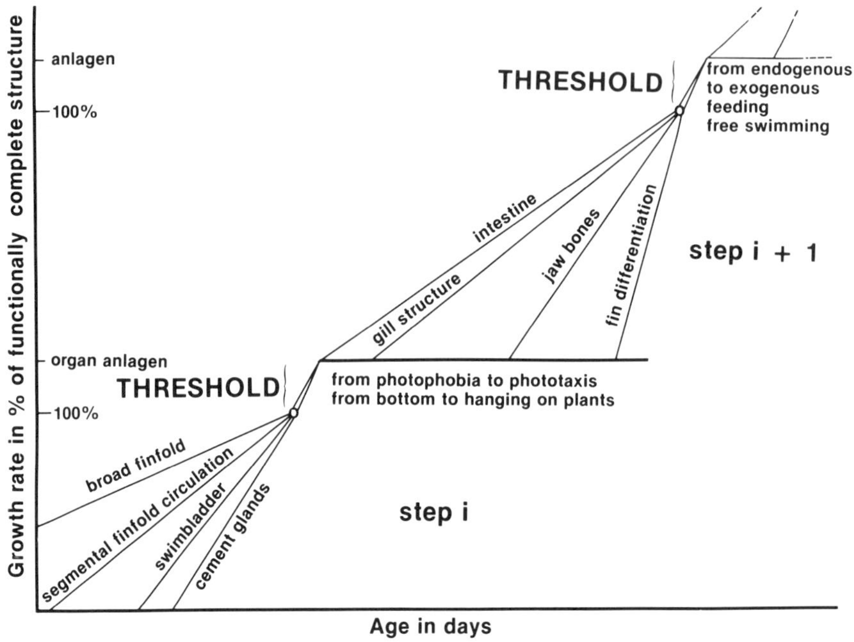

FIG. 1 Graphical presentation of a saltatory threshold of ontogeny. Several structures grow quantitatively at various rates achieving completion at the same time. Sudden change of rates marks the brief interval of a threshold during which a new integrative action and/or interaction with the environment (e.g. photoresponse) commences.

If we agree with Garstang (1922) and many others that evolution is but a succession of ontogenies, the same processes ought to be responsible for the patterns seen in both. I have reviewed all these ideas in detail elsewhere (Balon 1980, 1981a, 1983, 1984b) so let me state only that, while the patterns of saltation in evolution is currently a debated issue, it is recognized as rectangular (Stanley 1975) or punctuated equilibria (Eldredge and Gould 1972, Gould and Eldredge 1977). I consider the processes of saltatory ontogeny (Balon 1981b) responsible for these patterns. I also agree with Dodson and Hallam (1977) and Gould (1979) that Darwin's (1859) gradualism *a priori* precludes independent existence of evolutionary units — species — which I can perceive only as viable steady states in the evolutionary landscape, separated by saltatory nonviable thresholds.

First however, let me satisfy those who have recognized in the above nothing but the Hegelian dialectic. Gould (1977, p. 409) has put it admirably: "The reconciliation of our gradualistic bias with the appearance of discontinuity is a classical problem of intellectual history. We have sought to reduce the external phenomena of saltation to an underlying continuity of process — to reduce the qualitative to the quantitative. Philosophies of change and progress have wrestled with this dilemma and have tried to resolve it by formulating such laws as the 'transformation of quantity into quality' of the Hegelian dialectic: the addition of quantitative steps will lead eventually to a qualitative leap . . .''. Armed with the theory of saltatory ontogeny we are able to perceive individual stages of ontogeny (in a sense of instantaneous observable patterns) as entities of developmental dynamics — among variable developmental rates and of certain qualitative interrelationships with the epigenetic processes and environmental factors. The acceptance of arbitrary stages as a sufficient representation of ontogenetic changes is therefore utterly misleading and naive.

To construct a saltatory model of life history and to make it consistent and comparable, at least among all vertebrates, an unambiguous set of terms had to be defined first. Some of the earlier terms had to be abandoned in the process, partly because of their confused meaning but especially because of their limitation to ''egg and larval'' (sic) fish taxonomy, a specialized usage which failed to integrate with developmental biology. Otherwise, the illogical application of some terms, e.g. protolarva, mesolarva, metalarva (Snyder 1976) or prolarva and postlarva (Hubbs 1943) would have been obvious for some time, especially in cases when no larvae exist — a fact of great practical implication (Balon 1984b). Parturition and hatching would be viewed as less static (Noakes 1978, 1980a) and the superficial descriptive reasons for the recognition of the yolksac larva abandoned in favour of an ecophysiological understanding of what an embryo (a term frequently absent from egg and larval taxonomists' jargon) is all about (Balon 1975a).

If I may repeat myself again, the saltatory model of life history should be a hierarchical system, comprised of a series of intervals — periods, phases and steps — whose relationship may be compared to the relationship among days, hours and minutes. Stage refers to an instantaneous state of ontogeny and should not be confused with an interval. The entire life history, from the time ontogeny commences until death, normally consists of five periods: embryonic, larval, juvenile, adult and senescent. The thresholds which separate periods are stronger and much more significant than those separating the lowest saltatory intervals of life history called steps.

The embryonic period begins with activation (Balon 1984b). This period is characterized by an exclusively endogenous nutrition from the yolk of the ovum or, in the case of viviparous fishes, via special absorptive organs (placental analogues), from an ovarian secretion or embryonic cannibalism and/or a combination of yolk and secretion (Balon 1981c).

The larval period, a unique energy aquisition interval, commences with the transition to exogenous feeding (i.e. oral and intestine digestion, and/or whole body nutrient absorption) and usually lasts until metamorphosis, superficially often marked by the formation or ossification of the axial skeleton and the differentiated median finfold. This period is further characterized by the persistence of some temporary embryonic organs (respiratory plexuses) and by the development of special larval organs (e.g. respiratory vessels in fins, in gill covers, various temporary spines, flaps and filamentous appendages); i.e. transient features (Oppenheim 1981), which must be remodelled into different definitive organs of similar function, or disappear with the loss of their functional need. This period can be extremely long in some (e.g. eels) but is usually absent or truncated in fishes with large reproductive cost (Miller 1979, Bell 1980, Balon 1984a) and large offspring at the time of transition to exogenous feeding (e.g. as in cichlids and salmonids respectively).

The juvenile period begins after metamorphosis, e.g. when the fins are fully differentiated and most of the temporary organs are replaced by definitive organs. The transition from larva to young fish (Fig. 2) therefore involves extensive changes, often from an unfishlike appearance into an adultlike condition (e.g. Wilbur and Collins 1973, Smith-Gill and Berven 1979). Some organs develop only later in the juvenile period (scales, intromittent organs) and, conversely, some special organs persist from the larval period and disappear only later (I named such organs metamorphic, Balon 1980). The juvenile period lasts until the maturation of the first gametes (puberty) and is usually characterized by rapid growth, specific juvenile colouration and behaviour.

The adult period commences with maturation of the first gametes and is characterized by a spawning run, spawning concentration or by special behaviour accompanied by changes in colour, and/or morphology. Spawning — being a series of courtships, matings and gamete depositions — can be repeated in seasonal or circannual rhythms (iteroparous), in alternate years, or even less often (Bull and Shine 1979), or performed only once (semelparous). Somatic growth often decreases substantially, or in the case of one-time spawners, becomes completely arrested.

The senescent period is old age: the rate of reproduction of cells in most tissues has reached its limits (see also Oppenheim 1981, pp. 95-96). Growth slows considerably, stops permanently, or even becomes negative. Gametes, if produced, are inferior. This period can last for years (e.g. sturgeons) or days (e.g. *Oncorhynchus*).

Each period is divided into phases (Fig. 3) which are a convenient means of morphophysiological identification of developmental states but are of little saltatory relevance (Balon 1975a). The basic saltatory intervals are steps. Their

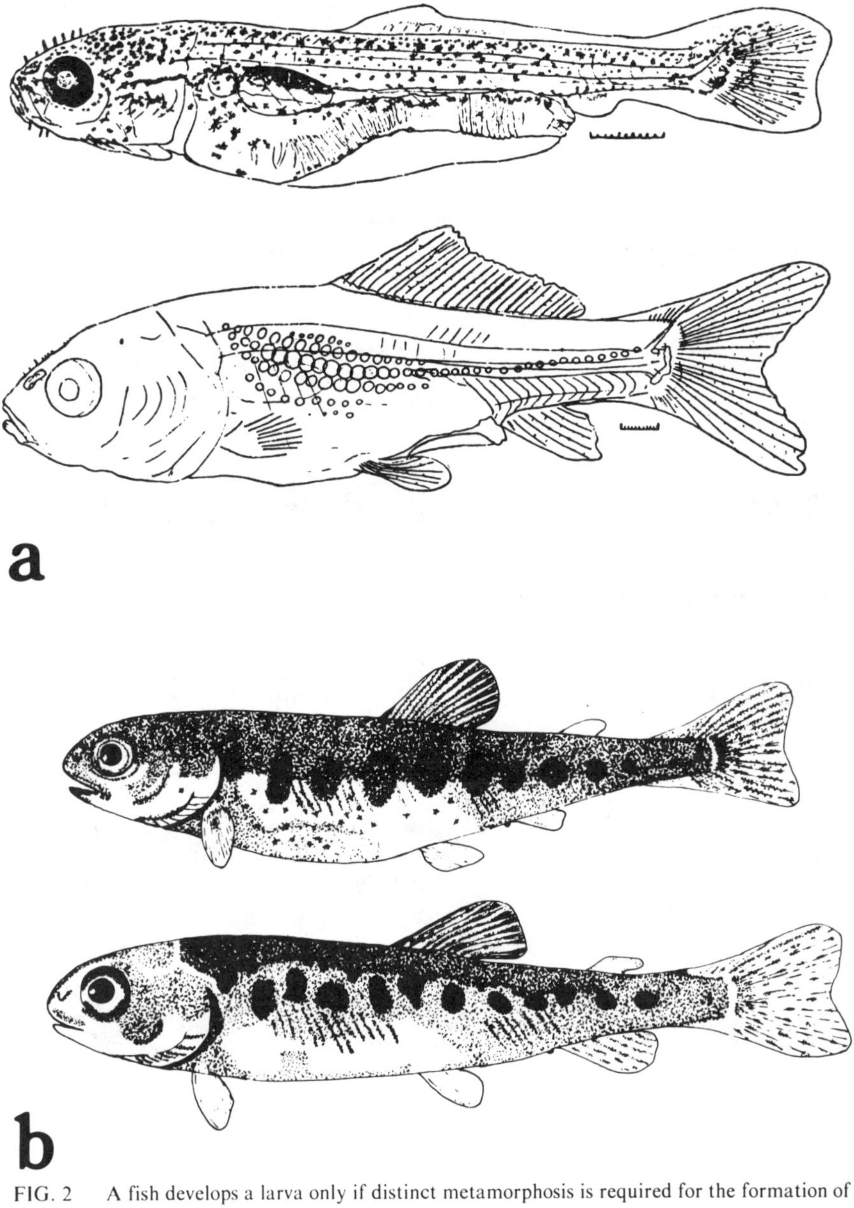

FIG. 2 A fish develops a larva only if distinct metamorphosis is required for the formation of a juvenile: *a* — the common carp, *Cyprinus carpio*, larva prior to and early juvenile after metamorphosis; *b* — alevin and juvenile of Arctic charr that lacks metamorphosis.

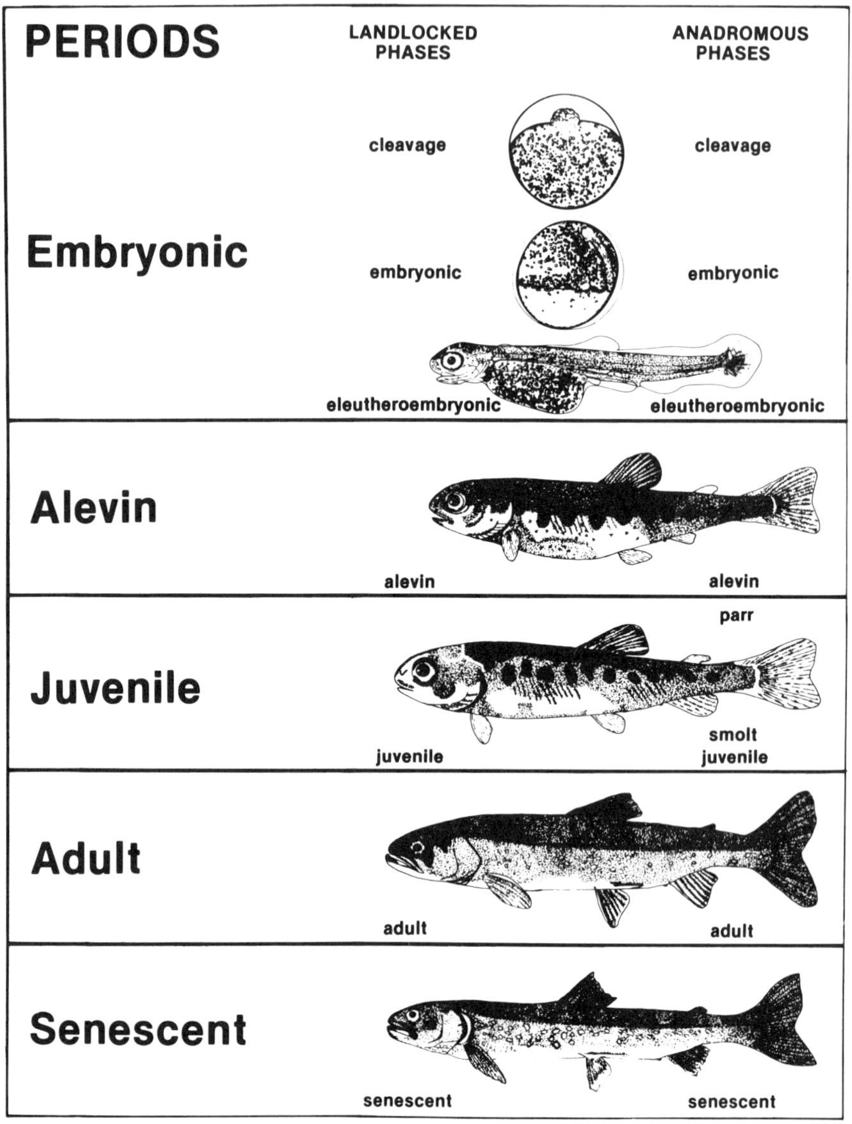

FIG. 3 Periods and phases of a life history model for landlocked and anadromous Arctic charrs.

sequence in the early ontogeny of an American landlocked Arctic charr (*Salvelinus alpinus oquassa*) is given in Table 1.

Table 1. Saltatory intervals of ontogeny in sunapee, *Salvelinus alpinus oquassa* (after Balon 1980).

Embryonic period	
The cleavage phase	
Step C^1 1.	Formation of the perivitelline space and bipolar differentiation.
Step C^2 2.	First multiplication of cells, cleavage, and the formation of a morula.
Step C^3 3.	Blastula, gastrula, the beginning of epiboly, the formation of the germ ring and embryonic shield.
The embryonic phase	
Step E^1 4.	Closure of the germ ring; brain, optic and auditory vesicle formation; separation of the tail-end.
Step E^2 5.	The beginning of trunk movements, heart beats, the circulation anlage and the appearance of first melanophores.
Step E^3 6.	Rapid expansion of the respiratory yolk plexus of subintestinal, and mixed hepatic origin.
Step E^4 7.	Function of the hepatic-vitelline and caudal respiratory plexuses, rapid expansion of pigment and hatching.
The eleutheroembryonic phase	
Step F^1 8.	Differentiation of the embryonic finfold, formation of first dermal rays in unpaired fins, development of first chondrified skeletal structures and growth of pelvic fins beyond the finfold edge.
Step F^2 9.	Onset and duration of strong photophobia, transfer to mainly branchial respiration and calcification of jaw tooth points.
Step F^3 10.	The active free embryo; complete absorption of yolk, calcification of skeletal structures and rapid expansion of iridocyte distribution on the body flanks.
Alevin period (a vestige of larval period in other fishes)	
Step A^1 11.	Mixed, endogenous and exogenous feeding; separation of all unpaired fins, and disappearance of remnants of preanal embryonic finfold.
Step A^2 12.	Swimbladder filled, calcification of axial skeleton, efficient exogenous feeding, rapid growth, increase in depth and pigmentation.
Juvenile period	
Adult period	
Senescent period	

Periods are the longest and most conspicuous intervals of life history, steps the shortest saltatory steady states (Balon 1984b). The processes which modify the step boundaries, i.e. thresholds, probably determine phenotypic changes or life history styles; the processes of epigenesis may elaborate and skew in a desired direction variations available for natural selection (Balon 1983).

EPIGENETIC PROCESSES: HETEROCHRONY AND JUVENILIZATION

Whether or not we accept that life history and evolutionary patterns are effects of saltatory processes of epigenesis, it is obvious that genetically instructed (see also Alberch et al. 1979), epigenetically formed and environmentally induced heterochronies in early ontogeny increase variation in a required direction and may accelerate selection of epiphenotypes (and subsequently via reshuffling of genes, canalization and genetic assimilation or transcription genotypes, see later) for better survival in new environments (Balon 1981b, 1983). When the consequences of heterochronies combine, they result and are responsible for the amplitude and timing of thresholds — in short are an important mechanism of changes.

If we agree with Nachtsheim that "the laws governing macroevolution are the same as those governing microevolution" (as cited by Hennig 1966, p. 225) — and in accord with my earlier opinion that evolution is nothing but a sequence of ontogenies, I do, in spite of Gould and Calloway (1980) — then the theories explaining processes and changes of ontogenetic development must also explain evolution. Therefore, attempts to separate microevolution from macroevolution (e.g. Stanley 1979) are probably misleading (Heberer 1958) and as Greenwood (1979) reflects "I wonder if we need the word 'macroevolution' at all?" The entire difference may rest in the "amplitude" of thresholds, which form as a consequence of essentially the same biological processes. The delay or acceleration of thresholds in early ontogeny modifies the formation of morphological, physiological and ethological characters with developmental, adaptive and/or evolutionary consequences. Selection may favour genotypes with enough variation to permit phenotypic shifts back and forth between two alternative stable states, but, to this I intend to devote the next section.

Let me now elucidate some aspects of heterochronies using the genus *Salvelinus* as an example (Balon 1980). Epiboly, as evidenced by the arrival of the germ ring at the equator and its subsequent closure, is slower in lake (*Salvelinus namaycush*) and brook (*S. fontinalis*) charrs than in Dolly Varden (*S. malma*), Arctic (*S. alpinus alpinus*) and sunapee (*S. a. oquassa*) charrs. Similarly, the median time to hatching is shortest in Dolly Varden and longest in brook charr. The cardiac pulse, however, begins at about the same time in all charrs.

Exogenous feeding begins earliest in Dolly Varden and Arctic charr, later in sunapee and lake charr, and latest in brook charr. The appearance of flank blotches (parr marks) follows the same sequence. Parr marks, the first character which can be explained directly in terms of environmental adaptation, appear slightly earlier in brook charr than in sunapee and lake charrs (Fig. 4).

By disruptive colouration, flank blotches probably facilitate concealment (Noakes 1980a). Fluvial charrs, which inhabit shallow water, will require such concealment earlier than lacustrine forms. The blotches are most important after emergence, when active life begins (= exogenous feeding). As would be expected, flank blotches form just before the onset of exogenous feeding. Although exogenous feeding begins latest in the brook charr, flank blotches form earlier than in sunapee and lake charrs. This is an adaptation to the fluvial environment. In the charrs that are most fluvial — Dolly Varden and brook charr — flank blotches form well in advance of feeding.

The filling of the swimbladder and perhaps the timing of calcification, both of which affect buoyancy, coincide only in the Arctic charr. In sunapee, calcification begins much earlier than in Arctic charr, but the swimbladder fills much later. The retention of oil globules for buoyancy and the longer ice cover explain this discrepancy in more specialized charrs. The need for denser bodies in the open fluvial habitats of brook charr explain the earlier calcification and relatively late filling of swimbladders (Balon 1980, 1981a).

Before I prove that salmonins in general, and charrs in particular, have juvenilized, let me repeat a parallel idea which has been accepted for the origin of chordates. Through paedomorphosis, a ciliated, auricularia-like larva became fish-like. The sessile lophophore-feeding adult interval was eliminated and the first pelagophilous fish appeared (Garstang 1928, 1962, Young 1978, but see Gutmann 1981, and Jollie 1982), its entire ontogeny adapted to planktonic life (Schmalhausen 1968). The planktonic fish-like chordates invaded fluvial (Berrill 1955) and reinvaded benthic habitats (similarly maybe to some sciaenids, Powles and Burgess 1978) and there they most likely evolved hiding, guarding and bearing reproductive styles (Moy-Thomas and Miles 1971).

Although the fossil record is poor for Salmoninae (as per Behnke 1972), at least one fossil, *Smilodonichthys rastrosus* Cavender and Miller 1972, points to planktonic feeding by an anadromous giant salmonid in the Pliocene. Were earlier salmonins also related to the oceanic pelagial by their reproductive style? All the evidence on the evolution of reproductive guilds (Balon 1975b, 1981c) supports this assumption. A predecessor of the salmonins, therefore, would be a nonguarding, open-substratum spawning pelagophil with relatively slow embryonic differentiation, a long larval period, and metamorphosis preceding the juvenile period. Consequently, the life history of extant salmonins, including

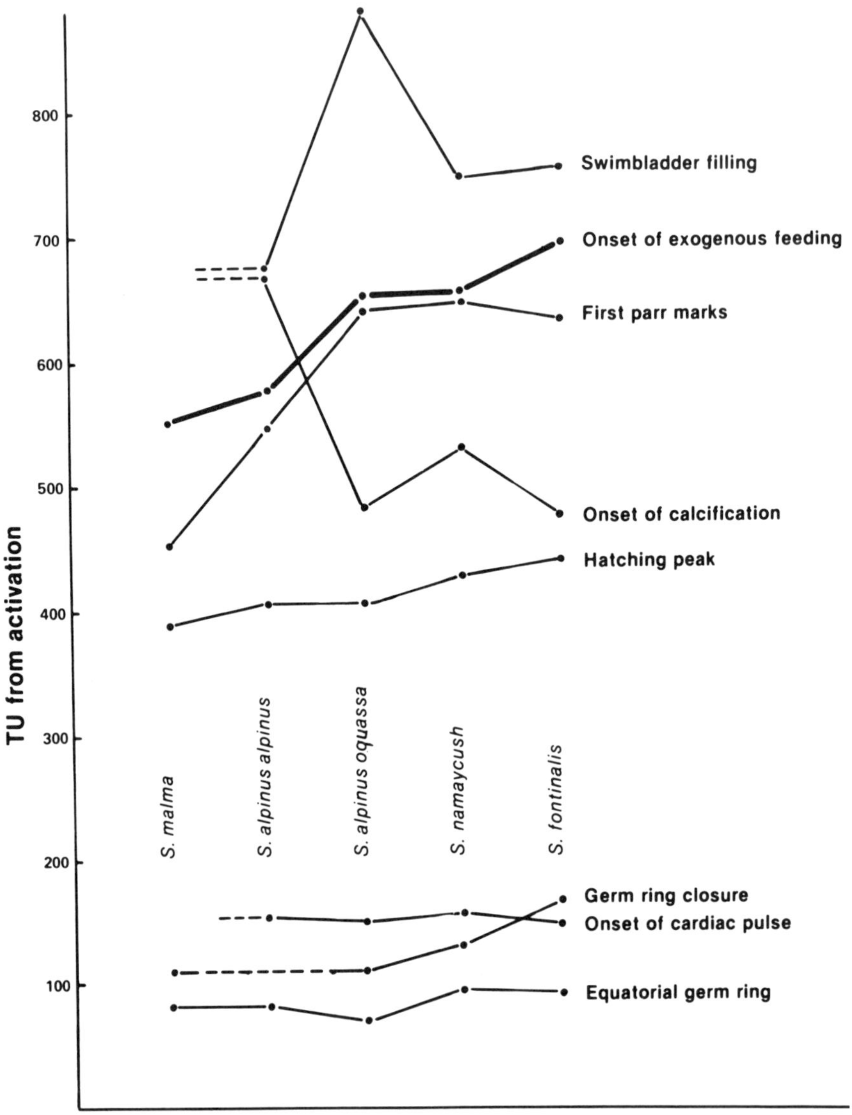

FIG. 4 Heterochronous thresholds for some structures and for the onset of some functions in five charrs raised under the same conditions (at temperature 4.4°C).

charrs, may have evolved from this ancestor by paedomorphosis and a subsequent dichotomy producing on one side specialized Pacific salmons (Balon 1968) and on the other side generalized *Salmo* or *Salvelinus* and secondarily juvenilized *Parasalmo* (Balon 1981b and later in this presentation). The specialists retained some traces of metamorphosis (secondary metamorphosis of Oppenheim 1981) now displaced towards the end of their ontogeny prior to spawning (Smirnov 1975). Initial adult allometries could resemble elongated fins and frontal humps on the snout, as in some labrids or cichlids. Such changes, while still evident in the Pacific salmons are reduced in other salmonins to thickening of the skin (Stoklosowa 1966, 1970, Noakes 1980b). Retention of juvenile allometries is obvious in all other extant forms which reproduce, so to speak, prior to the metamorphosis which had been eliminated by energy allocation into ova (Fig. 5). Reproducing as juveniles, free from the irreversible changes in form and function which accompanied secondary metamorphosis enabled multiple spawning — an adaptation of clear survival advantage (Gould 1977).

The evolution of protective hiding of eggs and embryos may have been a consequence of invasion into fluvial systems prior to juvenilization. It caused selection to favour higher yolk density and larger yolk size, which in turn enabled earlier differentiation of structures vital for better survival after emergence and elimination, in time, of vulnerable larva through reduction of the need for external energy aquisition. This way, the ground for juvenilization could be prepared as this energy shift and hiding permitted truncation of many structures known to develop later in the ontogeny of non-hiding fishes (Balon 1977). Some vital organs, characteristic of the larval period in other fishes, develop completely during the embryonic period in salmonins, the larval period being eliminated or retained only as a vestige (alevin, Fig. 3).

I would like to conclude this part by quoting from the Preface to the anthology on CHARRS (Balon 1980) which absorbed so many years of my life: "The evolution of charrs parallels in many aspects the evolution of man. The charrs decelerated from more specialized, planktonophagous ancestors of a warm climate, into paedomorphic, less specialized, eternal juveniles, as did man from their pithecine ancestors into neotenic hominids (Gould 1977b, . . .). The paedomorphic charrs were able to invade the most inhospitable habitats; so were paedomorphic men.

Due to their migratory habits, which enable utilization of resources from larger areas, charrs are able to keep their densities per unit area far above the carrying capacity of an occupied area. If landlocked by forces beyond their control charrs still persist, able to adapt and survive as altricial dwarfs. Early men behaved similarly and they are now repeating the physical example of the charrs on a meme level (Dawkins 1976)".

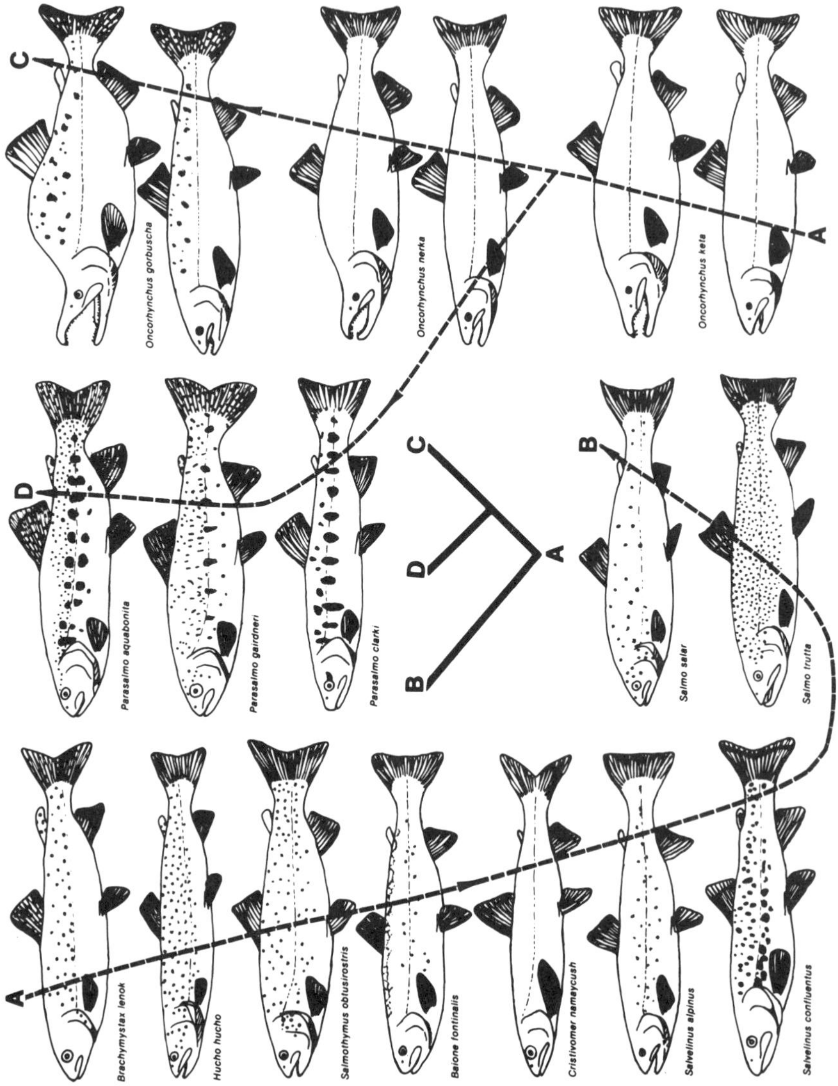

FIG. 5 In spite of relatively uniform appearance of salmonines due probably to the initial common paedomorphosis, males of the most specialized *Oncorhynchus* (right) undergo a secondary metamorphosis, and the adult allometries of others enable one to distinguish generalists (left) from the secondarily juvenilized *Parasalmo* (center). Specifically, after the initial paedomorphosis A (refer also to Fig. 8) the evolutionary trajectory of generalists (A - B) formed species of Huchonini (chromosome numbers 92 - 84) and Salmonini (82 - 58), whereas trajectory of specialists (A - C) led to the Oncorhynchini (74 - 52), the most specialized tribe, whose ancestors cleaved (probably via second paedomorphosis) into juvenilized (D) Parasalmonini (64 - 56).

DICHOTOMY OF LIFE HISTORY STATES

As usual in the panorama of evolution, we find that the single process is Janus-like, facing in two directions.

Gregory Bateson (1979)
in Mind and Nature

By consistently referring to Arctic charrs in the plural I hope to have prepared the ground for the looming doubts concerning the idea of a species complex, or any other taxonomic scenario, e.g. Linnean, biological, evolutionary, multispecies or terminal species concepts (Van Valen 1976, Løvtrup 1979, Wiley 1980). While the saltatory ontogeny, and its effect called evolution, permits the distinction of evolutionary units (stable states), I tend to agree with Rosen (1978) or Løvtrup (1979) that the species concept may be a ''convenience'' or ''fiction'' which imposes an unnecessary strait-jacket on our perception of the true character of this unit. Other such units in fishes, e.g. *Haplochromis* (Greenwood 1974, 1981) or *Cyprinodon* (e.g. Humphries and Miller 1981), present a similar ''bewildering complexity of morphologically and physiologically differentiated forms'' as Arctic charrs, ''both within and among water bodies . . .'' (Johnson 1980, p. 31).

Most of the Arctic charr forms come in pairs called ''normal'' and ''dwarf'' or any variant of these. If we survey the available data on life history styles of other organisms closely, a pattern of two alternating states often emerges. Populations or even demes of each organism are capable of alternating between the two states. Less perturbed and competitive conditions will favour selection of specialized states, more perturbed and colonizing conditions selection of generalized states. The generalized marsupials, for example, persist successfully in uncertain environment but are poor competitors with specialized placentals in stable and competitive environments (Low 1978). The processes responsible for these homeostatic states were since named the ''Chen Principle'' and the dichotomy formalized as the theory of ''Altricial $\rightleftharpoons$ Precocial Dynamic States'' (Balon 1983a).

Briefly, for the generalist states I have used the term altricial and for the specialists the term precocial, following the long accepted concept for birds (Portmann 1945, Ricklefs 1979). The most general consequences are: relatively small and incompletely developed young in the altricial and relatively large and completely developed young in precocial.

So, for example, if we compare the ''dwarf'' or altricial Arctic charrs (Balon 1980, p. 626) with the ''normal'' or precocial charrs (p. 602) the embryonic period of the former is shorter than the alevin period — 125 days versus 152 days or 550 TU (temperature units or degree days) versus 669 TU. In contrast, the embryonic and alevin periods of the precocial charr last 148 and 97 days respectively. In the altricial charr hatching and the onset of exogenous feeding occurs earlier, i.e. the onset of alevin period occurs earlier and the entire period lasts longer than in precocial charrs (Fig. 6). Embryos and eleutheroembryos of altricial charrs are always relatively smaller and less developed than those of precocial charrs (Fig. 7).

The onset of exogenous feeding long before the yolk reserves are exhausted is clearly an adaptation for uncertain trophic conditions in an unpredictably perturbed environment. In the case of charrs, smaller alevins can utilize smaller food particles and disperse more widely, having more time to do so. This liberates the parent from the expense of large, dense yolks, which is essential if the alevin (larval) period is to be reduced or eliminated. Yolk size and density also affect the extent of the respiratory plexuses on the yolk (Balon 1980, p. 626).

Consequently the life history features in both forms of Arctic charr display the expected trends in relative timing. In precocial forms the alevin period is truncated and the senescent period is lengthy. In altricial forms the early periods are prolonged and the senescent period is reduced or absent.

THE CONCEPT OF SYMPATRIC SPECIATION

If the formation of life history traits in response to environmental and community interactions is caused by the same epigenetic processes which are responsible for the formation of evolutionary units (speciation), then the strict requirement of a spatial separation (allopatry) makes no sense. I wish it to be clear, that my argument does not preclude allopatric speciation which can enhance the amplitude of a saltatory threshold or increase the evolutionary distance of sympatrically formed steady states which become spatially separated only later. In cases where spatial separation occurs first, the precocial form will usually be more appropriate for a founder state than the altricial form; the precocial form will survive only on rare occasions if the conditions for juvenilization are not present.

It is common knowledge (Johnson 1980) that in Arctic charrs, more than in any other fishes, distinct forms occur in sympatry (e.g. Smirnov 1956, Behnke 1972, 1980, Volubuyev 1977, Savvaitova 1980). Most speakers at this Symposium confirmed this, but few admitted the possibility of "sympatric speciation". If the reason is philosophical, in the sense of my opening paragraphs, it may prove more acceptable to speak from now on of the "formation of evolutionary units without spatial separation" or in a fashionable contraction of — FOREVOLUNITWISS. As already stated by Garstang (1928, p. 62) on a similar occasion: "To deny a fact is, of course, impossible; but I will try to show that

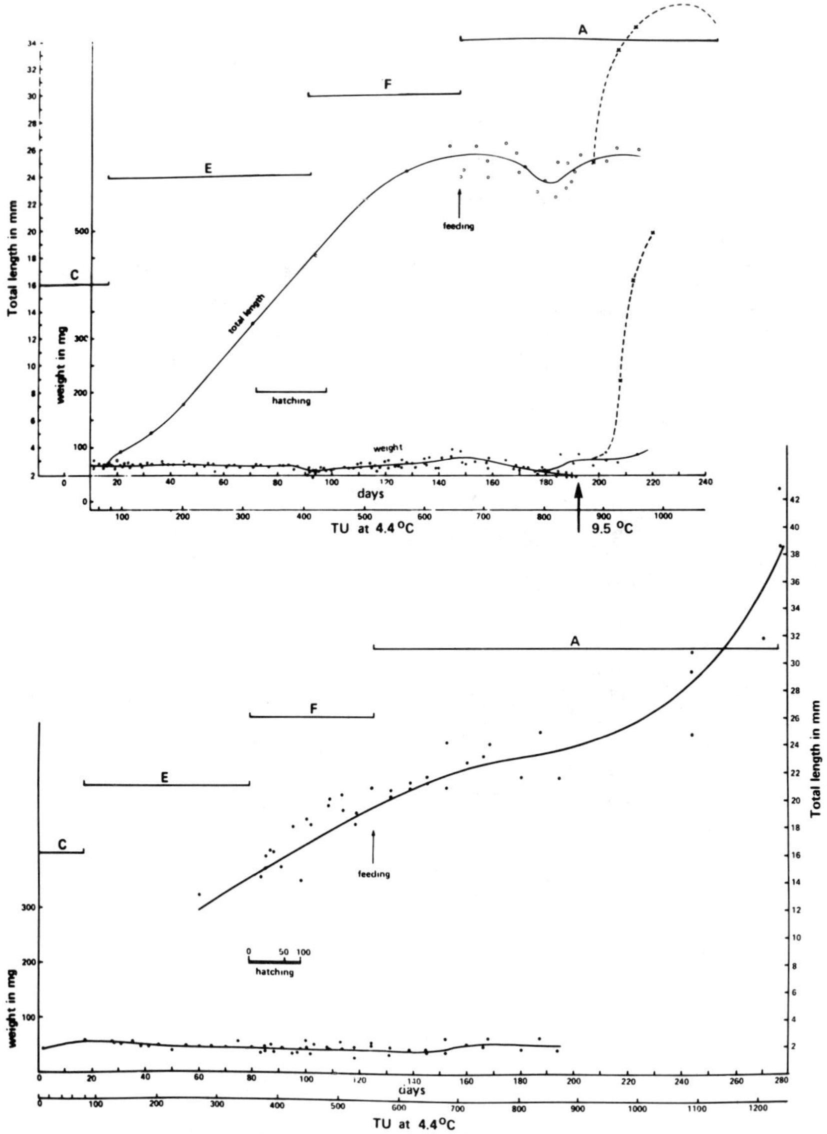

FIG. 6 Weight and length rates of growth in relation to hatching, onset of feeding and for phases in early ontogeny of precocial (above) and altricial (below) Arctic charrs. C, E, F, and A stand for the cleavage, embryonic, eleutheroembryonic and alevin phases, respectively.

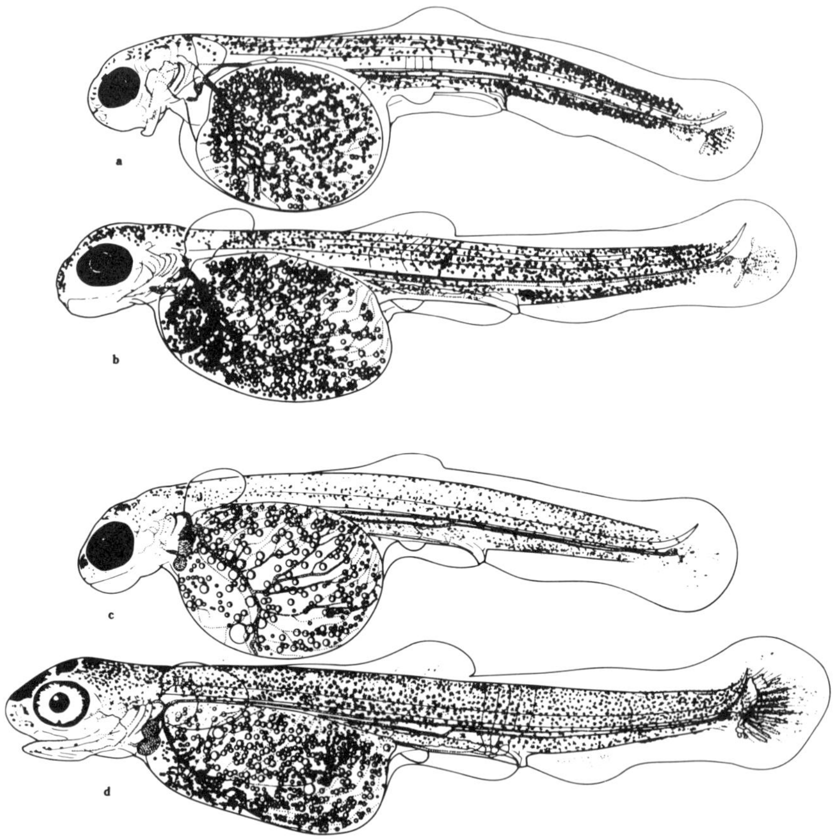

FIG. 7 Early (a,c) and late (b,d) hatched embryos of an altricial Arctic charr (a,b) 15.0 and 16.3 mm long, and a precocial Arctic charr (c,d) 15.9 and 19.8 mm long.

certain facts, which are demonstrable and undeniable, admit of, and in my judgment demand, a very different interpretation.''

The ''altricial ⇌ precocial states'' explain FOREVOLUNITWISS. In contrast to their precocial counterparts, the altricial charrs, being effective invaders, appear to accelerate certain functionable thresholds related to lower reproductive cost. They also seem to prolong some ontogenetic steps as a consequence of their small size. In essence, through epigenesis an Arctic charr may acquire invading abilities by forming altricial life history states. Such altricial charrs can then survive in a perturbed system. When the system stabilizes a certain part may form a ''normal'' precocial charr which may undergo further specialization and eventually form an evolutionary unit entirely different to the one from which the parental precocial form initially separated. The same evidence used earlier for allopatric speciation can now support FOREVOLUNITWISS (Lindsey et al. 1970, Nilsson and Filipsson 1971, Nyman 1972, Henricson and Nyman 1976, Dodson and Hallman 1977, Klemetsen and Grotnes 1980).

Once again to elucidate my points — when conditions become unpredictably preturbed, a precocial charr AA (normal or ordinary charr) establishes an altricial subpopulation or stock AB (dwarf or blattjen charr). The altricial AB is now capable of invading any perturbed accessible system or or surviving the period of perturbances in the original system. If a newly formed and invaded system eventually stabilizes, part of the altricial form AB finds favourable conditions for specialization and forms a separate precocial life history state BC (another normal charr or sunapee or Tiefseesaibling), and they in turn may produce a new altricial invader CD by juvenilization if environmental conditions so require. Therefore, an occasion there may be three sympatric forms in a single lake — one initial invader AB, one precocial BC and a new juvenilized altricial CD. Two of these FOREVOLUNITWISSed in accordance with the classical Hennigian paradigm. The esterase alleles presented by Nyman (1972) and others at this Symposium reflect this situation. The sympatric occurrence of two precocial forms AA and BC should be less common, even though it represents the double invasion of allopatrically differentiated units, also because precocial forms are less capable invaders (Behnke 1972, Savvaitova 1980).

As the earliest piscine glacial invaders, the Arctic charrs had to be adapted for feeding on aquatic and terrestrial invertebrates (e.g. Pechlaner et al. 1972a,b, Schulz 1975, Pechlaner 1979). The altricial charr presents such an adaptation. As dwarf altricial charrs increase in density or as their habitat is invaded by other fish species, conditions suitable for the FOREVOLUNITWISS of precocial states come to existence. The separation of a large precocial form follows stabilization of the abiotic environment and the formation of a fish taxocene capable of supporting a piscivorous life style. Later, other pelagic piscivores may arrive and eliminate the precocial form, leaving two altricial forms (Henricson and Nyman 1976). FOREVOLUNITWISS therefore easily accounts for all the phenomena previously explained by the independent invasion of allopatric forms and introgression (Svärdson 1958, 1961, 1970, Nyman 1972, Kornfield et al. 1981, but see White 1978b). Palaeogeographic speculation and the invention of physical barriers are replaced by biological processes, partly or wholly understood or at least falsifiable (Løvtrup 1974, Waddington 1975, Gould 1977).

Earlier explanations of FOREVOLUNITWISS had been constrained by the accepted genetic dogmas; therefore spatial isolation has been at most replaced by a reproductive, seasonable or habitat isolation (Kosswig 1947, 1963, Tauber and

Tauber 1977, McKaye 1978, 1980). Omnipresent dichotomy of life history states had never been seriously considered as the most likely mechanism of FOREVOLUNITWISS. No wonder that, in the absence of knowledge on convincing processes which could account for such a mode of speciation (Balon 1983), the observed patterns have been fitted as best as possible to the orthodox theories.

THE PREDICTION OF DWARFISM AS A TEST: PITFALLS OF PRACTICISM

Klemetsen and Grotnes (1980) conclude that their dwarf charr (i.e. altricial) is the one which actively invaded the mountain lakes; but in order to avoid the heresy of sympatric speciation, they claim that the normal (i.e. precocial) charrs, which occur in sympatry with the dwarfs, were stocked by man (p. 762). In contrast Pechlaner (1966, 1979) claims of the high-mountain lakes of the Alps, that "nearly all have acquired their charrs artificially" because all the 30 or so lakes in the Tyrol which contain fishes lack an "immigration pathway for fish" *today* (the last word mine). These lakes contain mainly the dwarfed altricial charr (Balon 1980, p. 792, Table 7).

The topographical situation in the Alps at the time of the glacial retreat, however, was certainly different from the present. Natural invasion of charr into the newly formed lakes was then a distinct possibility. Similar is the recent invasion of charrs and coregonins into glacier scoured lakes elsewhere in the holarctic (McPhail 1961, Lindsey et al. 1970) and their natural invasion into lakes discussed by Klemetsen and Grotnes (1980), many of which also lack an "immigration pathway" today.

Pechlaner's case for stocking in the 15th and 16th centuries is weak for other reasons also. To rech a high altitude alpine lake in the 16th century with a heavy load of water and fish would have been even more difficult and adventurous than nowadays (see Balon 1980, p. 776), for roads barely extended into the lower central valleys. Artificial propagation of salmonids was unknown then, so wet eggs could hardly have been transported. Charrs were not worth transporting beyond the range of easily accessible lakes, and if they were, such a project would surely have been recorded somewhere, as others were (e.g. Balon 1974a), by the company of chamois-hunting knights of the Archduke Sigismund and Kaiser Maximilian (Diem 1964). Fishing in these lakes illustrated by the ingenuous painting published in 1504 (Pechlaner 1979, p. 8) is as valid a proof of stocking charrs as the size of the fish depicted in the lake is proportional to the chamois and fishermen on the shore.

Frivolity aside, it is highly probably that many more glacial lakes in the Alps and elsewhere in the holarctic were originally invaded by altricial charrs than are currently inhabited by this species. Filling by silt, rock and a decrease in water supply has decided the fate of some lakes and their charrs. The evidence from Scottish lochs attests to this (e.g. Maitland, this volume).

Be that as it may, my theory of the altricial $\rightleftharpoons$ precocial dynamic states predicts that in lakes where conditions stabilized, the density of altricial charr could increase until competition induces the formation of the precocial charr. These large piscivorous charrs were probably encountered by the olden fishers, but with the increased number of human visitors soon became fished out. With this the conditions reversed to a less competitive and crowded state similar to early invasion, and the altricial dwarfs took over again.

Conventional fishery theory would predict the exact opposite to that which I have outlined above.Therefore, the practical consequences of misinterpretation are considerable: the demise of larger precocial fish and the takeover by dwarf altricial fish is not a result of underfishing (Pechlaner, this volume), on the assumption that the fish are somatically stunted from lack of sufficient food (e.g. Reimers 1979), but rather the condition results from overfishing. Therefore, adding food (e.g. in aquaculture conditions) could worsen the situation. The fish have responded to a vacant, low-competition habitat. Providing even more food (e.g. kitchen refuse in Dosener Lake, Balon 1980, p. 777) may simply reinforce that response. The problem is not one of inhibited somatic growth but one of prolonged early ontogeny and accelerated sexual maturation. Inadequate data on the true density of charr also contribute to this misinterpretation (Steiner and Pechlaner 1975, Chadwick 1976, Mahon 1980). As we explained earlier (Noakes and Balon 1982), if one wants large fish with late maturity and a long interval of somatic growth, one should do things to push the fish more towards a precocial life history state. This would mean that conditions should favour high intraspecific and/or interspecific competition, with relatively higher mortality rates for younger fish and a lower rate for older individuals. This in effect means that fishing, which usually selects for larger individuals and contributes to their higher mortality, must be relaxed. Limited food, stable conditions and high densities will favour the appearance of precocial charr.

The presence of only altricial charr may also be purely natural. Harsh climatic conditions with frequent environmental perturbations may preclude *a priori* the formation of precocial life history states. If the environmental perturbations increase after initial stabilization, the already formed precocial charr may perish or, given certain conditions, again become altricial by the process of juvenilization. We may never know the sequence of events by which the altricial or precocial forms in alpine lakes have been formed. However, we know that at present they occur most often as altricial forms (Reisinger 1953, Balon 1980).

The impoundment of a mountain lake inhabited by Arctic charrs creates a habitat which is perturbed and appropriate for invasion. "Dwarf" charr with altricial life history states will prevail in this situation, irrespective of the existence of a large precocial form in the watershed prior to impoundment (usually of low access for fishing). Misinterpretation of this situation leads to encouragement of fisheries which, as previously explained, may delay or preclude the formation or build-up of large precocial charrs, contrary to what is expected from such an action (Klemetsen, this volume).

Man is only one among many factors contributing to the high extinction rate of the specialized, large, precocial Arctic charr, which when landlocked may never produce enough available yield (sensu Balon 1974b) to support sustained harvest. In contrast, it is inherent in the life history state of the altricial charr to produce a large excess of available yield as a precaution against environmental perturbations (which man can harvest as a part of such perturbations). The safety margin of available yield in the anadromous charr may be a little more flexible because of the utilization and transport of resources from a larger area. Even in a fluvial system, however, altricial charrs are more resistant to fishing pressure, as evidenced by the cases of the precocial sea- or ferox trout (*Salmo trutta*) and its altricial form — the brown trout (Schäferna 1928, Berg 1948, Skrochowska 1953, Balon 1968, Campbell 1979, Went 1979). Whereas anadromous precocial forms became extinct in many systems, the altricial part of the species persist as dwarf brown trout (Baratch 1952, Balon 1964, 1968).

ALTRICIAL ⇌ PRECOCIAL DYNAMIC STATES (APDS) AS THE CAUSE OF EVOLUTIONARY PATTERNS NAMED "THE EFFECT HYPOTHESIS"

> Union of like and like was conformity and not strength but weakness; union of diversity and of opposites was the only real unity and source of strength.
>
> Laurens van der Post (1975)
> in A Mantis Carol

In the recent review of CHARRS Clark Hubbs (1981) complained that "no cladograms or phylogenetic trees are presented and their absence makes it difficult to understand the differences among the specialists." I presume there is a good reason why no one had the courage to express their idea graphically. Actually this is not entirely true, Reshetnikov (1980, p. 144) presents a phylogenetic tree of salmonids in general and coregonins in particular. In a book sent to me recently by Eizo Kimura entitled "Special on Freshwater Fishes: *Salvelinus*" (1980) T. Kubo presents a modified dendrogram (p. 11), originally constructed by Numashiri (1975) on the basis of salmonid isozymes. It is an interesting phylogenetic tree of salmons, trouts and charrs, with speciation events related to the timing of glacial episodes. It was a joy to compare it with the excellent colour plates that precede the Japanese text which I had some problem deciphering.

In viewing Figure 8, which I decided to present in spite of fear of ridicule, four points must be born in mind: 1) it is not intended as a phylogenetic tree of salmonines nor as a cladogram but as a graphical presentation of my ideas which may be useful in constructing the badly needed cladograms; 2) the number of extinct evolutionary units and the distances are entirely arbitrary; 3) the names of extant units (genera, subgenera and species) are neither complete nor meant as an endorsement of their validity or status; and finally, 4) the same applies to the numbers of chromosomes (Denton 1973, Reshetnikov 1980, p. 136, Gold 1977, Gold et al. 1980, Cavender, pers. comm.).

The importance of the two separately formulated theories (Balon 1983) — the Chen Principle and the APDS — for the understanding of Figure 8 cannot be overemphasized. As a biologist, I lack sufficient mastery, I confess, of the sophisticated geometry which would be required to produce a convincing illustration. It is the best I can do and if my effort stimulates someone to complete the story my purpose will be accomplished.

The lower part of the figure represents my original presentation of the APDS, based on the continuous tendency of organisms to form specialists (left to right), with the way back to generalists remaining open (paedomorphosis). The generalists with a highly heterogeneous gene pool, and an ability to survive a harsh environment with frequent unpredictable perturbations, are as a rule organisms of early maturation and/or long reproductive interval, delayed somatic differentiation, and low reproductive cost. Because early ontogenetic features of a highly general nature are retained in large numbers of offspring through most of the life history, the constant presence of an extremely broad range of variation capable of satisfying a flexible selection is maintained by the Chen Principle (Balon 1983). Given environmental stability and increasing competitive pressure, selection would favour specialization. After several numbers of ontogenies (= generations) along the specialization trajectory, the epiphenotypic variation will be considerably reduced, and some genetic heterogeneity ultimately lost. Age and size at maturity will be delayed and the higher reproductive cost will be distributed among a few precious zygotes. The lifespan will be prolonged, either by the senescent period, which is of no reproductive value, or terminated by semelparity. Fewer specialized early ontogenetic individuals result. This reduces variation and resutant flexibility so much that environmental perturbations which were previously accommodated now become totally destructive. For this process I have used the term gerontomorphosis — an initial hypermorphic sequence of ontogenies and evolution which is not applicable to salmonines, having been lost in times before smilodonichthyids.

Given that the only possible sequence of evolution was hypermorphic, every species would sooner or later have become extinct. However, each organismal unit reacts to unpredictable environmental perturbations by forming in turn, or at the same time, altricial and precocial life history states. These states may canalize into an evolutionary event as explained earlier. In short, juvenilization, in a strong threshold fashion, may change a specialist back into an altricial generalist, thus avoiding extinction. This would enable the sequence to be repeated along a new paedomorphic evolutionary trajectory. The entire evolution of salmonins, as we know it, appears to represent such primary and secondary paedomorphic trajectories (top of Figure 8).

The earliest recognizable (but maybe not the first) paedomorphosis (Fig. 5) initiated a dichotomy leading on the right to the precocial trajectory (A-C) of *Oncorhynchus* and *Parasalmo* and on the left to the altricial trajectory (A-B) representing remaining extant salmonins. Within the precocial trajectory *Oncorhynchus* spp. represent specialized terminal units of the most precocial primary sequence, whereas *Parasalmo* appears to represent a secondary paedomorphic sequence (A'-D) subsequent to juvenilization of an *Oncorhynchus*-like predecessor (*Rhabdofario*? or *S. australis* Cavender and Miller 1982). The altricial part of the initial dichotomy (A-B), along a trajectory of least specialization, led to the most generalized form, *Brachymystax*, and in order of increased specialization, *Salmothymus, Platysalmo, Cristivomer, Baione, Salvelinus, Salmo trutta*, and *Salmo salar*. Formation of distinct evolutionary units is more frequent along the precocial trajectories A-C as would be expected in the case of effect hypothesis (Vrba 1980).

The chromosome numbers added later corroborate the final scenario based on altricial $\rightleftharpoons$ precocial dynamic states and juvenilization, and may reflect the loss in genetic variation along the specialization trajectories. It is generally accepted that salmonines evolved from tetraploid ancestors (Ohno et al. 1969, Victorovsky 1976, Dorofeyeva 1977) via a regressive genome diploidization. The karyotype 92 of *Brachymystax lenok* is closest to the ancestral tetraploid type and on the other extreme are the diploid types of *Oncorhynchus gorbuscha* with the lowest number 52 coinciding with the most extreme secondary metamorphosis (Fig. 5),or of *Salmo salar* with another lowest number 58 coinciding with the most specialized form of the least specialization trajectory. It is tempting to speculate, therefore, that paedomorphosis may be associated with gene duplication and higher levels of DNA per cell achieved through such chromosomal rearrangements like tetraploidy and tandem gene duplication. Speciation and specialization are after all already suspected of association with gene fusion, deletion and lower levels of DNA per cell values (Dingerkus 1979, Schultz 1980). Altricial $\rightleftharpoons$ precocial dynamic states may be facilitated by correspond-

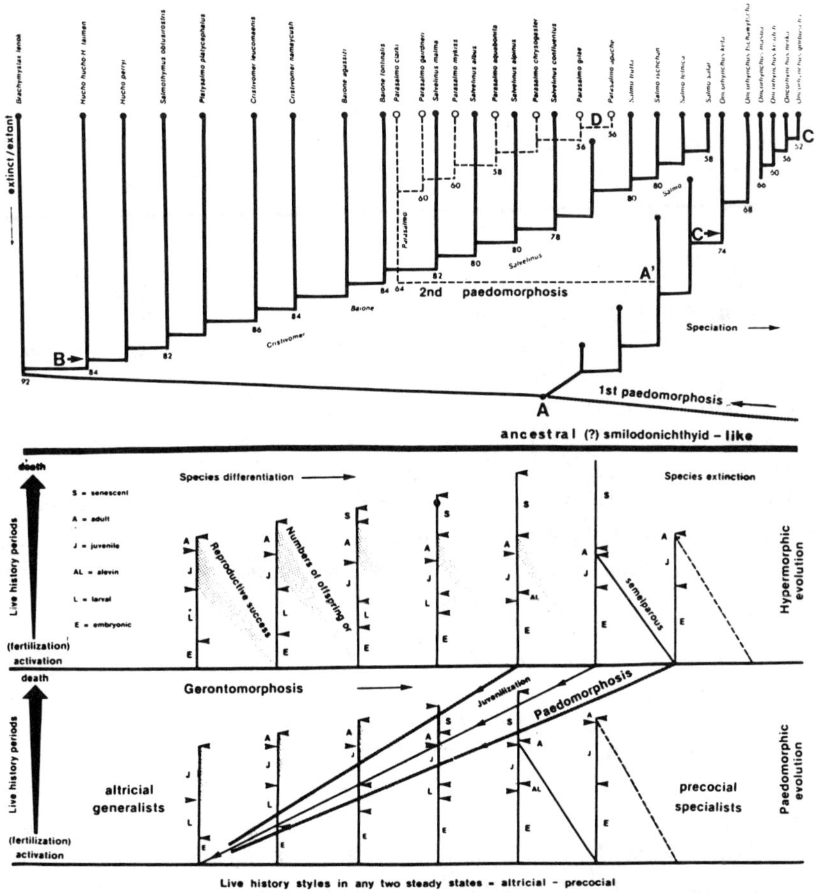

FIG. 8 Lower part: A simplified scheme of possible life history styles with changing duration of intervals (embryonic to senescent) from generalized altricial (left) to specialized precocial (right). Any two of these can represent at the same time the dualistic steady states named APDS. Given certain conditions an overspecialized precocial form can partially reinstate its lost variation by juvenilization, starting a new sequence of paedomorphic evolution, i.e. changing from precocial into altricial form. Upper part: Proposed scenario of the evolutionary relationships of salmonins, based primarily on the theories named the Chen Principle and APDS. The initial juvenilization of the ancestral pelagophilous "smilodonichthyid" cleaved into an altricial (A - B, left) and a precocial (A - C) trajectory (right) of which the most specialized forms are represented by the extant *Oncorhynchus* (C), and the secondarily juvenilized by *Parasalmo* (D). All the other extant salmonins (see also Fig. 5) are terminal taxa of the altricial trajectory of which *Brachymystax lenok* is least and *Salmo salar* the most specialized form. The numbers of diploid chromosomes (given at the base of each unit) are incidental to the central theme, but correspond surprisingly well with the scenario.

ing chromosomal rearrangements, size of which may be directly proportional to the amplitude of associated thresholds, i.e. affecting changes of life history only, or of intra or extra "specific" specialization. I hope that necessary techniques will soon be developed and reliable values on chromosome patterns will become available to test the relationship of altricial $\rightleftharpoons$ precocial dynamic states and cytogenetic features.

EPILOGUE

Because of the purposefully selective and invalid (according to the Rules of Zoological Nomenclature) speculation published after this paper's completion and presentation at the Symposium (Nyman et al. 1981), and the data reported by Holčik (1982) I feel encouraged to expand and explain my proposed phyletic classification of the Salmoninae (Fig. 5 and 8). The "altricial $\rightleftharpoons$ precocial dynamic state" hypothesis still accounts for all the available data. It predicts the dichotomous formation of altricial and precocial homeostatic states of charr phenotypes in response to environmental perturbations. Although these phenotypes are not separate evolutionary units, the same processes which cause this ecological dichotomy are probably responsible for the evolutionary patterns (Balon 1981a,b, 1983) in salmonids, and warrant a tentative cladogram as in Figure 9.

Huchonini encompasses the genera *Brachymystax* and *Hucho*, Salvelini the subgenera *Salvelinus, Cristivomer* and *Baione*, Salmonini *Salmo* and *Salmothymus*, Parasalmonini all western North American *Salmo (Parasalmo)*, and Oncorhynchini the Pacific salmons *Oncorhynchus*.

Within each subgenus charrs of the genus *Salvelinus* appear as evolutionary twin forms (altricial and precocial), i.e. *Cristivomer leucomaenis* and *C. namaycush, Baione fontinalis* and *B. agassizi, Salvelinus malma* and *S. confluentus, Salvelinus alpinus oquassa (S. a. salvelinus)* and *S. alpinus alpinus (S. a. profundus)*. Each species is a descendant of one or more juvenilization processes; if necessary it can maintain ecological flexibility through epigenesis as an altricial generalist or precocial specialist. Therefore forms (or stocks) like "dwarf" and "normal", sunapee and anadromous *Salvelinus alpinus*, lake charr and siscowet *S. namaycush*, and brook and aurora *S. fontinalis* recognized by ecologists are a source of confusion to conventional taxonomists. I hope my theories (Balon 1983) help to clarify some of the confusion, in spite of such powerful opponents as "belief" and "politics".

At the start I accused nearly everyone, including myself, of unforgivable biases — beliefs — which mark our conclusions. One is, however, influenced not only by the past. After a last minute reading I am tempted to restate my opening paragraphs by quoting from the keynote lecture given by Northrop Frye

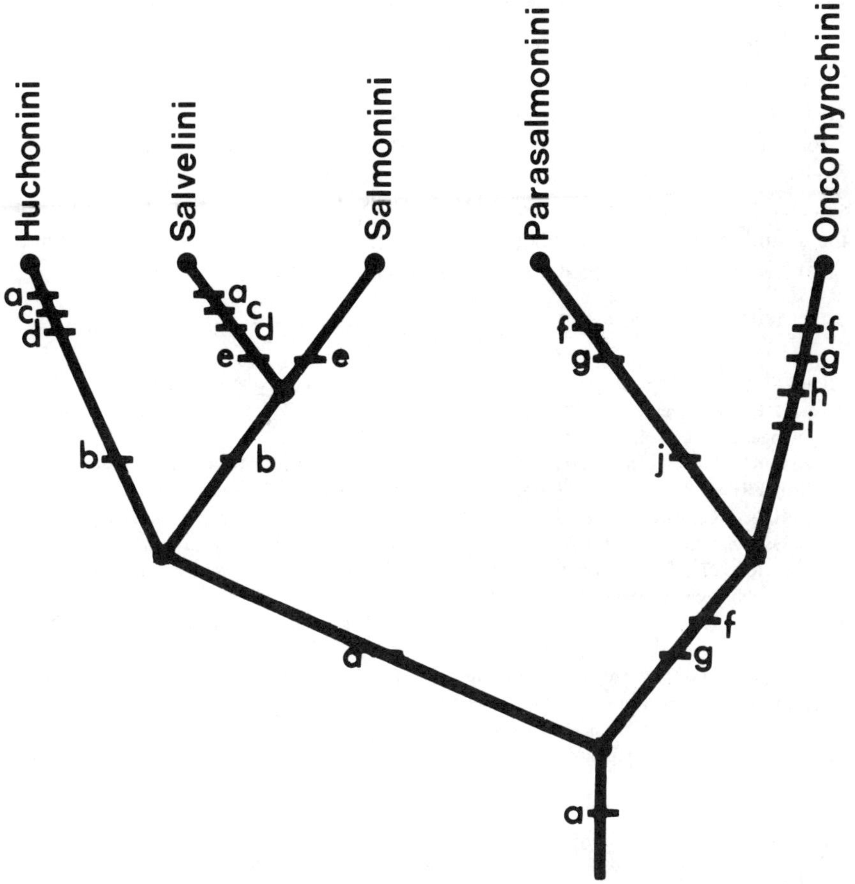

FIG. 9 Tentative cladogram of the evolutionary classification of salmonids, based in part on the conclusion drawn from epigenesis (a = the highest number of chromosomes (92), b = teeth on palatine bones and vomer in a continuous arched band, c = lateral line scales in the form of cantilevered eaves, d = short vomer, e = shaft or vomer with teeth, f = lowest number of chromosomes (74-52), g = in Asia and North America only, h = strongly developed sexual dichromatism and secondary metamorphosis, i = mostly anadromous, j = spring spawners.

(1981, p. 129) at the Toronto meeting of the AAAS, recently published in Science: "I speak of religious or political concerns rather than belief, because the conviction of its truth is less important than the sense of the social necessity of accepting it. (. . .)"

ACKNOWLEDGEMENTS

I wish to thank Robin Mahon who provided me with valuable references, was the first patient listener and opponent of the ideas, and finally read and corrected the hastily assembled drafts; Christine Flegler-Balon duplicated all along Robin's efforts adding her touch of precision in logics and was later joined by David Noakes; Marilyn White finished the figures and S. Yamashiro kindly translated selected parts from theJapanese ''Special on *Salvelinus*''. My studies that form the foundation for this synthesis were supported by grants from the Natural Sciences and Engineering Research Council of Canada

REFERENCES

Adolph, E.T. 1982. Physiological integrations in action. The Physiologist 25 (supplement): V, 1-67.

Alberch, P., S.J. Gould, G.F. Oster and D.B. Wake. 1979. Size and shape in ontogeny and phylogeny. Paleobiology 5:296-317.

Balon, E.K. 1964. A contribution to the knowledge of the ichthyofauna in the headwaters of the Poprad River. Sbornik prac o Tatranskom narodnom parku 7:140-164. (In Slovak).

Balon, E.K. 1968. Notes to the origin and evolution of trouts and salmons with special reference to the Danubian trouts. Věst. Čs. spol. zool. 32:1-21.

Balon, E.K. 1971. The intervals of early fish development and their terminology (review and proposals). Věst. Čs. spol. zool. 35:1-8.

Balon, E.K. 1974a. Domestication of the carp *Cyprinus carpio* L. ROM Life Sci. Misc. Publ., Toronto. 37 p.

Balon, E.K. 1974b. Fish production of a tropical ecosystem, p. 249-676. *In* E.K. Balon and A.G. Coche [eds.] Lake Kariba: a man-made tropical ecosystem in Central Africa, Monographiae Biologicae 25, Dr. W. Junk Publishers, The Hague.

Balon, E.K. 1975a. Terminology of intervals in fish development. J. Fish. Res. Board Can. 32:1663-1670.

Balon, E.K. 1975b. Reproductive guilds of fishes: a proposal and definition. J. Fish. Res. Board Can. 32:821-864.

Balon, E.K. 1976. A note concerning Dr. Richards' comments. J. Fish. Res. Board Can. 33:1254-1256.

Balon, E.K. 1977. Early ontogeny of *Labeotropheus* Ahl, 1927 (Mbuna, Cichlidae, Lake Malawi), with a discussion on advanced protective styles in fish reproduction and development. Env. Biol. Fish. 2:147-176.

Balon, E.K. 1979a. The theory of saltation and its application in the ontogeny of fishes: steps and thresholds. Env. Biol. Fish. 4:97-101.

Balon, E.K. 1979b. The juvenilization process in phylogeny and the altricial to precocial forms in the ontogeny of fishes. Env. Biol. Fish. 4:193-198.

Balon, E.K. [ed.] 1980. Charrs: salmonid fishes of the genus *Salvelinus*. Dr. W. Junk Publishers, The Hague. 928 p.

Balon, E.K. 1981a. Saltatory processes and altricial to precocial forms in the ontogeny of fishes. Amer. Zool. 21:573-596.

Balon, E.K. 1981b. About processes which cause the evolution of guilds and species. Env. Biol. Fish. 6:129-138.

Balon, E.K. 1981c. Additions and amendments to the classification of reproductive styles in fishes. Env. Biol. Fish. 6:377-389.

Balon, E.K. 1983. Epigenetic mechanisms: reflections on evolutionary processes. Can. J. Fish. Aquat. Sci. 40:2045-2058.

Balon, E.K. 1984a. Patterns in the evolution of reproductive styles in fishes, p. 35-53. *In* C.W. Potts and R.J. Wooton [eds.] Fish reproduction: strategies and tactics, Academic Press, London.

Balon, E.K. 1984b. Reflection on some decisive events in the early life of fishes. Trans. Amer. Fish. Soc. (In press).

Baratch, G.P. 1952. The significance of the brown trout on the resource origin of the Black Sea salmon. Zool. Zh. 31:906-915. (In Russian).

Bateson, G. 1979. Mind and nature. A necessary unity. E.P. Dutton, New York. 238 p. Behnke, R.J. 1972. The systematics of salmonid fishes of recently glaciated lakes. J. Fish. Res. Board Can. 29:639-671.

Behnke, R.J. 1980. A systematic review of the genus *Salvelinus*. pp.441-480. *In* E.K.

Balon [ed.] Charrs: salmonid fishes of the genus *Salvelinus*. Dr. W. Junk Publishers, The Hague.

Bell, G. 1980. The costs of reproduction and their consequences. Amer. Natur. 116:45-76.

Berg, L.S. 1948. Freshwater fishes of the U.S.S.R. and adjacent countries. Izd. AN SSSR, Moskva-Leningrad (transl. Nat. Sci. Found., Jerusalem 1962, Vol. 1, p. 504).

Berrill, N.J. 1955. The origin of vertebrates. Clarendon Press, Oxford. 257 p.

Bull, J.J. and R. Shine. 1979. Iteroparous animals that skip opportunities for reproduction. Amer. Natur. 114:296-303.

Campbell, R.N. 1979. Ferox trout, *Salmo trutta* L., and charr, *Salvelinus alpinus* (L.), in Scottish lochs. J. Fish Biol. 14:1-29.

Cavender, T.M. and R.R. Miller. 1972. *Smilodonichthys rastrosus* a new Pliocene salmonid fish from Western United States. Bull. Mus. Nat. Hist. Univ. Oregon (Eugene) 18:1-44.

Cavender, T.M. and R.R. Miller. 1982. *Salmo australis*, a new species of fossil salmonid from southwestern Mexico. Contr. Mus. Paleontol. Univ. Michigan 26:1-17.

Chadwick, E.M.P.1976. Ecological fish production in a small Precambrian shield lake. Env. Biol. Fish. 1:13-60.

Darwin, C. 1859. On the origin of species by means of natural selection, or the preservation of favoured races in the struggle for life. John Murray, London. (Facsimile of the 1st edition Harvard Univ. Press, Cambridge).

Denton, T.E. 1973. Fish chromosome methodology. Charles C. Thomas Publ., Springfield. 166 p.

Diem, H. 1964. Beiträge zur Fischerei Nordtirols. B. Bie Fisherei in den natürlichen Gewässern in der Vergangenheit. Veröff. Mus. Ferdinand. Innsbruck 43:7-132.

Dingerkus, G. 1979. Chordate cytogenetic studies: an analysis of their phylogenetic implications with particular reference to fishes and the living coelacanth, p. 111-127. *In* J.E. McCosker and M.D. Lagios [eds.] The biology and physiology of the living coelacanth, Occ. Pap. Calif. Acad. Sci., San Francisco.

Dodson, M.M. and A. Hallam. 1977. Allopatric speciation and the fold catastrophe. Amer. Natur. 111:415-433.

Dorofeyeva, E.A. 1977. Utilization of cariological characters for the solution of problems in systematics and phylogeny of salmonid fishes, p. 86-95. *In* Principles of Classification and Phylogeny of Salmonid Fishes, Acad. Sci. U.S.S.R., Leningrad. (In Russian).

Eigen, M. and P. Schuster. 1979. The hypercycle. A principle of natural self-organization. Springer-Verlag, Berlin. 92 p.

Eldredge, N. and S.J. Gould. 1972. Punctuated equilibria: an alternative to phyletic gradualism, p. 82-115. *In* T.J.M. Schopf [ed.] Models in Paleobiology, W.H. Freeman, San Francisco.

Frye, N. 1981. The bridge of language. Science 212: 127-132.

Futuyma, D.J. and G.C. Mayer. 1980. Non-allopatric speciation in animals. Syst. Zool. 29:254-271.

Garstang, W. 1922. The theory of recapitulation: a critical restatement of the biogenetic law. Linn. J. Zool. 35:81-101.
Garstang, W. 1928. The morphology of the Tunicata, and its bearing on the phylogeny of the Chordata. Quart. J. Microscop. Sci. 72:51-187.
Garstang, W. 1962. Larval forms with other zoological verses. Blackwell, Oxford. 77 p.
Gold, J.R. 1977. Systematics of western North American trout (*Salmo*), with notes on the redband trout of Sheepheaven Creek, California. Can. J. Zool. 55:1858-1873.
Gold, J.R., W.J. Karel and M.R. Strand. 1980. Chromosome formulae of North American fishes. Prog. Fish-Cult. 42:10-23.
Gould, S.J. 1977. Ontogeny and phylogeny. The Belknap Press of Harvard Univ. Press, Cambridge. 501 p.
Gould, S.J. 1979. A quahog is a quahog. Natural History 88:18-25.
Gould, S.J. and C.B. Calloway. 1980. Clams and brachiopods — ships that pass in the night. Paleobiology 6:383-396.
Gould, S.J. and N. Eldredge. 1977. Punctuated equilibria: the tempo and mode of evolution reconsidered. Paleobiology 3:115-151.
Greenwood, P.H. 1974. The cichlid fishes of Lake Victoria, East Africa: the biology and evolution of a species flock. Bull. Brit. Mus. Nat. Hist. (Zool.) Suppl. 6:1-134.
Greenwood, P.H. 1979. Macroevolution — myth or reality? Biol. J. Linn. Soc. 12:293-304.
Greenwood, P.H. 1981. The haplochromine fishes of the East African lakes. Kraus. Intern. Publ., München. 839 p.
Gutmann, W.F. 1981. Relationships between invertebrate phyla based on functional-mechanical analysis of hydrostatic skeleton. Amer. Zool. 21:63-81.
Heberer, G. 1958. Zum Problem der additiven Typogenese, p. 40-47. *In* O. Hedberg [ed.] Systematics of Today, Uppsala.
Hennig, W. 1966. Phylogenetic systematics. University of Illinois Press, Urbana. 263 p.
Henricson, J. and L. Nyman. 1976. The ecological and genetical segregation of two sympatric species of dwarfed char (*Salvelinus alpinus* (L.) species complex). Inst. Freshw. Res. Drottningholm Rep. 55:15-37.
Holčik, J. 1982. Towards the characteristics of the genera Hucho and Brachymystax (Pisces, Salmonidae). Folia Zoologica 31: 369-380.
Hubbs, C. 1981. Review of Charrs ed. by Eugene K. Balon. Copeia 1981:249-251.
Hubbs, C.L. 1943. Terminology of early stages of fishes. Copeia 1943:260.
Humphries, J.M. and R.R. Miller. 1981. A remarkable species flock of pup-fishes, Genus *Cyprinodon*, from Yucatan, Mexico. Copeia 1981:52-64.
Johnson, L. 1980. The Arctic charr, *Salvelinus alpinus*, p. 15-98. *In* E.K. Balon [ed.] Charrs: salmonid fishes of the genus *Salvelinus*, Dr. W. Junk Publishers, The Hague.
Jollie, M. 1982. What are the 'Calcichordata'? and the larger question of the origin of Chordates. Zool. J. Linn. Soc. 75: 167-188.
Klemetsen, A. and P.E. Grotnes. 1980. Coexistence and immigration of two sympatric Arctic charr, p. 757-763. *In* E.K. Balon [ed.] Charrs: salmonid fishes of the genus *Salvelinus*, Dr. W. Junk Publishers, The Hague.
Kornfield, J., K.F. Beland, J.R. Moring and F.W. Kircheis. 1981. Genetic similarity among endemic Arctic char (*Salvelinus alpinus*) and implications for their management. Can. J. Fish. Aquat. Sci. 38:32-39.
Kosswig, C. 1947. Selective mating as a factor for speciation in cichlid fish of East African lakes. Nature (London) 159:604-605.
Kosswig, C. 1963. Ways of speciation in fishes. Copeia 1963:238-244.

Lindsey, C.C., J.W. Clayton and W.G. Franzin. 1970. Zoogeographic problems and protein variation in the *Coregonus clupeaformis* whitefish species complex, p. 127-146. *In* C.C. Lindsey and C.S. Woods [eds.] Biology of coregonid fishes, Univ. of Manitoba Press, Winnipeg.

Løvtrup, S. 1974. Epigenetics — a treatise on theoretical biology. John Wiley and Sons, London. 548 p.

Løvtrup, S. 1979. The evolutionary species: fact or fiction? Syst. Zool. 28:386-392.

Low, B.S. 1978. Environmental uncertainty and the parental strategies of marsupials and placentals. Amer. Natur. 112:197-213.

Mahon, R. 1980. Accuracy of catch-effort methods for estimating fish density and biomass in streams. Env. Biol. Fish. 5:343-360.

Mayr, E. 1976. Evolution and diversity of life. Selected essays. The Belknap Press of Harvard Univ. Press, Cambridge. 721 p.

McKaye, K.R. 1978. Explosive speciation: the cichlid fishes of Lake Malawi. Discovery 13:24-29.

McKaye, K.R. 1980. Seasonality in habitat selection by the gold color morph of *Cichlasoma citrinellum* and its relevance to sympatric speciation in the family Cichlidae. Env. Biol. Fish. 5:75-78.

McPhail, J.D. 1961. A systematic study of the *Salvelinus alpinus* complex in North America. J. Fish. Res. Board Can. 18:793-816.

Miller, P.J. 1979. A concept of fish phenology. Symp. Zool. Soc. Lond. 44:1-28.

Moy-Thomas, J.A. and R.S. Miles. 1971. Palaeozoic fishes. W.B. Saunders, Philadelphia. 259 p.

Nilsson, N.A. and O. Filipsson. 1971. Characteristics of two discrete populations of Arctic char (*Salvelinus alpinus* L.) in a north Swedish lake. Inst. Freshw. Res. Drottningholm Rep. 51:90-108.

Noakes, D.L.G. 1978. Ontogeny of behavior in fishes: a survey and suggestions, p. 103-125. *In* G.M. Burghardt and M. Bekoff [eds.] The development of behaviour: comparative and evolutionary aspects, Garland STPM Press, New York.

Noakes, D.L.G. 1980a. Social behavior in young charrs, p. 683-701. *In* E.K. Balon [ed.] Charrs: salmonid fishes of the genus *Salvelinus*, Dr. W. Junk Publishers, The Hague.

Noakes, D.L.G. 1980b. Some histological observations on the skin of charrs, p. 423-433. *In* E.K. Balon [ed.] Charrs: salmonid fishes of the genus *Salvelinus*. Dr. W. Junk Publishers, The Hague.

Noakes, D.L.G. and E.K. Balon. 1982. Life histories of tilapias: an evolutionary perspective, p. 61-82. *In* R.S.V. Pullin and R. Lowe-McConnell [eds.] Biology and culture of tilapias, ICLARM Conf. Proc. 7, Manila.

Nyman, L. 1972. A new approach to the taxonomy of the "*Salvelinus alpinus* complex". Inst. Freshw. Res. Drottningholm Rep. 52:103-131.

Nyman, L., J. Hammar and R. Gydemo. 1981. The systematics and biology of landlocked populations of Arctic charr from northern Europe. Inst. Freshw. Res. Drottningholm Rep. 59: 128-141.

Ohno, S., J. Muramoto, J. Klein and N.B. Atkin. 1969. Diploid-tetraploid relationship in clupeid and salmonid fish, p. 139-147. *In* C.D. Darlington and K.R. Lewis [eds.] Chromosomes today, Vol. 2, Oliver and Boyd, Edinburgh.

Oppenheim, R.W. 1981. Ontogenetic adaptations and retrogressive processes in the development of the nervous system and behaviour: a neuroembryological perspective, p. 73-109. *In* K.J. Connolly and H.F.R. Prechtl [eds.] Maturation and development: biological and psychological perspectives, J.P. Lippincott, Philadelphia.

Pechlaner, R. 1966. Salmonideneinsätze in Hochgebirgsseen und-tümpel der Ostalpen. Verh. Internat. Verein. Limnol. 16:1182-1191.

Pechlaner, R. 1979. High-mountain lakes in Tyrol. *In* Tirol-Immer Einen Urlaub Wert, Heimatwerbung Ges., Innsbruck. 24 p. (German and English text).

Pechlaner, R., G. Bretschko, P. Gollmann, H. Pfeifer, M. Tilzer and H.P. Weissenbach. 1972a. The production processes in two high-mountain lakes (Vorderer and Hinterer Finstertaler See, Kühtai, Austria), p. 239-269. *In* Z. Kajak and A. Hillbricht-Ilkowska [eds.] Productivity problems of freshwaters, Proc. IBP-UNESCO Symp., Warszawa.

Pechlaner, R., G. Bretschko, P. Gollmann, H. Pfeifer, M. Tilzer and H.P. Weissenbach. 1972b. Ein Hochgebirgssee (Vorderer Finstertaler See, Kühtai, Tirol) als Modell der Energietransportes durch ein limmisches Ökosystem. Verhandlungsbericht der Deutschen Zool. Ges. 65:47-56.

Portmann, A. 1945. Die Ontogenese des Menschen als Problem der Evolutionsforschung. Verh. Schweiz. Naturf. Ges. 125:44-53.

Powles, H. and W.E. Burgess. 1978. Observations on the benthic larvae of *Pareques* (Pisces: Sciaenidae) from Florida and Colombia. Copeia 1978:169-172.

Regier, H.A. 1980. Epilogue: understanding of charrs and the need for a "charr watch", p. 889-893. *In* E.K. Balon [ed.] Charrs: salmonid fishes of the genus *Salvelinus*. Dr. W. Junk Publishers, The Hague.

Reimers, N. 1979. A history of a stunted brook trout population in an Alpine lake: a lifespan of 24 years. Calif. Fish and Game 65:196-215.

Reisinger, E. 1953. Zum Saiblingsproblem. Carinthia II, nat.-wiss. Beiträge zur Heimatkunde Kärntens 63:74-102.

Reshetnikov, J.S. 1980. Ecology and systematics of the whitefishes. Izd. Nauka, Moskva. 302 p. (In Russian).

Richards, W.H. 1976. Some comments on Balon's terminology of fish development intervals. J. Fish. Res. Board Can. 33:1253-1254.

Ricklefs, R.E.1979. Adaptation, constraint, and compromise in avian post-natal development. Biol. Rev. 54:269-290.

Rosen, D.E. 1978. Vicariant patterns and historical explanation in biogeography. Syst. Zool. 27:159-188.

Rosen, D.E. and D.G. Buth. 1980. Empirical evolutionary research versus neo-darwinian speculation. Syst. Zool. 29:300-308.

Savvaitova, K.A. 1980. Taxonomy and biogeography of charrs in the Palearctic, p. 281-294. *In* E.K. Balon [ed.] Charrs: salmonid fishes of the genus *Salvelinus*, Dr. W. Junk Publishers, The Hague.

Schäferna, K. 1928. Gigantic trout and a thought on fish growth. Rybarsky Věstnik 8:1-2 reprint pages. (In Czech).

Schmalhausen, I.I. 1968. The origin of terresterial vertebrates. Academic Press, New York. 314 p.

Schultz, N. 1975. Untersuchungen zur Biologie der Seesaiblinge (*Salvelinus alpinus*) (L.)) (Pisces: Salmonidae) im Achensee (Tirol, Österreich). Teil I: Nahrungsaufnahme. Ber. nat.-med. Ver. Innsbruck 62:139-151.

Schultz, R.J. 1980. Role of polyploidy in the evolution of fishes, p. 313-340. *In* W.H. Lewis [ed.] Polyoplidy: biological relevance, Plenum Publishing Corp., New York.

Settle, T. 1979. Popper on "when is a science not a science?" Syst. Zool. 28:521-529.

Skrochowska, S. 1953. Preliminary results in investigation of migrations in sea trout (*Salmo trutta* L.) and other salmonids reared in ponds. Pol. Arch. Hydrobiol. 1:89-135. (In Polish).

Smirnov, A.F. 1956. Arctic charr of Segozero. Trudy Karelsk. fil. AN. SSSR 5:119-130. (In Russian).

Smirnov, A.I. 1975. Biology, reproduction and development of Pacific salmons. Izd. Moskovskogo Univ., Moskva. 335 p. (In Russian).

Smith-Gill, S.J. and K.A. Berven. 1979. Predicting amphibian metamorphosis. Amer. Natur. 113:563-585.

Snyder, D.E. 1976. Terminologies for intervals of larval fish development, p. 41-58. *In* J. Boreman [ed.] Great Lakes fish egg and larvae identification: proceedings of a workshop, National Power Team, Power Plant Project FWS/OBS 76/23, Ann Arbor, Michigan.

Stanley, S.M. 1975. A theory of evolution above the species level. Proc. Nat. Acad. Sci. 72:646-650.

Stanley, S.M. 1979. Macroevolution: pattern and process. W.H. Freeman and Co., San Francisco. 332 p.

Steele, E.J. 1979. Somatic selection and adaptive evolution. The University of Chicago Press, Chicago. 134 p.

Steiner, V. and R. Pechlaner. 1975. Mark-recapture experiments with Arctic char (*Salvelinus alpinus*) in Austrian lakes. EIFAC Techn. Pap. 23 (Suppl. 1):672-688.

Stoklosowa, S. 1966. Sexual dimorphism in the skin of sea trout, *Salmo trutta trutta*. Copeia 1966:613-614.

Stoklosowa, S. 1970. Further observations on the sexual dimorphism in the skin of *Salmo trutta trutta* in relation to sexual maturity. Copeia 1970:332-339.

Svärdson, G. 1958. Specification in freshwater fish as illustrated by *Coregonus*. Proc. XVth Inter. Conf. Zool., Sect. 2:137-141.

Svärdson, G. 1961. Young sibling fish in northwestern Europe, p. 498-513. *In* W.F. Blair [ed.] Vertebrate speciation, Univ. Texas Press, Austin.

Svärdson, G. 1970. Significance of introgression in coregonid evolution, p. 33-59. *In* C.C. Lindsey and C.S. Woods [eds.] Biology of coregonid fishes, Univ. of Manitoba Press, Winnipeg.

Tauber, C.A. and M.J. Tauber. 1977. Sympatric speciation based on allelic changes at three loci: evidence from natural populations in two habitats. Science 197:1298-1299.

Van der Post, L. 1975. A mantis carol. The Hogarth Press, London. 166 p.

Van Valen, L. 1976. Ecological species, multispecies, and oaks. Taxon 25:233-239.

Victorovsky, R.M. 1976. Comparative cariology, evolution and speciation of the genus *Salvelinus*, p. 16-19. *In* Ecology and taxonomy of the salmonid fishes, Acad. Sĉi. U.S.S.R., Leningrad. (In Russian).

Volobuev, V.V. 1977. On the dwarf form of *Salvelinus neiva* (Salmoniformes, Salmonidae) from the Korral Lake (Okhota River basin). Zool. Zh. 56:405-411. (In Russian).

Vorobyeva, E.I., L.A. Sytina and V.V. Rubcov. 1981. Eco-morphological research on the early stages of fish ontogenesis, p. 153-157. *In* M. Peňaz and M. Prokeš [eds.] Topical problems of ichthyology, Czech. Acad. Sci., Brno.

Vrba, E.S. 1980. Evolution, species and fossils: how does life evolve? South Afr. J. Sci. 76:61-84.

Waddington, C.H. 1975. The evolution of an evolutionist. Edinburgh University Press, Edinburgh. 328 p.

Waddington, C.H. 1977. Tools for thought. Paladin, Frogmore. 250 p.

Went, A.E.J. 1979. 'Ferox' trout, *Salmo trutta* L. of Longhs Mask and Corrib. J. Fish Biol. 15:255-262.

White, M.J.D. 1978a. Modes of speciation. W.H. Freeman, San Francisco. 455 p.

White, M.J.D. 1978b. Chain processes in chromosomal speciation. Syst. Zool. 27:285-288.

Wilbur, H.M. and J.P. Collins. 1973. Ecological aspects of amphibian metamorphosis. Science 182:1305-1314.

Wiley, E.O. 1980. Is the evolutionary species fiction? — A consideration of classes, individuals and historical entities. Syst. Zool. 29:76-80.

Young, J.Z. 1978. The life of vertebrates. 2nd edition (7th Impression), Oxford Univ. Press, Oxford. 820 p.

Identification of Anadromous Arctic Charr Stocks in Coastal Areas of Northern Labrador

J. BRIAN DEMPSON

Department of Fisheries and Oceans, Fisheries Research Branch, Newfoundland Region, P.O. Box 5667, St. John's, Newfoundland A1C 5X1

DEMPSON, J.B. 1984. Identification of anadromous Arctic charr stocks in coastal areas of Northern Labrador, p. 143–162. *In* L. Johnson and B.L. Burns [eds.] Biology of the Arctic charr, Proceedings of the International Symposium on Arctic Charr, Winnipeg, Manitoba, May 1981. Univ. Manitoba Press, Winnipeg.

Samples of Arctic charr (*Salvelinus alpinus*) from various localities in northern Labrador were examined for nine meristic and thirteen morphometric characters. Univariate analyses showed that variation was slight with respect to meristic characters, although significant differences between samples indicated the discrete nature of the stocks. In general, higher mean values of meristic characters were found in charr from northerly areas. Multivariate analyses of morphometric characters revealed significant differences between all areas. Analysis of distribution patterns, resulting from tagging experiments carried out since 1975, indicated little mixing between populations in adjacent bays. Growth rate and maturation differences were also found. These results support the hypothesis that discrete anadromous Arctic charr stocks occur along the northern Labrador coast.

INTRODUCTION

Most mathematical models used in managing fish populations are based on the assumption that one well defined stock with its own characteristic distribution, growth rates, mortality rates and reproductive potential, is being exploited. In managing such a plastic species as the Arctic charr it is essential to know whether local stocks form discrete units, or whether they are part of one large intermingling population.

Many techniques have been used to differentiate stocks of fish, including tagging studies (Templeman 1974), biochemical analyses (Cross and Payne 1978), morphological comparisons (Sharp, Able, Leggett and Carscadden 1978; Carscadden and Misra 1980), taxonomy of parasites (Pippy 1969) and scale characteristics (Lear and Misra 1978). Several of these techniques have also been employed in examining systematic relationships within the *Salvelinus* species complex. For example, differences in meristic characteristics were examined by

Vladykov (1954), McPhail (1961), Qadri (1974), McCart and Craig (1971), Behnke (1972, 1980), Cavender (1980), McCart (1980) and Morrow (1980), while electrophoretic studies have been carried out by Saunders and McKenzie (1971), Nyman (1972) and Kornfield, Beland, Moring and Kircheis (1981). Similarly, Makhovenko (1972), Henricson and Nyman (1976) and Dick and Belosevic (1981) have used parasites as indicators of different forms of charr.

The charr is generally considered to be a localized species which remains in close proximity to its natal river (Nielsen 1961; Grainger 1953). Recent studies, however, have indicated that in certain cases marine migrations may be rather extensive (Jensen and Berg 1977; Johnson 1980), implying considerable mixing of populations at sea. Owing to the intense fishery that has developed in the last 4 years along the Labrador coast, from Anton's Point to Hebron Fiord, it was considered essential to determine whether anadromous charr harvested within this region represented one population or whether individual bays were characterized by their own discrete stocks with a minimal amount of interchange between arbitrarily defined geographical areas (Fig. 1) (Coady and Best 1976; Dempson 1978).

It is the purpose of this paper to examine the morphological and life history characteristics of anadromous charr populations in several of these neighbouring areas to determine whether there are differences significant enough to consider managing them as individual stocks.

MATERIALS AND METHODS

Morphological Data

A sample of 1007 Arctic charr was collected in 1978 from seven different marine areas for examination of meristic, morphometric and biological characteristics. The sample was frozen and transported to St. John's, Newfoundland, for examination. An additional sample of 1155 charr was collected from four areas during 1979 and 1980 for analysis of biological characteristics (Table 1, Fig. 1).

Nine meristic and 13 morphometric characters were analyzed. Counts and measurements were made following the criteria proposed by Hubbs and Lagler (1970). Morphometric measurements to the nearest mm, taken on the left side of the fish, included: total length (TL), fork length (FL), standard length (ST), opercular girth (GOP), maximum girth (GMX), and predorsal length (PRDL). Dorsal length to the adipose fin (DADL), head length (HL), postorbital length (PORBL), pectoral fin length (PCTL), snout length (SNL), orbit length (ORBL), and length of upper jaw (UJL) were measured to the nearest 0.1 mm, using a set of Helios vernier calipers.

Table 1. Number of specimens of Arctic charr collected from 1978-1980 for analysis of meristic, morphometric and biological characteristics.

Area	Date	Gear Mesh Size (mm) 38	51	64	76	89	102	114	127	140	152	Total
Ramah Bay (58°52'N, 63°15'W)	1978 Aug. 7		13			45		34	77	28	24	221
Saglek Fiord (58°29'N, 63°15'W)	1978 Aug. 8-9		15			72		29	23	9	1	149
	1979 Aug. 4-5		50	11	2			122	5			190
	1980 Aug. 14-15	26		13	7		42	119			2	209
	Total	26	65	24	9	72	42	270	28	9	3	548
Hebron Fiord (58°09'N, 62°45'W)	1978 Aug. 10-11		28			34		30				92
	1979 Aug. 6		21	23	15			96				155
	1980 Aug. 16				99			95			50	244
	Total		49	23	114	34		221			50	491
Napartok Bay (58°01'N, 62°19'W)	1978 Aug. 11							60	33			93
	1979 Aug. 7				6			55	24			85
	1980 Aug. 17	1			51		44	54			8	158
	Total	1			57		44	169	57		8	336
Okak Bay (57°28'N, 62°20'W)	1978 Aug. 12		5			20		138	36			199
	1979 Aug. 8							76				76
	1980 Aug. 19						13	21			4	38
	Total		5			20	13	235	36		4	313
Tikkoatokak Bay (56°42'N, 62°12'W)	1978 Aug. 1-5							124				124
Voisey Bay (56°15'N, 61°50'W)	1978 Aug. 1-5							129				129

The first gill arch on the left side of the fish was removed and all rakers on both limbs, including rudiments, were counted. Vertebral and fin ray counts were taken from radiographs. Charr specimens were filleted to expose the vertebral column and increase the clarity of the radiographs. Similarly, fins were removed and spread out on X-ray trays to allow for easier counting. All vertebral centra were counted, including those articulating with the fourth and fifth hypurals but excluding the urostyle, as illustrated by Vladykov (1954). Fin ray counts also included rudimentary rays.

Biological Characteristics

The age of each charr was determined from the sagittal otoliths, viewed under reflected light, following the method of Andrews and Lear (1956). Sex and maturity were determined by gross inspection of the gonads. The numbers of immature and mature fish in each age-class were recorded, and the age and length at which 50 percent of the sample was mature was derived from probit analysis (Bliss 1952), following the technique outlined by Fleming (1960).

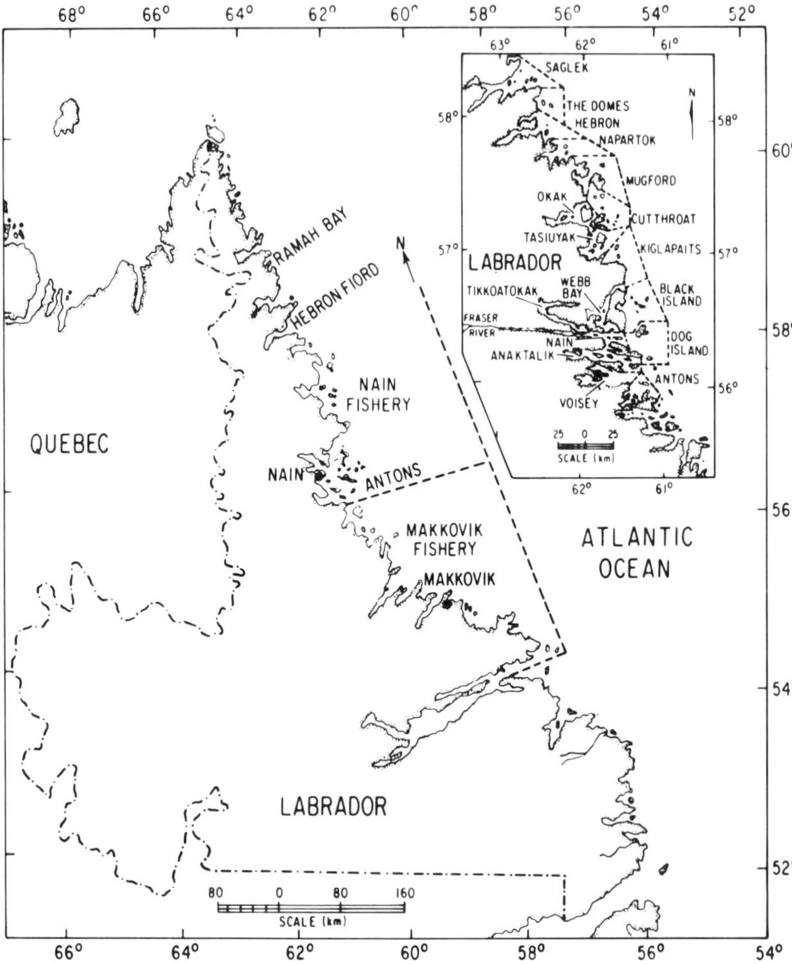

FIG. 1 Location of the Nain and Makkovik Arctic charr commercial fishing regions in northern Labrador. Insert illustrates the geographical stock separation within the Nain fishing region.

Samples from the period 1978-1980 were pooled to provide adequate numbers for the statistical analysis.

Samples of Arctic charr from the commercial fishery were used in the analysis of growth differences between areas since selectivity of various gear, or different gear size, was not a source of bias due to consistent use of 114 mm and 127 mm mesh gillnets. Data from Voisey Bay (n = 1400), Tikkoatokak Bay (n = 1566), and Okak Bay (N = 1246) were pooled for the years 1977-1980, while

Hebron Fiord data (N = 468) from 1977 and 1980 were combined to ensure that all ages were adequately represented.

Tagging Studies

Tagging experiments were carried out in conjunction with analyses of morphological and biological characteristics, in order to examine distribution patterns and determine the extent of intermingling of coastal populations. Locations and dates of tagging experiments are listed in Table 2.

Tagging operations at the Fraser River were carried out on charr migrating upstream through a counting fence. In the remaining locations charr were either seaward migrants, tagged during the spring, or individuals tagged during the marine phase of their migration. Numbered Carlin tags with double stainless steel thread were used throughout. Recaptures were obtained from the commercial fishery and the local recreational fishery, and from charr returning back through the Fraser River counting fence.

Statistical Analyses

Statistical analyses of meristic and morphometric data were confined to samples from the following four areas: Voisey Bay, Tikkoatokak Bay, Okak Bay and Hebron Fiord (Fig. 1).

The significance of each meristic variable and of growth in length at age between areas was tested by analysis of variance (ANOVA). Statistical analyses were performed on square root transformations of the meristic data (Sokal and Rohlf 1969). However, the standard single classification ANOVA, is sensitive to lack of homogeneity of variances within groups (Brown and Forsythe 1974a). Therefore, the Biomedical Computer Program BMDP7D (Dixon and Brown 1979) was used to calculate a modified F-statistic, the Brown-Forsythe statistic (F^*), which does not assume equality of variances (Brown and Forsythe 1974a). Multiple range tests (Brown and Forsythe 1974b) were conducted to identify which of the four areas differed significantly at the $P < 0.05$ level. Multivariate techniques were not applied to meristic data since discrete variables do not generally conform to the assumption of multivariate normality required in the analyses (Misra, pers. comm).

In analyzing morphometric data the Biomedical Computer Program BMDP7M (Stepwise Linear Discriminant Analysis, Jennrich and Sampson 1979) was utilized, which initially performs a multivariate analysis of variance (MANOVA) to determine the equality of character mean vectors (group population centroids). However, measurement of fork length, standard length and total length were omitted from the analysis. The multivariate test employed Wilk's lambda statis-

Table 2. Summary of Arctic charr tag recaptures by area and year in northern Labrador.

Location	Date	No. tagged	Recaptures year	Recaptures No.	Area of recapture* 1	2	3	4	5	5A	6	7	8	9	10	11	12	13	14	15	16	Unknown
Voisey Bay	1979 June	39	1979	6	2	4																
			1980	2							1		1									
	1980 July	100	1980	8	2	5					1											
	TOTAL	139		16	4	9					2		1									
Anaktalik Bay	1979 June	29	1979	1			1															
			1980	4			2						1									1
	1980 June	91	1980	9			3		1				3									2
	TOTAL	120		14			6		1				4									3
Main Bay	1979 June-July	94	1979	7						6	1											
			1980	15					2		9		4									
	1980 June	99	1980	22	2		1	1		1	7		9									1
	TOTAL	193		44	2		1	1	2	7	17		13									1
Fraser River	1975 July-August	287	1976	38					22	10	5											1
			1977	16					6	6	4											
			1978	14					4	1	9											
			1979	13						4	8											1
			1980	5							2		1									2
	1976 August	100	1977	11					8	3												
			1978	4					2	1	1											
			1979	5						3	2											
			1980	1							1											
	1977 August	200	1978	15			2		6		7											
			1979	17					1	5	11											
			1980	8	1				1	1	2		3									
	1978 August	31	1979	9	1				1	3	4											
			1980	1				1														
	1979 August	190	1979	5						5												
			1980	38			2	4	7	1	10		9	2								3
	TOTAL	808		200	2		4	5	58	43	66		13	2								7
Tikkoatokak Bay	1979 May	72	1979	24					1		21		1	1								
			1980	10					1		9											
	1979 June-July	84	1979	10				1	1	1	7											
			1980	3							2											1
	1980 June	110	1980	26							15		11									
	TOTAL	266		73				1	3	1	54		12	1								1

tic which has an approximate F distribution (Rao 1952, Morrison 1967). The relative success of the method can be judged by the subsequent discriminating procedure which classifies individuals into groups and is analogous to a multivariate multiple range test (Pimental 1979). The program employs a jackknife validation procedure which has been shown to reduce bias (Lachenbruch and Mickey 1968) by classifying each case into a group according to the classification information derived from all data except the specimen being classified (Jennrich and Sampson 1979). Homogeneity of within group variance-covariance

Table 2. (Cont'd).

Location	Date	No. tagged	Recaptures		Area of recapture*																	
			year	No.	1	2	3	4	5	5A	6	7	8	9	10	11	12	13	14	15	16	Unknown
Okak Bay	1978 August	50	1979	1												1						
			1980	1										1								
	1979 August	19	1979	1												1						
	1980 August	100	1980	0																		
	TOTAL	169		3										1		2						
Napartok Bay	1978 August	50	1979	1															1			
	1979 August	53	1979	1															1			
	1980 August	123	1980	2															2			
	TOTAL	226		4															4			
Hebron Fiord	1978 August	100	1980	1																		1
	1979 August	50	1980	1																1		
	1980 August	149		0																		
	TOTAL	299		2																1		1
Saglek Fiord	1978 August	100	1979	1																	1	
			1980	1																1		
	1979 August	52		0																		
	1980 August	150		0																		
	TOTAL	302		2																1	1	

* 1 - Anton's, 2 - Voisey Bay, 3 - Anaktalik Bay, 4 - Dog Island, 5 - Nain Bay, 5A - Fraser River, 6 - Tikkoatokak Bay, 7 - Webb Bay, 8 - Black Island, 9 - Kiglapaits, 10 - Tasiuyak, 11 - Okak Bay, 12 - Cutthroat, 13 - Mugford, 14 - Napartok Bay, 15 - Hebron Fiord, 16 - Saglek Fiord

matrices was examined by Box's test utilizing the SPSS Discriminant Analysis program (Klecka 1975).

In order to reduce the effect of different mean fish size on intergroup comparisons, yet still maintain the biological integrity of the data, charr samples selected for morphometric analyses were those captured in 114 mm mesh gillnets and therefore subject to the same degree of gear selectivity. Additionally, these data were stratified by selecting charr of fork length 440-599 mm.

RESULTS

Morphological Data

Results of univariate analyses of meristic variables are shown in Table 3. Interpopulation variation in meristic characters was slight; however, of the 9 variables examined, 6 differed significantly ($P < 0.01$) in their means. Multiple

Table 3. Univariate statistics for meristic and morphometric characters from northern Labrador Arctic charr. Analysis of variance results including Levene's test of heteroscedasticity and multiple comparison tests of meristic characters. Statistical analyses based on transformed ($\sqrt{x + 0.5}$; Sokal and Rohlf 1969) meristic data. Significance levels are $P < 0.05$(*) and $P < 0.01$(**). Underlined areas are not significantly different. DR = dorsal rays; AR = anal rays; PCTR = pectoral rays: PELR = pelvic rays; BR = branchiostegal rays; V = vertebrae; UGR = upper gillrakers; LGR = lower gillrakers; TGR = total gillrakers.

Meristic characters	Voisey Bay (V)				Tikkoatokak Bay (T)				Okak Bay (O)				Hebron Fiord (H)				Levene's test	df	F*		Multiple comparison test
	N	$\bar{x}$	S.D.	Range	N	$\bar{x}$	S.D.	Range	N	$\bar{x}$	S.D.	Range	N	$\bar{x}$	S.D.	Range					
DR	91	14.65	0.69	13-16	90	14.71	0.74	13-16	133	14.50	0.74	12-17	68	14.77	0.95	12-17	NS	3,281	2.06	NS	
AR	91	13.74	0.83	12-16	91	13.46	0.67	12-15	135	13.36	0.82	11-15	68	13.26	0.75	12-15	NS	3,349	0.90	NS	
PCTR	93	13.57	0.68	12-16	90	13.31	0.77	12-15	136	13.66	0.86	11-16	68	13.91	0.88	10-16	NS	3,323	7.46	**	TVOH
PELR	93	9.07	0.41	8-10	91	9.05	0.31	8-10	135	9.10	0.40	8-10	68	9.16	0.37	9-10	NS	3,346	1.27	NS	
BR	128	10.62	0.78	8-12	124	10.53	0.64	9-12	190	10.72	0.63	9-12	91	10.96	0.73	9-13	**	3,439	6.97	**	TVOH
V	122	65.05	0.97	62-68	121	64.93	1.03	62-67	176	64.81	0.93	63-68	89	65.54	1.29	63-72	*	3,370	9.42	**	OTVH
UGR	124	9.47	0.83	7-11	120	9.33	0.77	7-11	187	9.43	0.84	7-13	88	10.06	0.76	8-12	**	3,476	16.21	**	TOVH
LGR	124	14.85	0.88	13-17	120	14.67	1.00	12-17	187	14.47	1.00	12-17	88	15.16	1.23	12-18	**	3,379	9.39	**	OTVH
TGR	125	24.31	1.29	21-28	123	24.01	1.35	21-27	187	23.91	1.44	20-28	89	25.21	1.62	22-30	NS	3,422	18.07	**	OTVH
Morphometric characters																					
TL	119	53.40	3.69	47.0-62.4	99	54.09	4.52	46.0-63.4	98	51.31	3.65	46.5-62.6	25	55.27	4.46	47.1-61.9					
FL	119	50.48	3.52	44.1-58.9	99	51.17	4.35	44.1-59.8	98	48.43	3.51	44.0-59.3	25	52.36	4.48	44.1-59.7					
ST	119	45.97	3.19	40.4-53.4	99	46.55	4.01	39.9-54.6	98	43.85	3.19	34.4-53.9	25	47.21	3.93	39.5-53.3					
GOP	118	25.65	2.28	20.4-31.2	99	25.41	2.81	20.2-32.8	86	24.07	2.15	20.3-29.4	25	25.31	2.37	21.4-29.2					
GMX	119	30.98	3.01	24.4-39.7	99	29.87	3.28	24.8-41.0	86	28.98	2.36	24.4-35.0	25	29.38	2.65	25.3-33.9					
PRDL	119	21.08	1.64	17.6-24.7	99	20.99	1.79	17.7-25.2	86	19.96	1.44	17.4-23.6	25	21.83	2.07	17.9-25.2					
DADL	119	11.09	1.16	8.7-13.4	99	10.99	1.23	8.7-14.4	86	10.87	0.96	8.7-13.4	25	12.09	1.29	9.8-14.3					
HL	119	9.73	0.93	8.0-14.6	99	9.87	0.98	8.0-12.0	87	9.53	0.87	8.0-11.9	25	10.29	1.06	8.5-12.7					
PORBL	119	5.53	0.47	4.5-6.9	99	5.68	0.54	4.7-6.9	85	5.38	0.56	4.3-6.9	25	5.74	0.48	5.0-6.7					
PCTL	119	6.39	0.72	4.9-8.0	99	6.68	0.96	3.7-8.9	86	6.68	0.88	2.7-8.8	25	6.97	0.97	5.1-9.0					
SNL	119	2.70	0.38	1.9-4.1	99	2.88	0.48	2.1-4.2	86	2.67	0.43	1.2-4.0	25	2.91	0.44	2.2-4.0					
ORBL	119	1.38	0.12	1.1-1.8	99	1.40	0.15	1.1-1.8	86	1.42	0.20	1.0-1.9	25	1.49	0.20	1.3-1.9					
UJL	119	5.04	0.65	3.9-7.3	99	5.16	0.73	3.9-7.4	86	4.99	0.67	4.0-7.1	24	5.37	0.86	4.2-7.6					

Table 4. Multivariate analysis of variance of Arctic charr morphometric characters (above diagonal), and percent correctly classified by discriminant analysis (below diagonal). Significance levels are P < 0.05(*) and P < 0.01(**).

	Voisey	Tikkoatokak		Okak		Hebron	
		df	F	df	F	df	F
Voisey		4,212	18.70**	3,199	31.75**	4,137	19.30**
Tikkoatokak	75			5,178	11.14**	4,118	10.93**
Okak	77	73				3,105	18.57**
Hebron	78	78		78			

range tests indicated that Voisey Bay and Tikkoatokak Bay charr did not differ significantly in any of the meristic characters examined. Voisey Bay charr differed from Okak charr in lower gillraker counts, while Tikkoatokak Bay charr differed from Okak charr in pectoral fin ray counts. Arctic charr from Hebron Fiord were clearly separate, differing significantly in at least 4 characters from fish from all other areas (Table 3). Highest meristic counts were observed in 8 of 9 variables from charr collected from the Hebron Fiord, the most northerly area examined in this study.

Univariate statistics for morphometric characters are summarized in Table 3. Morphometric multivariate analyses indicate that there were significant differences ($P < 0.01$) in the character mean vectors between all areas (Table 4).

The stepwise procedure selected characters which were most useful in discriminating between groups. In order of importance, these were: maximum girth, dorsal length to the adipose fin, predorsal length, left pectoral fin length, and snout length. Approximately 77 percent of individuals were correctly classified into their respective areas of origin by the discriminant analysis. Four of six area comparisons, Voisey-Okak, Voisey-Hebron, Tikkoatokak-Okak and Tikkoatokak-Hebron, had heterogeneous variance-covariance matrices.

Biological Characteristics

Significant differences were observed between areas in comparisons of growth at most ages (Table 5). Voisey Bay charr had the fastest growth rate (Fig. 2), averaging 632 mm at age 12, while Tikkoatokak, Hebron and Okak charr averaged 594, 573 and 568 mm respectively at this age. Between ages 7 and 12, Voisey Bay Arctic charr were significantly larger than charr from all other areas.

Table 5. Results of analysis of variance of the length (cm) at age between commercial samples of Arctic charr from four areas in northern Labrador. Significance levels are $P < 0.05$(*) and $P < 0.01$(**).

	Voisey Bay			Tikkoatokak Bay			Okak Bay			Hebron Fiord					
Age	N	Mean	S.D.	N	Mean	S.D.	N	Mean	S.D.	N	Mean	S.D.	df	F*	
6	46	45.6	4.8	36	42.2	4.0	5	44.0	4.8				2, 14	5.6	*
7	283	49.7	4.1	248	46.9	4.1	76	46.8	3.9	9	45.0	4.1	3, 64	26.0	**
8	491	52.7	4.8	524	50.2	4.5	286	49.7	3.7	47	49.3	3.5	3, 598	46.4	**
9	290	56.6	5.2	389	53.7	5.1	265	51.6	4.6	99	51.5	4.5	3, 756	57.8	**
10	134	59.2	5.2	206	56.7	5.3	221	52.9	5.5	75	54.1	5.0	3, 494	46.0	**
11	72	61.1	6.2	90	57.2	5.6	126	54.1	5.6	50	54.7	3.9	3, 291	29.3	**
12	36	63.2	4.6	51	59.4	6.1	104	56.8	6.4	64	57.3	5.9	3, 221	12.6	**
13	27	62.8	5.6	11	62.0	6.1	62	56.9	4.8	38	59.4	5.5	3, 56	8.1	**
14	12	66.3	4.9	8	59.2	3.5	33	56.7	5.7	28	60.7	6.2	3, 64	11.1	**
15	6	64.9	10.9	1	61.6		36	56.8	7.0	26	59.4	5.6	2, 10	2.5	NS
16							15	56.5	6.2	11	59.2	3.8	1, 23	1.9	NS
17							9	56.3	6.0	10	59.6	3.8	1, 13	2.1	NS
18							3	57.7	10.4	3	60.6	8.3	1, 4	0.1	NS

Table 6. Age and length (cm) at sexual maturity of Arctic charr from different areas in northern Labrador.

	Male							Female						
		Age			Length				Age			Length		
	N	Range	Age_{50}	S.E.	Range	Lth_{50}	S.E.	N	Range	Age_{50}	S.E.	Range	Lth_{50}	S.E.
Fraser River	403	1-13	5.2	0.1	4.4-72.8	24.5	0.1	519	1-18	6.9	0.1	4.1-66.5	38.1	0.0
Tikkoatokak Bay	52	6-13	7.2	0.4	40.8-66.2	44.4	0.3	70	6-15	7.9	0.2	39.4-71.0	46.3	0.2
Okak Bay	130	6-18	6.6	0.2	34.3-60.7	39.6	0.2	165	6-18	8.5	0.1	32.9-60.8	42.0	0.1
Napartok Bay	136	5-20	6.3	0.3	15.7-66.4	33.0	0.3	191	6-19	8.5	0.1	30.3-62.5	42.0	0.1
Hebron Fiord	235	4-21	8.0	0.1	21.7-75.5	40.8	0.1	245	4-21	9.6	0.1	22.1-74.8	45.9	0.1
Saglek Fiord	236	3-17	7.3	0.1	16.5-67.0	40.1	0.1	280	3-21	10.0	0.1	18.2-59.7	47.3	0.1
Ramah Bay	108	5-17	7.4	0.2	21.0-71.9	39.3	0.2	110	6-16	10.2	0.1	33.6-58.9	49.9	0.1

Growth of Okak and Hebron charr were similar for ages 7 to 12, after which the Hebron fish achieved a larger size. No significant differences, however, occurred in any length-at-age comparison between these two areas. Okak charr were significantly smaller than charr from Tikkoatokak Bay at ages 9 to 11. Similarly, Hebron charr were significantly smaller than charr from Tikkoatokak at the same ages.

Arctic charr from the Fraser River had mean ages at sexual maturity of 5.2 and 6.9 years for males and females respectively. These charr were obtained from the counting fence and seine nets and, therefore, are not directly comparable with charr from the other areas. Age at sexual maturity of charr in the remaining areas varied from 6.3-8.0 years in males (Table 6). In females, there was a definite trend for a greater mean age at sexual maturity with increase in latitude, ranging from 7.9 years at Tikkoatokak Bay to 10.2 years at Ramah (Table 6). Size at sexual maturity in females also increased similarly, ranging from 420 mm at Okak Bay to 499 mm at Ramah (Table 6).

Distribution

Since 1975, a total of 1,526 tags was applied to charr from the Fraser River, Voisey Bay, Anaktalik Bay, Tikkoatokak Bay and Nain Bay. Recaptures totalling 347 (23 percent) are summarized by area and year up to 1980 in Table 2. Approximately 80 percent have been recovered within 50 km of the release site, or in the case of Fraser River charr, within 50 km of the freshwater marine transition area at the head of Nain Bay.

Recaptures of charr tagged in Nain Bay, Tikkoatokak Bay and at the Fraser River are illustrated in Figure 3. Except for the interchange of charr between Nain Bay (Fraser River) and Tikkoatokak Bay, migrations to outer offshore areas appear more common than interchange of charr between inner bays. The latter accounts for 4.5 percent of the tag recoveries, if Nain and Tikkoatokak Bay are considered as one area and recaptures in offshore areas are excluded. However, offshore movements occurred in 1980 only. It is assumed that late ice conditions prevailing within the bays during the spring of 1980 influenced the abundance of capelin (*Mallotus villosus*), the predominant food item for Arctic charr in the Nain region (Andrews and Lear 1956; Dempson, LeDrew and Nicholls 1979), resulting in a directed offshore feeding migration. Except for one Fraser River charr caught in the Anton's region in 1979, all other distant recaptures occurred during 1980 (Table 2). This offshore movement was similarly observed in the 1980 recaptures of charr from Nain, Taikkoatotak, and Anaktalik bays (Table 2). It should be noted that Nain Bay has been closed to commercial fishing since 1978 and recaptures of tags in that area are made only in the spring domestic fishery or during recreational fishing in the summer.

Recaptures also showed that seaward migrants tagged in Tikkoatokak Bay during the springs of 1979 and 1980, spent an average of 40 days at sea. For seaward migrants tagged in Nain Bay in 1980, the average period was 54 days; the longest time spent at sea was 98 days. Most rapid movement was observed in a charr tagged in Nain Bay in 1979 and recovered 5 days later at the Fraser River counting fence, a distance of 83 km, representing a minimum speed of 16.4 km per day.

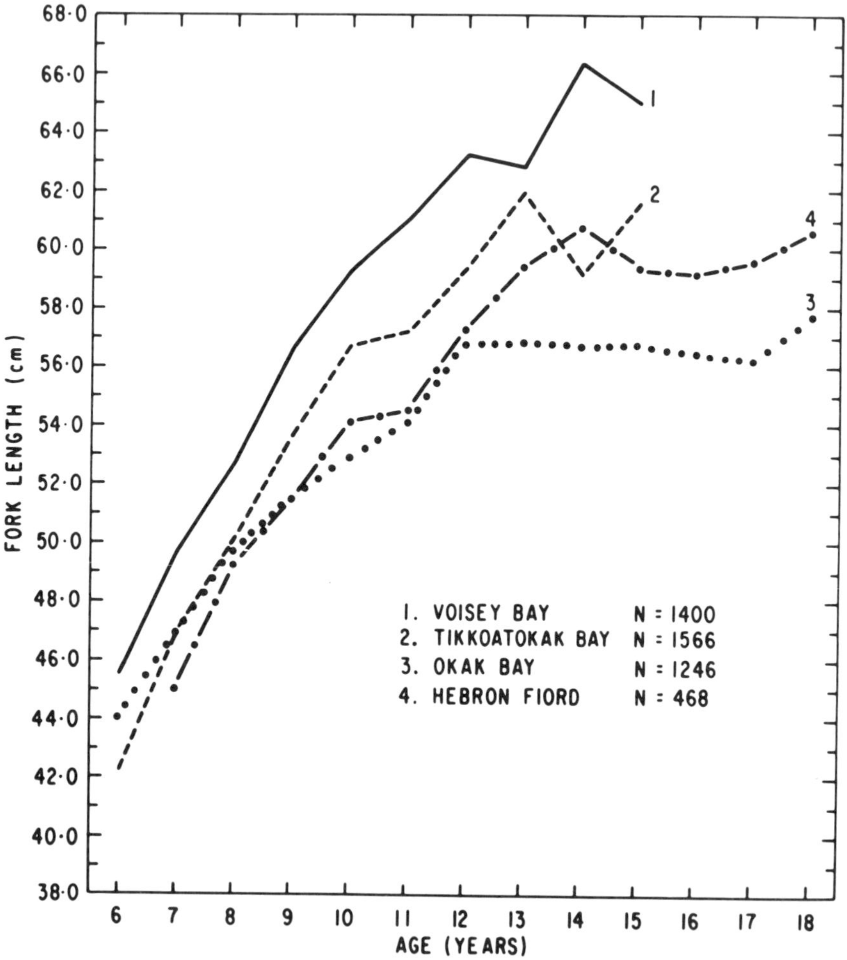

FIG. 2 Comparison of mean length at age of four Arctic charr populations from northern Labrador.

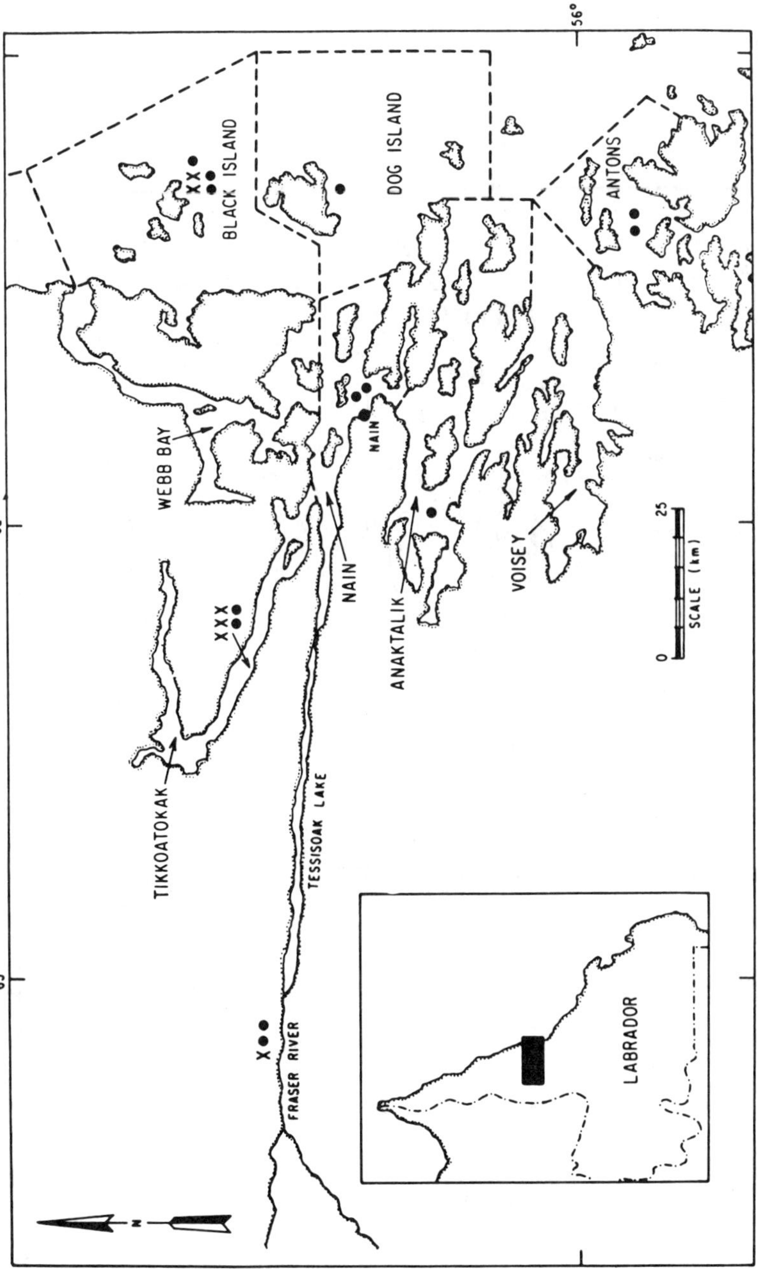

FIG. 3 Recaptures of charr tagged in Nain Bay during 1979 and 1980. (X) = five recaptures, (●) = one recapture.

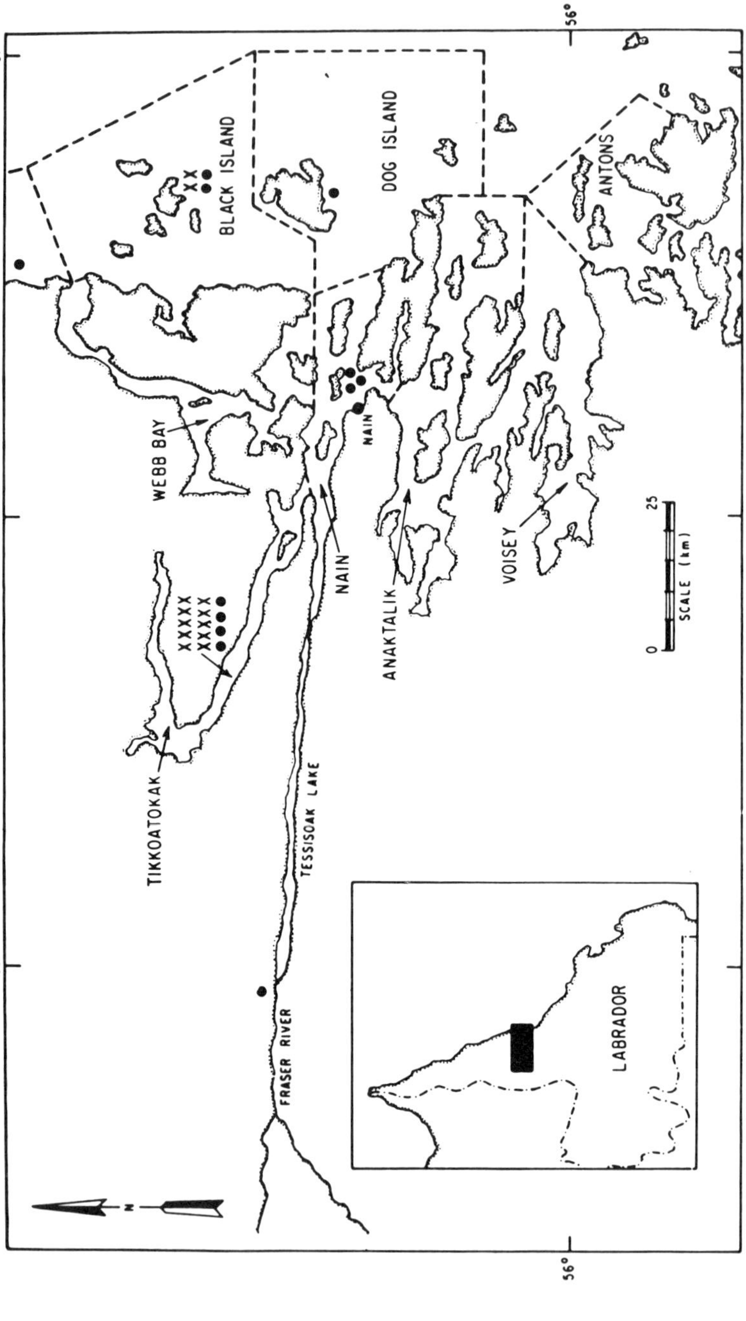

FIG. 4 Recaptures of charr tagged in Tikkoatokak Bay during 1979 and 1980. (X) = five recaptures, (●) = one recapture.

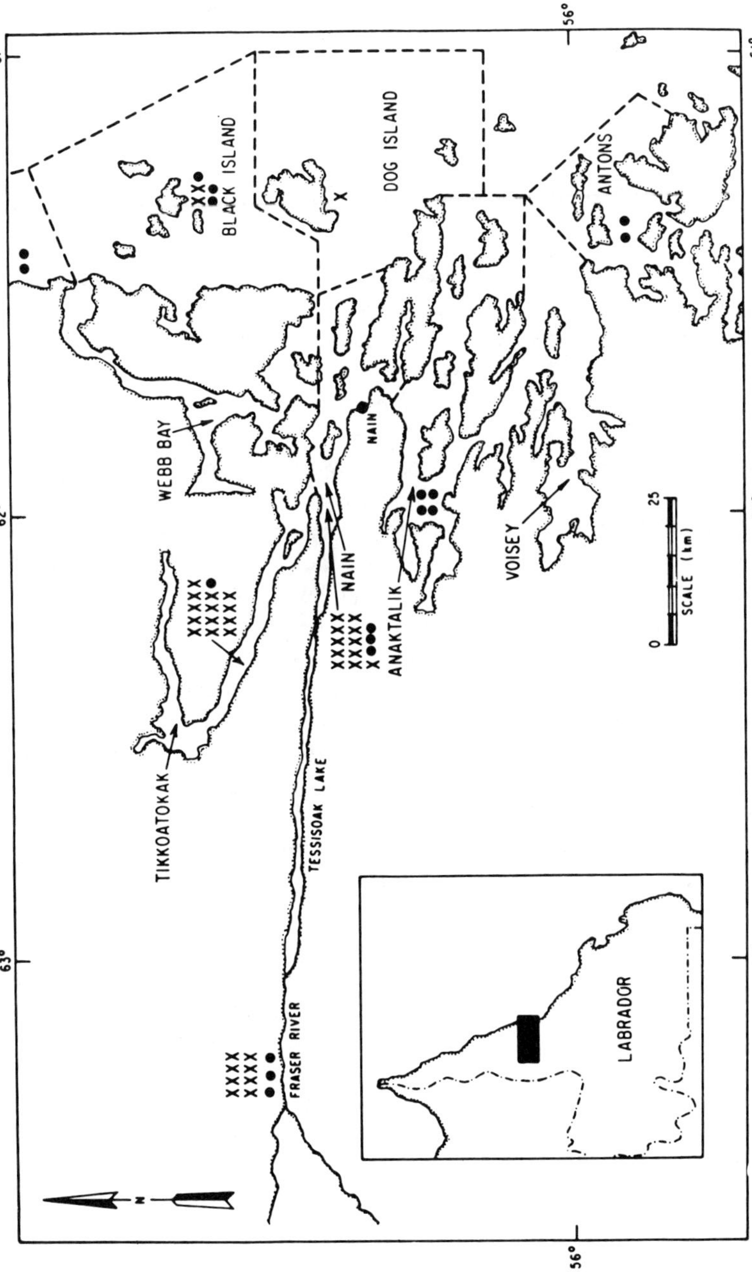

FIG. 5 Recaptures of Arctic charr tagged at the Fraser River from 1975 to 1979. (X) = five recaptures, (●) = one recapture.

Recaptures of Arctic charr tagged in the northern areas of Okak-Saglek (Table 2) were too few for any conclusions to be drawn.

DISCUSSION

Analyses of morphometric, meristic and biological characteristics and recaptures of tagged fish indicate that there are a number of discrete stocks of charr along the coast of Labrador.

Numerous studies have examined the variation in meristic characteristics of different charrs in order to clarify the systematic relationships within the genus *Salvelinus*. Literature pertaining to the use of morphological and biological data for the purpose of stock discrimination required for proper management is generally lacking, although Dick and Belosevic (1981) have recently used parasites and morphological characteristics to separate anadromous and non-migratory components in the commercial Arctic charr fishery from the Nettilling Lake-Koukdjuak River system on Baffin Island. Analysis of meristic characters, which are known to be influenced by genetic and environmental parameters (Tåning 1952; Barlow 1961; Garside 1966), indicated that there were significant differences between charr from separate areas along the northern Labrador coast, with the exception of those from Voisey Bay and Tikkoatokak Bay. Andrews and Lear (1956) found no significant differences in meristic characters between Nain and Okak charr, but concluded that minimal overlap occurs in extreme areas.

Multivariate analysis of morphometric data revealed significant differences ($P < 0.01$, Table 4) between all areas, providing further evidence for the existence of separate charr stocks along the coast. Analyses were performed on raw untransformed data since several studies have indicated that corrections to remove effects of size and particularly size-related shape often produce meaningless results (Marr 1955; Atchley et al. 1976; Oxnard 1978). Morphometric data, however, were also analyzed after adjusting for size by, a) logarithmic transformation of data, and b) using body proportions. Results were virtually identical with analyses of untransformed data. The fact that covariance matrices differed significantly between several groups is acceptable if Klecka (1975), Lachenbruch (1975) and Pimentel (1979) are correct in stating that the discriminant analysis technique is robust enough to disregard this assumption as long as departures are not 'too' severe.

Differences in growth rate, age at sexual maturity, and distribution pattern corroborate the results of morphological analyses and support the hypothesis that discrete charr stocks occur along this limited section of the Labrador coast. The slowest rate of growth was found in Okak Bay charr, as previously observed by Andrews and Lear (1956), and there was a general trend for the growth rate to be inversely related to latitude (Coady and Best 1976; Dempson, LeDrew and

Nicholls 1979). Although charr from Voisey Bay and Tikkoatokak Bay are more closely related geographically than charr from other areas examined, growth differences and distribution patterns clearly illustrate the discreteness of these two populations.

Grainger (1953) reported that charr from Sylvia Grinnell River, Frobisher Bay, have a tendency to remain in close proximity to their home river. Moore (1975), on the basis of length-frequency data, suggested that there was little intermingling of charr stocks from different rivers in the Cumberland Sound area. Glova and McCart (1978) stated that Arctic charr from Nachvak Fiord, Labrador, were most abundant within 25-30 km of the Nachvak and Palmer rivers, while few charr were captured at the fiord mouth. However, tagging studies by Jensen and Berg (1977) and Johnson (1980) show that Arctic charr marine migrations may be rather extensive. Results presented in this paper suggest that minimal interchange occurs between inner bays, but environmental conditions may occasionally influence distribution patterns of food organisms, resulting in a directed movement of charr offshore to feeding areas. Considering the intensity and distribution of the commercial Arctic charr (Dempson 1978) and Atlantic salmon fishery along the Labrador coast north of Hamilton inlet it is likely that, if charr were undertaking extensive migrations, then some recaptures from distant areas would have occurred.

Charr from Fraser River (Nain Bay) and Tikkoatokak Bay are defined as one stock, distinctly different from charr from Voisey Bay, Okak Bay and Hebron Fiord which should also be treated as discrete populations. Growth comparisons between charr from Voisey Bay and adjacent Anaktalik Bay yielded no significant differences (Dempson, unpublished data); however, tag recaptures during the next few years will help to clarify the relationship of charr from these two areas, as well as indicating the relative contributions of charr from the inner bays to the offshore regions.

In conclusion, the morphological and biological characteristics examined in this paper suggest that discrete populations of anadromous Arctic charr do exist along this limited section of the northern Labrador coast. Consequently, in order to ensure the proper management and exploitation of this valuable resource, these discrete populations should be monitored and managed on an individual basis.

ACKNOWLEDGEMENTS

I thank Michael Chadwick, Henry Lear and Rex Porter for their comments and suggestions concerning the manuscript. Dr. Raj Misra and Don Stansbury kindly provided statistical and computing advice. Appreciation is also extended to Herb Mullett for preparing the figures and to Karen Harding and Janice Lannon for typing the report.

REFERENCES

Andrews, C.W. and E. Lear. 1956. The biology of Arctic char (*Salvelinus alpinus* L.) in northern Labrador. J. Fish. Res. Board Can. 13:843-860.

Atchley, W.R., C.T. Gaskins and D. Anderson. Statistical properties of ratios. I. Empirical results. Syst. Zool. 25:137-148.

Barlow, G.W. 1961. Causes and significance of morphological variation in fishes. Syst. Zool. 25:105-117.

Behnke, R.J. 1972. The systematics of salmonid fishes of recently glaciated lakes. J. Fish. Res. Board Can. 29:639-671.

Behnke, R.J. 1980. A systematic review of the genus *Salvelinus*, p. 441-481. *In* E.K. Balon [ed.] Charrs: salmonid fishes of the genus *Salvelinus*. Dr. W. Junk Publishers, The Hague.

Bliss, C.I. 1952. The statistics of bioassay with special reference to the vitamins. Academic Press, New York. 184 p.

Brown, M.B. and A.B. Forsythe. 1974a. The small sample behaviour of some statistics which test the equality of several means. Technometrics 16(1):129-132.

Brown, M.B. and A.B. Forsythe. 1974b. The ANOVA and multiple comparisons for data with heterogeneous variances. Biometrics 30:719-724.

Carscadden, J.E. and R.K. Misra. 1980. Multivariate analysis of meristic characters of capelin (*Mallotus villosus*) in the Northwest Atlantic. Can. J. Fish. Aquat. Sci. 37:725-729.

Cavender, T.M. 1980. Systematics of *Salvelinus* from the north Pacific Basin, p. 295-322. *In* E.K. Balon (ed.) Charrs: salmonid fishes of the genus *Salvelinus*. Dr. W. Junk Publishers, The Hague.

Coady, L. and C.W. Best. 1976. Biological and management investigations of the Arctic char fishery at Nain, Labrador. Canada, Fish. Mar. Serv. Tech. Rep. 624:103 p.

Cross, T.F. and R.H. Payne. 1978. Geographic variation in Atlantic cod, *Gadus morhua*, off eastern North America: a biochemical systematics approach. J. Fish. Res. Board Can. 35:117-123.

Dempson, J.B. 1978. Biological assessment of Arctic char (*Salvelinus alpinus* L.) stocks and summary of the Atlantic salmon (*Salmo salar* L.) fishery in northern Labrador. Canada, Fish. Mar. Serv. Tech. Rep. 817:54 p.

Dempson, J.B., L. LeDrew and T. Nicholls. 1979. Review of biological information on Arctic char (*Salvelinus alpinus*) in Labrador and present status of the commercial fishery, p. 60-74. *In* Symposium on research in the Labrador coastal and offshore region. Memorial University of Newfoundland Printing Services.

Dick, T.A. and M. Belosevic. 1981. Parasites of Arctic charr *Salvelinus alpinus* (Linnaeus) and their use in separating sea-run and non-migrating charr. J. Fish Biol. 18:339-347.

Dixon, W.J. and M.B. Brown. 1979. Biomedical computer programs P-series. University of California Press, Berkeley. 880 p.

Fleming, A.M. 1960. Age, growth and sexual maturity of cod (*Gadus morhua* L.) in the Newfoundland area, 1947-1950. J. Fish Res. Board Can. 17(6):775-809.

Garside, E.T. 1966. Developmental rate and vertebral number in salmonids. J. Fish. Res. Board Can. 23(10):1537-1551.

Glova, G.J. and P.J. McCart. 1978. Studies of Arctic char populations in Nachvak Fiord, Labrador. Report to Esso Resources Canada Limited, Aquitaine Co. of Canada Ltd., and Canada-Cities Service Ltd., Calgary. 116 p.

Grainger, E.H. 1953. On the age, growth, migration, reproductive potential and feeding habits of the Arctic char (*Salvelinus alpinus*) of Frobisher Bay, Baffin Island. J. Fish. Res. Board Can. 10:326-370.

Henricson, J. and L. Nyman. 1976. The ecological and genetical segregation of two sympatric species of dwarfed char (*Salvelinus alpinus* (L.) species complex). Inst. Freshw. Res. Drottningholm Rep. 55:15-37.

Hubbs, C. and K.F. Lagler. 1970. Fishes of the Great Lakes region. University of Michigan Press, Ann Arbor, Mich. 213 p.

Jennrich, R. and P. Sampson. 1979. Stepwise discriminant analysis, p. 711-733. *In* W.J. Dixon and M.B. Brown [ed.] Biomedical computer programs P-series. University of California Press, Berkeley.

Jensen, K.W. and M. Berg. 1977. Growth, mortality and migrations of the anadromous char, *Salvelinus alpinus* L., in the Vardnes River, Troms, Northern Norway. Inst. Freshw. Res. Drottningholm Rep. 56:70-80.

Johnson, L. 1980. The Arctic charr, *Salvelinus alpinus*, p. 15-98. *In* E.K. Balon [ed.] Charrs: salmonid fishes of the genus *Salvelinus*. Dr. W. Junk Publishers, The Hague.

Klecka, W.R. 1975. Discriminant analysis, p. 434-467. *In* N.H. Nie et al. [eds.] Statistical package for the Social Sciences. McGraw-Hill, New York.

Kornfield, I., K.F. Beland, J.R. Moring and F.W. Kircheis. 1981. Genetic similarity among endemic Arctic char (*Salvelinus alpinus*) and implications for their management. Can. J. Fish. Aquat. Sci. 38:32-39.

Lachenbruch, P.A. 1975. Discriminant analysis. Hafner Press, London. 128 p.

Lachenbruch, P.A. and M.R. Mickey. 1968. Estimation of error rates in discriminant analysis. Technometrics 10:1-11.

Lear, W.H. and R.K. Misra. 1978. Clinal variation in scale characters of Atlantic salmon (*Salmo salar*) based on discriminant function analysis. J.Fish. Res. Board Can. 35:43-47.

Makhovenko, E.T. 1972. The parasite fauna of different forms of the Arctic char, *Salvelinus alpinus* L., from Kamchatka. Parazitologiya 6(4):369-375. Can. Trans. Fish. Aquat. Sci. No. 4650, 1980.

Marr, J.C. 1955. The use of morphometric data in systematic, racial and relative growth studies. Copeia 1:23-31.

McCart, P.J. 1980. A review of the systematics and ecology of Arctic char, *Salvelinus alpinus*, in the western Arctic. Can. Tech. Rep. Fish. Aquat. Sci. 935:89 p.

McCart, P.J. and P. Craig. 1971. Meristic differences between anadromous and freshwater-resident Arctic char (*Salvelinus alpinus*) in the Sagavanirktok River drainage, Alaska. J. Fish. Res. Board Can. 28:115-118.

McPhail, J.D. 1961. A systematic study of the *Salvelinus alpinus* complex in North America. J. Fish. Res. Board Can. 18(5):793-816.

Moore, J.W. 1975. Distribution, movements and mortality of anadromous Arctic char, *Salvelinus alpinus* L., in the Cumberland Sound area of Baffin Island. J. Fish Biol. 7:339-348.

Morrison, D.F. 1967. Multivariate statistical methods. McGraw-Hill, New York. 338 p.

Morrow, J.E. 1980. Analysis of the Dolly Varden charr, *Salvelinus malma*, of northwestern North America and northeastern Siberia, p. 323-338. *In* E.K. Balon [ed.] Charrs: salmonid fishes of the genus *Salvelinus*. Dr. W. Junk Publishers, The Hague.

Nielsen, J. 1961. Preliminary results of tagging experiments with char, *Salvelinus alpinus* (L.), in Greenland, p. 24-48. *In* Contributions to the biology of the Salmonidae in Greenland. Medd. Gronl. 159(8).

Nyman, L. 1972. A new approach to the taxonomy of the "*Salvelinus alpinus* species complex". Inst. Freshw. Res. Drottningholm Rep. 52:103-131.

Oxnard, C.E. 1978. One biologist's view of morphometrics. Ann. Rev. Ecol. Syst. 9:219-241.

Pimental, R.A. 1979. Morphometrics: The multivariate analysis of biological data. Kendall/Hunt Publishing Co., Iowa. 276 p.

Pippy, J.H. 1969. *Pomphorhynchus laevis* (Zoega) Muller, 1776 (Acanthocephala) in Atlantic salmon (*Salmo salar*) and its use as a biological tag. J. Fish. Res. Board Can. 26:909-919.

Qadri, S.U. 1974. Taxonomic status of the *Salvelinus alpinus* complex. J. Fish. Res. Board Can. 31:1355-1361.

Rao, C.R. 1952. Advanced statistical methods in biometric research. John Wiley and Sons, New York. 390 p.

Saunders, L.H. and J.A. McKenzie. 1971. Comparative electrophoresis of Arctic char. Comp. Biochem. Physiol. 38B:487-492.

Sharp, J.C., K.W. Able, W.C. Leggett and J.E. Carscadden. 1978. Utility of meristic and morphometric characters for identification of capelin (*Mallotus villosus*) stocks in Canadian Atlantic waters. J. Fish. Res. Board Can. 35:124-130.

Sokal, R.R. and F.J. Rohlf. 1969. Biometry. W.H. Freeman and Co., San Francisco. 776 p.

Tåning, A.V. 1952. Experimental study of meristic characters in fishes. Biological Reviews 27:169-193.

Templeman, W. 1974. Migrations and intermingling of Atlantic cod (*Gadus morhua*) stocks of the Newfoundland area. J. Fish. Res. Board Can. 31:1073-1092.

Vladykov, V.D. 1954. Taxonomic characters of the eastern North American chars (*Salvelinus* and *Cristivomer*). J. Fish. Res. Board Can. 11:904-932.

The Interactive Segregation of Landlocked Arctic Charr (*Salvelinus alpinus*) from Lake Charr (*S. namaycush* and Brook Charr (*S. fontinalis*) in Two Lakes of Subarctic Quebec, Canada

NOEL C. FRASER[1] and GEOFFREY POWER

Department of Biology, University of Waterloo, Waterloo, Ontario, Canada

FRASER, N.C. AND G. POWER. 1984. The interactive segregation of landlocked Arctic charr (*Salvelinus alpinus*) from lake charr (*S. namaycush*) and brook charr (*S. fontinalis*) in two lakes of subarctic Quebec, Canada, p. 163–181. *In* L. Johnson and B.L. Burns [eds.] Biology of the Arctic charr, Proceedings of the International Symposium on Arctic Charr, Winnipeg, Manitoba, May 1981. Univ. Manitoba Press, Winnipeg.

The diets and population structures of two lake-resident Arctic charr populations (*Salvelinus alpinus*) in subarctic Quebec were compared. One population co-existed with the threespine stickleback (*Gasterosteus aculeatus*), the other with two congeneric species, lake charr (*S. namaycush*) and brook charr (*S. fontinalis*) as well as threespine stickleback.

A diet analysis showed that the allopatric charr fed on a wide variety of food items including sticklebacks and small charr, aquatic insect larvae and pupae, gastropods, gammarids and zooplankton. The sympatric charr, however, were not piscivorous and consumed a much higher proportion of gastropods, gammarids and zooplankton. The lake charr were mainly piscivorous and the brook charr insectivorous; zooplankton were negligible in both their diets. The results of a multivariate analysis of variance indicated that the diets of all four charr populations were significantly different.

The allopatric population of *S. alpinus* grew at a significantly higher rate than the sympatric population but had a lower age-specific length. The former had a much higher frequency of smaller fish and a higher catch per unit effort; fish first matured at a younger age (4+ males and 5+ females vs. 8+ males and females) and at a smaller size (15-20 cm vs. 30-40 cm), and carried a significantly higher internal parasite load (mainly *Diphyllobothrium ditremum*, *D. dendritecum* and *Eubothrium salvelini*).

Based on the diet analysis we concluded that Arctic charr interactively segregate from sympatric lake charr and brook charr in the study lake. The potential effects of the segregated diet on the growth and reproduction of the Arctic charr may be modified by the influences of different predation pressures and parasite infections.

---

1 Present address: Ontario Ministry of Natural Resources, Fisheries Branch, Queen's Park, Toronto, Ontario, Canada M7A 1W3.

INTRODUCTION

"Interactive segregation", a term adapted to fisheries work by Nilsson (1967), refers to a mechanism by which ecologically similar species with overlapping ranges segregate as a result of the magnification of small differences, usually related to behaviour. This mechanism may be effected directly, through interference (predation, food fighting, territoriality) or indirectly, through more efficient resource and/or habitat exploitation. Because direct evidence is rarely recorded, evidence for interactive segregation is largely based on differences in diet and distribution between allopatric and sympatric populations of fish species (Nilsson 1967, Svärdson 1976).

In North America, lake charr, *Salvelinus namaycush*, is of outstanding importance for its interactive association with Arctic charr, *S. alpinus*. Evidence for interactive segregation of the two species, however, relies solely on differences in their local distributions. In larger lakes, lake charr either eliminate Arctic charr or relegate it to a subordinate position, the two rarely co-occurring in the absence of other fish species. Arctic charr occurs alone only in smaller lakes, where it may also co-exist with the ninespine stickleback (*Pungitius pungitius*) (Johnson 1980).

Evidence for interactive segregation of Arctic charr and another congeneric, brook charr, *S. fontinalis*, is also based mainly on differences in their respective distributions. Throughout eastern Canada and northern New England, brook charr virtually replaced Arctic charr after the last glaciation (Behnke 1980), leaving only relict populations of the latter (Saunders and Power 1969). In the northern extreme of the brook charr's distribution, however, both charr species frequently co-exist in inland lakes (Power 1980).

The purpose of this paper is to document the interactive segregation of *S. alpinus* from co-existing *S. namaycush* and *S. fontinalis* in a subarctic lake, by comparing the diet of the sympatric Arctic charr with that of an allopatric Arctic charr population in a smaller lake in the region. The potential effects of the interaction on growth, length-weight relationships, fecundity and other characteristics of the Arctic charr were also investigated.

STUDY SITE

Lac Ducreux (57°47′N 69°34′W) and Lac Noname (58°15′N 69°08′W) (Fig. 1) are located on Laurentian Shield granite-gneiss rock, in a transition zone of taiga to tundra vegetation. Lac Ducreux (213 m asl.; 378 ha.; mean depth 9.8 m) contains only Arctic charr and threespine stickleback (*Gasterosteus aculeatus*), whereas Lac Noname (183 m asl.; 407 ha.; mean depth 2.0 m) has Arctic charr, threespine stickleback, lake charr, brook charr and a few burbot (*Lota lota*). Both lakes are oligotrophic, slightly acidic and low in conductivity (L. Ducreux: pH = 6.8, cond. = 36.0 umhos·cm^{-1}; L. Noname: pH = 6.3, cond. = 26.2 umhos·cm^{-1}). Due to the isolation and pristine nature of the lakes, it was possible to carry out the study free from the influence of commercial or sport fisheries, fish stocking, water level manipulations or pollution.

METHODS

The two lakes were sampled in six consecutive alternating sessions over eleven weeks during the summer of 1979. Fish were caught in nylon gillnets (stretch mesh sizes 38, 51, 64, 76, 102 and 114 mm) set from shore both on the surface and on the bottom. The fork length, fresh whole weight, sex, stage of maturity and a qualitative evaluation of the extent of internal parasitization were recorded. A subsample of mature female gonads were preserved in 4 percent formaldehyde; the egg diameters were measured and the number of eggs counted. Otoliths were stored dry, immersed in glycerin and aged under reflected light. Criteria for the identification of otolith annuli were similar to those used by Nordeng (1961).

The stomach contents were preserved and pooled according to sampling session and fish length (Borgeson 1963). The contents of the grouped stomachs were sorted into specific food types and their volumes determined by water displacement. The percentage of each food type in the total volume of the combined stomach contents in each length interval was calculated. A version of multivariate analysis of variance (SAS 1979) was employed to discern differences among the diets of the *Salvelinus* species within Lac Noname, and between the Arctic charr diets of the two lakes.

RESULTS

Diet

The widespread piscivory of the Lac Ducreux Arctic charr is evident from Table 1. Trichoptera larvae and pupae, gastropods and zooplankton were important food items at more specific times for fish of certain length intervals. The large charr (> 40 cm) showed a consistent near-exclusive preference for fish in both sampling periods, whereas the smaller fish changed from predominantly bottom feeding to more pelagic feeding (fish and zooplankton) in the later period. Larger charr were cannibalistic; smaller charr ate proportionately more sticklebacks. Dominant zooplankton in the diet were *Holopedium gibberum* and *Bosmina longirostris*.

In contrast, the Lac Noname Arctic charr were essentially non-piscivorous. They ingested mainly gastropods, *Gammarus lacustris*, zooplankton and all life

Table 1. Diet of allopatric *S. alpinus* in Lac Ducreux and *S. alpinus* sympatric with *S. namaycush* and *S. fontinalis* in Lac Noname, northern Quebec (% volume measurement)

SAMPLING PERIOD	LENGTH INTERVAL (cm)	STOMACHS WITH FOOD	STOMACHS EMPTY	*Gasterosteus aculeatus*	*Salvelinus* sp.	Zooplankton	*Valvata* sp.	*Lymnaea* sp.	*Gammarus lacustris*	Trichoptera larvae & pupae	Insect pupae & larvae	Trichoptera adults	Insect adults	Others	Debris
L. Ducreux															
1-5 Aug 79	15-20	86	5	0	0	6	1	0	2	24	12	1	7	9	40
	21-30	98	18	3	1	1	2	2	1	59	4	1	0	12[a]	14
	31-40	34	14	9	1	0	19	18	0	37	0	0	0	3	13
	41-50	24	26	3	97	0	0	0	0	0	0	0	0	0	0
	>50	5	3	0	84	0	0	0	0	0	0	0	0	14[b]	2
23-28 Aug 79	15-20	69	20	5	0	47	0	0	0	1	0	1	24	0	22
	21-30	67	58	39	10	13	0	0	3	17	0	1	4	1	12
	31-40	28	24	32	36	0	20	8	1	1	0	0	0	2	0
	41-50	31	28	10	82	0	3	4	0	0	0	0	0	0	0
	>50	7	5	5	95	0	0	0	0	0	0	0	0	0	0
L. Noname															
16-24 Jul 79	15-20	0	0	-	-	-	-	-	-	-	-	-	-	-	-
	21-30	7	0	1	0	4	9	0	0	0	60	0	18	0	8
	31-40	17	2	0	0	0	4	31	9	2	27	0	6	0	21
	41-50	21	19	0	0	0	64	14	7	4	5	0	0	0	5
	>50	4	5	0	0	0	81	17	0	0	0	1	0	0	1
9-17 Aug 79	15-20	8	0	0	0	0	0	0	0	5	3	1	52	0	40
	21-30	10	2	0	0	63	0	0	0	7	1	4	15	1	8
	31-40	10	5	0	0	8	1	33	52	0	0	0	2	2	3
	41-50	35	83	1	0	0	23	41	26	1	0	0	0	2	5
	>50	3	26	0	0	0	0	96	0	1	0	0	0	0	3
3-10 Sep 79	15-20	2	1	0	0	100	0	0	0	0	0	0	0	0	0
	21-30	2	0	0	0	100	0	0	0	0	0	0	0	0	0
	31-40	9	4	0	0	15	1	0	52	0	1	0	0	22[c]	10
	41-50	49	31	0	0	0	1	51	20	0	0	0	0	19[c]	7
	>50	9	8	0	0	0	1	66	19	1	0	0	0	7[c]	5

[a]8% *Nostoc* [b]13% rodent [c]*Salvelinus* *spp.* eggs

stages of insects other than trichoptera. The larger charr (> 40 cm) ate predominately gastropods but consumed proportionately more *G. lacustris* and fish eggs later in the season. Fish less than 30 cm in length shifted from a diet of insects (other than trichoptera) in mid-July to zooplankton by September, whereas charr 30-40 cm in length shifted from insect larvae and pupae (other than trichoptera), and gastropods in early summer to gammarids, fish eggs, and some zooplankton in autumn. *H. gibberum* and *Daphnia* sp. were the dominant zooplankton food items.

In mid-July, the diet of *S. namaycush* consisted of trichoptera and other insects of all life stages, sticklebacks and other charrs. By September, however, they were nearly exclusively piscivorous (Table 2). *S. fontinalis* were consistently insectivorous, consuming all life stages of trichoptera and other insects in

Table 2. Diet of *S. namaycush* and *S. fontinalis* sympatric with *S. alpinus* in Lac Noname, northern Quebec. (% volume measurement)

SAMPLING PERIOD	LENGTH INTERVAL (cm)	STOMACHS WITH FOOD	STOMACHS EMPTY	*Gasterosteus aculeatus*	*Salvelinus* sp.	Zooplankton	*Valvata* sp.	*Lymnaea* sp.	*Gammarus lacustris*	Trichoptera larvae & pupae	Insect pupae & larvae	Trichoptera adults	Insect adults	Others	Debris
S. namaycush															
16-24 Jul 79	20-30	0	0	-	-	-	-	-	-	-	-	-	-	-	-
	31-40	5	0	30	17	0	0	0	0	8	17	7	1	0	19
	41-50	22	2	15	0	0	0	0	10	7	15	15	0	7[a]	30
	>50	24	7	0	97	0	0	0	0	0	0	0	0	2	1
9-17 Aug 79	20-30	2	0	87	0	0	1	0	0	8	0	0	0	0	4
	31-40	5	0	87	0	0	0	0	3	3	0	0	0	1	6
	41-50	21	10	57	28	0	0	0	1	2	0	0	2	6[c]	2
	>50	20	12	1	93[b]	0	0	0	0	0	0	0	1	3[c]	0
3-10 Sep 79	20-30	1	0	0	96	0	0	0	0	4	0	0	0	0	0
	31-40	4	2	97	2	0	0	0	1	0	0	0	0	0	0
	41-50	21	10	87	7	0	0	1	3	1	0	0	1	1	0
	>50	16	10	0	100[d]	0	0	0	0	0	0	0	0	0	0
S. fontinalis															
16-24 Jul 79	15-20	1	0	0	0	0	0	0	0	27	27	0	0	0	46
	21-30	16	0	0	0	0	0	8	12	16	41	0	10	0	13
	31-40	5	0	0	0	0	0	2	0	21	58	0	6	1	11
9-17 Aug 79	15-20	4	1	1	0	0	0	3	0	23	25	0	25	1	22
	21-30	15	3	0	0	0	3	66	0	12	5	1	5	0	9
	31-40	0	0	-	-	-	-	-	-	-	-	-	-	-	-
3-10 Sep 79	15-20	4	2	0	0	0	3	0	0	7	85	0	0	4	1
	21-30	39	10	1	0	0	1	38	3	27	3	2	3	2	20
	31-40	15	8	1	0	0	0	36	0	9	40	2	4	9[e]	0

[a]rocks, vegetation [b]51% *Lota lota* [c]Rodentia [d]42% *Lota lota* [e]epiphytic algae

all study periods. *Lymnaea* sp. became more important in the diet of the larger brook charr in the latter two sampling sessions (Table 2).

Clear differences in diet existed between the allopatric and sympatric Arctic charr populations. The allopatric charr over 40 cm in length were almost exclusively piscivorous whereas similar-sized charr from the sympatric charr population consumed only gastropods, *Gammarus lacustris*, and fish eggs. For smaller fish, common food items were restricted to gastropods (for fish 30-40 cm), or zooplankton and insects (for fish 15-30 cm).

A multivariate analysis of variance was used to compare the diets of the Arctic charr populations in the two lakes. The analysis indicated that there was a significant difference between the diets of these two charr populations (Wilk's criterion F-tests; $p=0.0500$). This analysis, however, did not account for the covariance of food consumption with predator length. When length was accounted for, the differences became more pronounced.

In Lac Ducreux the consumption of fish was positively related to the length of the predatory Arctic charr ($p = 0.0001$) whereas the consumption of trichoptera ($p = 0.0006$) and other insects ($p = 0.0012$) was negatively related. There was no significant relationship with length for consumption of zooplankton ($p = 0.0536$), gammarids ($p = 0.5058$) or gastropods ($p = 0.7296$). For the sympatric Lac Noname Arctic charr the consumption of gastropods ($p = 0.0001$) and gammarids ($p = 0.0429$) was positively related to length whereas the consumption of insects (other than trichoptera; $p = 0.0156$) and zooplankton ($p = 0.0025$) was negatively related to length. There was no significant relationship with length for consumption of fish ($p = 0.9636$) and trichoptera ($p = 0.9438$).

A multivariate analysis of variance was similarly used to compare the diets of the three *Salvelinus* populations in Lac Noname. It indicated that there were highly significant differences among the three diets ($p = 0.0002$). With the exception that trichoptera consumption was negatively related to the length of lake charr ($p = 0.0409$), the consumption of various food types was not significantly related to the length of either brook charr or lake charr.

There was negligible overlap between the diets of Arctic charr and lake charr in Lac Noname, but there were some similarities between those of brook charr and smaller Arctic charr in the earlier period (i.e. insect larvae, pupae and adults). By September, however, the overlap became minimal.

Growth, Length and Age Frequencies, Catch/Unit Effort

The allopatric Arctic charr population had a significantly higher growth rate than that of the sympatric Arctic charr (t-test; $p = 0.0001$):

Lac Ducreux
$$\log FL = \underset{(\pm 0.0177\ \text{S.E.})}{1.7236} + \underset{(\pm 0.0193\ \text{S.E.})}{0.7823} \log A \qquad (1)$$
$$n = 901 \qquad r^2 = 0.65 \qquad p < 0.0001$$

Lac Noname
$$\log FL = \underset{(\pm 0.0139\ \text{S.E.})}{2.1115} + \underset{(\pm 0.0144\ \text{S.E.})}{0.5379} \log A \qquad (2)$$
$$n = 340 \qquad r^2 = 0.80 \qquad p < 0.0001$$
$$FL = \text{fork length (mm)} \qquad A = \text{age (years)}$$

There was no significant difference in growth between sexes (L. Ducreux $0.50 > p > 0.25$; L. Noname $0.10 > p > 0.05$); therefore the data were pooled.

Although the Lac Ducreux charr were growing faster, their mean length was smaller than that of the Lac Noname Arctic charr (Table 3). The sympatric Arctic charr exhibited a typical smooth growth curve in contrast to the allopatric charr which displayed an irregular pattern of growth (Fig. 2).

The length frequency distributions were fundamentally different (Fig. 3). The allopatric distribution was bimodal (major mode: 15-20 cm; minor mode: 40-45

Table 3. Age-specific mean lengths (mm ± st. dev.) of allopatric *S. alpinus* taken from Lac Ducreux and *S. alpinus* sympatric with *S. namaycush* and *S. fontinalis* taken in Lac Noname, northern Quebec.

	LAC DUCREUX			LAC NONAME		
AGE (years)	N	Mean Fork Length (mm)	S.D.	N	Mean Fork Length (mm)	S.D.
2	2	121.5	16.3	3	191.7	2.5
3	1	195.0	--	8	206.6	19.5
4	35	191.8	31.3	17	253.1	44.3
5	101	195.1	10.7	15	310.2	27.4
6	144	206.4	28.4	10	350.6	36.9
7	145	221.6	49.4	17	383.7	37.9
8	92	271.5	81.5	28	416.3	43.8
9	91	322.5	91.2	64	440.9	35.6
10	68	341.8	77.2	47	469.0	30.9
11	55	336.5	70.8	57	471.0	34.2
12	46	402.6	96.0	31	480.2	37.3
13	37	409.1	73.8	23	478.4	23.7
14	28	446.2	76.7	11	491.2	33.8
15	28	465.3	69.9	7	478.4	20.7
16	11	435.8	34.5	0	--	--
17	9	480.0	60.1	1	536.0	--
18	5	499.8	107.1	1	505.0	--
19	3	520.0	82.3	0	--	--

cm) whereas that of the sympatric Arctic charr showed only one mode (45-50 cm).

Both age frequency distributions, however, were unimodal (Lac Ducreux Arctic charr: 6-7 years; Lac Noname Arctic charr: 9-11 years, Fig. 4).

The catch per unit effort (unit effort = 61 m of net per night) was calculated as an indicator of the relative abundance of fish vulnerable to gill-netting:

	Numbers ± S.D.	*Biomass ± S.D.*	*n*
allopatric *S. alpinus*	31.8 ± 24.5	10.75 ± 6.35	29
sympatric *S. alpinus*	9.4 ± 6.4	9.32 ± 7.92	40
sympatric *S. namaycush*	5.0 ± 4.0	6.95 ± 6.75	40
sympatric *S. fontinalis*	3.2 ± 4.1	0.67 ± 0.83	40
total sympatric *Salvelinus* sp.	17.5 ± 10.7	16.93 ± 13.37	40

Allopatric Arctic charr were caught significantly more often than both the sympatric Arctic charr and the three sympatric *Salvelinus* sp. grouped together (t-test; $p<0.0010$ and $p<0.0020$, respectively). The biomass per unit effort for the

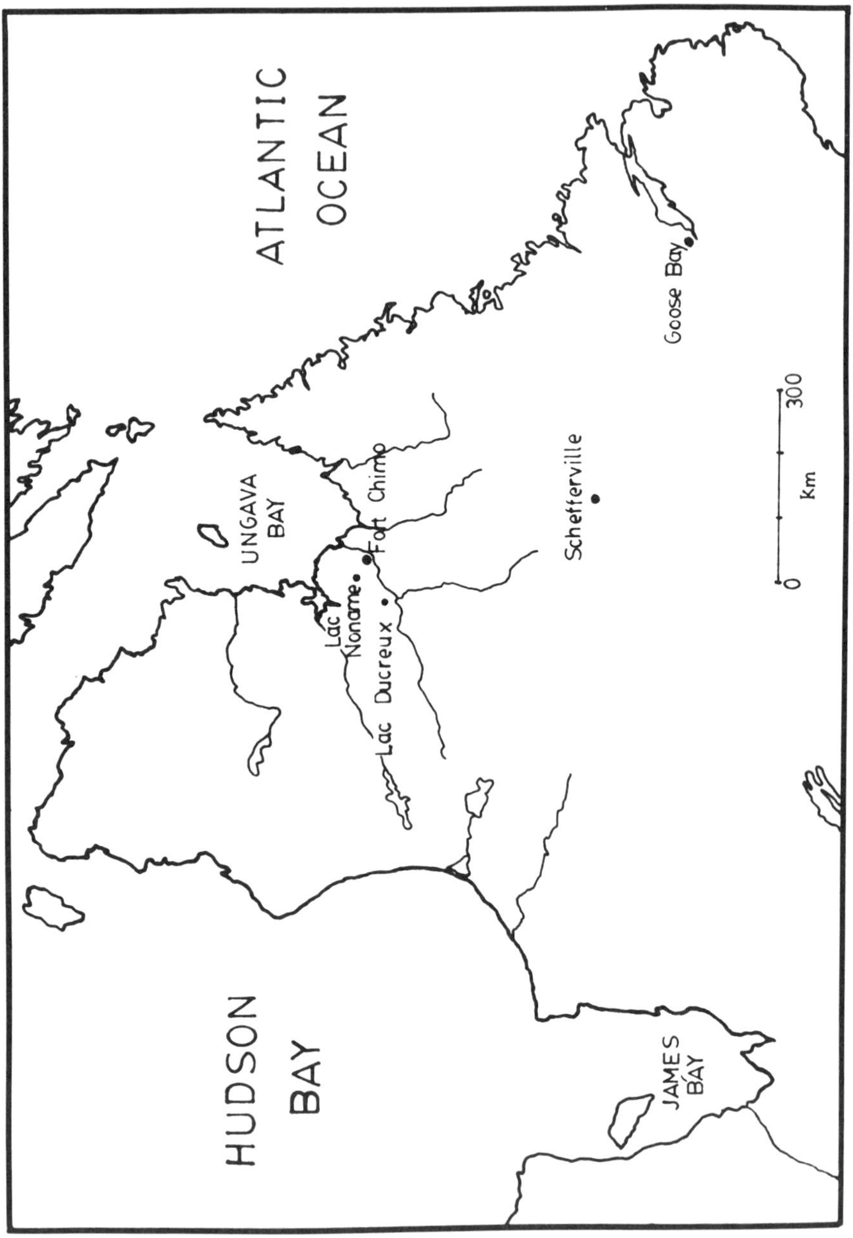

FIG. 1 Map of Quebec-Labrador showing location of Lac Ducreux and Lac Noname.

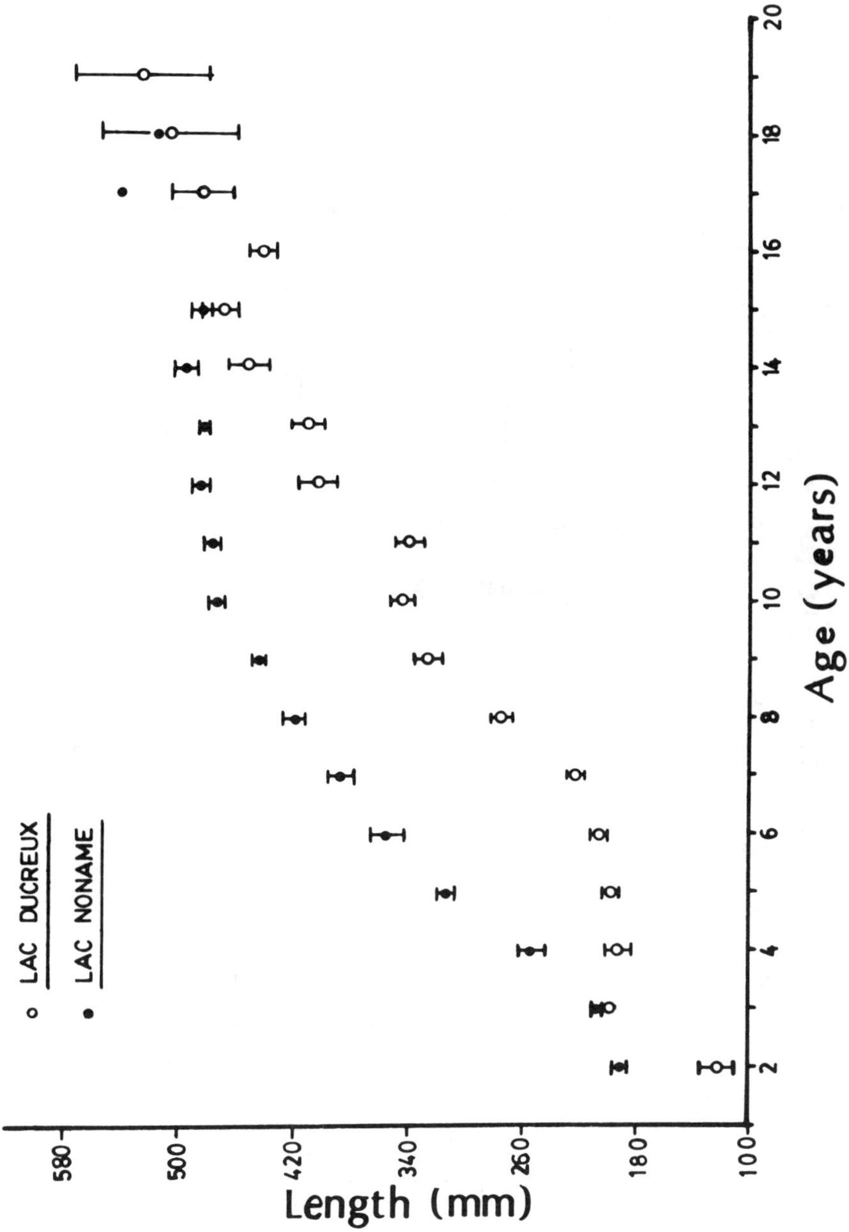

FIG. 2 Growth of allopatric *S. alpinus* in Lac Ducreux and *S. alpinus* sympatric with *S. namaycush* and *S. fontinalis* in Lac Noname, northern Quebec. (Bars indicate standard error of the mean.)

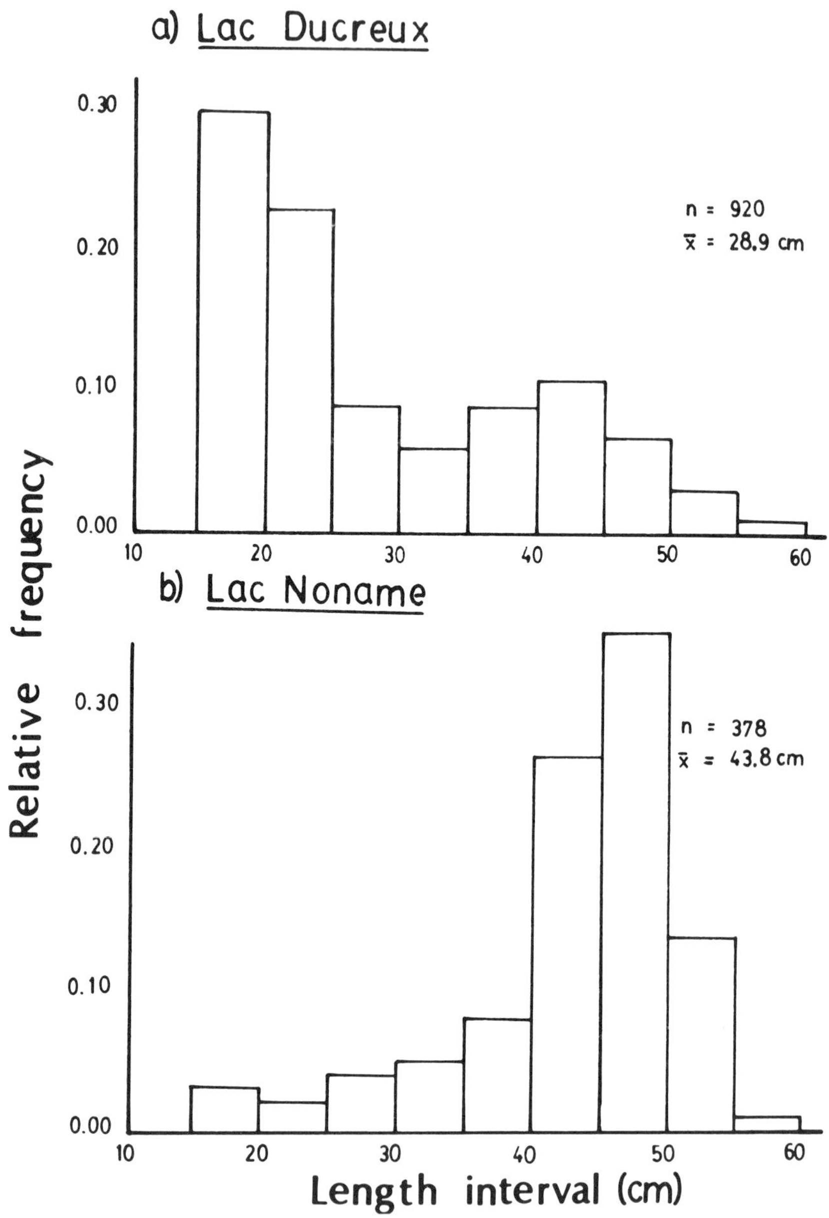

FIG. 3 Length frequency distributions of allopatric *S. alpinus* in Lac Ducreux and *S. alpinus* sympatric with *S. namaycush* and *S. fontinalis* in Lac Noname, northern Quebec.

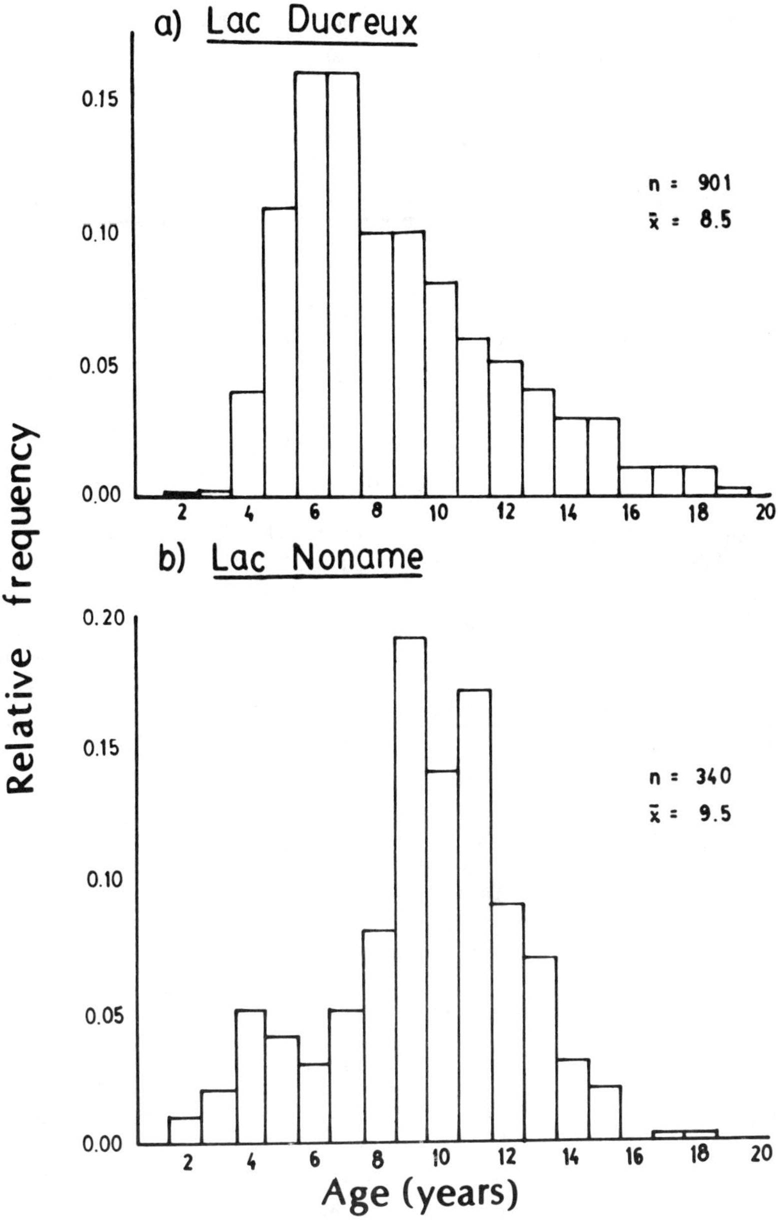

FIG. 4 Age frequency distributions of allopatric *S. alpinus* in Lac Ducreux and *S. alpinus* sympatric with *S. namaycush* and *S. fontinalis* in Lac Noname, northern Quebec.

allopatric charr, however, was not significantly different from that of the sympatric Arctic charr ($p>0.2000$) and was in fact significantly lower than the three sympatric charr species grouped together ($0.0500>p>0.0200$).

Length-Weight Relationships

The slope of the length-weight regression for the allopatric charr population was lower than that of the sympatric Arctic charr (t-test, $p<0.0010$):

Lac Ducreux

$$\log W = -1.9597 + 2.9621 \log FL \quad (3)$$
$$(\pm 0.226 \text{ S.E.}) \quad (\pm 0.0157 \text{ S.E.})$$
$$n=916 \quad r^2=0.98 \quad p<0.0001$$

Lac Noname

$$\log W = -2.1445 + 3.1140 \log FL \quad (4)$$
$$(\pm 0.0441 \text{ S.E.}) \quad (\pm 0.0270 \text{ S.E.})$$
$$n=376 \quad r^2=0.97 \quad p<0.0001$$

FL = fork length (cm) W = weight (g)

No significant differences in slope were found between males and females of each population; therefore, the data were pooled (L. Ducreux: $p>0.20$; L. Noname: $p>0.20$).

After the potential bias of including maturing, ripe, spent or highly-parasitized charr was removed, no significant difference in length-weight relationships existed between the fish less than 32.5 cm in length in either lake (t-test; $p = 0.50$). However, amongst the larger fish, the Lac Noname Arctic charr still had a significantly higher slope than that of the Lac Ducreux charr ($p<0.02$):

Lac Ducreux

$$\log W = -1.6057 + 2.7489 \log FL \quad (5)$$
$$(\pm 0.0979 \text{ S.E.}) \quad (\pm 0.0608 \text{ S.E.})$$
$$n=173 \quad r^2=0.92 \quad p<0.0001$$

Lac Noname

$$\log W = -1.9464 + 2.9933 \log FL \quad (6)$$
$$(\pm 0.1370 \text{ S.E.}) \quad (\pm 0.0837 \text{ S.E.})$$
$$n=139 \quad r^2=0.90 \quad p<0.0001$$

Maturation, Fecundity and Sex Ratios

The allopatric Arctic charr of Lac Ducreux matured first at an earlier age and at a smaller size than the sympatric Arctic charr of Lac Noname (Table 4). Allopatric males and females were maturing at ages 4+ and 5+ repectively, whereas the sympatric population generally remained immature until age 8+. Greater than 80 percent of the fish in the smaller length interval sampled in Lac Ducreux were mature, but in Lac Noname only charr greater than 30 cm long

Table 4. Frequency of fish maturing[a] in each age group of *S. alpinus* in Lac Ducreux and Lac Noname, northern Quebec.

	LAC DUCREUX[b]				LAC NONAME[c]			
	Females		Males		Females		Males	
AGE (years)	Total Numbers	% Mature	Total Numbers	% Mature	Total Numbers	% Mature	Total Numbers	% Mature
2-4	3	0	5	60	7	0	11	0
5-7	42	79	90	91	14	7	8	13
8-10	41	37	54	61	46	67	64	67
11-13	33	33	25	56	29	72	68	79
>14	16	19	19	32	5	100	11	91
TOTAL	135		193		101		162	

a "Mature" defined as gonads mature, ripe or spent
b Sampling period 23-28 August 79
c Sampling periods 9-17 August 79 and 3-10 September 79

were mature (Table 5). The sympatric fish, however, maintained a higher proportion of spawning individuals in the older year classes and larger length intervals.

The mean diameter of the preserved eggs and the mean number of eggs. 100 g^{-1} were essentially identical for Arctic charr from both lakes:

	Egg Diam. (mm) + S.D.	*No. Eggs · 100g^{-1} + S.D.*	*n*
Lac Ducreux	4.83 ± 0.51	154.75 ± 40.33	16
Lac Noname	4.74 ± 0.17	162.02 ± 21.23	16

Parasite Load

The parasitization of the internal organs of the allopatric Arctic charr was significantly more prevalent and intense than that of the sympatric Arctic charr (chi-square analysis; $p = 0.0010$). The frequencies of Arctic charr with each level of infection intensity were:

	Absent	*Light*	*Heavy*	*Total*
Lac Ducreux	163	630	125	918
Lac Noname	84	294	0	378

Heavy parasitization was more frequent in the larger fish; 87 percent of the heavily-infected Ducreux fish were greater than 40.0 cm in length. A chi-square

Table 5. Frequency of fish maturing[a] in each length interval of *S. alpinus* in Lac Ducreux and Lac Noname, northern Quebec.

	LAC DUCREUX[b]				LAC NONAME[c]			
	Females		Males		Females		Males	
LENGTH (cm)	Total Numbers	% Mature	Total Numbers	% Mature	Total Numbers	% Mature	Total Numbers	% Mature
15-20	34	82	64	91	4	0	5	0
20-30	40	58	76	89	7	0	7	0
30-40	28	21	24	42	19	16	11	18
>40	35	17	36	22	80	79	164	76
Total	137		200		110		187	

a "Mature" defined as gonads mature, ripe or spent
b Sampling period 23-28 August 79
c Sampling periods 9-17 August 79 & 3-10 September 79

analysis indicated that the allopatric Arctic charr feeding on fish were more likely to be heavily parasitized ($p=0.0001$). In addition, the probability of a charr maturing in Lac Ducreux was related to its parasite load (chi-square analysis; $p=0.0001$); the more intense the infection, the less likely was the fish to mature.

The Arctic charr were parasitized by the cestodes *Diphyllobothrium ditremum, D. dendriticum, Eubothrium salvelini, Proteocephalus* sp., and *Cyathocephalus truncatus*; the acanthocephalan *Echinorhynchus leidyi*; the trematode *Schistocephalus solidus*; and unidentified free-living nematodes. The Lac Noname lake charr and brook charr had light parasite infections, similar to that of the sympatric Arctic charr (chi-square analysis; $0.50>p>0.25$).

DISCUSSION

The interactive segregation of Arctic charr from sympatric lake charr and brook charr in Lac Noname is clearly demonstrated by the differences in diet of the four charr populations in this study. The diet of the allopatric Arctic charr of Lac Ducreux was significantly different from that of the sympatric Arctic charr in Lac Noname. The three sympatric charr populations had significantly different diets as well. We deduced from this evidence that the shift in feeding behaviour of the sympatric Arctic charr from that of the allopatric charr was a result of the interspecific interactions of the three congeneric species. This pattern of resource partitioning meets Nilsson's (1960, 1965, 1967) criteria for identifying interactive segregation of fish species.

The interactive segregation was more evident between lake charr and Arctic charr than between brook charr and Arctic charr. Behnke (1980) described the lake charr as being an evolutionarily stable species with a highly specialized predatory mode of life and narrow environmental tolerances. Martin and Olver (1980) typified the feeding of adult lake charr as being mainly piscivorous, and stated that interspecific food competition was not likely a significant factor governing the feeding behaviour of this fish. Such descriptions suggest that adult lake charr would be more likely to maintain their piscivorous habit than relinquish it in the face of interspecific interactions. The adult lake charr in Lac Noname were almost exclusively piscivorous, whereas Arctic and brook charr maintained fishless diets. The Arctic charr in Lac Ducreux, however, in the absence of lake charr were heavy fish-eaters. Thus, the apparent exclusion of fish from the diet of the sympatric Arctic charr by the lake charr is interpreted as evidence for strong interactive segregation between the two charr species in Lac Noname.

Brook charr were described by Behnke (1980) as the most generalized and adaptable of the three charrs. Power (1980) emphasized their opportunism and the wide variety of natural prey in their diet. Brook charr in Lac Noname consumed mainly Trichoptera, other insects and *Lymnaea* sp. Their diet overlapped somewhat with that of the sympatric Arctic charr. However, the brook charr were not eating zooplankton, *Valvata* sp. or *Gammarus lacustris* as were the sympatric Arctic charr and as may brook charr in other environments (Power 1980). Moreover, the Arctic charr were not consuming Trichoptera as they had been in Lac Ducreux. Thus, interactive segregation, although incomplete, was probably occurring between these two species.

The diet of the Lac Ducreux Arctic charr resembled that of the Lac Noname lake charr much more closely than that of the brook charr. The diet of the sympatric Arctic charr therefore appears to represent a greater change away from that of the lake charr than from that of the brook charr in Lac Noname, suggesting a more intense interactive segregation between the sympatric Arctic charr and lake charr.

The diet differences cannot be attributed to qualitative differences in the potential prey communities of Lac Ducreux and Lac Noname. The zooplankton of both lakes were similar (six out of seven major copepods and cladocerans in common) as were the prey organisms found in fish from both lakes (Fraser 1981). Based on the general omnivory and opportunism of the charrs, we assumed that the types of food in the diet of such a large sample of fish would accurately reflect the types of prey organisms available in both lakes. However, we would not expect to find quantitative similarities between the two potential prey communities. Since different predator communities exert different pressures on their prey communities, the relative proportions of each prey item in each lake would be unique to that lake (Larkin 1979; Brooks 1979).

Although interactive segregation was demonstrated through the diet differences of the four charr populations in this study, the actual mechanism of diet segregation was not observed. One factor influencing the segregation of food may be the variations among species of the degree of aggression connected with feeding and/or territoriality (Nilsson 1967).

The Arctic charr was described by Johnson (1980) as an unaggressive fish, able to avoid interspecific confrontations to some extent by changing its diet. In this study, it is possible that the Arctic charr abandoned its piscivorous habit in response to aggression from the co-habiting lake charr.

Behnke's (1980) assertion that brook charr are the most generalized and adaptable of the three charrs may be interpreted to mean that brook charr are the least aggressive of the three charrs. However, little is known of the relative aggression levels of brook charr and Arctic charr during interspecific encounters and the diet results in this study do not indicate which species may have been more aggressive.

Differences in feeding efficiencies for particular food types and related morphological differences among the three species are also important components of food segregation (Nilsson 1967; Werner and Hall 1979). Werner and Hall (1979) observed that allopatric populations of three congeneric sunfishes (*Lepomis* sp.) displayed a preference for the same type of habitat, though having different foraging efficiencies in that habitat. They showed that in sympatry, habitat switching and ultimate segregation occurred as a result of the different foraging efficiencies of the three species in conjunction with decreasing prey densities in the three available habitats. As the initially excessive prey density declined in the preferred habitat, the profitability ranking of that prey item declined at a different rate for each fish species. The least efficient forager was the first to abandon the preferred habitat but was also the first to take advantage of the food resources in the nextpreferred habitat.

The three *Salvelinus* species in Lac Noname could be co-existing in a similar fashion. For example, of the three charr species, the lake charr have the most appropriate and by implication, the most efficient body size and morphology for capturing prey fish. They are, therefore, least likely to switch from fish and will remain predominately piscivorous even at low prey densities. The less appropriately equipped predator, *S. alpinus* may find it profitable to include fish in its diet only when prey fish are abundant. When prey fish are rare, however, the Arctic charr may find it energetically inefficient to pursue them and consequently drop them from their diet.

The charr length-frequency distributions (including brook and lake charr (Fraser 1981)) suggest that densities of small forage fish in Lac Noname are low relative to Lac Ducreux. This difference in prey density may account for the fact that lake charr were piscivorous in Lac Noname and Arctic charr were piscivorous in Lac Ducreux. We suggest that, as with the three sympatric *Lepomis* species, the most efficient forager, *S. namaycush*, remained in the preferred habitat whereas the two less efficient foragers, *S. alpinus* and *S. fontinalis*, took advantage of alternative food resources in alternative habitats.

The divergent population structures and reproductive strategies of the two Arctic charr populations could be a response to their different diets and interactions related to foraging (Nilsson 1963, 1965). However, to paraphrase S. Utida (in Larkin 1956), the regulatory action of inter- or intraspecific interactions can be transformed or out-weighed by such higher order interspecific interactions as predation and parasitism.

Predation pressure on small charrs is probably greater in Lac Noname than in Lac Ducreux because lake charr were more consistently piscivorous than were Arctic charr. The threat of predation probably confined the small fish in Lac Noname to the unsampled inlets and broad, rocky shallow areas of the lake, thereby explaining the paucity of specimens less than 25 cm in our samples. Werner et al. (1977) hypothesized the same type of size-regulated predation to explain why smaller fish were found only in habitats of dense cover in the Michigan lakes they studied. In Lac Ducreux, Arctic charr between 15 and 20 cm were abundant, implying that the larger charr were exerting a less potent predation force on them.

The large difference in early age-specific mean lengths indicates that the Lac Noname Arctic charr were growing faster than the Lac Ducreux charr during the first few years of life. This may also be a response to greater predation pressure, in this case selecting against slow-growing juveniles which remain vulnerable for a longer period of time.

The greater length and age of maturation in the Lac Noname Arctic charr could also be a response to the predatory habits of the co-existing lake charr. The reproductive behaviour of charr results in the aggregation of spawning fish at specific times and places in the lake, making them more vulnerable to piscivores. We hypothesize that the Lac Noname Arctic charr have evolved toward a larger size at first maturation to avoid heavy lake charr predation during spawning and that this delay is compensated by the increased number of eggs released by each spawning female.

In contrast, the Lac Ducreux Arctic charr mature first between 15 and 20 cm or less, at the early age of four or five. The greater proportion of energy channelled towards gonadal rather than somatic growth may account for the age-specific lengths of these fish.

The role of parasites in the interactive segregation of the three *Salvelinus* species does not appear significant. However, the allopatric population of Arctic charr suffered extensively from internal cestode infestations. Body length, lowered fecundity, lowered slope in the length-weight relationship, and the consumption of prey fish were all related to the intensity of infection. Plerocercoids in the prey fish, *Gasterosteus aculeatus*, were the main source of infection. On the other hand, the Lac Noname charr were relatively free of parasites, probably because of the absence of sticklebacks from their diet (Curtis 1983). In this instance, the interactive segregation of food items appeared to have indirectly benefitted the Arctic charr.

In conclusion, the interactive segregation of *S. alpinus* from *S. namaycush* and *S. fontinalis* is supported by the observed lack of overlap in their diets when in sympatry, and by the shift in diet of the Arctic charr from that observed when in allopatry. The potential effects of the changed diet on the growth and reproduction of the Arctic charr may be modified by the influences of different predation pressures and parasite infections.

ACKNOWLEDGEMENTS

The financial and logistical help of the Centre d'Etudes Nordiques, Universite Laval; the National Science and Engineering Research Council of Canada (for grant to G.P.); the President's Northern Studies Committee, University of Waterloo; the Dean of Graduate Studies, University of Waterloo; and the finance committee of this symposium are gratefully acknowledged.

The mechanical, biological and culinary expertise of Mr. Chris Hart helped make our field season both a success and a pleasure. Mark Curtis identified the charr parasites; and Patricia Chow-Fraser kindly reviewed the manuscript.

REFERENCES

Behnke, R.J. 1980. A systematic review of the genus *Salvelinus*, p. 441-480. *In* E.K. Balon [ed.] Charrs: salmonid fishes of the genus *Salvelinus*. Dr. W. Junk Publishers, The Hague.

Borgeson, D.P. 1963. A rapid method for food-habits studies. Trans. Am. Fish. Soc. 93:434-435.

Brooks, J.L. 1979. Multilevel systems affects of predation, p. 77-79. *In* R.H. Stroud and H. Clepper [eds.] Predator-prey systems in fisheries management. Sport Fishing Institute, Wash., D.C..

Curtis, M.A. 1983. *Diphyllobothrium* spp. and Arctic charr: parasite acquisition and its effects on a lake-resident population. Paper presented at Int. Symp. on Arctic charr, Winnipeg, 1981 (This volume).

Fraser, N.C. 1981. The interactive segregation of landlocked Arctic charr from lake charr and brook charr in southern Ungava, Quebec. M.Sc. thesis. University of Waterloo, Waterloo, Ontario, Canada.

Johnson, L. 1980. The Arctic charr, *Salvelinus alpinus*, p. 15-98. *In* E.K. Balon [ed.] Charrs: salmonid fishes of the genus *Salvelinus*.Dr. W. Junk Publishers, The Hague.

Larkin, P.A. 1956. Interspecific competition and population control in freshwater fish. J. Fish. Res. Board Can. 13:327-342.

Larkin, P.A. 1979. Predator-prey relations in fishes — an overview of the theory, p. 13-22. *In* R.H. Stroud and H. Clepper [eds.] Predatory-prey systems in fisheries management. Sport Fishing Institute, Wash., D.C.

Martin, N.V. and C.H. Olver. 1980. The lake charr, *Salvelinus namaycush*, p. 205-277. *In* E.K. Balon [ed.] Charrs: salmonid fishes of the genus *Salvelinus*. Dr. W. Junk Publishers, The Hague.

Nilsson, N.-A. 1960. Seasonal fluctuations in the food segregation of trout, charr and whitefish in 14 north-Swedish lakes. Inst. Freshw. Res. Drottningholm Rep. 41:185-205.

Nilsson, N.-A. 1963. Interaction between trout and charr in Scandinavia. Trans. Am. Fish. Soc. 92:276-285.

Nilsson, N.-A. 1965. Food segregation between salmonid species in north Sweden. Inst. Freshw. Res. Drottningholm Rep. 46:58-78.

Nilsson, N.-A. 1967. Interactive segregation between fish species, p. 295-313. *In* S.D. Gerking [ed.] The biological basis of freshwater fish production. Blackwell Scientific Publications, Oxford and Edinburgh.

Nordeng, H. 1961. On the biology of charr (*Salmo alpinus* L.) in Salangen, North Norway. I. Age and spawning frequency determined from scales and otoliths. Nytt. Mag. Zool. (Oslo). 10:67-123.

Power, G. 1980. The brook charr, *Salvelinus fontinalis*, p. 141-203. *In* E.K. Balon [ed.] Charrs: salmonid fishes of the genus *Salvelinus*. Dr. W. Junk Publishers, The Hague.

SAS Institute. 1979. SAS User's Guide, 1979 Edition, Gary, North Carolina, 494p.

Saunders, L.H. and G. Power. 1969. The Arctic charr, *Salvelinus alpinus* (Linnaeus), of Matamek Lake, Quebec. Nat. Can. 96:919-934.

Svärdson, G. 1976. Interspecific population dominance in fish communities of Scandinavian lakes. Inst. Freshw. Res. Drottingholm Rep. 55:144-171.

Werner, E.E. and D.J. Hall. 1979. Foraging efficiency and habitat switching in competing sunfishes. Ecol. 60:256-264.

Werner, E.E., D.J. Hall, D.R. Laughlin, D.J. Wagner, L.A. Wilsmann and F.C. Funk. 1977. Habitat partitioning in a freshwater fish community. J. Fish. Res. Board Can. 34:360-370.

Nucleolytic Enzymes in White-Spotted Charr (*Salvelinus leucomaenis*)

MINEO SANEYOSHI and SHUNJI IZUTA

Faculty of Pharmaceutical Sciences, Hokkaido University, Sapporo 060, Japan

SANEYOSHI, M. and S. IZUTA. 1984. Nucleolytic enzymes in white-spotted charr (*Salvelinus leucomaenis*), p. 183–191. *In* L. Johnson and B.L. Burns [eds.] Biology of the Arctic charr, Proceedings of the International Symposium on Arctic Charr, Winnipeg, Manitoba, May 1981. Univ. Manitoba Press, Winnipeg.

White-spotted charr (*Salvelinus leucomaenis*) is a common charr in Japan closely related to Arctic charr (*Salvelinus alpinus*) and Dolly Varden (*Salvelinus malma*). In northeastern parts of the Japanese Islands, especially Hokkaido, this species is mainly anadromous. In the mountainous regions of central and southwest Japan the only stocks of this species are confined to freshwaters. As a possible approach to the differentiation of species within the charr complex, various nucleolytic enzymes were examined in samples of the liver, testes and pyloric caeca of the white-spotted charr. Using DEAE-cellulose and phospho-cellulose column chromatographic techniques, a 3′ exonuclease, with similar specific activity, was found in all tissues. However, endonucleolytic acid deoxyribonuclease (DNase) activity was found to be strongest in testes extracts. Acid DNase was separated into six activity peaks by DEAE-cellulose column chromatography. Acid ribonuclease activity in pyloric caeca extracts was also strong and was separated into three peaks on the same chromatographic system. The distinctive enzymatic activity of each tissue may vary among the species of charrs and related salmonids and may be useful in biochemical taxonomy. 3′-exonuclease will be useful for this purpose, because this enzyme can cleave non-fluorescent 3′-umbeliferyl nucleotide, generating a highly fluorescent umbeliferon. This reaction should be applicable to the micro-determination of 3′-exonuclease isozymes on gels after electrophoresis.

INTRODUCTION

As a contribution to the biochemical systematics of charrs and other salmonid fishes the distribution in tissues and the characterization of several nucleolytic enzymes was studied in white-spotted charr (*Salvelinus leucomaenis*).

Two recognized species of *Salvelinus* occur in Japan: Dolly Varden, *Salvelinus malma* and white-spotted charr. Dolly Varden is the more northerly of the two and is restricted to Hokkaido and the northern islands; white-spotted charr is widely distributed throughout the country. In the northern part of Japan, including Hokkaido, white-spotted charr is usually anadromous; by contrast in the southern regions it occurs as a landlocked or resident species in cold headwater lakes in mountainuous regions. Morphological variation is common in southern stocks particularly with respect to the number and colour pattern of the red spots on the body.

Nucleolytic enzymes have an important function in nucleic acid metabolism; a considerable number have been isolated from various tissues. It is believed that analysis of these enzymes will prove to be useful in the characterization of charrs and related salmonids. The only investigations on nucleolytic enzymes in fish previously reported are those of Smith and Dixon (1969) and Menon and Smith (1970) on 3′-exonuclease derived from Chinook salmon testes.

In this paper we describe the distribution in various tissues of white-spotted charr of the enzymes: 3′- and 5-′exonuclease, acid and neutral deoxyribonuclease (DNase) and acid and neutral ribonuclease (RNase). The activity of each enzyme was analyzed by DEAE-cellulose column chromatography. The 3′-exonuclease was then further purified using phospho-cellulose column chromatography. Acid DNase and acid RNase from white-spotted charr testes was then compared with that from the testes of salmonids in different genera: rainbow trout (*Salmo gairdneri*) and cherry salmon (*Oncorhynchus masou*).

MATERIALS AND METHODS

Three year old male white-spotted charr were kindly provided by Professor Tatsuro Kubo, of the Fishery Science Faculty, Hokkaido University, Hakodate; three year old male rainbow trout were purchased from the Sapporo Rainbow Trout Co. Ltd., Sapporo, and two year old cherry salmon were collected from the Asuta River under special permit from the Governor of Hokkaido.

The state of maturity was determined microscopically by examining spermatogonia to ensure the presence of 30-40 percent primary and secondary spermatocytes and 50-60 percent spermatids. Tissue from selected fishes was immediately frozen using solid carbon dioxide or liquid nitrogen and stored at -80°C until required for use.

Substrates

Salmon sperm DNA was purchased from P.L. Biochemicals. Yeast RNA was obtained from Kojin Co. Ltd.. The 2,4-dinitrophenyl thymidine 3′-phosphate (2,4-d-Tp) and 2,4-dinitrophenyl thymidine 5′-phosphate (2,4-D-pT) were synthesized chemically according to the process described by von Tigerstrom and Smith (1969).

Table 1. Assay of nucleolytic enzymes in white-spotted charr.

Enzyme	Buffer (pH)	Substrates
3'-Exonuclease	0.04 M potassium acetate (5.5)	2,4-D-Tp (0.1 μM)
5'-Exonuclease	0.04 M Tris-HCl (8.5)	2,4-D-pT (0.1 μM)
Acid DNase	0.04 M potassium acetate (5.5)	denatured DNA (0.25 mg/ml)
Neutral DNase	0.02 M Tris-HCl (7.5)	denatured DNA (0.25 mg/ml)
Acid RNase	0.04 M potassium acetate (5.5)	yeast RNA (0.25 mg/ml)
Neutral RNase	0.04 M Tris-HCl (7.5)	yeast RNA (0.25 mg/ml)

Assay of Nucleolytic Enzymes

The assay mixtures, made up to a total volume of 0.5 ml, were prepared as in Table 1.

Exonucleases

The preparations were incubated at 25°C for 25 minutes; 0.5 ml of distilled water was added and the absorbance then measured at 358 nm.

RNases and DNases

Mixtures were incubated for the same period of time as for exonucleases but 0.2 ml of 0.3N HCl was added. The precipitated non-digested DNA or RNA was removed by centrifugation (1000 × g, 15 min) after storage at 0°C for 3 hr; 0.3 ml of distilled water was added to the supernatant and its UV absorbance at 260 nm was measured.

Preparation of Tissue Extracts

All procedures were carried out at 4°C. Amounts of 10g of the frozen tissues were thawed and homogenized in a Waring blender for one minute with 5 volumes of buffer 'B' containing 0.05M Tris-HCl, pH 7.5, 0.01M MgCl2, 0.01M 2-mercaptoethanol, 5 percent glycerol and 0.3M NaCl. The homogenization vessel was surrounded by an ice bath; the procedure was repeated five times with the same time period on each occasion.

The homogenates were centrifuged (3500 × g, 20 min) and to the resulting supernatant liquid was added sufficient solid ammonium sulfate to achieve 70 percent saturation. The mixture was allowed to stand for 10 min at 0°C then the precipitates were collected by centrifugation (6000 × g, 20 min). Precipitates were dissolved in buffer 'A' containing 0.5M Tris-HCl, pH 7.5, 0.01M MgCl2, 0.01M 2-mercaptoethanol and 5 percent glycerol, then dialyzed against one litre

Table 2. Tissue distribution of the nucleolytic enzymes in white-spotted charr (as specific activity, units/g of tissues).

Enzyme	Liver	Testis	Pyloric caeca
3'-Exonuclease*	1.1	2.3	2.0
5'-Exonuclease*	0.19	0.15	0.45
Acid DNase**	3.9	44.0	32.0
Neutral DNase**	11.0	3.5	61.0
Acid RNase**	19.0	13.0	35.0
Neutral RNase**	2.7	2.4	21.0

* 1 unit = liberated 1 nmole of 2,4-dinitrophenol per 20 min at 25°. specific activity = units/ g tissues.

** 1 unit = formed 0.05 mg of oligonucleotides in acid soluble fraction per 25 min at 25°.

of the same buffer (5 hr × 3). Small amounts of insoluble material were removed by centrifugation and the resulting supernatant used as the enzyme preparation.

RESULTS AND DISCUSSION

Tissue Distribution of Enzyme Activity

The exonucleolytic activities of the two enzymes classified as 3′-exonuclease and 5′-exonuclease were first measured using two synthetic substrates. Both 2,4-D-Tp and 2,4-D-pT liberated 2,4 dinitrophenol (λmax 358 nm) when 3′-exonuclease or 5′-exonuclease, respectively, was present in the tissue extract. The activity of 3′-exonuclease was found to be similar in testis and pyloric caeca but much less than in liver (Table 2). By contrast 5′-exonuclease (present in snake venom and the intestine of mammals) had low activities in all tissues.

The four endonucleolytic DNase and RNase activities were then measured in the tissue extracts. Testes had the highest activity of acid DNase among the three tissues examined; pyloric caeca also exhibited significant activity but liver extracts yielded low values. Neutral DNase activity could be detected only in pyloric caeca extract; testis and liver activities were very low. The activity of endonucleolytic acid and neutral RNase was also examined. In acid media, RNase (pH 5.5) digests RNA to oligonucleotides and neutral RNase (pH 7.5) digests RNA to mononucleotides. The activity of acid RNases was highest in pyloric caeca extracts (35 units × g^{-1} of tissue), 0.6 and 0.5 of this activity in liver and testis respectively. Neutral RNase activity was only found at significant levels in pyloric caeca (21 units × g^{-1} of tissue); liver and testis had only one tenth of this activity.

DEAE-Cellulose Column Chromatographic Analysis of Nucleolytic Enzymes in Tissue Extracts

In order to characterize, at least partially, the enzymes described above, DEAE-cellulose column chromatographic analysis was performed. The extract obtained from 10g of tissue was applied to a column of DEAE-cellulose (2.5 cm × 25 cm). The column had been previously equilibrated with buffer 'A' and washed well with the same buffer until no absorbance at 280 nm could be observed. Elution was performed in a linear gradient of NaCl from 0 to 0.3 M in buffer 'A' (total volume 600 ml) at a flow rate of 0.5 ml × min^{-1}. The results show (Fig. 1) that of the enzyme activities in liver extract only 3′-exonuclease

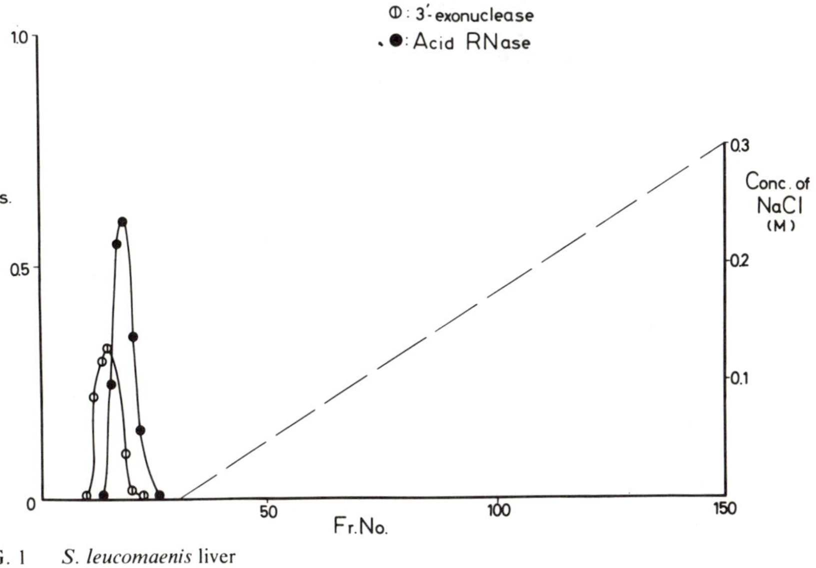

FIG. 1 *S. leucomaenis* liver

and acid RNase could be detected on the chromatogram. It is not yet quite clear but this could result from protease contamination. In the case of testis extract (Fig. 2) 3′-exonuclease activity appeared as a single peak at the same position as that of liver enzyme. On the other hand, acid DNase activity was clearly resolved into at least six peaks (designated as DNase I to DNase VI). Acid RNase activity was detected in this column chromatography as a single symmetrical peak. The

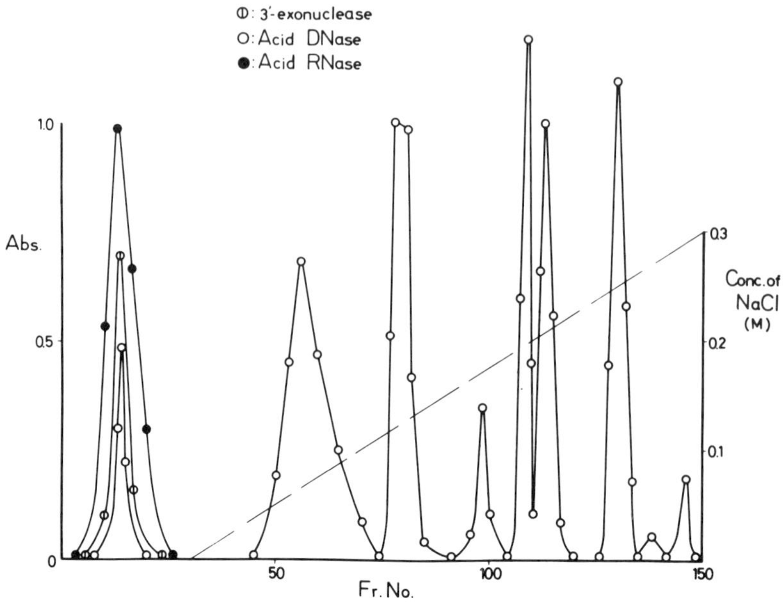

FIG. 2 Testes

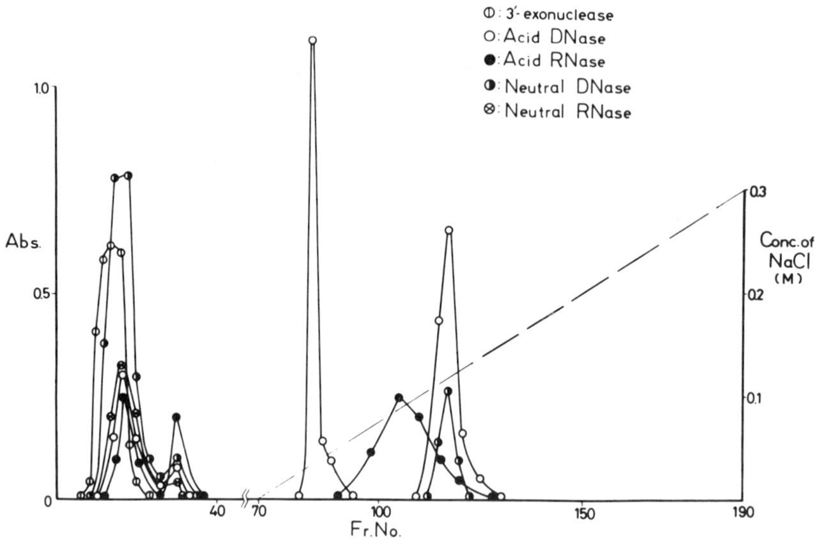

FIG. 3 Pyloric caeca

enzyme activities from pyloric caeca extract were resolved as shown in Figure 3; 3′-exonuclease activity appeared as a single peak in the unadsorbed fractions, the same as with testis and liver preparations. It is important to note that a neutral DNase was eluted as a strong single peak in unadsorbed fractions which was not detected in other tissue extracts. Acid DNase activities were eluted as three separate peaks at identical positions to those of DNase I, II and V of the testis DNase. Acid RNase activities were also recognized as three separate peaks.

3′-Exonuclease Activities in Phospho-Cellulose Column Chromatography

As described in the previous section, 3′-exonuclease activity was eluted as a single peak on a DEAE-cellulose column in all the tissues examined. To characterize this enzyme in more detail the fractions indicating 3′-exonuclease activity were analyzed by cationic ion-exchanger. After recovering the protein by precipitation with ammonium sulfate (70 percent saturation) and dialysis, it was examined in a phospho-cellulose column. Elution was performed in a linear salt gradient after washing the column with buffer. At 0.2 to 0.25 M salt concentrations, 3′-exonuclease showed the same profile of activity for all extracts independent of tissue from which they were obtained (Fig. 4). This result suggests that 3′-exonuclease may be common to the three tissues examined.

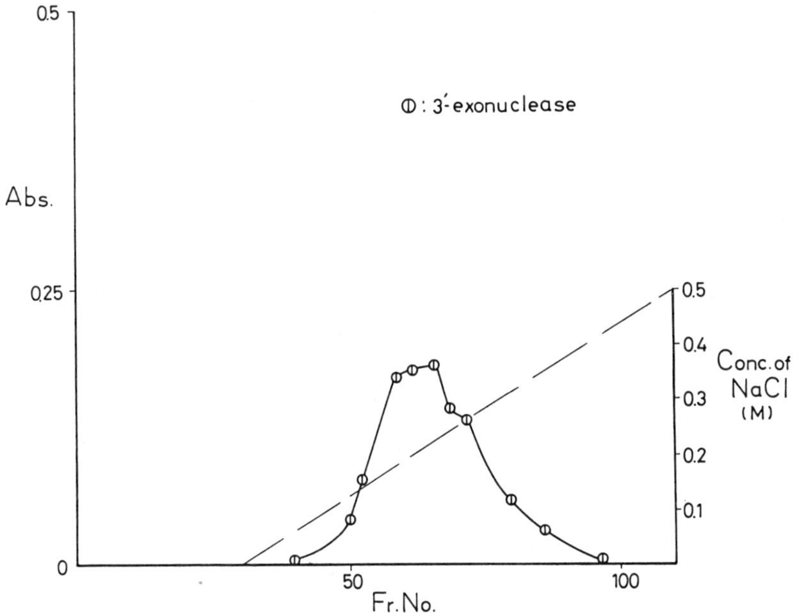

FIG. 4 Elution profile of 3′-exonuclease activity on phospho-cellulose

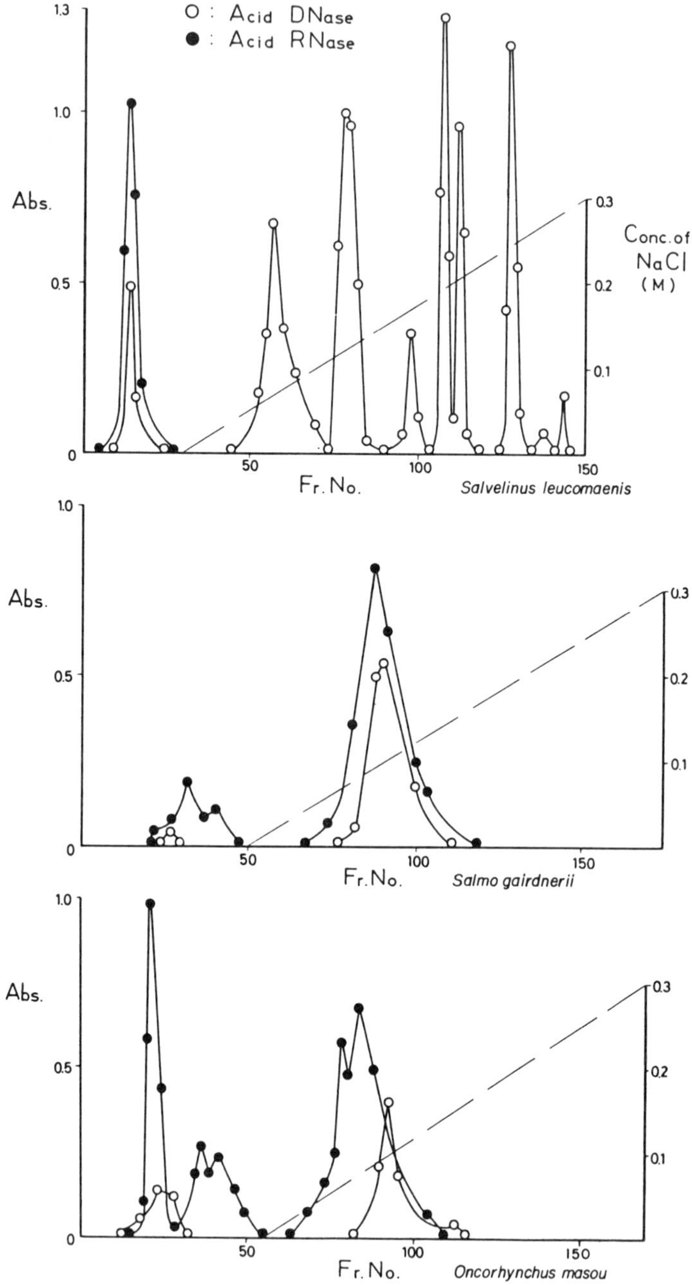

FIG. 5 Comparison of elution profiles of three genera.

Comparison of Acid DNase and RNase Activities in White-Spotted Charr, Cherry Salmon and Rainbow Trout

In the preceding sections, the distribution and specificity of several nucleolytic enzymes in white-spotted charr have been partially characterized. In this section we describe a preliminary characterization of two enzyme activities (acid DNase and RNase) from salmonids of distinct genera: cherry salmon (*Oncorhynchus masou*) and rainbow trout (*Salmo gairdneri*).

Acid DNase activities in testis extract of white-spotted charr, rainbow trout and cherry salmon were determined to be 44 units, 14 units and 7.7 units per gram of tissue respectively. By contrast acid RNase activities in the same species were 13, 16 and 23 units respectively. It is believed that as DNase and RNase activities vary significantly both with respect to tissue and species it will be a useful tool for the biochemical discrimination of salmonids.

To obtain further quantitative information on enzyme activity, DEAE-cellulose chromatography was employed. Acid DNase activity of rainbow trout was eluted at a single peak at 0.1M salt concentration and this was identical to that in white-spotted charr (Fig. 5); on the other hand acid DNase activities of cherry salmon testes were resolved into two separate peaks corresponding to DNase I and DNase III of white-spotted charr; in the case of acid RNase, rainbow trout had one peak with small shoulders in the unadsorbed fractions and one big peak on the gradient. Interesting results were obtained from acid RNase activities from cherry salmon tissues: two well-separated peaks were found in unadsorbed fractions, closely corresponding to two peaks on the gradient fractions. Whether these separate peaks of activity correspond to ''isozymes'' or different enzymes which have very similar isoelectric points remains to be elucidated. We are currently working on these problems.

ACKNOWLEDGEMENTS

We are indebted to Professor Tatsuro Kubo for providing white-spotted charr and valuable discussions. We thank the Governor of Hokkaido for the special collection permit for cherry salmon.

REFERENCES

Dixon, G.H. and M. Smith. 1968. Nucleic acids and protamines in salmon testes. Prog. Nucl. Acid Res. and Mol. Biol. 8:9.

Menon, K.M.J. and M. Smith. 1970. Exonuclease (Phosphodiesterase) from the testes of chinook salmon (*Oncorhynchus tshawytscha*). Biochemistry 9:1584-1592.

Von Tigerstrom, R.G. and M. Smith. 1969. Preparation of the 2,4-dinitrophenyl esters of thymidine 3′- and thymidine 5′-phosphate and their use as substrates for phosphodiesterases. Biochemistry 8:3067-3070.

The Status and Biology of Arctic Charr, *Salvelinus alpinus* (L.), In Scotland

P.S. MAITLAND
Institute of Terrestrial Ecology, 78 Craighall Road, Edinburgh, Scotland

R.B. GREER
Freshwater Fisheries Laboratory, Faskally, Pitlochry, Scotland

R.N. CAMPBELL
Nature Conservancy Council, 9 Culduthel Road, Inverness, Scotland

and the late G.F. FRIEND
Department of Zoology, The University, Edinburgh, Scotland

MAITLAND, P.S., R.B. GREER, R.N. CAMPBELL AND G.F. FRIEND. 1984. The status and biology of Arctic charr, *Salvelinus alpinus* (L.), in Scotland, p. 193–215. *In* L. Johnson and B.L. Burns [eds.] Biology of the Arctic charr, Proceedings of the International Symposium on Arctic Charr, Winnipeg, Manitoba, May 1981. Univ. Manitoba Press, Winnipeg.

This paper reviews available information on the Arctic charr in Scotland and presents new data on its biology. This fish is much commoner than formerly supposed and occurs in a large number of lochs, particularly in the north and west of the country. All known sites are listed and analyses of habitat characteristics presented. Extinct populations are discussed. Meristic data on samples of charr from a number of habitats are reviewed. There is considerable morphological variation among populations. Ecological data on food, growth, parasites and reproduction are presented and considered in relation to previous studies. The conservation of charr and aspects of future research are discussed. A bibliography of major references to Scottish charr is included.

INTRODUCTION

The Arctic charr, *Salvelinus alpinus* (Linnaeus 1758) is a stenothermic, holarctic species occurring widely in lakes and some rivers in northern and alpine Europe, Asia and North America. In Scotland, the species has been largely ignored as far as fisheries are concerned. Charr appear to be sensitive to pollution and other stresses, and a number of populations — particularly in lakes on the southern edge of the geographic distribution of the species — have become extinct in the last 200 years.

In the British Isles there are well over two hundred lakes containing charr populations. The majority of these are in Scotland (Maitland 1972), but there are a considerable number in Ireland, about twelve in the Lake District of England and three in Wales. Systematically, the British charrs (like those in other parts of the world) are interesting because their local races are often quite recognizably different from one another.

Inevitably this has led to confusion in their taxonomy for though Day (1887) regarded all British charr as belonging to the single species *Salvelinus alpinus*, Gunther (1880) and Regan (1911) between them distinguished at least 14 species of charr in the British Isles alone. It is accepted now that all native charr in the British Isles do, in fact, belong to the plastic *Salvelinus alpinus* — though it must be remembered that there are several thriving populations of introduced American brook charr, *Salvelinus fontinalis* (Mitchill 1815) in various lakes. Recently, stock from migratory charr from Norway have been introduced to waters in Scotland (D. Brown, pers. comm.).

Apart from the interesting series of studies on the population of charr in Windermere (Frost 1952, Kipling and Frost 1978, etc), information on the status and biology of charr in the British Isles is fragmentary and generally inadequate.

The present paper is a synthesis of the data currently available on the charr of Scotland. The work of the four authors has been carried out independently on different aspects of the biology of this species though there has been collaboration and exchange of material from time to time. Much of the research was carried out by Mr. G.F. Friend over a period of about 25 years. During this period he published two short papers on charr (Friend 1956; 1959) but unfortunately his death in 1979 took place before he was able to prepare the bulk of his material for publication. It is included in this paper along with the results of recent studies by the other three authors.

DISTRIBUTION

The indigenous freshwater fish communities of Scotland date from the retreat of the Pleistocene ice-cap about 10,000 years ago (Maitland 1974). Initial recolonization could only be carried out by species able to live in both salt and fresh water; therefore the number of native species now common throughout Scotland, and virtually the only species in the north and west, is small: sea lamprey (*Petromyzon marinus*), river lamprey (*Lampetra fluviatilis*), brook lamprey (*Lampetra planeri*), Atlantic salmon (*Salmo salar*), trout (*Salmo trutta*), charr (*Salvelinus alpinus*), European eel (*Anguilla anguilla*), threespine stickleback (*Gasterosteus aculeatus*), and tenspine stickleback (*Pungitius pungitius*).

These species probably recolonized Scotland by moving round the coast as it became progressively free from ice, penetrating river and loch systems. Other, purely freshwater, species gradually dispersed from the south of Britain northward but some have apparently not yet reached Scotland, while most of those which have (eg roach, *Rutilus rutilus*) are largely restricted to the south of the country.

Most of the existing populations of charr are natural, but some have undoubtedly been created, intentionally and unintentionally by man. Harvie-Brown and Buckley (1888) refer to MacLaine of Lochbuie stocking his lochs on Mull with alpine charr and brook charr (which are still there), while Maxwell (1926) has given an account of his introduction of charr to Loch Ossian (though it seems highly probable that they were there already). The charr (and brook charr) of the Uaine Lochs were introduced during the 1890's (Harvie-Brown and Buckley 1888). At Howietoun in 1892 these two species were hybridized successfully (Maitland 1887) and it has been suggested (A.F. Walker, pers. comm.) that if such hybrids were introduced to waters (such as the Uaine Lochs) the stock could have segregated into the original parental types: this would explain the present odd communities in these waters. More recently similar hybrids have been introduced to several Scottish lochs.

Charr arrived in the newly-made Loch Errochty in the late 1950's by way of the tunnel which picks up water from the upper River Garry — the fish originating from Loch Garry. Similarly, in the new Cruachan Reservoir — part of the Cruachan pumped-storage hydroelectric scheme at Loch Awe — charr are now quite common, having presumably been pumped up through the 1400 m tunnel to an altitude of 365 m.

The information on charr in Scottish lochs has been carefully documented by the authors, based on published data (eg. Hardie 1940, Campbell 1979, etc), unpublished communications from various people and personal fishing records. The master list is regularly up-dated and the current version now includes 168 different lochs, the distribution of which is shown on Figure 1. Only a few populations occur in the southern part of Scotland, the great majority being confined to the north-west. No existing populations are found in the south-east, and though there are many charr lochs throughout the Western Isles, there is only one population on Orkney and one on Shetland (Swan 1954, 1964).

HABITAT

Standing Waters

The general conditions in known charr lochs from which data are available are given in Table 1. The great majority of Scottish lochs, especially upland

Table 1. Physico-chemical characteristics of Scottish charr lochs from which data are available.

CHARACTER	MINIMUM	MEAN	MAXIMUM	NO. LOCHS
Area (km^2)	0.026	4.54	38.46	102
Length (km)	0.3	5.02	35.2	100
Altitude (m)	10.0	163.41	483.0	76
Catchment (km^2)	59.0	10034.21	81953.0	103
Distance sea (km)	0.3	17.38	97.7	101
Depth: mean (m)	1.4	18.55	132.0	84
Depth: max (m)	2.1	51.48	344.0	85
pH	4.3	6.47	7.4	35
Ca (mg/l)	0.5	4.63	30.4	16
Alkalinity (mg/l)	0.6	9.69	109.0	30
Na (mg/l)	2.1	5.44	20.2	16
Conductivity	31.0	47.87	130	15

ones, are glacial in origin (Murray and Pullar 1910). Many are glacial basins, some of great depth, gouged from acid, weather-resistant rocks. Others are ponded back by glacial moraines. All such lochs are oligotrophic in character, many being dystrophic because of extensive areas of peat in their catchments. In a large number (eg. Loch Sionascaig) little sedimentation has taken place (Pennington et al. 1972). The physical and chemical conditions in Scottish charr lochs indicate that, in general, they are large and deep in areas of moderate altitude, at varying distances from the sea, with an acid to neutral water low in nutrients. Data on conductivity, sodium and chloride show that much of the area concerned is oceanic with significant amounts of cyclic salts in many waters (Holden 1966).

Only a small fraction (less than 3 percent) of the 3788 lochs in Scotland greater than 5 ha (Smith and Lyle 1979) are known to contain charr; on the whole they are the larger waters (Table 2) which are also the deepest. One of the best known areas of Scotland limnologically is the Tayside (Maitland et al. 1981) where comparisons of the known charr lochs with those of the area as a whole are of considerable interest (Fig. 2). The data show quite clearly that the charr habitat in Tayside tends to be the large lochs (six out of the seven largest lochs contain charr) in areas of moderate (but never high) altitude and base poor geology. It is perhaps significant that the only base rich charr loch in the catchment (Loch

Table 2. A comparison of the size of all Scottish lochs (from Smith & Lyle, 1979) and Scottish charr lochs from which data are available.

AREA CLASSES (km^2)	ALL SCOTTISH LOCHS		SCOTTISH CHARR LOCHS		
	NO	%	NO	%	% OF EACH CLASS
0-0.25	2973	78.5	7	7.1	0.2
0.25-0.5	417	11.0	15	15.3	3.6
0.5-1	209	5.5	20	20.4	9.6
1-2	83	2.2	11	11.2	13.3
2-4	54	1.4	19	19.4	35.2
4-8	27	0.7	12	12.2	44.4
8-16	12	0.3	5	5.1	41.7
16-32	9	0.2	7	7.1	77.8
32+	4	0.1	2	2.0	50.0
TOTALS	3788	99.9	98	99.8	-

Leven) is one in which the species has become extinct. In other parts of Scotland different criteria may apply.

Charr are inexplicably absent from a number of large and apparently suitable lochs in Scotland (eg. Loch Shiel) and their presence cannot be assumed as it can with trout. Toivonen (1972) considers that shallowness and staining of the water by humic colloids, together with the presence of some other fish species (eg. whitefish) are contributing factors responsible for the absence of charr from many otherwise suitable Fennoscandian lakes. These factors do not however apply to all Scottish lochs without charr.

Running Waters

Although, as previously discussed, there are no native anadromous charr in Scotland, all existing populations occur predominantly in lochs, although not infrequently fish are found in running waters. Some of the more recent records are listed in Table 3; these appear to fall into several categories:

1) Fish accidently entrained into power station systems or swept over dams and forced downstream (eg. fish in the Rivers Kingie and North Esk which pass through the dams on Lochs Quoich and Lee, respectively).

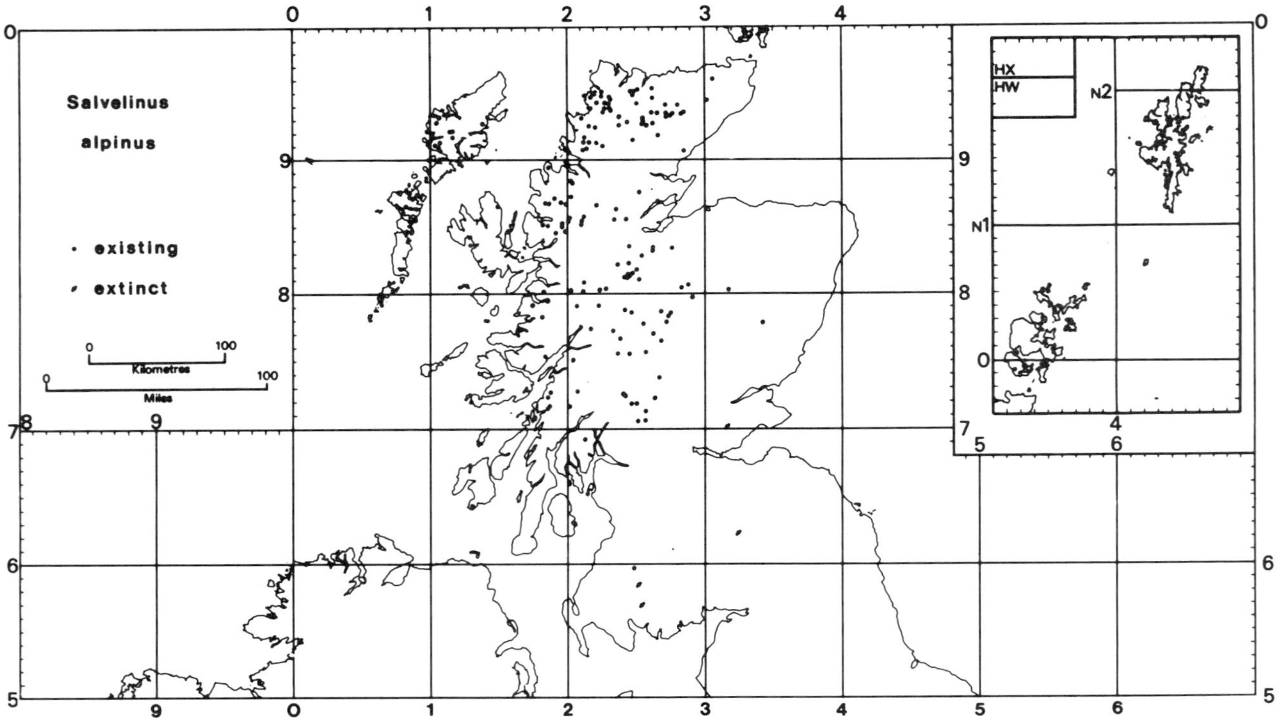

FIG. 1 An outline map of Scotland, showing the distribution of known charr lochs. Sites where charr are extinct, or thought to be extinct, are shown also.

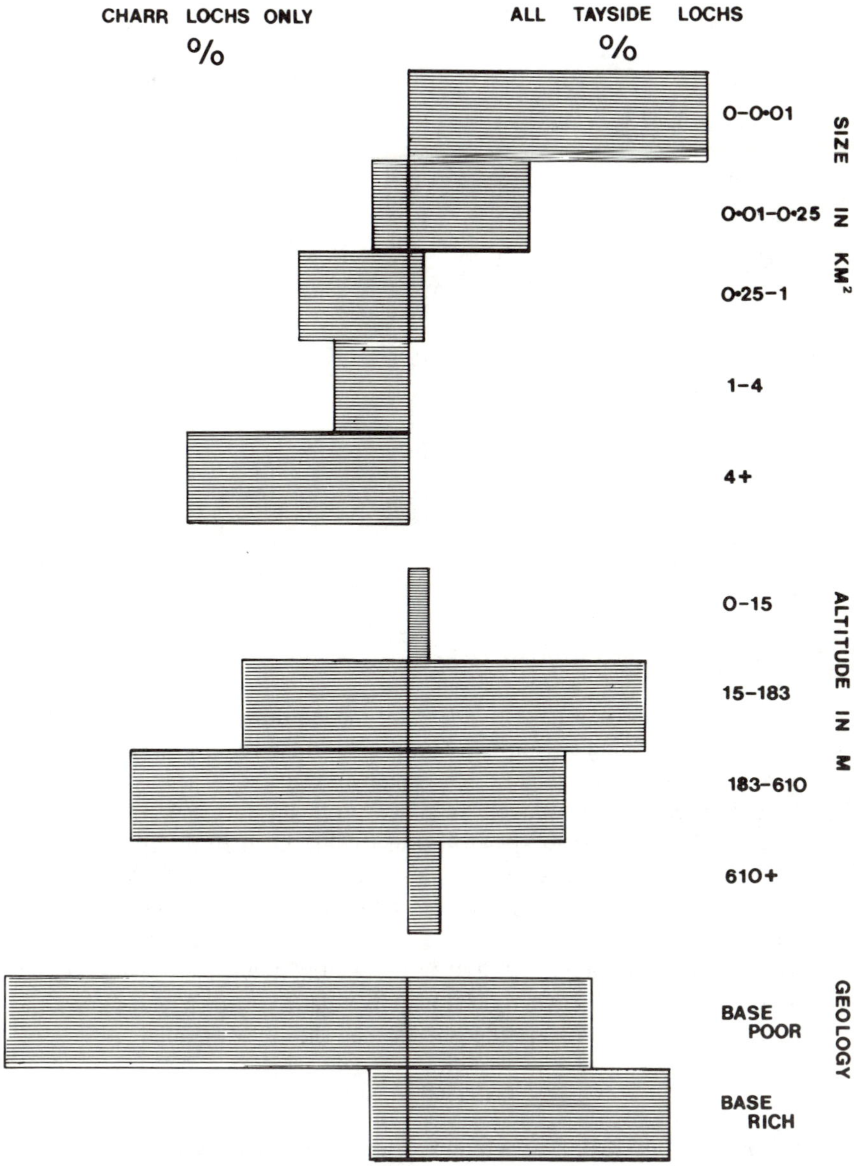

FIG. 2 A comparison of the distribution of sizes, altitudes and base richness between all lochs and charr lochs only in the Tayside Region of Scotland.

Table 3. Unpublished records of charr from running waters in Scotland.

RUNNING WATER	YEAR	NEAREST LOCH
River Borgie	1973	Brora
River Tummel	1960	Dunalistair
River Earn	1979	Earn
Inflow burn	1978	Garry (Perth)
River Spey	1958	Insh
River North Esk	1960	Lee
Outflow burn	1975	Meallt
Inflow burn	1975	Meallt
River Ness	1979	Ness
River Kingie	1954	Quoich
River Tummel	1978	Rannoch
River Shin	1960	Shin
River Shin	1973	Shin

2) Ripe fish which may have accidentally moved downstream into outflows during the spawning season (eg. fish caught in the River Ness below Loch Ness: recorded by Traquhair (1898) and by one of the present authors).
3) Ripe fish which have actively migrated upstream during the autumn into suitable inflow burns to spawn. Formerly the Loch Insh population was thought to be the only one with this habit, but it appears that the same situation exists at Loch na Sealga, possibly Loch Meallt, and probably other lochs.
4) Young fish (probably 0+ and 1+) which, like trout, use suitable inflow burns as nursery areas. The Loch Meallt population is the only one known to have this habit, although further survey may show other similar stocks.

EXTINCTIONS

It is quite certain that charr are now extinct in some lochs where they were previously abundant. The best documented case is that of Loch Leven (Burns-Begg 1874) where charr were at one time very common and taken regularly in nets. However, in 1830 the loch level was lowered by some 1.4 m and its surface area reduced by 546 ha (29 percent) as part of a local land drainage and reclamation project. Charr seemed to disappear shortly after this and have not been recorded since 1837 in this well fished and studied water. The population formerly present in St. Mary's Loch (Maitland 1974) also appears to have disappeared though the reasons are less clear.

In recent years it appears that acid precipitation may have changed the water chemistry of many lochs in granite unbuffered areas. It is well known that this effect in Scandinavia has caused the extinction of many populations of charr (Maitland et al. 1981) and other species. Certainly, recent fishing (gillnetting and angling) in several lochs likely to be affected in this way has failed to produce any charr. The waters concerned so far are Lochs Grannoch, Dungeon, Achray and Venacher.

Charr have never been recorded in several Scottish lochs where their presence on historical grounds seems likely. Examples of such waters are Lochs Lomond, Shiel and Tummel. Loch Lomond is one of only two Scottish lochs to contain whitefish (*Coregonus lavaretus*) and it is known from Scandinavia (Hoglund et al. 1975) that, where this species is introduced to waters containing charr, the latter often disappear, apparently through competition. However, in Loch Eck, both whitefish and charr co-exist — though the latter are unusual morphologically (see below). In shallow waters such as Loch Tummel, it has been suggested that pike (*Esox lucius*) and perch (*Perca fluviatilis*) may eliminate charr and prevent them recolonizing from upstream.

MORPHOLOGY

Meristic Characters

A detailed study of meristic characters in samples of adults from 26 Scottish lochs has been carried out. The logistics and methodology of this work have been described in a preliminary paper (Friend 1959) and the results form the basis of a more detailed paper to be presented elsewhere (Maitland and Friend 1982). It is clear from this that there are considerable and consistent differences in the morphology of various populations. How much of this is due to ecotypic and how much to phenotypic variation is uncertain and considerable further research is warranted in this field, especially using modern biochemical techniques.

Two examples of the variation in meristic characters are given in Figure 3. The number of gillrakers ranges from 13 to 17; as in the whitefish *Coregonus* spp. the numbers may have evolved differentially in relation to feeding habits (Svärdson 1979). Thus a low mean gillraker number is found in populations such as those at Lock Eck (Site 9) where the fish are small and inconspicuous in

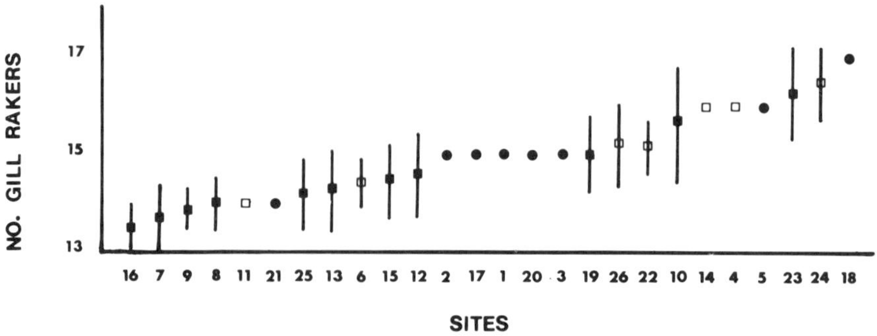

FIG. 3 Examples of the morphological variation among various populations of Scottish charr. The means and range of gillraker number and dorsal fin lengths are shown for each loch, identified by a site number.

colour, apparently having been forced into a deep benthic habit by competition from whitefish and trout (*cf* Nilsson 1965). The charr of Loch Stack, on the other hand, have a much higher mean gillraker number appropriate to their pelagic, plankton-feeding niche. Considerable differences occur too in other characters, such as the length of the dorsal fin (Fig. 3) where long fins appear to be commonly associated with dull-coloured benthic races and short fins with more brightly-coloured pelagic forms.

Walker and Greer (1981) discovered, in 1975, two morphometrically distinct sympatric forms of charr in Loch Rannoch. These two forms also show differences in their ecology and analysis of heavy metal concentrations in viscera and muscle tissue also revealed substantial differences. This is the first recorded instance of two sympatric forms of charr occurring in a Scottish loch and apart from the two sympatric races in Windermere described by Frost (1965) is the only record of this phenomenon in Britain. It is of interest to note that the famous Rannoch charr (originally described as *S. struanensis*) was almost certainly the pelagic form: the type specimens were collected by sweep net in shallow water.

Colour

One of the attractions of charr is the variability and sometimes surprising brightness of their colour. Although there are undoubtedly major differences among many Scottish populations, it must be remembered that within any population there may be variations related to size, sex, season and sexual maturity. An analysis of the adult colour and colour patterns in 14 populations of Scottish charr has been carried out using objective criteria; three reasonably consistent characters were found to be of value in characterizing colour forms.

These were the persistence of parr marks (which may be clearly marked or quite absent in adults of different populations), the pattern of spotting (spots may be absent, present on the sides only, or present on the sides and on the back, where they may elongate to form vermiculate patterns) and the degree of brightness of red pigment, especially on the belly. The latter was categorized using standard colour charts and ranged from pale cadmium yellow through mandarin and blood reds to deep magenta.

Considerable variation in colour exists but there is a tendency for the dull, poorly spotted forms to have persistent parr marks (eg. Loch Ness) whereas the brighter, better spotted forms do not have parr marks as adults (eg. Loch Lee). There is no detailed information on the colour changes in any population related to age, season or sex. The colour of the two forms existing in Loch Rannoch is quite different, the benthic form having persistent parr marks, no spots and a dull colour whereas the pelagic form has no parr marks, faint pink spots on the sides and a general dark claret colour on the body. One of the clearest colour differences in mature fish is in the pectoral fins: the benthic form having orange-yellow fins, the pelagic form having dark grey-black fins.

FOOD

Very little information has been published on the food of charr in Scotland and virtually nothing is known of the diet of young fish. Table 4 summarizes much of the available information. The food of adult fish consists almost entirely of aquatic invertebrates: sometimes it is dominated by crustacean zooplankton and sometimes by zoobenthos; often both these items form an important part of the diet. Seasonal data from several waters, eg. Loch Garry (Greer 1981), indicate that, as with whitefish (Slack 1957), plankton may dominate at one time of the year (summer) and benthos at another (winter). Uncommonly, fish (usually sticklebacks) form part of the diet — as at Loch Borralie (Darling and Boyd 1964) and Loch Meallt (Campbell 1976).

The situation at Loch Rannoch is of particular interest (Walker and Greer 1981). Here, the two sub-populations of charr have quite distinct feeding habits. The diet of the pelagic form is dominated by crustacean zooplankton (usually about 95 percent of the contents), whereas that of the benthic form is made up almost entirely of zoobenthos. Thus the differences in morphology are reflected also in the ecology of these two forms.

Although the stomachs of many charr from Lochs Garry and Rannoch have been examined at spawning time, no eggs have ever been found. This contrasts with the situation in Windermere charr (Frost 1977) and whitefish in Loch Lomond (Slack 1957) where adult fish eat their own eggs in large numbers during the spawning season.

Table 4. The food of charr in various Scottish lochs.

LOCH	YEAR	% COMPOSITION OF FOOD			NUMBER OF FISH	DOMINANT FOOD ITEMS
		PLANKTON	BENTHOS	FISH		
a Bharpa	1978	100	0	0	1	Cladocera
Borralie	1975	0	100	0	4	Ephemeroptera
Fada	1975	78	22	0	25	Cladocera
Garry (Perth)	1977	50	50	0	309	Cladocera, Mollusca
Luichart	1959	48	52	0	18	Diptera, Cladocera
Lurgain	1977	100	0	0	10	Cladocera
Meallt	1975	3	90	7	94	Mollusca, Hemiptera
Morar	1966	100	0	0	12	Cladocera
Moy	1960	100	0	0	20	Cladocera
nam Brathain	1977	50	50	0	12	Cladocera, Diptera
Ness	1977	100	0	0	30	Cladocera
Rannoch (P)	1977	90	10	0	50	Cladocera
Rannoch (B)	1977	5	94	1	50	Mollusca, Trichoptera
Shin	1977	100	0	0	10	Cladocera
Uachdair	1980	50	50	0	2	Cladocera, Diptera

PREDATORS

The principal predators of charr in Scottish lochs appear to be other species of fish. In Loch Tay, Calderwood (1930) recorded a charr in the stomach of a kelt salmon, and in the same loch at the time of writing (March 1981) a fresh-run salmon of about 13 kg contained a very large charr. Mills (1964) showed that pike (*Esox lucius*) in Loch Luichart feed mainly on other fish but charr feature in their diet between April and August. Pike in Loch Rannoch have also been found to contain charr on various occasions. In an extensive study of the food of large brown trout (*Salmo trutta*) in Scottish lochs, Campbell (1979) lists eight lochs where charr have been found in trout stomachs. He notes that small trout may take small charr, but stresses that the large ferox trout may be heavily dependent on charr as a main source of food and indeed that this form of trout rarely occurs where charr are absent.

As with whitefish (*Coregonus albula* and *C. lavaretus*), charr are very prone to attack by eels (*Anguilla anguilla*) when caught in gillnets, and in fact are clearly preferred to trout. How relevant this is to the natural situation is uncertain, but it seems likely that injured, dead or dying charr must be vulnerable to eels. Cannibalism also occurs among charr, and a 375 mm charr caught in Loch Rannoch was found to contain another of 180 mm.

PARASITES

Ectoparasites

A total of some 324 adult charr from 26 different lochs have been examined for ectoparasites. The majority of fish have none, and only two species have been recorded from others. The details of positive records are given in Table 5.

In charr from several lochs, the base of the tail shows small whitish cysts which contain large numbers of the protozoan parasite *Heneguya*.

The most obvious ectoparasite of Scottish charr is *Salmincola edwardsii* a parasitic copepod usually found attached to the walls of the gill cavity (never the gill filaments), and occasionally on the fins (Fryer 1981). This parasite is known to occur in only five of the Scottish lochs (Doon, Lee, Moy, Stack and Tay), but up to 60 percent of the charr in these populations may be infested. In some lochs (eg. Loch Lee) about 15 percent of the brown trout, *Salmo trutta fario* carry quite another parasitic copepod, *Salmincola gordoni*, which is also found in about one percent of the sea trout *Salmo trutta trutta* in the River North Esk.

SPAWNING

Little previous information is available on the spawning of charr in Scotland. A summary of information at present available is given in Table 6. All the

Table 5. Mean density (on infested fish) and percentage occurrence of *S. edwardsii* and occurrence of Heneguya on charr where ectoparasites were recorded.

LOCH	Salmincola edwardsii		Heneguya
	DENSITY	OCCURRENCE	
an Dubh	-	-	+
Assynt	-	-	+
Doine	-	-	+
Doon	1.8	42	+
Lee	1.3	60	+
Moy	2.2	50	+
Stack	1.0	33	-
Tay	1.5	33	-
Voil	-	-	+

populations studied so far appear to spawn in the autumn and there is no evidence of spring spawning as occurs in Windermere (Frost 1965) in England. The only detailed information available to date is from Loch Luichart, where Mills (pers. comm.) caught numbers of charr while gillnetting for pike during the autumns of 1958 and 1959 (Mills 1964). The data on gonad condition (Fig. 4) indicate that in 1958 spawning probably took place from late October to late November and in 1959 from mid October to mid November.

More recently, samples of spawning charr were collected in Loch Garry by Greer (1981) in early October 1973 and 1975 and in late September 1974. These samples were taken at a site 50 m offshore, in 7-10 m of water, at the eastern end of the loch. Ripe and running fish of both sexes were present in both October samples, but in the sample of September 1974 only males were ripe and running. The mature gravid females in this sample did not appear to have reached full spawning condition although they were obviously very near to it.

The sample of September 1974 comprised 69 male charr (56 sexually mature, 13 immature) and 35 female charr (26 sexually mature, 9 immature). A similar proportion of males to females was noted in the October samples. The sexes were almost equally represented in samples taken in April, June and July. All the mature fish of both sexes were at least 2 years old apart from one male found to be mature at 1+ years. Immature fish consisted of 1+ year old fish apart from two 2+ year old female charr.

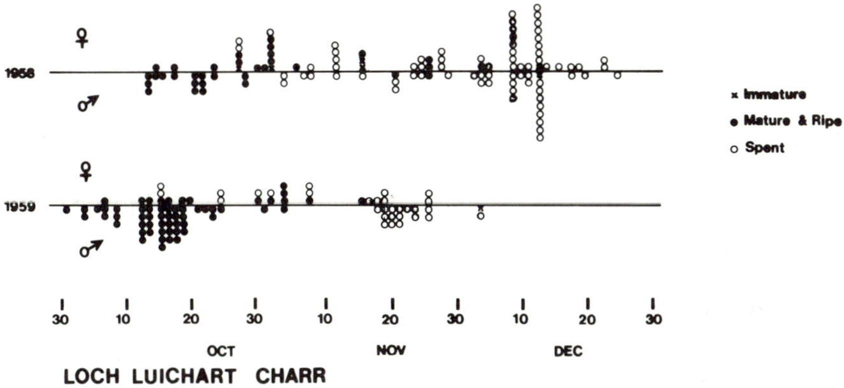

FIG. 4 Gonad conditions of individual charr caught in Loch Luichart in 1958 and 1959.

Although it is probable that the majority of Scottish charr spawn in the lochs themselves — usually along wave-washed shores or over off-shore reefs (e.g. Loch Earn) — it is clear that there are a number of populations which migrate into running water during the spawning season (Table 6). The stream chosen may be the inflow (as at Loch Insh and Loch na Sealga) or the outflow (as at Loch Tarff) of the loch concerned. There appear to be no detailed observations of the spawning act or the actual site of spawning in Scottish waters.

Information on the fecundity of charr in Loch Luichart is available from the samples taken by Mills (pers. comm.) and shown in Figure 4. In 1958, females with a mean length of 210 mm (range: 197-220) had a mean fecundity of 319 eggs (range: 257-406). In 1959, females with a mean length of 233 mm (range: 192-293) had a mean fecundity of 308 eggs (range: 200-437). The relationship between fecundity and length or weight was poor, but the average number of eggs per gram body weight of fish was 2.42.

The fecundity of 20 female charr taken from Loch Garry in September 1974 was examined. These fish had a mean length of 200 mm (range: 170-225 mm), a mean fecundity of 190 eggs (range: 124-261) and an average of 1.63 eggs per gram of body weight (range: 128-206).

SIZE, AGE AND GROWTH

Adult charr in Scotland vary greatly in size from one loch to another. For instance, those of Lochs Doon and Eck are small, with mean lengths of 174 and 200 mm, and weights of 49 and 89 g respectively, whereas those of Lochs Borralie and Insh are much larger with mean lengths of 284 and 365 mm, and weights of 285 and 598 g. The reason for these differences is uncertain. The

Table 6. Information on the spawning of charr in Scotland.

LOCH	SEASON	SITE	SOURCE
Achilty	Autumn	Littoral	Hardie (1940)
an Sealga	Oct-Nov	Inflow	Campbell: 1979
Dochart	Autumn	Littoral	Brown: 1975
Earn	Autumn	Littoral	Barbour: 1980
Garry (Perth)	Sep-Oct	Littoral	Greer: 1975
Girlsta	October	Littoral	Friend: 1960
Insh	Autumn	Inflow	Hardie (1940)
Killin	September	Littoral	Hardie (1940)
Lee	October	Littoral	Friend: 1964
Leven	Autumn	Inflow	Hardie (1940)
Luichart	Oct-Nov	Littoral	Mills: 1959
St Mary's	Autumn	Inflow	Maitland (1974)
Tarff	Autumn	Outflow	Hardie (1940)

largest known Scottish charr of which full details are available was a male 432 mm in length and 814 g in weight that came from Loch Insh; the largest rod-caught fish (801 g) also came from the same waters (Malloch 1910).

The age of a large number of Scottish charr from many locations has been determined using both scales and otoliths (see Swan 1954). The adult populations of many lochs were usually dominated by fish of 3, 4 or 5 years of age (eg. Lochs Garry, Rannoch and Awe) but in some waters (eg. Loch Insh) much older charr occur and fish of 8 and 9 y are not uncommon (Table 7, Fig. 5).

The growth of charr is variable (Table 7). In some lochs (eg. Lochs Garry and Awe) growth in length is slow, fish taking some 4 to 5 yr to reach 200 mm. In other waters (eg. Loch Insh) growth is much more rapid and fish are usually well over 300 mm after 4 or 5 yr. The growth of the two races of charr in Loch Rannoch is of special interest. It appears that although the benthic form may grow more slowly than the pelagic one at first, after the age of 4 yr growth in the benthic race is better and most of the larger (and older) fish belong to this group.

With few exceptions (eg. Loch Awe) charr grow fast during their first year of life and compare favourably with loch-grown trout of the same age (see Campbell 1971). Thereafter the growth of charr is usually much slower than that of trout.

Table 7. Mean length in mm of charr of various age groups from different Scottish lochs.

	AGE GROUPS									
	1+	2+	3+	4+	5+	6+	7+	8+	9+	10+
Awe	-	144	162	178	191	180	-	-	-	-
Borralie	84	163	229	265	340	-	-	-	-	-
Garry (Perth)	123	158	185	203	199	-	225	-	-	-
Insh	-	-	-	336	345	328	350	385	370	-
Lurgain	-	152	175	170	-	-	-	-	-	-
Meallt	-	-	158	168	-	-	-	-	-	-
Ness	-	151	163	215	241	-	-	-	-	-
Rannoch (P)	-	159	174	184	196	195	193	-	-	-
Rannoch (B)	-	158	168	197	199	221	229	224	250	-
Shin	125	163	205	-	-	-	-	-	-	-

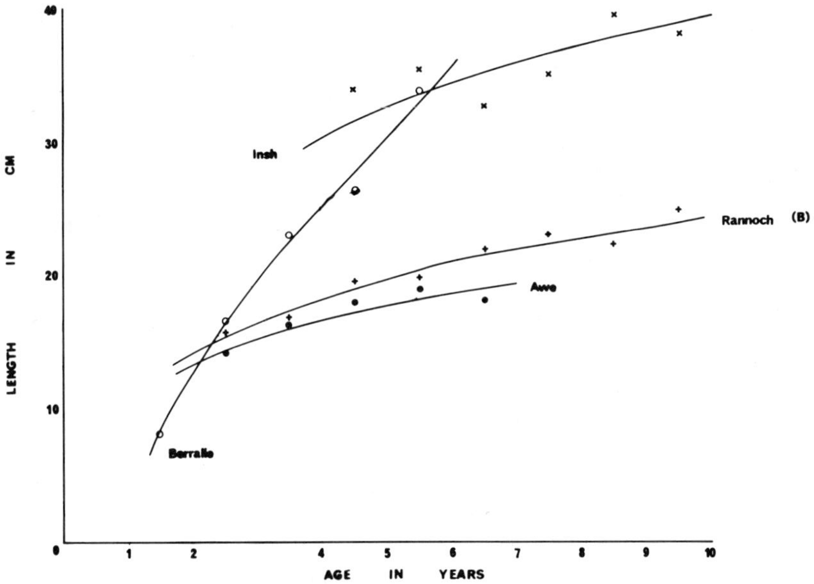

FIG. 5 Growth curves of charr in selected Scottish lochs.

ANGLING

Charr in Scotland are considered by many anlgers to be difficult to catch on rod and line and to be rather rare and elusive fish. In fact, many anglers do not even know of the existence of this species.

The results of a rather limited survey (Table 9) however, indicated that charr are caught (usually incidentally) by angling in a large proportion of the lochs where they are known to occur (40 of 168 listed in the present paper). Further, it is likely that this number is an underestimate as anglers may fail to record their catch of charr and in some cases fail to identify charr correctly. For instance it was found that anglers fishing in Loch Garry (Perthshire) had wrongly identified charr as juvenile rainbow trout (*Salmo gairdneri*, Richardson).

Charr may be caught by accident in the course of fishing for trout, but in a number of lochs are fished for deliberately, (eg. Lochs Bhrodain, Ericht, Garry, Lubnaig, Maree, an t Seilich, Voil). Most of the angling methods used to catch charr are those practiced in normal trout fishing using bait, fly or spinning lures. Anglers specializing in angling for charr may further adapt these trout angling methods specifically for catching charr.

Walker (pers. comm. 1979) noted that while fishing in Loch Garry (Perthshire) he successfully caught charr by allowing his fly line to sink deeper than in the course of normal fishing. A line fished normally was rarely successful in catching charr. McLaren (pers. comm. 1979) found that fly fishing was frequently successful in catching charr in Loch Maree and in a number of other lochs in the Northern Highlands. Spinning and trolling with small spinners is also a successful method of catching charr (Burns, Cairney pers. comm. 1979). One of the authors has caught charr by using tackle similar to that described by Frost (1955) for catching charr in Windermere.

Bait fishing using mainly worms and maggots is commonly used for angling charr. The current rod caught record in Scotland (a charr of 801 gm from Loch Insh) was taken using a bottom fished earthworm.

A paternoster tackle rigged with multiple hook droppers and baited with maggots is frequently used by anglers fishing for charr in the Trossachs area. This is a successful method in Loch Garry (Perthshire) particularly if the tackle is retrieved very slowly along the bottom. Williams (1980) gave a full description of methods for charr fishing in a recent article in the popular angling press.

Large invididual catches are reported. Greer (1981) mentions a catch of over 200 charr, taken by three anglers in Loch Garry (Perthshire) during an afternoon and evening. McLaren (pers. comm. 1979) stated that catches of over 20 charr in a single outing were taken in Loch Maree. He recorded one catch of over 100 charr taken in an evening's fly flishing. Cairney (pers. comm. 1979) has taken up to 40 charr in Loch Bhrodain and up to 80 charr in Loch an Seilich in single outings.

The size of charr taken by angling varies considerably according to locality. In Loch Ericht, Craib (pers. comm. 1973) few charr in excess of 23 cm (ca 100 g) length are taken by angling. Malloch (1910), however, caught a charr of approximately 1 kg in weight in this loch. Walker caught a charr of 30.5 cm in Loch Garry in 1975 but normally few fish greater than 25 cm in length are caught here. In most lochs a charr of 30 cm is considered to be a large fish. Much larger charr are however commonly taken in Loch Insh; anglers fishing for salmon in this loch frequently catch charr over 500 g in weight (McBain pers. comm. 1978).

A more detailed survey of angling catches of charr in Scotland will be necessary before a full picture can be drawn up but it is evident from the existing information that charr are more frequently fished for and larger catches taken than previously supposed.

DISCUSSION

The charr in Scotland has been much neglected in the past by local inhabitants, anglers and research workers. The reason for its infrequent use as a resource was probably the fact that this species, owing to its cryptic habits, was seldom encountered and difficult (by contemporary methods) to harvest. Only in the few locations in Scotland where charr migrate up-stream to spawn would this be possible to any extent. It is interesting to note that there appear to be no Gaelic names for lochs that refer to the presence of charr, while those referring to salmon, trout and pike (also to various birds and mammals) are common.

Sport-fishing did not get underway in the Highlands until the late 18th century but in the prolific writings on angling that have been produced subsequently, very little attention has been paid to charr; usually it was regarded as an oddity and not a worthy target species for anglers. In more recent times little research has taken place on Scottish charr as the emphasis on fishery research in Scotland has been on the main commercial species: Atlantic salmon and migratory (sea) trout and, to a lesser extent, the main anglers' species, brown trout. University research has been on similar lines.

The identification and recording, in recent years, of the many charr lochs in Scotland have produced new horizons for research workers, as the considerable number of charr populations, evolving in isolation since the retreat of the Quaternary Ice Cap (about 10,000 years BP), offer ideal material for the evolutionary biologist. This is in contrast to the many brown trout populations existing in conditions of apparent genetical isolation, for it is impossible, in Scotland, to state with certainty that any brown trout population has not been adulterated by introductions from other geographical areas. From the middle of last century on, most trout lochs and river systems were stocked by the landowners, and even before that, trout were introduced into many fish-less lochs by the local inhabitants as a source of food for the time when they were attending cattle and sheep on high ground during the summer.

Charr were never used to any extent for stocking purposes although some very limited experimental stockings were carried out at the turn of the century. A further attraction to research on charr is in connection with energy-flow studies on the large, deep, recently-glaciated oligotrophic lochs, where the standing crop of fish seems to be high in relation to the chemical poverty of the water, lack of allochthonous organic material and the poor climate.

Charr, currently held in low regard as a target species, may grow in popularity as a sports fish much in the same way that coarse fish, previously largely disregarded by anglers in Scotland, have, in the last decade or so, become popular with local and visiting anglers from other areas of Scotland and abroad. In recent decades the popularity of charr as a sports fish in Norway has increased greatly from a comparatively minor sports fish to a major one (Aass, pers. comm. 1974).

When considering the future potential of charr as a resource in Scotland it should be borne in mind that both recreational and commercial fisheries for charr are common in the Scandinavian countries and Iceland. Many of these fisheries are based on populations living in oligotrophic lakes and reservoirs which consist of charr of small individual size and poor growth rate. A pilot test on the market potential of small sized charr (about 200 mm) revealed that hoteliers would accept such fish and could sell them to tourists. So perhaps the small size of most Scottish charr need not be a complete deterrent against their use as a resource. Natural resources in the Highlands have often, in the past, been unwisely used and if any development of charr fisheries takes place it should be carefully monitored so that misuse does not occur.

As several important charr populations have now become extinct special consideration must be given to the conservation of this species. A number of points are relevant here. Firstly, the majority of Scottish charr populations are probably safe (as far as one can predict anything these days). Secondly, possibly the greatest danger is from (a) introductions of alien fish species, either intentionally or by escapes from aquaculture facilities and (b) 'new' diseases or parasites originating from such operations. In certain places pollution and eutrophication have affected charr lochs and in other regions acid rain has been detrimental to charr populations. Hydroelectric installations, which often favour the same large highland lochs as charr, do not seem to have harmed charr populations. In Scandinavia there is some evidence that regulation for hydro-generation can actually increase the spawning potential of some waters (Aass 1972).

At the present time there are four major charr lochs which are wholly or partly within National Nature Reserves managed by the Nature Conservancy Council in Scotland while several more, as well as a number of smaller ones, are similarly situated within notified Sites of Special Scientific Interest. The wildlife on land carrying either of these official designations is protected to some extent from adverse developments and/or introductions.

A computer listing of all charr waters in Scotland is available on request from the authors.

ACKNOWLEDGEMENTS

We are grateful to Dr. P.F. Friend for putting all his father's research notes and material at our disposal. Dr. D.H. Mills was kind enough to allow us access to his data from Loch Luichart, and we have received many useful notes and pieces of information from several people notably: Mr. A.E. Joyce, Mr. A.F. Walker, Mr. J. Burns and Mr. P. Cairney. Some of the data analysis was carried out by Ms. R. MacKenzie and Miss A.E. Sneddon and Mr. C. McEvoy prepared most of the figures. Mr. Friend's later research was supported by a grant from the Leverhulme Trust. The manuscript was typed by Mrs. M.S. Wilson.

REFERENCES

(including bibliography of major works on charr in Scotland)

Aass, P. 1972. The winter migrations of char, *Salvelinus alpinus* L., in the hydroelectric reservoirs Tunhovdfjord and Pålsbufjord, Norway. Inst. Freshw. Res. Drottningholm Rep. 50:5-44.

Burns-Begg, R. 1874. The Loch Leven angler. Barnett, Kinross.

Calderwood, W.L. 1930. Salmon and sea trout. Arnold, London.

Campbell, R.N. 1971. The growth of brown trout, *Salmo trutta* L. in northern Scottish lochs with special reference to the improvement of fisheries. J. Fish. Biol. 3:1-28.

Campbell, R.N. 1979. Ferox trout, *Salmo trutta* L., and charr, *Salvelinus alpinus* (L.), in Scottish lochs. J. Fish. Biol. 14:1-29.

Campbell, R.N.B. 1976. Examinations of, and comparisons between, the food of two Hebridean loch populations of charr (*Salvelinus alpinus*, Linnaeus 1758) one allopatric, the other sympatric with trout (*Salmo trutta*, Linnaeus 1758). B.Sc. Thesis, University of Aberdeen.

Darling, F.F. and J.M. Boyd. 1964. The highlands and islands. Collins, London.

Day, F. 1887. British and Irish Salmonidae. Williams and Norgate, London.

Friend, G.F. 1956. A new sub-species of char from Loch Eck. Glasg. Nat. 17:219-220.

Friend, G.F. 1959. Subspeciation in British chars. System. Ass. Publ. 3:121-129.

Frost, W.E. 1952. Predators on the eggs of char in Windermere. Salmon Trout Mag. 136:193-197.

Frost, W.E. 1955. An historical account of char in Windermere. Salmon Trout Mag. 143:15-24.

Frost, W.E. 1965. Breeding habits of Windermere char, *Salvelinus willughbii* (Gunther) and their bearing on speciation in those fish. Proc. R. Soc. Lond. B. 163:232-284.

Frost, W.E. 1977. The food of char, *Salvelinus willughbii* (Gunther), in Windermere. J. Fish. Biol. 11:531-547.

Fryer, G. 1981. The copepod *Salmincola edwardsii* as a parasite of *Salvelinus alpinus* in Britain and a consideration of the so-called relict fauna of Ennerdale Water. J. Zool. Lond. 193:253-268.

Greer, R.B. 1983. Observations of the biology of charr in Loch Garry, Tayside, Scotland. (In preparation).

Gunther, A.C.L.G. 1880. An introduction to the study of fishes. Adam and Black, Edinburgh.

Hardie, R.P. 1940. Ferox and char in the lochs of Scotland. Oliver and Boyd, Edinburgh.

Harvie-Brown, J.A. and T.E. Buckley. 1888. A vertebrate fauna of the Outer Hebrides. Douglas, Edinburgh.

Hoglund, L.B., A. Bohman, and N.-A. Nilsson. 1975. Possible odour responses of juvenile Arctic char (*Salvelinus alpinus* (L.)) to three other species of subarctic fish. Inst. Freshw. Res. Drottningholm Rep. 54:21-35.

Holden, A.V. 1966. A chemical study of rain and stream waters in the Scottish highlands. Freshwat. Salm. Fish. Res. 37:1-17.

Kipling, C. and W.E. Frost. 1978. The weight-length relationship of charr, *Salvelinus willughbii* (Gunther), in Windermere, and their use for the determination of age and growth. J. Cons. Perm. Int. Explor. Mer. 38:208-215.

Maitland, J.R.G. 1887. The history of Howietoun. Guy, Stirling.

Maitland, P.S. 1972. A key for the identification of the freshwater fishes of the British Isles with notes on their ecology and distribution. Sci. Publ. Freshw. Biol. Ass. 27:1-139.

Maitland, P.S. 1974. The conservation of freshwater fishes in the British Isles. Biol. Conserv. 6:7-14.

Maitland, P.S. and G.F. Friend. 1983. A morphological analysis of Arctic charr, *Salvelinus alpinus* (L.) from various Scottish lochs (In preparation).

Maitland, P.S., H.A. Regier, G. Power and N.-A. Nilsson. 1981. A wild salmon, trout and charr watch: an international strategy for salmonid conservation. Submitted to STOCS.

Maitland, P.S., K.H. Morris and A.A. Lyle 1981. The fresh waters of the Tayside area 1. Introduction and map analysis. Unpublished ITE Report.

Malloch, P.D. 1910. Life history of the salmon, sea trout and other freshwater fish. Adam and Black, London.

Maxwell, H. 1926. The char of Loch Ossian. Scot. Nat. 1926:175-176.

Mills, D.H. 1964. The ecology of the young stages of the Atlantic salmon in the River Bran, Ross-shire. Freshwat. Salm. Fish. Res. 32:1-58.

Murray, J. and L. Pullar. 1910. Bathymetrical survey of the Scottish freshwater lochs. Challenger, Edinburgh.

Nilsson, N.-A. 1965. Food segregation between salmonoid species in North Sweden. Inst. Freshw. Res. Drottningholm Rep. 46:58-78.

Pennington, W., E.Y. Haworth, A.P. Bonny and J.P. Lishman. 1972. Lake sediments in northern Scotland. Phil. Trans. R. Soc. Lond. B. 264:191-294.

Regan, C.T. 1911. British freshwater fishes. Methuen, London.

Slack, H.D. 1957. Studies on Loch Lomond 1. Blackie, Glasgow.

Smith, I.R. and A.A. Lyle. 1979. Distribution of fresh waters in Great Britain. Institute of Terrestrial Ecology, Cambridge.

Svärdson, G. 1979. Speciation of Scandinavian *Coregonus*. Inst. Freshw. Res. Drottningholm Rep. 57:1-95.

Swan, M.A. 1954. Difficulties encountered in reading char scales. Salmon Trout Mag. September: 505-512.

Swan, M.A. 1964. The Girlsta char. One of Shetland's oldest natives. New. Shetl. Mag. 1964:23-26.

Toivonen, J. 1972. The fish fauna and limnology of large oligotrophic glacial lakes in Europe (about 1800 AD). J. Fish. Res. Board Can. 29:629-637.

Traquhair, R.H. 1898. On a peculiar charr from Inverness-shire. Ann. Scot. Nat. Hist. 78-79.

Walker, A.F. and R.B. Greer. 1983. Two morphologically distinct forms of charrs in Loch Rannoch. (In preparation).

Williams, P. 1980. Catching charr. Rod and Line. October 1980.

Habitat Selection and Segregation of Parr of Arctic Charr (*Salvelinus alpinus*), Brown Trout (*Salmo trutta*) and Atlantic Salmon (*Salmo salar* L.) in Two Streams in North Norway

TOR G. HEGGBERGET

Directorate for Wildlife and Freshwater Fish, Elgeseter gt. 10, N-7000 Trondheim, Norway

HEGGBERGET, T.G. 1984. Habitat selection and segregation of parr of Arctic charr (*Salvelinus alpinus*), brown trout (*Salmo trutta*) and Atlantic salmon (*Salmo salar* L.) in two streams in North Norway, p. 217–231. *In* L. Johnson and B.L. Burns [eds.] Biology of the Arctic charr, Proceedings of the International Symposium on Arctic Charr, Winnipeg, Manitoba, May 1981. Univ. Manitoba Press, Winnipeg.

Habitat selection in sympatric populations of presmolt Atlantic salmon (*Salmo salar* L.), brown trout (*Salmo trutta* L.) and Arctic charr (*Salvelinus alpinus* L.) in two streams, Kåfjordelva and Skibotnelva, North Norway, was examined by electrofishing.

Highly significant differences exist between the distribution of salmon and charr and salmon and trout, while there are only small differences in habitat selection between trout and charr. Salmon were located further from the river banks, in deeper water and in higher water velocities than the other two species. In Skibotnelva, 70, 50 and 30 percent of charr, trout and salmon respectively were caught within 1 m of the banks. In Kåfjordelva the corresponding numbers were 60, 60, and 8 percent. In relation to water depth, 94 percent of the charr, 90 percent of the trout and 60 percent of the salmon in Skibotnelva were located at depths between 0 and 20 cm. In Kåfjordelva the corresponding numbers were 92, 90 and 52 percent. 70 percent of the charr, 50 percent of the trout and 20 percent of the salmon were caught in water velocities between 0 and 10 $cm.sec^{-1}$ in Skibotnelva, while 85 percent of the charr, 75 percent of the trout and 23 percent of the salmon in Kåfjordelva were caught within the same water velocity range. At the sampling localities having the lowest water velocities, charr were the most abundant species. Charr also seemed to be located further from the banks in the pools rather than in the fast flowing localities.

There is a strong interdependence of the examined parameters, but the results indicate that water velocity is the most important in segregating these species. Each species is forced into the microhabitat to which it is best adapted.

The waterflow of these rivers is now being reduced due to hydroelectric development. In an attempt to maintain the water level in the rivers, it has been decided to establish weirs. To maintain the original fish fauna, a new type of weir is being tested. In contrast to the traditional type, this new weir will not create large areas of still or slow-flowing water. The decision for constructing this new type of weir in these streams is based on the results of the present research, which demonstrate a preference by young trout and charr for slow-flowing water.

INTRODUCTION

Anadromous Arctic charr (*Salvelinus alpinus* L.) are distributed in Norway along the coast from Finnmark south to mid-Norway (65°N) (Nordeng 1961). In the rivers populated by Arctic charr, there are usually sympatric populations of brown trout (*Salmo trutta* L.) and Atlantic salmon (*Salmo salar* L.) (Berg 1965).

Social behaviour and competition in young stages of brown trout and Atlantic salmon have been described by several workers (Lindroth 1955, Kalleberg 1958, Keenleyside 1962, Hartman 1963, Heggberget 1974, Karlstrøm 1977), but studies of social behaviour in young charr have been relatively infrequent; most studies have dealt with adult fish or comparisons of charr with various other salmonids (Noakes 1980). These workers point out that Atlantic salmon, brown trout and Arctic charr are all territorial, assuming the definition of territory as being "any defended area" (Noble 1939). The three species do not only defend territories against individuals of their own species (Noakes 1980), but also against fish of related species (Kalleberg 1958).

There is little information about competition amongst sympatric stocks of young Arctic charr, Atlantic salmon and brown trout. Competition and agonistic behaviour are described for sympatric populations of Atlantic salmon and brown trout by Kalleberg (1958) and Karlstrøm (1977), while the social behaviour in Arctic charr is described by Noakes (1980). Competition between rainbow trout (*Salmo gairdneri* Richardson) and brook trout (*Salvelinus fontinalis*) is describedby Stringer and Hoar (1955) and Newman (1956). The preceding studies show that young stream dwelling salmon, brown trout and Arctic charr are biologically similar and that these species have similar environmental demands. Thus, when the three species occur sympatrically, as they do in north Norway, there is great potential for competition amongst them.

THE STUDY AREA

Two small rivers, Kåfjordelva and Skibotnelva (69°N) were chosen for this study (Fig. 1). The catchment areas of the two rivers are 288 km^2 (Kåfjordelva) and 784 km^2 (Skibotnelva). Kåfjordelva was first regulated for hydroelectric purposes in 1971 and Skibotnelva in 1979, after this study was carried out. Skibotnelva is a wider and deeper stream than Kåfjordelva.

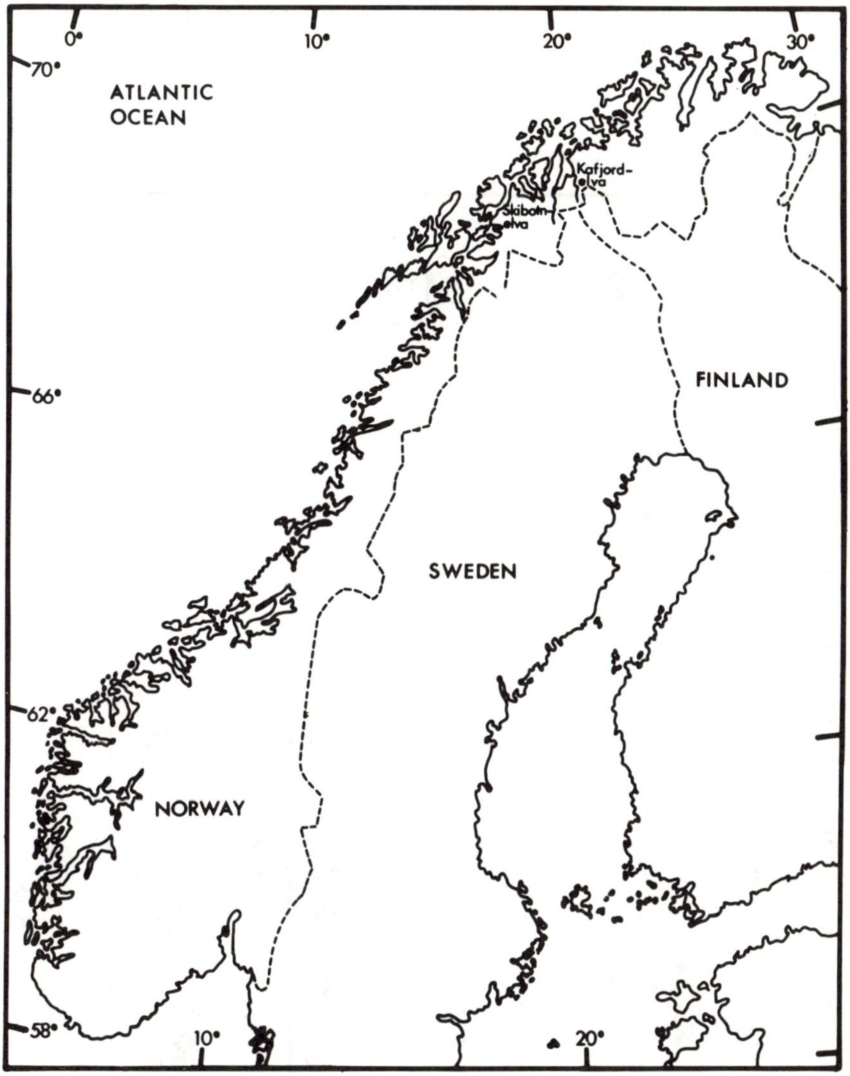

FIG. 1 Map showing location of the rivers chosen for study.

MATERIALS AND METHODS

The fish were sampled by electrofishing (D.C. type). The approximate location of each fish caught was determined by measuring the distance from the river bank, the water depth and water velocity. Possible differences in distribution of these species in relation to these parameters were tested by using a chi-square test (Guttman and Wilks 1967, p. 197).

RESULTS

The results were grouped into four categories, depending on the composition of the fish species at each sampling locality. Category one consisted of fish from the sampling stations where at least 5 specimens of each species were caught.

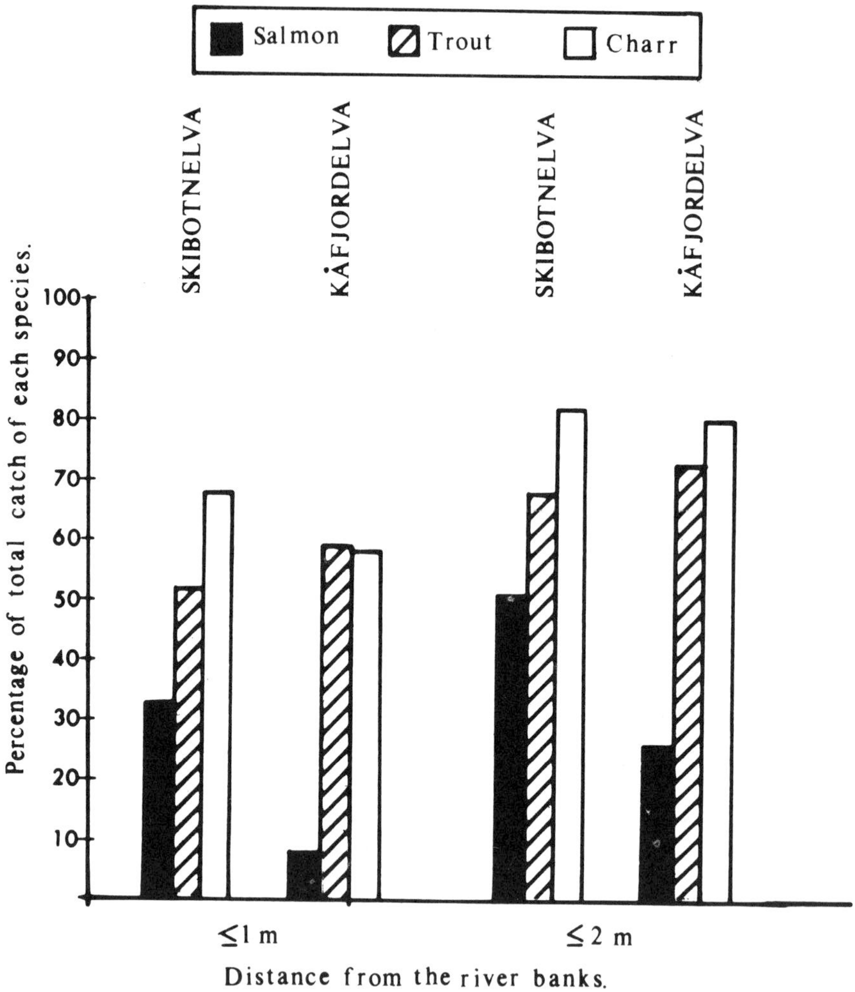

FIG. 2 Distribution of salmon, trout and charr in relation to distance from the river banks.

Table 1. Differences in distribution of salmon, trout and charr in sympatric populations of the three species in Kåfjordelva and Skibotnelva.
S = salmon; T = trout; C = charr
$p < 0.05$ * significantly different distribution
$p < 0.01$ * * highly significantly different distribution
$p > 0.05$ ns not a significantly different distribution

River	Species	Parameter	Intervals		
		Distance from the river banks	0-1 m		0-2 m
Skibotn	S-T		**		**
	S-C		**		**
	T-C		ns		*
Kåfjordelv	S-T		**		**
	S-C		**		**
	T-C		ns		ns
		Water Depth	0-10 cm	0-20 cm	0-30 cm
Skibotn	S-T		**	**	**
	S-C		**	**	**
	T-C		**	ns	ns
Kåfjordelv	S-T		**	**	**
	S-C		**	**	**
	T-C		ns	ns	ns
		Water velocity	0 $cm \cdot s^{-1}$	0-10 $cm \cdot s^{-1}$	0-20 $cm \cdot s^{-1}$
Skibotn	S-T		**	**	**
	S-C		**	**	**
	T-C		ns	**	**
Kåfjordelv	S-T		**	**	**
	S-C		**	**	**
	T-C		ns	ns	ns

The number of fish from these stations totalled 329 salmon, 97 trout and 271 charr. Category two consisted of samples having only salmon and charr or fewer than 5 specimens of trout. The numbers of salmon and charr at these stations were 46 and 43, respectively. Category three had charr and trout (fewer than 5 salmon), the numbers being 105 and 62 respectively. The fourth category consisted of salmon and trout (fewer than 5 charr), and the numbers of these species were 166 and 21, respectively.

The total number of fish of each species caught in this study were: 541

salmon, 180 trout and 419 charr. The length of fish ranged from 50 to 150 mm. The densities of fish are presented as the total numbers of fish caught after three electrofishing runs.

Comparison of Distribution at Stations with Three Species

When relating the distribution of salmon, trout and charr to their distance from the river banks, salmon seemed to be segregated from trout and charr. Some 50-70 percent of the trout and charr were observed within 1 m of the banks, while only 10-30 percent of the salmon were located in the same area (Fig. 2). The segregation was significantly different, while there were smaller differences in the distribution of trout and charr (Table 1). In total, 70-80 percent of the trout and charr were found within 2 m of the river banks, while only 30-50 percent of the salmon were caught within 2 m of the banks.

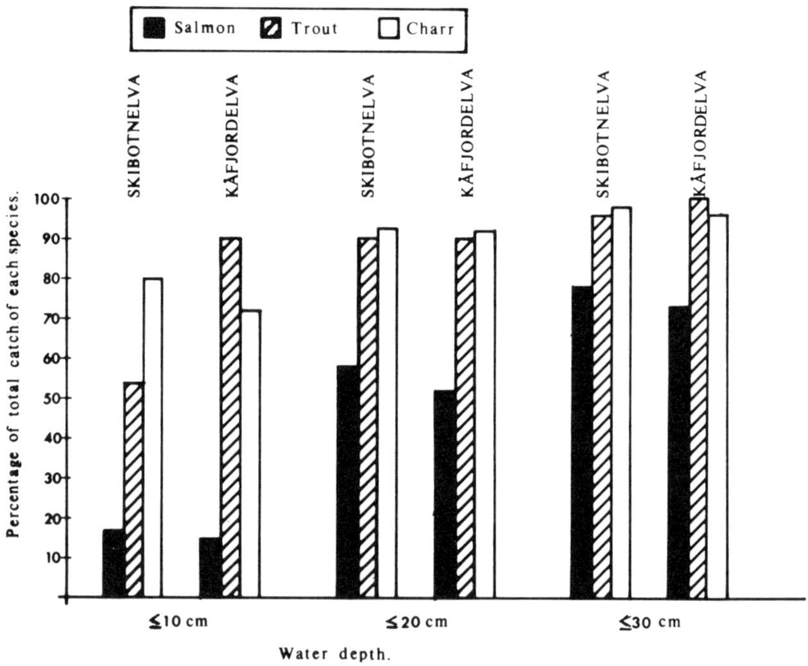

FIG. 3 Distribution of salmon, trout and charr in relation to water depth.

A comparison of the distribution of these species in relation to water depth showed that the salmon were found in deeper water than both the trout and charr (Fig. 3). In less than 10 cm of water in Skibotnelva the distribution of trout and charr was significantly different. About 80 percent of the charr were observed at depths less than 10 cm while about 50 percent of the trout were observed in the same depth range. At depths up to 20 and 30 cm there was no significant

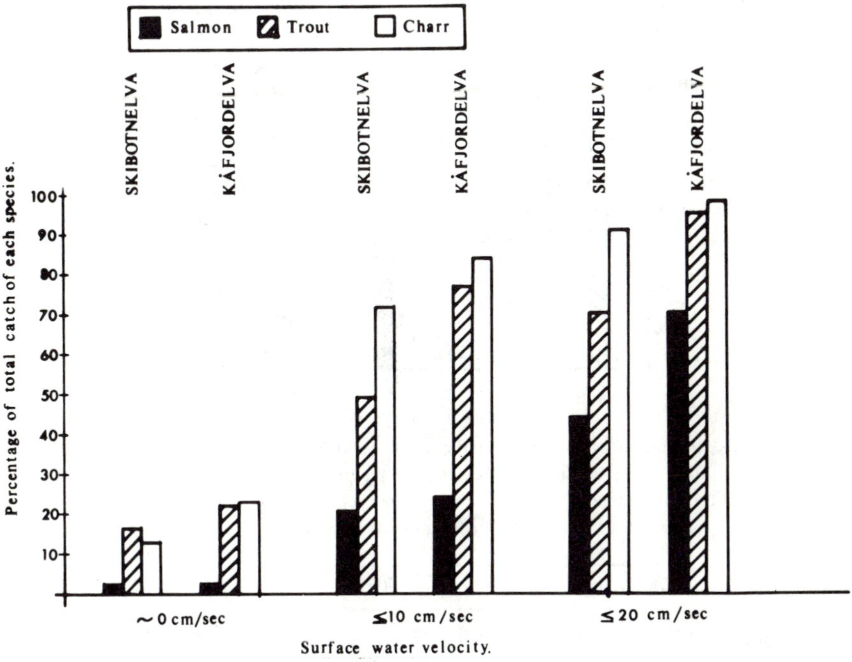

FIG. 4 Distribution of salmon, trout and charr in relation to water velocity.

difference in the distribution between the trout and charr in both Skibotnelva and Kåfjordelva. However, in both rivers, there were highly significant differences in the distribution of salmon compared to that of trout and charr.

In relation to water velocity, the distribution of salmon compared to that of trout and charr was found to be highly significantly different in all cases, both in Skibotnelva and Kåfjordelva (Table 1). Concerning trout and charr, the charr were found to be more frequent in slow-flowing water (Fig. 4), while the distribution of these species was significantly different only at water velocities between 0-10 and 0-20 cm $\times$ sec^{-1} in Skibotnelva (Table 1).

Comparison of Distribution at Stations Where Only Two Species were Present

At localities having only trout and charr, the trout were found to be more abundant near the river banks (Fig. 5) and in shallow water (Fig. 6), while the charr were more frequent at low water velocities (Fig. 7).

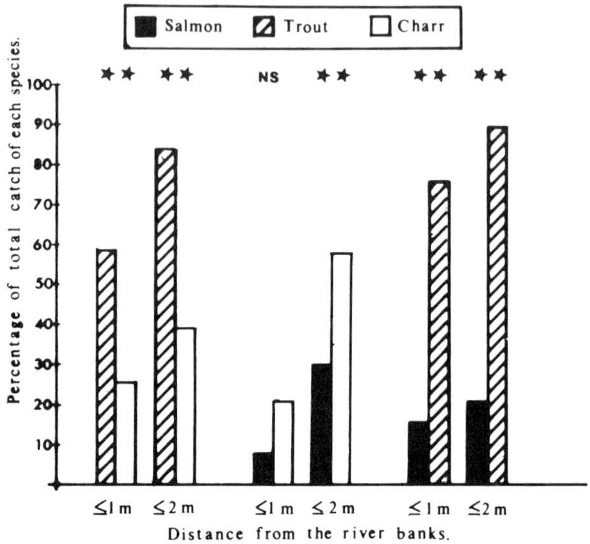

FIG. 5 Distribution of salmon, trout and charr in relation to distance from the river banks at localities where two of the three species are present. * * = $p < 0.01$ highly significantly different distribution. * = $p < 0.05$ significantly different distribution. – = $p > 0.05$ not a significantly different distribution.

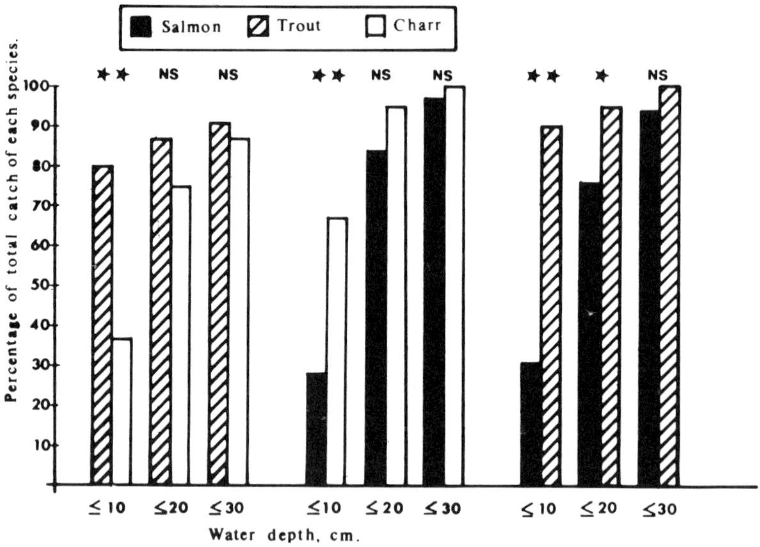

FIG. 6 Distribution of salmon, trout and charr in relation to water depth at localities where two of the species are present. (For explanation of the symbols, see Fig. 5)

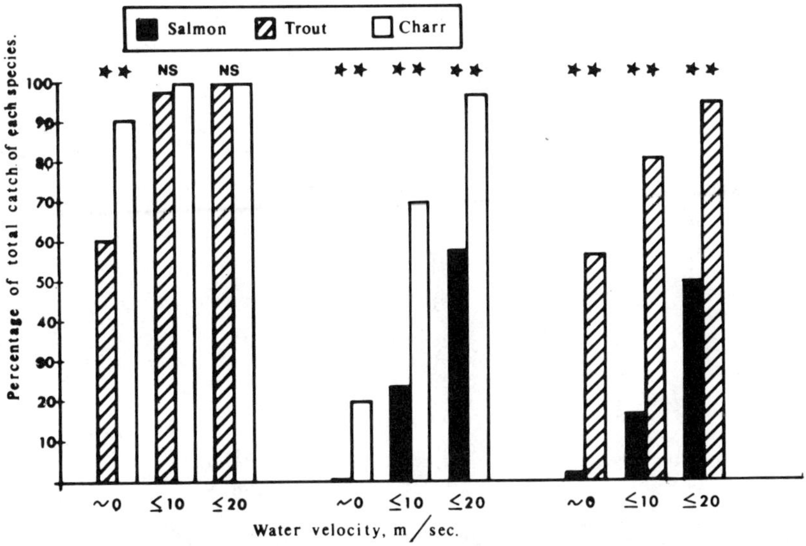

FIG. 7 Distribution of salmon, trout and charr in relation to water velocity at localities where two of the species are present. (For explanation of the symbols, see Fig. 5)

At stations where salmon and charr were present, there were significantly different distributions at $\leq$ 2 m from the river banks, at water depths equal to or less than 10 cm and at water velocities of $\simeq$ 0, 10 and 20 cm $\times$ sec^{-1}. In all cases, the salmon were located further from the banks, in deeper water and at higher water velocities than the charr.

Where only trout and salmon were present, there were highly significant differences in distribution in all cases except for depths more than 10 cm. The salmon were observed further from the banks, at greater water depths and at higher water velocities than the trout.

Segregation at Different Stations of the River

Sampling of fish was carried out at 11 different stations in Kåfjordelva and 6 stations in Skibotnelva (Table 2).

In Kåfjordelva, charr were present at all the localities, while only three of the 6 sampling stations in Skibotnelva contained charr. Salmon were absent from the three localities in Kåfjordelva which were dominated by pools and slow-flowing water. The three other stations were dominated by fast-flowing water (› 30 cm $\times$ s^{-1}), and only in a narrow zone along the banks was the water slow-flowing.

In Kåfjordelva there is little longitudinal zonation of the charr distribution. In the lower part of the river (< 5 km from the sea) charr constituted about 60 percent,

Table 2. Composition of species and densities of salmon, trout and charr in Kåfjordelva and Skibotnelva.
S = salmon, T = trout, C = charr

River	Locality	Distance from the sea km	Composition of species (%) S	T	C	Total density N/100 m^2
Kåfjordelva	1	< 1	9	29	62	15.2
	2	< 1	48	15	37	5.5
	3	< 1	0	31	69	16.4
	4	3	0	23	77	7.3
	5	4	0	59	41	17.3
	6	5	50	6	44	35.1
	7	6	36	15	49	22.9
	8	6	39	0	61	10.4
	9	7	62	3	35	26.5
	10	7	79	5	16	4.3
	11	8	14	31	65	14.1
Skibotnelva	1	2	57	14	29	7.0
	2	6	41	0	59	9.4
	3	7	98	2	0	33.3
	4	8	65	4	31	8.7
	5	9	85	15	0	6.3
	6	9	94	6	0	12.8

while in the upper reaches (5-8 km from the sea) they made up about 45 percent. The trout were also more abundant in the lower stretches, while salmon were more frequent in the upper reaches of the river.

In Skibotnelva charr were the most abundant in the lower part, while salmon were the most abundant in the upstream sections of the river. There were also generally lower water velocities at the sampling stations situated in the downstream part of Skibotnelva.

Comparison of Distribution Between the Rivers

Table 3 shows that the same species had a different distribution in Kåfjordelva and Skibotnelva in about one third of the cases. In relation to distance from the

Table 3. Comparison of distribution of salmon, trout and charr between the rivers Kåfjordelva and Skibotnelva.
S = salmon, T = trout; C = charr
$p < 0.05$ * significantly different distribution
$p < 0.01$ * * highly significantly different distribution
$p > 0.05$ ns not a significantly different distribution

River	Species	Parameter	Intervals		
		Distance from the river banks	0-1 m	0-2 m	
Kafjordelv-Skibotnelv	S-S		**	**	
	T-T		ns	ns	
	C-C		ns	ns	
		Water depth	0-10 cm	0-20 cm	0-30 cm
Kafjordelv-Skibotnelv	S-S		ns	ns	ns
	T-T		**	ns	ns
	C-C		ns	ns	ns
		Water velocity	0 $cm \cdot s^{-1}$	0-10 $cm \cdot s^{-1}$	0-20 $cm \cdot s^{-1}$
Kafjordelv-Skibotnelv	S-S		ns	ns	**
	T-T		ns	*	*
	C-C		*	ns	*

river banks, the salmon was found closer to the shores in Skibotnelva than in Kåfjordelva. Trout in Kåfjordelva were found in shallower water than in Skibotnelva and all three species were found at lower water velocities.

DISCUSSION

Observations on alevins of Arctic charr in stream enclosures or aquarium tanks (Fabricius 1953, Fabricius and Gustafson 1954, Frost 1965) showed an almost total lack of aggressiveness and the absence of any behaviour indicating defence of territories. Frost (1965) observed that only in one instance was a wandering alevin chased away by an occupant of a station. However, Nilsson (1963) showed that when alevins of Arctic charr and brown trout were kept together in aquaria, they tended to fight interspecifically, trout being the more aggressive. Fabricius (1953) found that aggressive behaviour in adult charr occurred in both sexes and at all times of the year, and in young specimens it was seen at least toward the end of their first summer of life. Noakes (1980) compared behaviour of one year old *S. alpinus, S. namaycush* and *S. fontinalis* in aquaria. He observed a higher level of agonistic behaviour in *S. alpinus* than in *S. namaycush* and *S. fontinalis* in allopatric samples of these species. Concerning intraspecific behaviour, Noakes (op. cit.) observed territoriality among *S. alpinus* and *S. fontinalis*. Only areas near the substrate were defended, and the largest fish were dominant.

For Atlantic salmon and brown trout, a few researchers have shown that the presmolt stages of these species are territorial and show agonistic behaviour, both inter- and intraspecifically (Lindroth 1955, Kalleberg 1958, Karlstrøm 1977).

The relatively few studies of fish populations where Arctic charr or other charr species were present in sympatry with salmon and/or trout indicate that salmon, trout and charr are all territorial and that they exhibit intra- and/or interspecific agonistic behaviour (Newman 1956, Chapman 1962, Keenleyside and Yamamoto 1962, Symons 1968). Thus, in Kåfjordelva and Skibotnelva, where salmon, trout and charr are all present, it is reasonable to hypothesize that these species are competing interspecifically.

Salmon were found to be located further from the river banks, in deeper water and at higher water velocities than the other two fish species investigated in this study. This is in agreement with the results of other investigations of habitat selection in sympatric populations of salmon and trout (Lindroth 1955, Heggberget 1974, Karlstrøm 1977). These studies demonstrated a preference by salmon for fast-flowing water.

At localities where only two of the species were present, the salmon were found further from the river banks and at higher water velocities than either the trout or charr. With relation to water depth, salmon were located in deeper water than both charr and trout, but the differences were not significant in all cases.

The differences in the distribution of sympatric trout and charr were smaller than for sympatric salmon and trout or salmon and charr. The trout were more frequently found near the river bank and in shallow water than the charr, while the charr were more abundant at low water velocity.

There is a strong interdependence of the parameters measured in this research. In general, depth and water velocity increase with distance from the banks. However, at the localities in Kåfjordelva where salmon were absent (loc. 3, 4, 5) there were large areas of still or very slow-flowing water extending far from the banks. A possible explanation of the observations at these localities is that charr are more aggressive than trout in still, or slow-flowing water, forcing the trout from the deeper areas of the pool into the shallow areas near the banks. At the localities where salmon, trout and charr were present, there were only minor differences of distribution between the trout and charr. At these localities, both trout and charr may be forced into the zone near the river bank due to the more aggressive behaviour of salmon in fast-flowing water (Kalleberg 1958). The methods used in the present research are too imprecise to unveil possible minor differences in distribution of the fish when they are together in this narrow zone.

Gibson (1966) and Gibson and Keenleyside (1966) observed habitat segregation of *S. fontinalis* and Atlantic salmon in sympatric populations. They found that the salmon were more frequent in the fast-flowing areas and in midstream, while the brook charr were located in the pools and near the river banks. Interaction between Arctic charr and brown trout in lakes has been investigated by several authors (Nilsson 1963, 1967, Svärdson 1976). They conclude that charr usually inhibit trout when existing sympatrically in lakes. Curry-Lindahl (1957) reported that the upper regions of North-Swedish rivers were occupied only by charr, while the numbers of charr in the lower regions of the rivers decreased due to interactions with other fish species. According to Beamish (1980) charr are rather poor swimmers compared to other salmonids.

The longitudinal distribution of charr observed in Kåfjordelva and Skibotnelva differs from the model proposed by Power (1973). He observed that charr generally occupied the upper, unproductive headwaters in North-Norwegian rivers, while trout occupied the slightly warmer, more productive middle sections, salmon occupying the lower regions. Between each section there was considerable overlapping but only in a few places did each of the three species contribute to the total fish production with any degree of equality (Power 1973). Both in Kåfjordelva and Skibotnelva, charr and salmon were the dominant species in the lower regions. According to Johnson (1980), one of the species in sympatric populations of brook charr, Arctic charr and Atlantic salmon almost invariably predominates. The Leaf River in Northern Quebec is the only river known where Arctic charr and Atlantic salmon both contribute with about equal proportions (Jessop et al. 1970). The rivers Kåfjordelva and Skibotnelva are two more examples of Arctic charr and Atlantic salmon coexisting in about equal proportions.

When two or more ecologically similar species occupy the same geographical area, they segregate completely by selecting different habitats (Grinnell 1904) or imperfectly by choosing habitats that overlap, attempting to share certain resources through interactive processes (Brian 1956). This second form of segregation, interactive segregation (Brian, op. cit.), implies that small ecological differences are magnified in competing species. Interactive segregation in stream dwelling young salmon, trout and charr has been demonstrated by Curry-Lindahl (1957), Kalleberg (1958), Gibson (1956) and Heggberget (1974).

There were some differences in distribution of the same species in the two rivers (Table 3). However, the relative distributions were principally the same, the salmon being found further from the banks, in deeper water and in higher water velocities than charr and trout in both rivers (Figs. 3-4). The between-river differences might be caused by small physical differences of Kåfjordelva and

Skibotnelva. Skibotnelva is a wider and deeper stream than Kåfjordelva, hence it was possible to carry out electrofishing in Kåfjordelva further from the banks.

Due to hydroelectric development, the waterflow of the rivers Kåfjordelva and Skibotnelva is reduced. The traditional way of mitigating damage due to reduced waterflow in Norwegian rivers is to establish weirs which create large flooded areas of still or slow-flowing water (Mellquist 1976). Due to the preference by trout and charr for still or slow-flowing water, weirs of the traditional type are thought to favour the populations of charr and trout but inhibit the populations of salmon. A new type of weir to enhance salmon production which does not create large areas of still water is therefore now being constructed in the rivers Kåfjordelva and Skibotnelva. These weirs are a socalled "Syvde-type", having two projecting piers from each side of the river, covering about two thirds of the river bed (Beheim et al. 1977)

REFERENCES

Beamish, F.W.H. 1980. Swimming performance and oxygen consumption of the charrs, p. 739-749. *In* E.K. Balon [ed.] Charrs: salmonid fishes of the genus *Salvelinus*. Dr. W. Junk Publishers, The Hague.

Beheim, E., K.W. Jensen, P. Mellquist and Ø. Vasshaug. 1977. Biotopforbedring i regulerte og uregulerte laksevassdrag. VN-Rapport nr. 3, 29 p. (In Norwegian).

Berg, M. 1965. North Norwegian salmon rivers. J. Grundt Tanum Forlag. 298 p. (In Norwegian with English summary).

Brian, M.W. 1956. Segregation of the species of the ant genus Myrmica. J. Anim. Ecol. 25: 319-337.

Chapman, D.W. 1962. Aggressive behaviour in juvenile coho salmon as a cause of emigration. J. Fish. Res. Board Canada 19: 1047-1080.

Curry-Lindahl, K. 1957. The occurrence of the char in running waters in Arctic and high boreal areas in Sweden. Kungl. Fysiogr. Sallsk. Lund Forh. 27(12): 161-172.

Fabricius, E. 1953. Aquarium observations on the spawning behaviour of the char, *Salmo alpinus*. Inst. Freshw. Res. Drottningholm Rep. 34: 14-48.

Fabricius, E. and K.J. Gustafson. 1954. Further aquarium observations on the spawning behaviour of the char, *Salmo alpinus* L. Inst. Freshw. Res. Drottningholm Rep. 35: 58-104.

Frost, W.E. 1965. Breeding habits of Windermere char, *Salvelinus willugbii* (Gunther) and their bearing on speciation in these fish. Proc. R. Soc. Edinb. B. 163: 232-284.

Gibson, R.J. 1966. Some factors influencing the distribution of brook trout and young Atlantic salmon. J. Fish. Res. Board Canada 23 (12): 1977-1980.

Gibson, R.J. and M.H.A. Keeleyside. 1966. Responses to light of young Atlantic salmon (*Salmo salar*) and brook trout (*Salvelinus fontinalis*). J. Fish. Res. Board Canada 23 (7): 1007-1024.

Grinnell, J. 1904. The origin and distribution of the chestnut-backed chikadee. Auk 21: 364-382.

Guttman, I. and S.S. Wilks. 1967. Introductory engineering statistics. John Wiley and Sons Inc. New York, London, Sydney. 340 p.

Hartman, G.F. 1963. Observation on behaviour of juvenile brown trout in a stream aquarium during winter and spring. J. Fish. Res. Board Canada 20 (3): 769-787.

Heggberget, T.G. 1974. Habitatvalg hos yngel av laks (*Salmo salar* L.) og ørret (*Salmo trutta* L.). K. nosrke Vidensk. Selsk., Mus. Rapport Zool. Ser. 1974-12. (In Norwegian).

Jessop, B.M., R.L.G. Lee and G. Power. 1970. Observations on the fish fauna of the Leaf River, Ungava, Canada, Can. Field-Nat. 84: 365-367.

Johnson, L. 1980. The Arctic charr, *Salvelinus alpinus*, p. 15-98. *In* E.K. Balon [ed.] Charrs: salmonid fishes of the genus *Salvelinus*. Dr. W. Junk Publishers, The Hague.

Kalleberg, H. 1958. Observations in a stream tank of territoriality and competition in juvenile salmon and trout (*Salmo salar* L. and *S. trutta* L.). Inst. Freshw. Res. Drottningholm Rep. 39: 55-99.

Karlstrøm, Ø. 1977. Habitat selection and population densities of salmon and trout parr in Swedish rivers. Inform. Inst. Freshw. Res. Drottningholm 6, 72 p. (Mimeographed in Swedish with English summary).

Keenleyside, M.H.A. 1962. Skin-diving observations of Atlantic salmon and brook trout in the Miramichi River, New Brunswick. J. Fish. Res. Board Canada 19(4): 625-634.

Keenleyside, M.H.A. and F.T. Yamamoto. 1962. Territorial behaviour of juvenile Atlantic salmon (*Salmo salar* L.). Behaviour XIX (1-2): 139-169.

Lindroth, A. 1955. Distribution, territorial behaviour and movements of sea trout fry in the river Indalsalven. Inst. Freshw. Res. Drottningholm Rep. 36: 104-119.

Mellquist, P. 1976. Informasjon om terskelprosjektet. NVE — Vassdragsdirektoratet. (In Norwegian with English summary).

Newman, M.A. 1956. Social behaviour and interspecific competition in two trout species. Physiol. Zool. 29(1): 64-81.

Nilsson, N.-A. 1963. Interaction between trout and char in Scandinavia. Trans. Am. Fish. Soc. 92: 276-285.

Nilsson, N.-A. 1967. Interactive segregation between fish species, p. 295-313. *In* S.D. Gerking [ed.] Biological basis of fish production. Blackwells, Oxford.

Noakes, D.L.G. 1980. Social behaviour in young charrs, p. 683-703. *In* E.K. Balon [ed.] Charrs: salmonid fishes of the genus *Salvelinus*. Dr. W. Junk Publishers, The Hague.

Noble, G.K. 1939. The role of dominance in the social life of birds. Auk 56: 263-273.

Nordeng, H. 1961. On the biology of char (*Salmo alpinus* L.) in Salangen North Norway. 1. Age and spawning frequency determined from scales and otoliths. Nytt Mag. Zool. 10: 67-123.

Power, G. 1973. Estimates of age, growth, standing crop and production of salmonids in some North Norwegian rivers and streams. Inst. Freshw. Res. Drottningholm Rep. 53: 78-111.

Stringer, G.E. and W.S. Hoar. 1955. Aggressive behaviour of underyearling Kamloops trout. Can. J. Zool. 33 (3): 148-160.

Svärdson, G. 1976. Interspecific population dominance in fish communities in Scandinavian lakes. Inst. Freshw. Res. Drottningholm Rep. 55: 144-171.

Symons, P.E.K. 1968. Increase in aggression and in strength of the social hierarchy among juvenile Atlantic salmon deprived of food. J. Fish. Res. Board Canada 25(11): 2356-2385.

Life History Patterns of the Miyabe Charr in Shikaribetsu Lake, Japan

KOJI MAEKAWA

Department of Oral Anatomy, School of Dentistry, Hokkaido University, N 13 W 7, Kita-ku, Sapporo, 060 Japan

MAEKAWA, K. 1984. Life-history patterns of the miyabe charr in Shikaribetsu Lake, Japan, p. 233–250. *In* L. Johnson and B.L. Burns [eds.] Biology of the Arctic charr, Proceedings of the International Symposium on Arctic Charr, Winnipeg, Manitoba, May 1981. Univ. Manitoba Press, Winnipeg.

Morphology and life history of the landlocked miyabe charr, *Salvelinus malma miyabei* (Oshima), was investigated in Shikaribetsu Lake and its inlet stream in central Hokkaido, Japan. Characteristic of this charr is that the number of gillrakers is the most numerous of all the Dolly Vardens studied across their geographical range. The development of gillrakers is completed during a transitional period of growth as the juvenile fish migrate into the lake from the stream in which they were spawned; migration takes place from July to November with a peak in October. The charr start feeding on plankton as they enter the lake and then pass the winter there. There is a remarkable point of inflection in the relative growth of body parts in the sea-run type of Dolly Varden; this occurs at the length at which smoltification takes place and it is not observable in either miyabe charr or in river-resident Dolly Varden. It is concluded that miyabe charr do not smoltify completely as do the river-resident (or landlocked) Dolly Vardens of Hokkaido. Like the river-resident type, miyabe charr regain their parr marks at maturity. The miyabe charr attain maturity at a later and more advanced growth stage than the river-resident type, but at an earlier stage than the sea-run type. Moreover, a large number of miyabe charr and river-resident Dolly Varden live to spawn more than twice. On the other hand, the miyabe charr resemble the Dolly Varden in some characteristics; i.e. they both return to spawn in the stream from which they originated, between August and November, with a peak in September. Also both forms produce river-resident types and have a prolonged lower jaw at maturity. We conclude that the miyabe charr has intermediate characteristics of development between the sea-run and the river-resident types of Dolly Varden. The important role of precocity in the process of developing the landlocked form of charr, the relationship between progenesis and r-selection in the life history strategy and the evolution of the miyabe charr from the original Dolly Varden is discussed.

INTRODUCTION

The landlocked miyabe charr, *Salvelinus malma miyabei*, in Shikaribetsu Lake, central Hokkaido, Japan, was described for the first time in 1938 by

Oshima as *S. miyabei*; he noted differences in pectoral fin length, the number of scales along the lateral line, and the muscle colour between the miyabe charr and the Dolly Varden of Hokkaido. Later, Yoshiyasu (1973) found a difference in the electrophoretic pattern between the hemoglobins of these charr groups. Behnke (1972), from data on the number of gillrakers (21-26 in the miyabe charr) reported by Oshima (1961), suspected that the miyabe charr might be a relict of *S. alpinus*. Most of the recent Japanese taxonomists, however, have treated *S. malma miyabei* as a subspecies or as a local population of the Dolly Varden (Aoyagi 1957; Kubo 1967; Ishigaki 1969; Miyaji et al. 1972, 1976), and, more recently, Behnke (1979).

Several fragmentary reports have appeared on the ecology and morphology of the miyabe charr (Oshima 1962; Inukai and Sato 1943; Hada and Tomita 1949; Inamura and Nakamura 1962; Kubo 1967; Yoshiyasu 1973; Kimura 1976).

In this paper I describe supplemental data obtained in the course of ecological and morphological studies on the sub-speciation of the miyabe charr (Maekawa 1977a, 1977b, 1978) and its characteristic specializations compared with stream-resident and anadromous types of the Dolly Varden.

LOCALITY OF SHIKARIBETSU LAKE

Shikaribetsu Lake is situated in a mountainous area at about 800 m a.s.l. in central Hokkaido, Japan, at a latitude of 45°N. The lake has three small inlet streams and one outlet; it is about 3 km long with a maximum depth of about 101 m. The outlet stream was dammed in 1953 and lake levels regulated. The lake is ice covered for about 6 months of the year; the water temperature in summer rises to about 18°C on the surface but maintains a temperature of about 4°C below a depth of 10 m.

The following fish species were collected during a period from 1971 to 1975: *Salvelinus malma miyabei, Salmo gairdnerii, Hypomesus transpacificus, Tribolodon hakonensis, T. ezoe, Barbatura toni toni, Carassius spp.* and *Pungitius pungitius tymemsis. S. malma miyabei* is the only native fish in the lake system; all other species have been introduced since 1955.

LIFE HISTORY

Like the anadromous Dolly Varden which ascends coastal rivers to spawn, the miyabe charr returns to spawn in the small inlet of Shikaribetsu Lake known as the Yambetsu stream (about 6 km length). The miyabe charr includes both lake-dwelling and stream-resident types, comparable with the anadromous Dolly Varden in having sea-run and stream-resident types. The life history of the miyabe charr, however, is different from that of the anadromous charr. The diagram in Figure 1 shows the life cycles of three types of the Dolly Varden

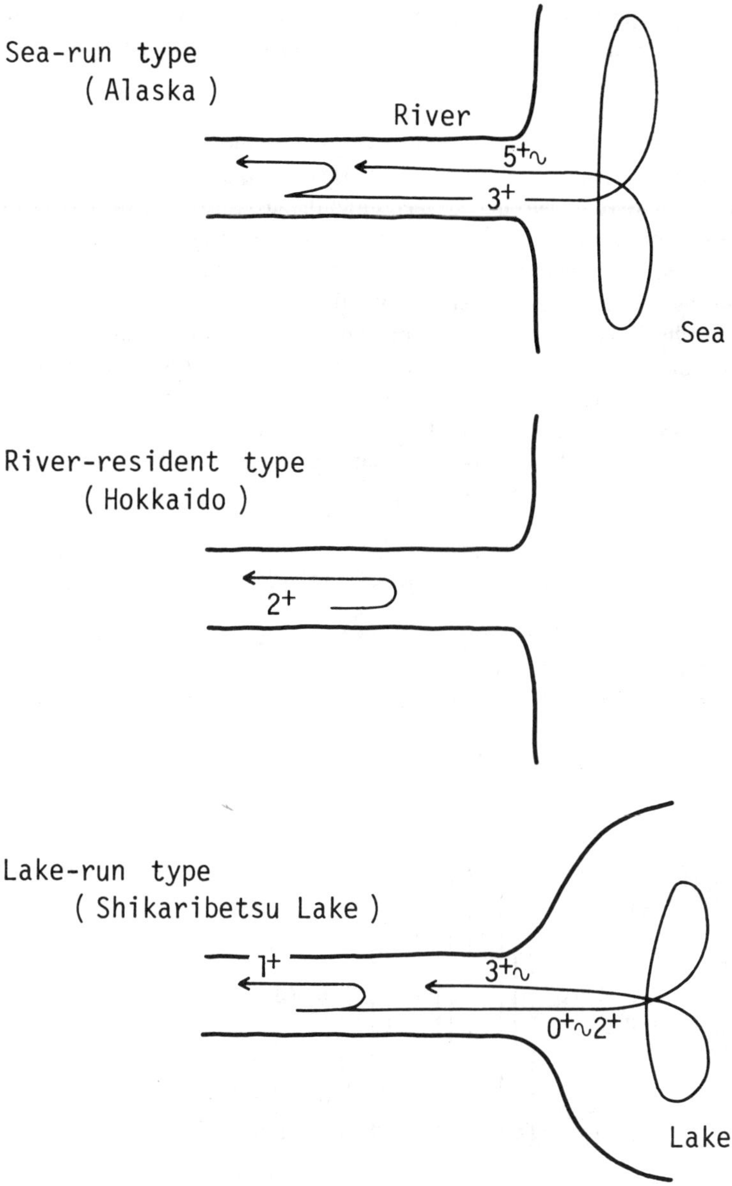

FIG. 1 Diagrams of the life cycle in each type of Dolly Varden. Note the age difference between the miyabe charr and the Dolly Varden during downstream and upstream migrations.

(drawn from reports by Ishigaki 1969; Armstrong and Morrow 1979; Maekawa 1977a). One clear difference between the miyabe charr and the anadromous Dolly Varden is that juveniles of the miyabe charr are relatively young fish (0+ with a mode of about 65 mm in fork length and 1+ with a mode of about 90 mm in fork length) which migrate downstream into the lake between July and November with a peak in October (Fig. 2). The fish are parr-like, although a few smolts (2+ or over in age) may descend into the lake in June and July just as the anadromous ones descend into the sea in June (Armstrong 1970). It is suggested that these fish, which are referred to as the 'fall migrating group', may be using the lake as a wintering area since, of all available water sources, the lake maintains the highest water winter temperature (Maekawa 1977a).

Few stream-resident Dolly Varden from the headwaters of the major river systems in central Hokkaido or from the headwaters to the mouth of the river in eastern Hokkaido have been found to descend to the sea (Hikita 1962; Ishigaki 1969; Maekawa 1973). We believe that the miyabe charr probably did not descend to the sea from the lake even before the outlet stream was artificially dammed in 1953.

Feeding Habits in the Stream

Insects are the most readily available food item at all seasons for both the miyabe charr and the stream-resident type of the Dolly Varden. The seasonal composition of the food organisms of the juvenile miyabe charr coincides with

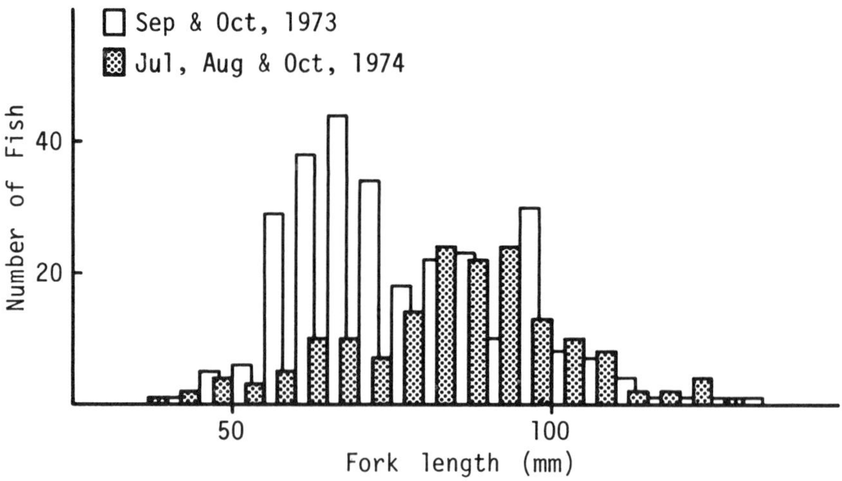

FIG. 2 Size composition and the total number of downstream migrating miyabe charr (from Maekawa 1977a).

seasonal changes in the composition of the drift as captured by a Surber-sampler (Maekawa 1977a). However, benthic organisms may occur more frequently in Dolly Varden when they co-exist with other salmonids (Ishigaki 1969).

Feeding Habits in the Lake

Investigating 20 fish, Hada and Tomita (1949) found the lake inhabiting miyabe charr to be highly planktivorous (about 100 percent occurrence in all individuals); according to Kubo (1967), off-shore migrants feed on zooplankton but the in-shore dwelling population near the river mouth feeds on benthic and drifting insects (Yamashiro, pers. comm.). Maekawa (1977a) reported that the zooplankters (*Bosmina* spp. and *Daphnia cucullata*) and insects were the most frequently occurring food items, with an occurrence rate of about 50 percent and 60 percent, respectively; fish of the introduced species were second in occurrence (about 15 percent). Therefore, miyabe charr may yet be highly planktivorous.

Reproduction

Data on reproduction were collected from Shikaribetsu Lake, the inlet stream, and the upper reaches of the Tokoro River in eastern Hokkaido during the period of 1971-1975 by three methods: handnets, castnets, and gillnets. In particular, investigations were conducted in the Tokoro River in 1972, 1973 and 1975.The lake-dwelling type of miyabe charr matures at age 3+ and about 180 mm fork length, whereas the stream-resident type matures at age 1+ and 95 mm fork length (Maekawa 1978). Figure 3 shows the size and age composition of the mature stream-resident type of Dolly Varden in the Tokoro River. Otoliths were used for age determination. These fish reach maturity at age 2+ when more than 100 mm in fork length. Possibly the male charr matures sexually at a smaller size than the female.

The fork length-fecundity curve (Fig. 4) and body weight-fecundity curve (Fig. 5) of the miyabe charr and stream-resident type Dolly Vardens appear to be similar. However, the range of egg number is somewhat more variable for the miyabe charr than for the other types. A comparison of egg size showed the miyabe charr egg to be substantially larger (mean diameter 5.2 mm) than that of the stream-resident type (mean 4.6 mm); the significance of this is not yet clear.

Comparing the monthly changes in the gonad-index between the various types of Dolly Varden, it appeared that stream-resident types of both the miyabe charr and Dolly Varden begin to develop sexually earlier than the lake-run and the sea-run types.

The spawning period for the lake-run miyabe charr is from September to November with a peak in October. Before spawning, the charr return to the inlet

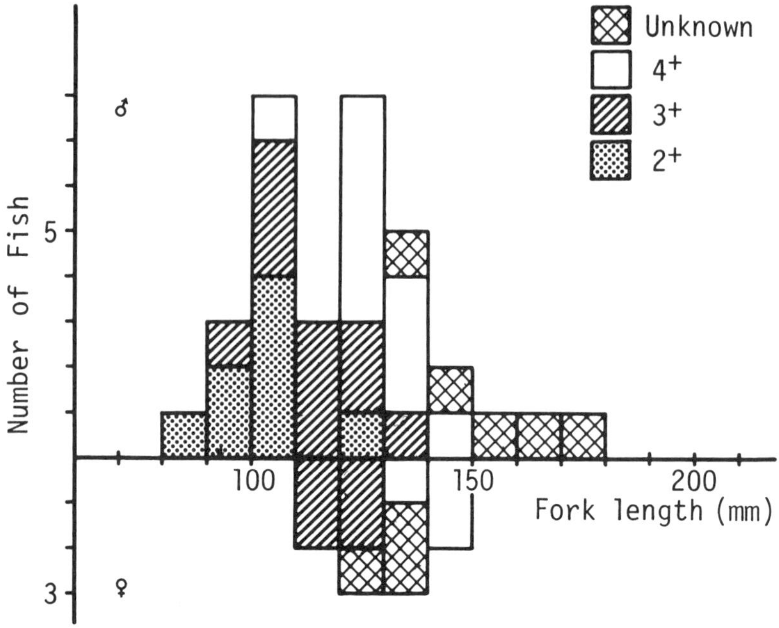

FIG. 3 Size and age composition of mature stream-resident Dolly Varden collected in the Tokoro River of Hokkaido during 1972 and 1975.

streams from the lake; they were collected by traps in the wood weir across the stream during 1972 and 1973. The migration upstream occurs only at night between late August and December with a peak in September (Fig. 6 and 7). Traps set in 1976 showed that stream-resident Dolly Varden also migrate to the upstream reaches of the Tokoro River.

Traps across the inlet stream showed that most of the lake-run types descend downstream soon after spawning. It is possible that most of the lake-run types spawn twice or more, since a few fish, tagged after spawning, returned to the stream the following year. It is unknown whether the lake-run miyabe charr spawns annually or at less frequent intervals.

SOME MERISTIC CHARACTERISTICS OF MIYABE CHARR COMPARED WITH THOSE OF OTHER TYPES

The meristic characteristics of the miyabe charr, with the exception of gillrakers, are similar to those of the Hokkaido Dolly Varden. An obvious ''geocline'' showing an increase in the number of vertebrae, gillrakers, dorsal rays and pored scales has been found in the Dolly Varden (McPhail 1967;

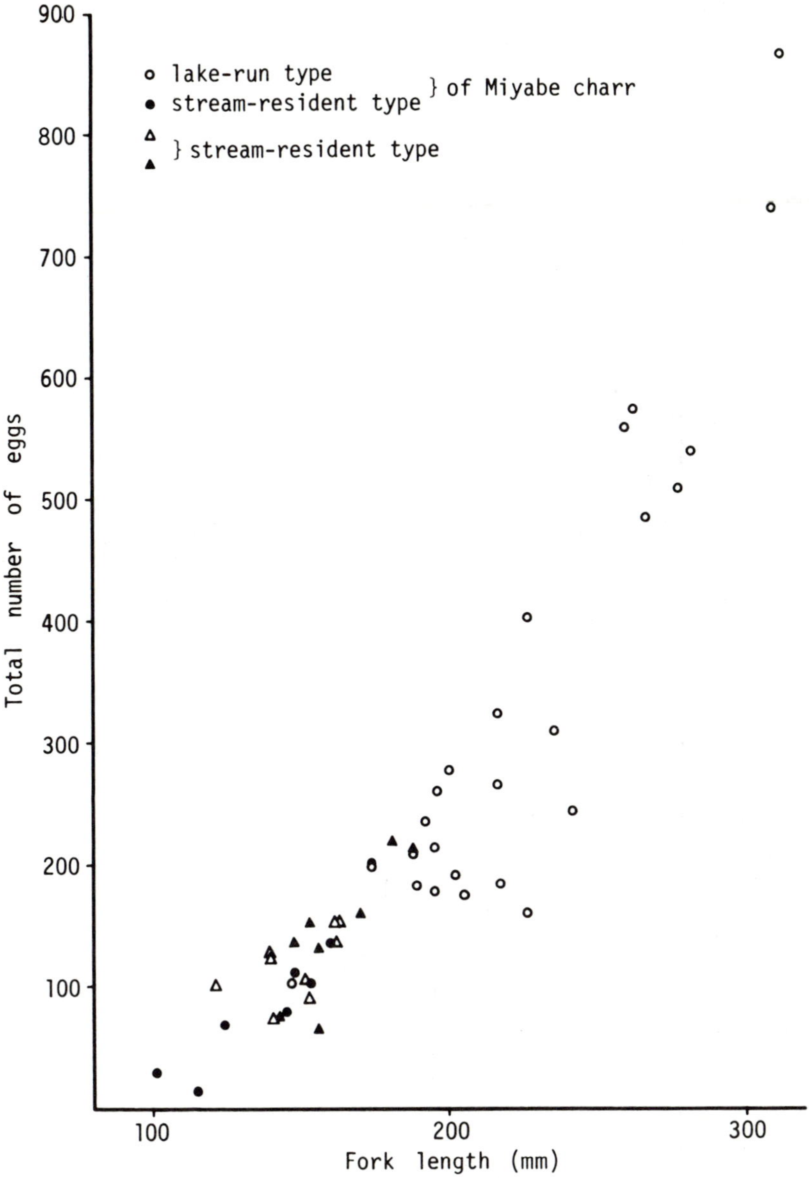

FIG. 4 Relationship between fecundity and fork length of the miyabe charr (○,●) and the stream-resident type of Dolly Varden in the Tokoro River (△) and the Nukabira River (▲) of Hokkaido.

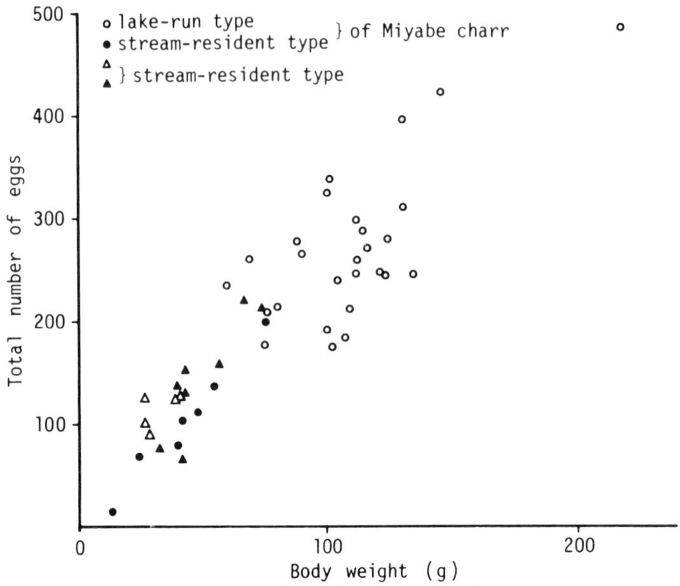

FIG. 5 Relationship between fecundity and body weight of the miyabe charr (○,●) and the stream-resident type of Dolly Varden in the Tokoro River (△) and the Nukabira River (▲) of Hokkaido.

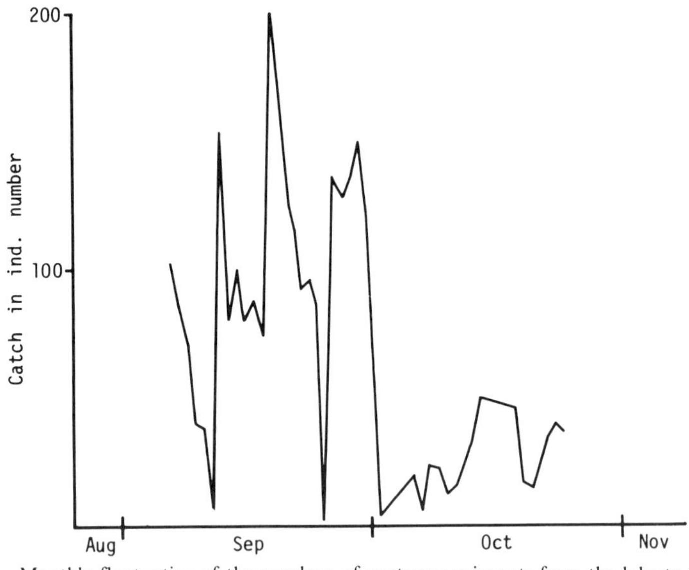

FIG. 6 Monthly fluctuation of the numbers of upstream migrants from the lake to the inlet stream for spawning.

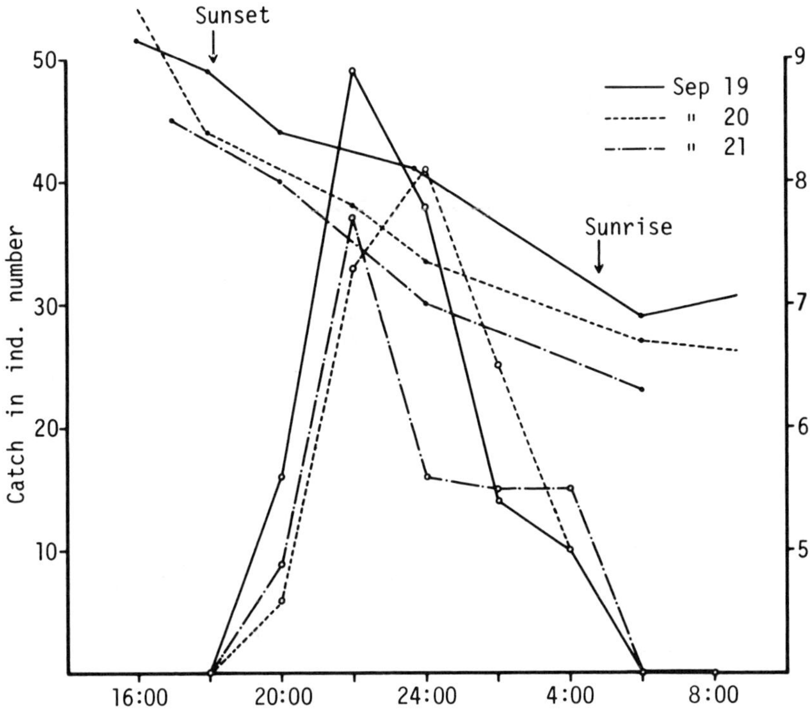

FIG. 7 Nocturnal fluctuation in the number of upstream migrating fish.

Maekawa 1977b). A striking characteristic of the miyabe charr is the large number of gillrakers and accessory gillrakers on the medial surface of the gill arch. The gillrakers of the miyabe charr (n = 23-39, mean: 26) are the most numerous among all the Dolly Varden stocks investigated: Nome (21 on the average); the Bering sea (about 22); Anchorage (about 18); Juneau (17); Seattle (16); the Okhotsuku Sea (about 22), and Hokkaido (about 21).

The gillrakers were observed morphologically and a comparison was made between those of the Dolly Varden types and the miyabe charr (Fig. 8). The gillrakers of the miyabe charr are slighter and longer than those of the Dolly Vardens; the gillrakers in the south-east type of Dolly Varden are, according to McPhail (1962), the thickest of all.

Figure 9 shows the relationship between the number of gillrakers on the first arch of the left side and fork length in the miyabe charr. The number of gillrakers increases rapidly up to a length of 600-700 mm thereafter remaining constant. No consistent differences were observed in gillraker count between males and females.

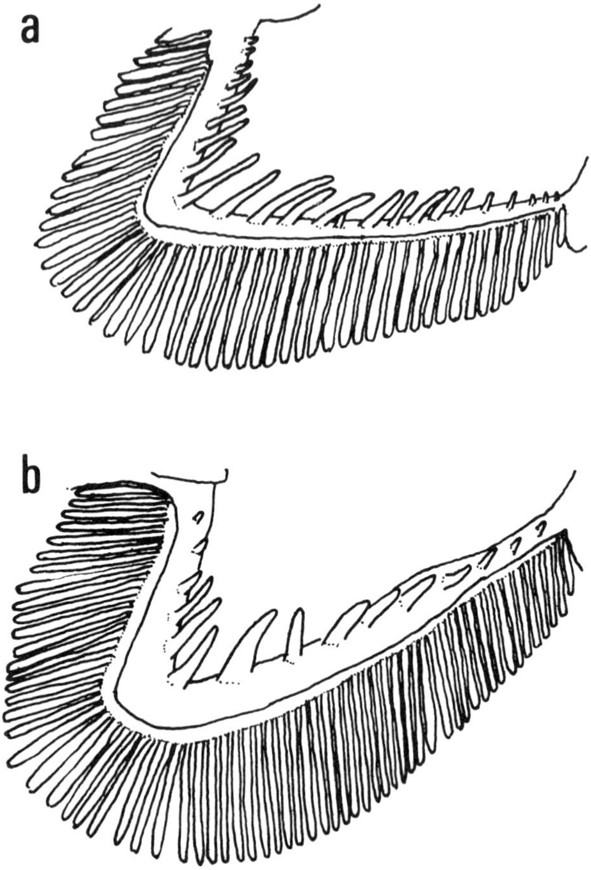

FIG 8 Left first gill arch of the miyabe charr (a) and the Dolly Varden (b) collected near Juneau.

GROWTH AND DEVELOPMENT

Three types of the Dolly Varden, the miyabe charr, and the stream-resident and sea-run types were examined with special reference to their growth and development. The growth rate of the miyabe charr closely resembled the early stages of growth of the stream-resident type, but a remarkable difference in body size appears at the age of 2 + when the former descends to the lake and thereafter begins to grow rapidly.

The relative growth pattern of all types is shown in Figure 10 (from Maekawa 1978). A number of inflection points were observed in the relative growth lines

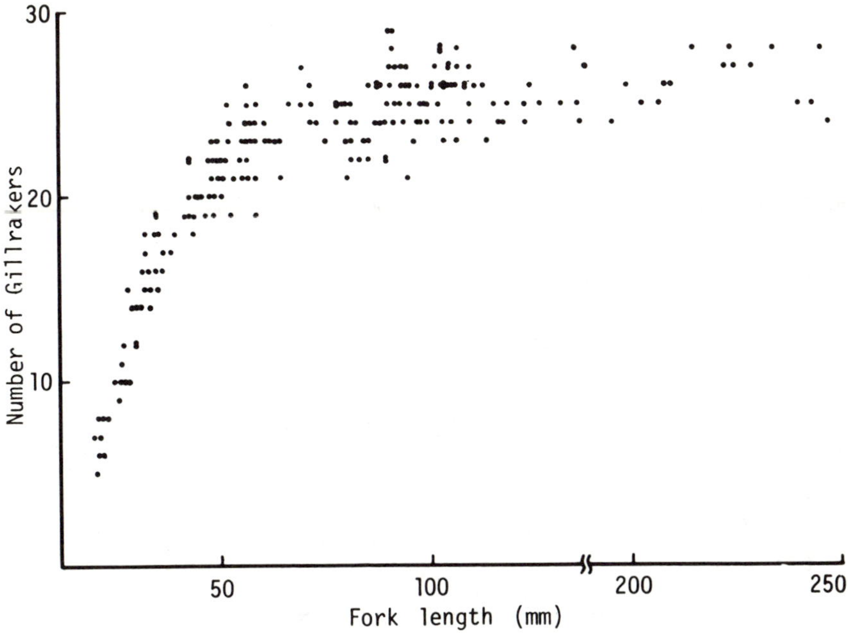

FIG. 9 Relationship between fork length and number of gillrakers.

of the Dolly Varden. The first inflection observed in some body parts occurred at about 30 mm in fork length. A further inflection appeared between 140 mm and 160 mm in fork length only in the sea-run type. After this inflection, differences in the relative size of body parts between these three types become clear.

Sexual dimorphism is apparent in the spawning adults of the miyabe charr; females develop a swollen ovipositor, and the male develops a pronounced kype. The stream-resident male, however, only shows a small projection at the tip of the lower jaw. The most striking characteristic of the lake-run type of miyabe charr is that, remarkably, the parr marks appear again at maturity, just like those of the juveniles and the adults of the stream-resident type. Except for the difference in colouration which occurs during the spawning season, the miyabe charr in the lake is basically silver with small, light-coloured pink spots as in the anadromous Dolly Varden.

DISCUSSION

It has been thought that miyabe charr might be a relict population of typical anadromous Dolly Varden which became landlocked in Shikaribetsu Lake due to volcanic eruptions during the Wurm glacial age (Tokachi Research Group 1972).

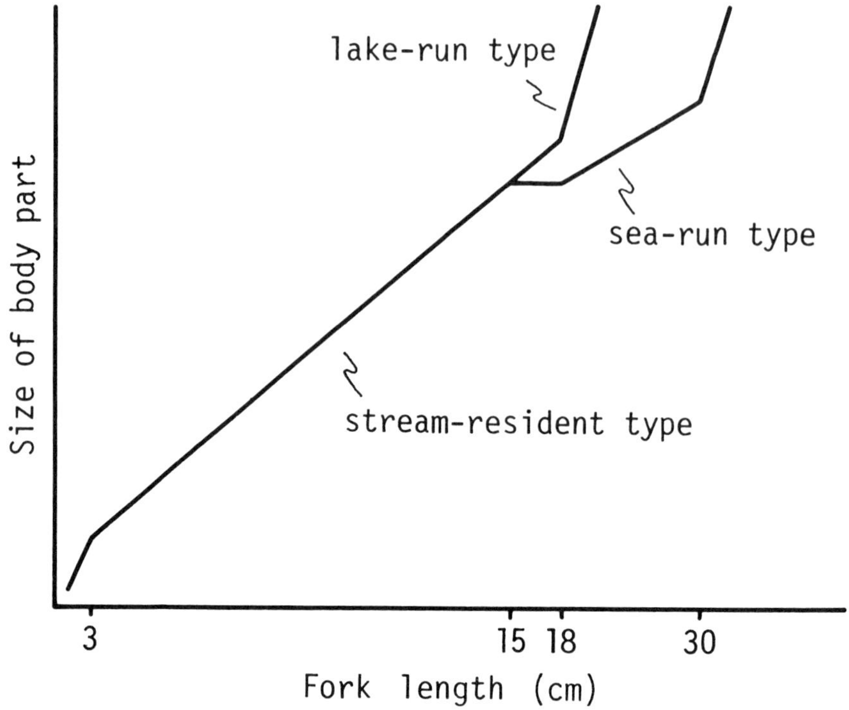

FIG. 10 Diagram of relationship between body parts and fork length in three types of Dolly Varden, as shown by logarithm (redrawn from Maekawa 1978).

This suspicion has been supported by studies of the life cycle of the stream-resident type, the silvery body colour, and the similarity of the sexual dimorphism in miyabe charr and the anadromous Dolly Varden. Maekawa (1978), however, has conjectured that the process of evolution of the miyabe charr might be very similar to that of the kokanee in relation to the landlocking mechanism in sockeye salmon, *Oncorhynchus nerka* described by Ricker (1940), since the miyabe charr possesses intermediate characteristics between the anadromous and stream-resident types of Dolly Varden and is more precocious, having more juvenilized characteristics than the anadromous form of Dolly Varden. The supplemental data of the present paper reinforces the hypothesis that the landlocking process has resulted from precocity and juvenilization, not merely from physical landlocking. Armstrong (1970) reported that a smolt downstream migration of the Dolly Varden occurrs in May and June in river systems fed by a lake, observing samples caught in weirs built across the outlet; this shows that smolts of the Dolly Varden may possess a physiological tendency to descend to the sea from the lake. It is interesting to consider why the miyabe charr does not descend from the lake to the sea. Landlocked processes or differentiation of the Salmonidae and closely related species have been discussed in relation to their precocity by many authors (Ohno 1933; Ricker 1938, 1940; Azuma 1973; Kubo 1974; Utoh 1976, 1977). Recently, in their study of masu salmon *Oncorhynchus masou*, whose life form is very similar to that of the Dolly Varden, Kubo (1980) and Utoh (1976) have reported that transformation of the smolt either contends with sexual maturation, or sexual maturation competes with the transformation. According to Kubo (1974, 1980), three distinct stages can be recognized in the smolt: presmolt, mid-smolt and post-smolt. The salinity tolerance of smolts becomes higher as smoltification progresses; particularly, the silver post-smolt can withstand abrupt transition from freshwater to sea-water, and in spring, negative rheotaxis in smolts appears as fish reach the climax of external metamorphosis (post-smolt).

In this study, the inflection point was not observed at 150 mm fork length, in the relative growth line of the miyabe charr, although it appeared at this fork length in the sea-run type; this almost coincides with that found by Armstrong (1970) for smolts. Therefore, the miyabe charr may grow to sexual maturity without transformation to a complete smolt (probably post-smolt), or the number of such transformed individuals might be very few. When the masu salmon smolts are confined to freshwater, a regressive transformation may often occur (Kubo 1980).

Similar to the stream-resident type of the Hokkaido Dolly Varden, the miyabe charr may have acquired a truncated ontogeny by the early attainment of sexual maturity in the process of landlocking. Dolly Varden living in recently dammed lakes have been found to change to a silvery body colour, although they retained juvenile characteristics such as a reduced lower jaw (Fig. 11). According to Kubo (1967), the silvery body colour could be due to a change in the quality of ambient light (such as the blue colour of the lake (Kubo 1967)), the colour change occurring without transformation to the complete smolt.

The miyabe charr and the stream-resident type of Dolly Varden may have been juvenilized via the processes of heterochrony (Gould 1977), giving rise to an altricial form (Balon 1979). Progenesis is associated with r-strategies (MacArthur and Wilson 1967; Gould 1977): the fish either occupy habitats with persistent unpredictable fluctuations or invade habitats before they are stabilized. Attributes of r-selected organisms include high fecundity, early maturation, short life span, rapid development and others (Pianka 1971). The high intrinsic rate of natural increase (r) can be obtained by producing a few offspring early in life; that is, precocious reproduction is all important (cited from Gould 1977). According to the above-mentioned findings, the miyabe charr and the stream-resident Dolly

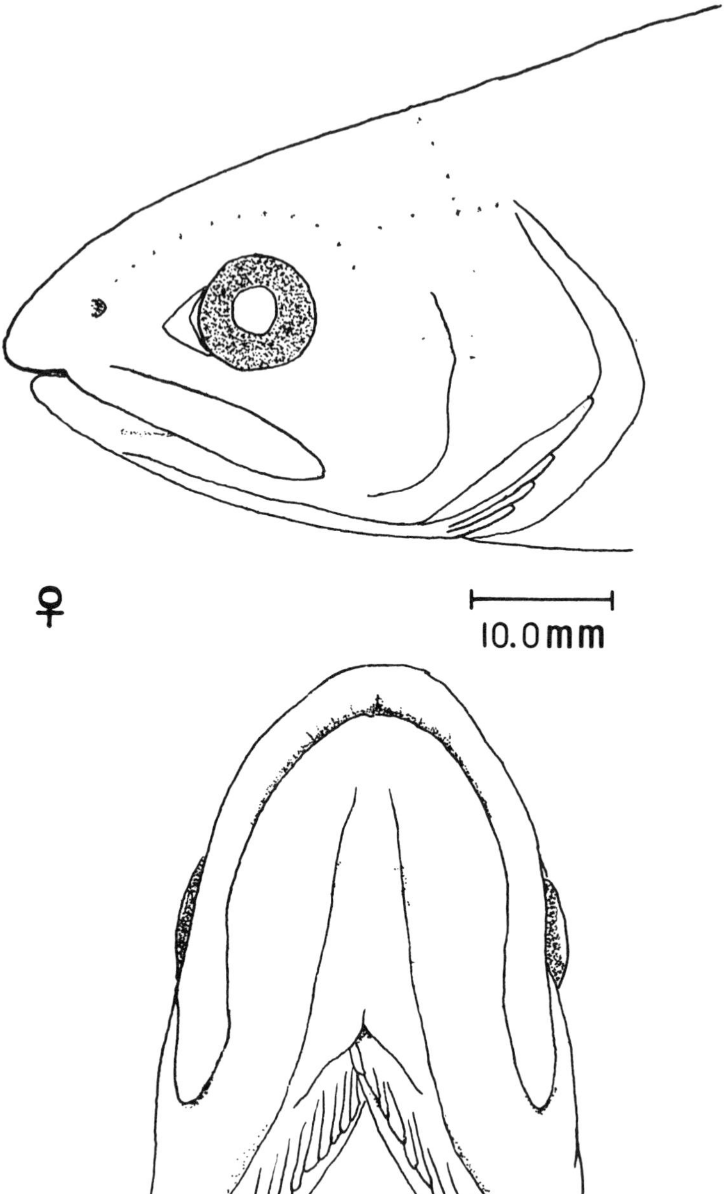

FIG. 11 Head aspect of the Dolly Varden collected in Niikappu Lake which is artificially dammed.

Varden are relatively r-selected. From the geological point of view, it is suspected that lakes and streams tend to produce more unstable and unpredictable environmental fluctuations than does the sea. Therefore, progenesis in the miyabe charr and the stream-resident Dolly Varden might "evolve as adaptations to strikingly different ecological conditions; progenesis with its acceleration of maturity is the life history strategy for r-selection" (Gould 1977).

The number of gillrakers in the miyabe charr exceeds that of all Dolly Varden collected previously in the southern parts of the Bering Straits (there may be a few Arctic charr also included in the samples used in this study). All other meristic characters in both the Dolly Varden of Hokkaido and the miyabe charr are the southern type in "geocline". Behnke (1971) reviewed the systematics of various species or populations of sympatric *Salvelinus* having more gillrakers than other groups. McCart (1971) reported that the Arctic charr in the Sagavanirktok River has more gillrakers (24-33) than other anadromous populations (19-27).

However, populations with numerous gillrakers have, generally, other meristic characters such as vertebrae, pyloric caeca and scale pores that are also more numerous. Martin and Sandercock (1967) reported that a correlation exists between the number of gillrakers, their length, and the amount of plankton eaten by lake trout *Salvelinus namaycush*, as well as with the number of accessory gillrakers. Compared to the piscivorous Dolly Varden (e.g. Ross 1959), the planktivorous sockeye salmon, *Oncorhynchus nerka*, has more gillrakers than all other species of *Oncorhynchus* (Hikita 1962); therefore, the aquisition of this characteristic might be due to plankton feeding (Maekawa 1977a). Maekawa (1977a) divided preliminary development in miyabe charr into seven intervals, based on the changes in life history (habitats, distribution, social behaviour, and morphology); these stages he regarded as an expression of saltation in development (Balon 1979). It has been reported that thresholds occur at the beginning and end of each step reflected in distinct functional changes (Vasnetzov 1949; Balon 1975). In this sense, a threshold should be defined as an abrupt functional change during ontogeny (Balon 1979). If this hypothesis is applied, a transitional sequence of steps, called a "critical period in development" by Tokuda (cited from Azuma 1973), may fluctuate in their response to changes in the environment. Here it is interesting to note that the development of gillrakers is completed during a transitional period (Maekawa 1977a), as the juvenile fish are migrating into the lake to begin feeding on plankton for the first time. We conclude that natural selection may be involved during this transitional period, allowing adaptation of the miyabe charr to the biotic environment of the lake.

ACKNOWLEDGEMENTS

I wish to express my hearty thanks to Drs. Hans Mori and Hisashi Abe of the Faculty of Agriculture of Hokkaido University for their valuable advice during the course of this work. It is also a pleasure to thank Mr. Robert H. Armstrong of the Alaska Department of Fish and Game and Mr. Masao Mitsuzuka for the supply of materials and data on stream-resident types of Dolly Varden. Thanks are also due to Miss Junko Tarazaki of the School of Dentistry, Hokkaido University, for her invaluable assistance in preparing the manuscript.

REFERENCES

Aoyagi, H. 1957. General notes on the freshwater fishes of the Japanese Archipelago. Seisensha, Osaka. 272 p. (In Japanese).

Armstrong, R.H. 1970. Age, food, and migration of Dolly Varden smolts in southeastern Alaska. J. Fish. Res. Board Can. 27:991-1004.

Armstrong, R.H., and T.E. Morrow. 1979. The Dolly Varden charr, *Salvelinus malma*, p. 16-98. *In* E.K. Balon [ed.] Charrs: salmonid fishes of the genus *Salvelinus*. Dr. W. Junk Publishers, The Hague.

Azuma, M. 1973. Studies on the variability of the landlocked ayu-fish, *Plecoglossus altivelis* T. et S., in Lake Biwa. IV. Considerations on the grouping and features of variability. Jap. J. Ecol. 23:255-265. (In Japanese with English summary).

Balon, E.K. 1975. Terminology of intervals in fish development. J. Fish. Res. Board Can. 32:1663-1670.

Balon, E.K. 1979. The theory of saltation and its application in the ontogeny of fishes steps and thresholds. Env. Biol. Fish. 4:97-101.

Behnke, R.T. 1972. The systematics of salmonid fishes of recently glaciated lakes. J. Fish. Res. Board Can. 29:639-671.

Behnke, R.T. 1979. A systematic review of the genus *Salvelinus*, p. 441-480. *In* E.K. Balon [ed.] Charrs: salmonid fishes of the genus *Salvelinus*. Dr. W. Junk Publishers, The Hague.

Gould, S.T. 1977. Ontogeny and phylogeny. The Belknap Press of Harvard Univ. Press, Cambridge, Massachusetts. 501 p.

Hada, Y., and M. Tomita. 1949. Plankton and fishes in some lakes of Hokkaido. I. *Salvelinus* in the Lake Shikaribetsu. Seibutsu 4:21-26. (In Japanese).

Hikita, T. 1962. Ecological and morphological studies of the genus *Oncorhynchus* (Salmonidae) with particular consideration on phylogeny. Sci. Rep. Hokkaido Salmon Hatchery 17:1-97.

Inamura, A. and M. Nakamura. 1962. Distribution and variation of fishes in the genus *Salvelinus* in Japan. Contributions Res. Institute for Nat. Resources 58:65-78. (In Japanese with English summary).

Inukai, T. and S. Sato. 1943. On breeding habits of the miyabe charr (*Salvelinus malma* Oshima) in the Lake Shikaribetsu. Dobutsugaku Zasshi 55:78-79. (In Japanese).

Ishigaki, K. 1969. Ecology and morphology of genus *Salvelinus* in Hokkaido. Doctor Thesis, Hokkaido Univ., Sapporo, Hokkaido, 249 p. (In Japanese with English summary).

Kimura, S. 1976. On the adult, alevin and fry of the lacustrine charr, *Salvelinus miyabei* Oshima. Sci. Bull. Fac. Agr. Kyushu Univ. 30:191-197. (In Japanese with English summary).

Kubo, T. 1967. Ecological and physiological studies on the Dolly Varden charr (*Salvelinus malma*) in Lake Shikaribetsu, Hokkaido. Sci. Rep. Hokkaido Salmon Hatchery 21:11-33. (In Japanese with English summary).

Kubo, T. 1974. Notes on the phase differentiation and smolt transformation of juvenile masu salmon. Sci. Rep. Hokkaido Salmon Hatchery 28:9-26. (In Japanese with English summary).

Kubo, T. 1980. Studies on the life history of the "Masu" salmon (*Oncorhynchus masou*) in Hokkaido. Sci. Rep. Hokkaido Salmon Hatchery 34:195. (In Japanese with English summary).

MacArthur, R.H. and O. Wilson. 1967. The theory of island biogeography. Princeton Univ. Press, Princeton, N.Y. 203 p.

Maekawa, K. 1973. On a silvery smolt of the Dolly Varden, *Salvelinus malma,* collected from the Shiretoko Peninsula, Hokkaido. Jap. J. Ichthyol. 20:245-247. (In Japanese with English summary).

Maekawa, K. 1977a. Studies on the variability of the landlocked miyabe charr, *Salvelinus malma miyabei* Oshima. I. Development and early life history. Jap. J. Ecol. 27:91-102. (In Japanese with English summary).

Maekawa, K. 1977b. Studies on the variability of the landlocked miyabe charr, *Salvelinus malma miyabei*. III. Geographical variations of the Dolly Varden, *Salvelinus malma*, and morphological characters of the miyabe charr. Jap. J. Ichthyol. 24:49-56. (In Japanese with English summary).

Maekawa, K. 1978. Growth and development of *Salvelinus malma miyabei* compared with other forms of *S. malma*. Jap. J. Ichthyol. 25:9-18.

Martin, N.U. and F.K. Sandercock. 1967. Pyloric caeca and gillraker development in lake Trout, *Salvelinus namaycush*, in Algonquin Park, Ontario. J. Fish. Res. Board Can. 24:965-974.

McCart, P. 1971. Meristic differences between anadromous and freshwater-resident Arctic charr (*Salvelinus alpinus*) in the Sagavanirktok River drainage, Alaska. J. Fish. Res. Board Can. 28:115-118.

McPhail, J.D. 1961. A systematic study of the *Salvelinus alpinus* complex in North America. J. Fish. Res. Board Can. 18:793-816.

Miyaji, D., H. Kawanabe, and N. Mizuno. 1972. Coloured illustrations of the freshwater fishes of Japan. 275 p. Hoikusha, Tokyo. (In Japanese).

Miyaji, D., H. Kawanabe, and N. Mizuno. 1976. Coloured illustrations of the freshwater fishes of Japan. 462 p. Hoikusha, Tokyo. (In Japanese).

Ohno, I. 1938. Life history of masu salmon in Hokkaido. Sake Masu Iho 5(2):15-26. (In Japanese).

Oshima, M. 1961. Studies on charrs found in Japanese waters. Japan Wildlife Bull. 18:3-70. (In Japanese with English summary).

Pianka, E.R. 1970. On r- and K- selection. Amer. Nat. 104:592-597.

Ricker, W.E. 1938. "Residual" and kokanee salmon in Cultus Lake. J. Fish. Res. Board Can. 4:192-218.

Ricker, W.E. 1940. On the origin of kokanee, a fresh-water type of sockeye salmon. Trans. Roy. Soc. Canada 34:121-235.

Ross, T.F. 1959. Feeding habits of the Dolly Varden, *Salvelinus malma* (Walbaum), at Chignik, Alaska. Trans. Am. Fish. Soc. 88:253-260.

Tokachi Research Group. 1972. The Pumice-fall deposits of the Lake Pleustocene in the Tokachi Plain, Hokkaido. Quaternary Res. 11:219-227. (In Japanese with English summary).

Utoh, H. 1976. Study of the mechanism of differentiation between the stream resident form and seaward migratory form in masu salmon, *Oncorhynchus masou* Brevoort. I. Growth and sexual maturity of precocious masu salmon parr. Bull. Fac. Fish. Hokkaido. Univ. 26:321-326. (In Japanese with English summary).

Utoh, H. 1977. Study of the mechanism of differentiation between the stream resident form and the seaward migratory form of masu salmon, *Oncorhynchus masou* (Brevoort). II. Growth and sexual maturity of precocious masu salmon parr (2). Bull. Fac. Fish. Hokkaido Univ. 28:66-73. (In Japanese with English summary).

Vasnetzov, V.V. 1946. Divergence and adaptation in ontogeny. J. Zool. 25: 185-199. (In Russian).

Yoshiyasu, K. 1973. Starch-gel electrophoresis of hemoglobins of freshwater salmonid fishes in Northeast Japan. Bull. Jap. Soc. Sci. Fish. 39:449-459.

Ecophysiological Adaptations of the Retina in the Arctic Charr

M.A. ALI and MARY ANN KLYNE

Département de Biologie, Université de Montréal, Montréal, Que., H3C 3J7

GUDMUNDUR EINARSSON

Department of Physiology, University of Iceland, Reykjavik, Iceland

ALI, M.A., M.A. KLYNE AND G. EINARSSON. 1984. Ecophysiological adaptations of the retina in the Arctic charr, p. 251–261. *In* L. Johnson and B.L. Burns [eds.] Biology of the Arctic charr, Proceedings of the International Symposium on Arctic Charr, Winnipeg, Manitoba, May 1981. Univ. Manitoba Press, Winnipeg.

The circumpolar distribution of the Arctic charr (*Salvelinus alpinus*) has been influenced by geological events of the Pleistocene period. The Arctic charr can thus be found in alpine and temperate lakes. These habitats vary enormously in light intensity, quality and regime and also in photoperiod.Thus, retinal characteristics of the Arctic charr must have the capacity to encompass a wide range of photic conditions. The presence of well-developed retinal pigment epithelium, photoreceptors (rods and cones), nuclear layers, plexiform layers and ganglion cells, coupled with the occurrence of retinomotor responses suggests that the Arctic charr is quite competent to meet this demand. This paper discusses, and attempts to place in perspective, some remarkable structural and functional features of the retina of the Arctic charr which enable it to occupy these various habitats and to succeed in them.

INTRODUCTION

Geological events of the Pleistocene and post-Pleistocene period have greatly influenced the distribution of the Arctic charr (*Salvelinus alpinus* L.); most freshwater forms are found in lakes formed during this period. As its name implies the Arctic charr is circumpolar in distribution. The anadromous form is found as far north as Discovery Harbour on the northeast coast of Ellesmere Island (82°34'N), and as far south as latitude 49°N — in the Gander River of Newfoundland (Bigelow 1963). The freshwater form on the other hand is found in most islands of the north Atlantic, the eastern coast of North America, the

European Alps and central Siberia. A recent comprehensive review of the Arctic charr is given by Johnson (1980).

From its geographical distribution it is clear that Arctic charr is exposed to great daily and seasonal changes of the light-dark cycle and as such it must possess a degree of physiological adaptation far surpassing that of other freshwater fishes. No unique physiological attributes have however been observed. No unusual trends were noted in the respiration rate (Holeton 1973, 1974) or the effect of temperature and oxygen on the growth rate (Swift 1964) of this charr. Its retina too was found (Ali 1965, Ali and Wagner 1980) to be remarkably similar to that of other salmonids.

Arctic charr from Thingvellirvatn (64°N, 21°W) in Iceland provide interesting material for study. Not only are these charrs subjected to the extreme conditions of high latitudes, but Thingvellirvatn offers several niches for this species. Fridricksson (1939) and Frost (1965) described four categories of Arctic charr in this lake — 'real' (normal), 'delpa', 'murta', and 'svart murta'. Fridricksson's classification was based on: size, age at maturity, feeding habits, shape of jaw and number of vertebrae. Further information on the history of these charrs may be obtained in the report of Snorrason (1981) and Gydemo (this volume).

The present study is part of a preliminary investigation on the retinal structure of the Arctic charr inhabiting three different niches in Thingvellirvatn, Iceland.

MATERIALS AND METHODS

Samples of Arctic charr (*Salvelinus alpinus* L.) were obtained from Thingvellirvatn (64°N, 21°W), Iceland, during the summer of 1980. Retinas of fish in their natural state (i.e. without prior adaptation) were used in this study. The environmental conditions were as follows: 7°C, water temperature and $\sim 10^4$ lux, light intensity at the water surface (light-adapted). Samples were obtained from crevices and different depths of the lake (< 20 m, 40 m). Under 'natural' dark conditions the light intensity was no less than 5 lux, while 'total' dark-adapted retinas refer to those fixed in complete darkness except for a dim red light (Ilford S-902).

Retinas in the primary fixative (1 percent glutaraldehyde + 1 percent paraformaldehyde in 0.1 M cacodylate buffer pH 7.2) were transported to the laboratory where they were processed further. After three rinses in cacodylate buffer (0.1 M, pH 7.2) containing 7 percent sucrose the retinas were postfixed in 1 percent buffered osmium tetroxide. Dehydration was through a graded series of ethanol followed by two changes of propylene oxide prior to infiltration with Spurr's resin.

Semi-thin (0.5 - 1.0 μm) and ultra-thin (70-90 nm) sections were cut on a Reichert OmU 3 ultramicrotome. Sections for light microscopic studies were stained with basic methylene blue (1 percent), while those for electron microscopic studies were stained with 2 percent uranyl acetate and basic lead citrate (Venable and Coggeshall 1965).

RESULTS

Light-microscope observation of light- and dark(total)-adapted retinas of 'murta' Arctic charr inhabiting the 40 m deep waters of Thingvellirvatn show that the retina is well-developed and capable of photomechanical movements (movement of both photoreceptors and retinal pigment epithelium (RPE)). However, when retinas are dark(natural)-adapted during the summer period there is relatively no photomechanical movement; there is only a slight movement of the pigment granules of the RPE. At the light-microscope level the chief notable difference of light-adapted retinas is the extent of photoreceptor outer segments; outer segments (OS) of the murta variety (40 m deep) extend right up to Bruch's membrane while OS of charrs from the crevices and shallow waters ($<$ 20 m) do not. Further studies were conducted with the electron-microscope.

Light-adapted retinas of 'murta' from deep waters (40 m) and 'real' charr from shallow waters ($<$20 m) of the lake were used as analogues of the charr occupying lakes at high northern latitudes. They were compared to retinas of charrs from crevices and also to dark-adapted retinas (natural and total).

Low magnification electron-micrographs show that there is apparently no structural difference between the retinal photoreceptors of these fishes (Figs. 1, 2). As in other teleosts, rods are elongated and cones contracted in the light-adapted retinas and the ellipsoid mitochondria of cones show differences in size and structure (width of the intermembrane space) while those in the rods do not (Fig. 1). Tangential sections through the cones show that single and twin cones are arranged in a square mosaic (Fig. 2 A,B), and that this is carried over into the outer plexiform layer (OPL).

Tangential sections through the cone and rod pedicels show that while spinules (finger-like extensions of the horizontal cells of cone pedicels) are abundant in cones they are relatively absent from rods (Fig. 2 C,D). Transverse sections augment this observation (Fig. 3 A,B). Both tangential (Figs. 2 C,D; 3 C,D) and transverse sections (Fig. 3 A,B) clearly illustrate that the OPL of the Arctic charr is well-developed and relatively similar in structure to light-adapted retinas of other charrs. No difference was apparent in the retinal structures of those charrs favouring different depths. However, when light(natural)-adapted cone pedicels are compared with cone pedicels of dark (total)-adapted charrs a notable reduction in the number of spinules is noted (Fig. 4 A,B). No reduction in

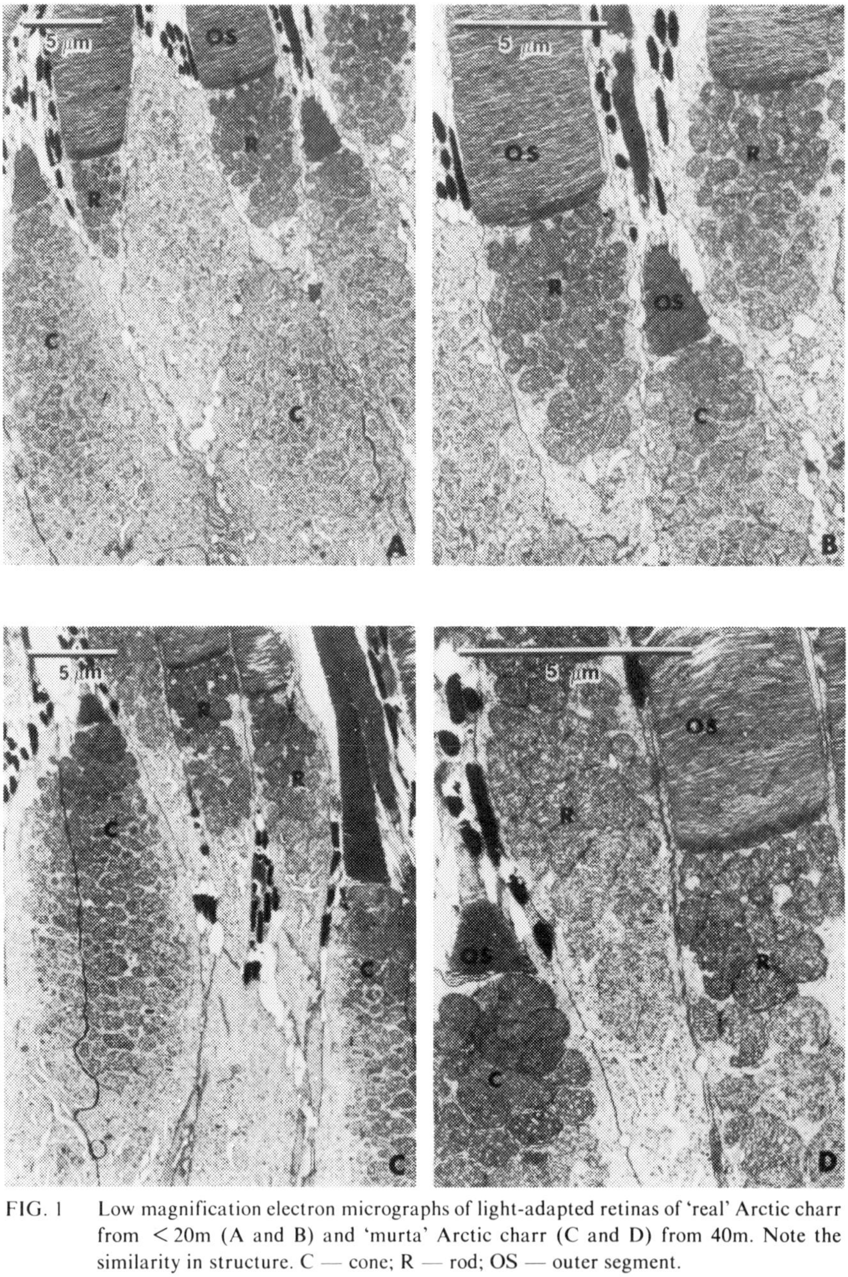

FIG. 1 Low magnification electron micrographs of light-adapted retinas of 'real' Arctic charr from <20m (A and B) and 'murta' Arctic charr (C and D) from 40m. Note the similarity in structure. C — cone; R — rod; OS — outer segment.

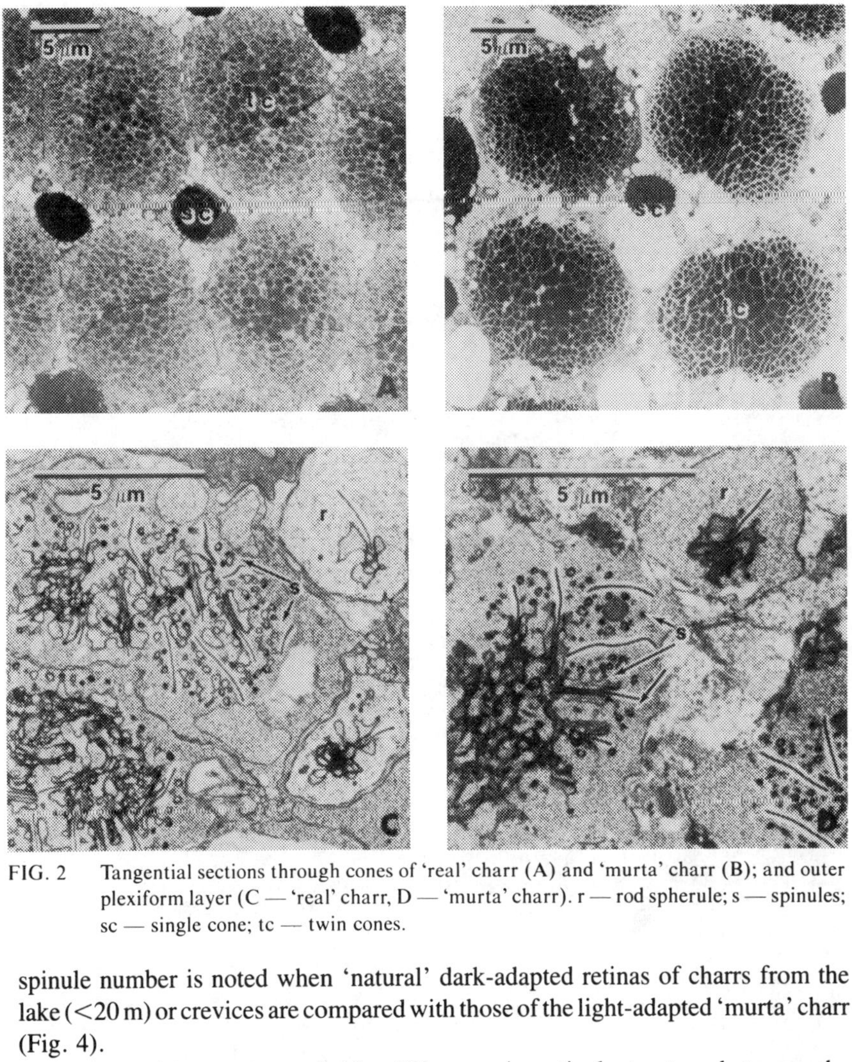

FIG. 2 Tangential sections through cones of 'real' charr (A) and 'murta' charr (B); and outer plexiform layer (C — 'real' charr, D — 'murta' charr). r — rod spherule; s — spinules; sc — single cone; tc — twin cones.

spinule number is noted when 'natural' dark-adapted retinas of charrs from the lake (<20 m) or crevices are compared with those of the light-adapted 'murta' charr (Fig. 4).

Perhaps the most remarkable difference in retinal structure between the 'murta' Arctic charr of deep (40 m) waters and 'real' Arctic charrs of shallow (<20 m) waters is in the pigment epithelium. Photoreceptor outer segments extend almost to Bruch's membrane in light-adapted retinas of the murta, consequently the nuclei of these cells appear squeezed into positions between the OS (Fig. 5A). When adapted to total darkness the OS recede vitread and the nuclei appear just vitread of Bruch's membrane (Fig. 5B), as in the light- and dark(natural)-adapted

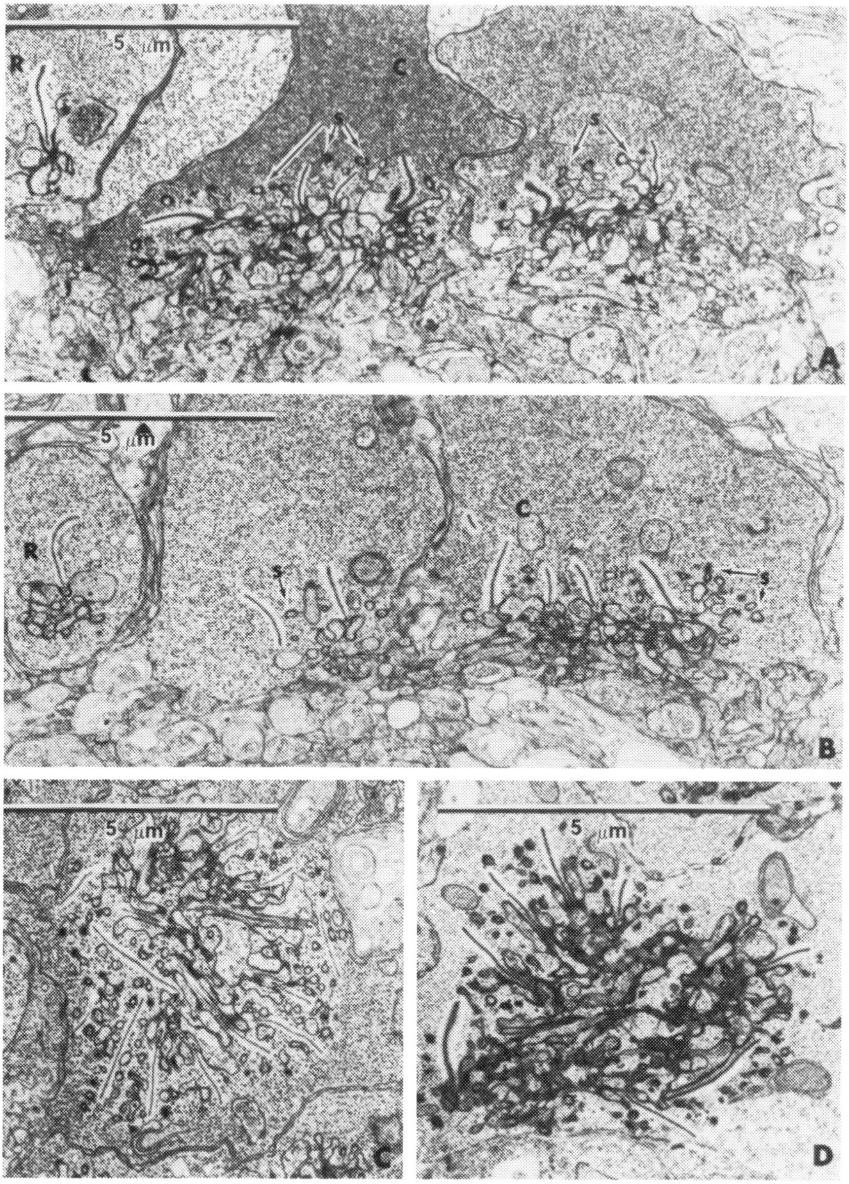

FIG. 3 Transverse (A and B) and tangential (C and D) sections through cone pedicels of the retinas of light-adapted 'real' and 'murta' charr, respectively. Note the presence of the large number of spinules (s). C — cone pedicel; R — rod spherule.

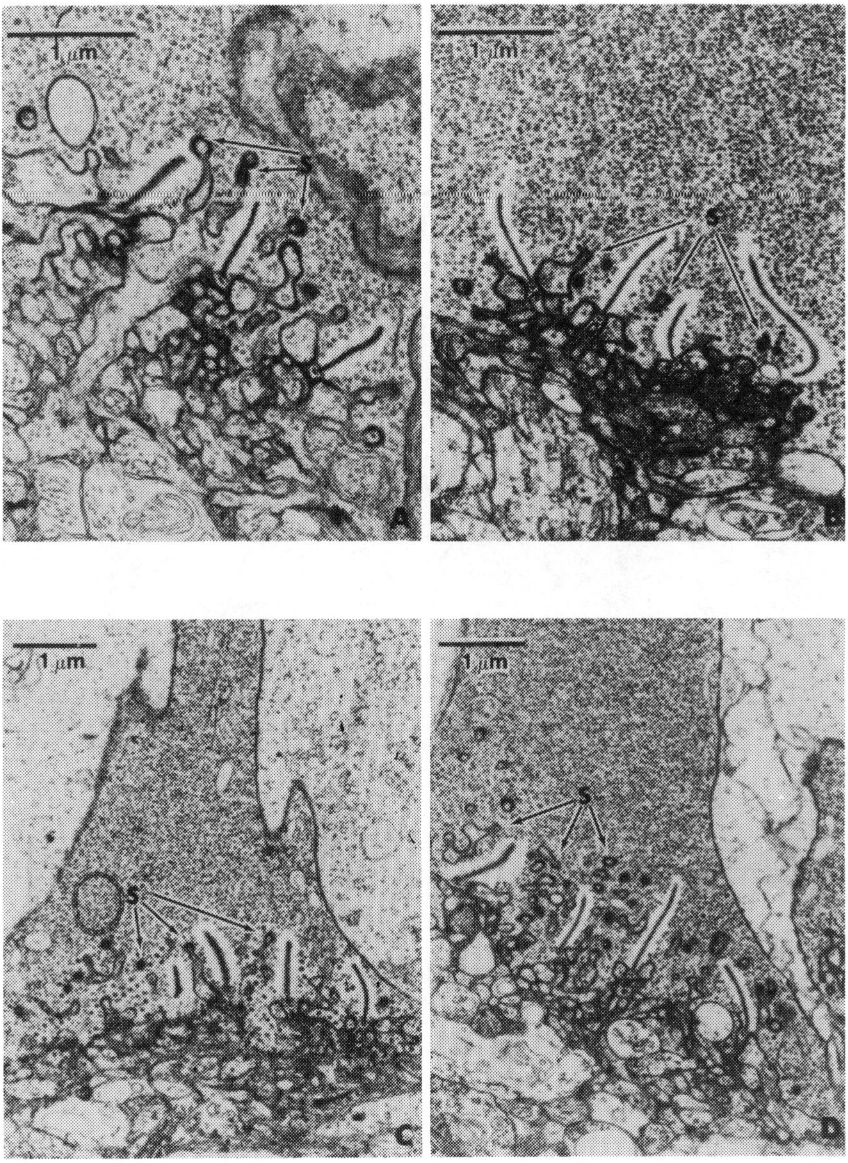

FIG. 4 Transverse section through cone pedicels of light-adapted (A) and dark(total)-adapted (B) retina of 'murta' (40 m), as well as dark(natural)-adapted 'real' charr from < 20 m (C) and crevices (D). Note the greater number of spinules (s) in the cone pedicels of light- and dark(natural)-adapted charrs.

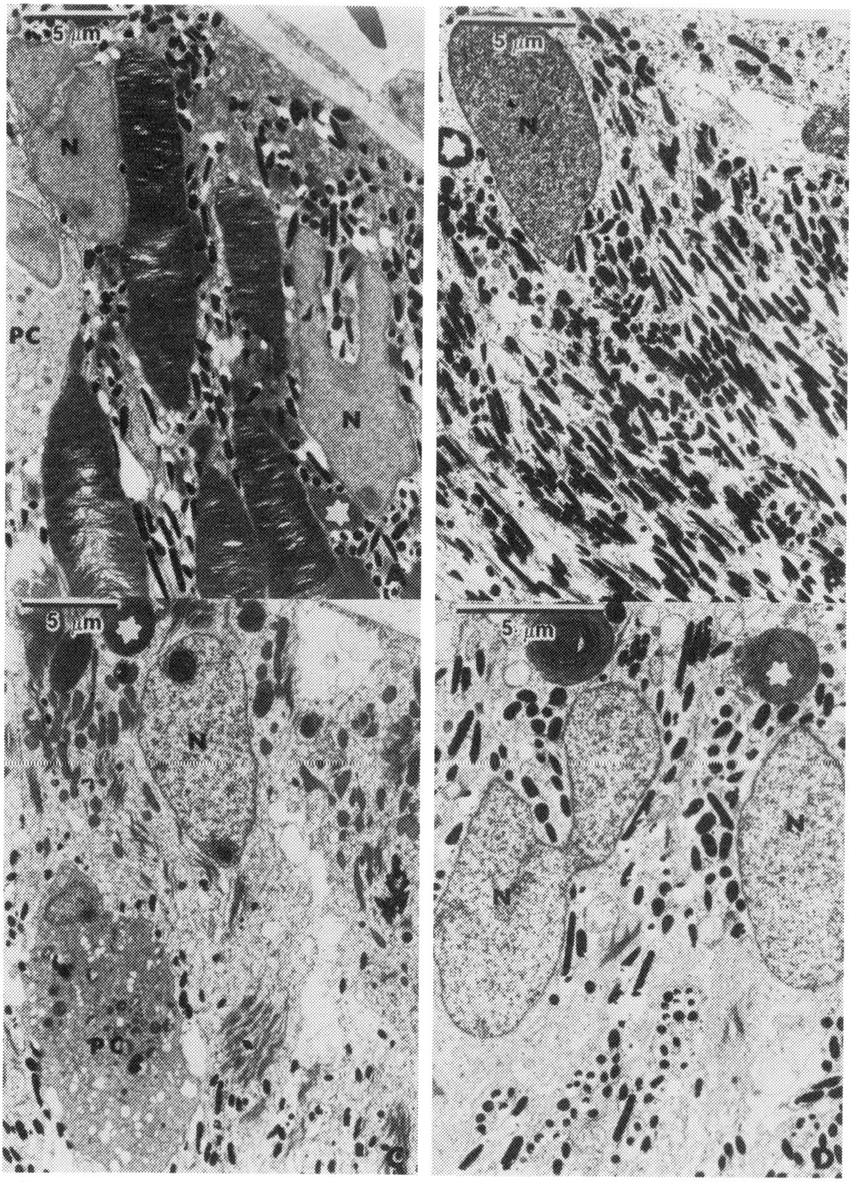

FIG. 5 Retinal pigment epithelium of light (A) — and dark(total) (B)-adapted retina of 'murta' charr (40 m); and light (C)- and dark(natural) (D)-adapted 'real' charr (< 20 m). Note location of nuclei (N) and presence of wandering phagocytes (PC) and phagosomes (star).

retinas of the real charr from shallow (< 20 m) waters (Fig. 5 C,D). It is interesting to note the presence of wandering phagocytes (macrophages) in the light-adapted retinas of both charrs (Fig. 5 A,C). Although wandering phagocytes are also present in the natural, dark-adapted retina they are fewer in number, while in the retina subjected to total darkness this structure is almost absent. Except for wandering phagocytes, light-and dark(natural)-adapted retinas the RPE appear very similar; only a slight sclerad contraction of pigment granules is observed (Fig. 5 C,D). Phagosomes are apparent in all the retinas observed (Fig. 5).

DISCUSSION

The retinal structure of the Arctic charr is similar to that of other salmonids (Ali 1965, Ali and Wagner 1980) and is a good example of a multipotential teleost retina with a high density of rods and cones, and a well-developed neural area. Although there are no specialized features, adaptations to a specific environment, the Arctic charr through its innate flexibility with respect to the circadian system is able to survive under conditions of continuous light (summer) and darkness (winter) in high northern latitudes.

Circadian rhythms in photoreceptors as well as spinules appear to be totally eliminated during the continuous light of summer when our samples were obtained. Pigment granules of the RPE also remained more or less extended (light-adapted state) during the summer. Extended pigment granules shield the light sensitive OS of rods, while contracted cones are exposed. Cone activity is evident from the large number of spinules present throughout the 24 hr period. Electrophysiological recordings imply that spinules mediate chromatic feedback into cones (Raynauld et al. 1979); however their exact function has yet to be resolved. Although the spinules of goldfish were shown to be cyclic in nature (Raynauld et al. 1979, Wagner 1980), they clearly do not exhibit this periodicity in the Arctic charr from high latitudes during the summer months. The Arctic charr nevertheless maintains its photomechanical capacity throughout the summer, evidenced when the charr was subjected to total darkness. This arhythmicity or desynchronization has also been noted for other fishes of high northern latitudes (Muller 1978).

An important observation is the presence of large number of phagosomes as well as wandering phagocytes within the RPE of Arctic charr during the summer. Shedding of OS is thought to be initiated by light (Basinger et al. 1976, Besharse et al. 1977, Hollyfield and Basinger 1978, Hollyfield et al. 1977, O'Day and Young 1978, Young 1977, 1978) and the greater degree of OS turnover during the continuous light of summer most likely resulted in increased amounts of wandering phagocytes in an attempt to assist in OS removal. The presence of these phagocytes does not represent an abnormal condition as they have been previously noted within the RPE of other teleosts including the Arctic charr (Braekevelt 1980). This renewal process nonetheless must place a constraint on retinal activity.

Thus our results indicate that the Arctic charr, irrespective of its niche, has adapted to high latitudes, not by specific adaptations of retinal structure, but rather by flexibility in its circadian system. Specific retinal adaptations would limit the charr to a particular environment, whereas the flexibility in rhythmic pattern means that the charr is adaptable not only to continuous darkness or continuous light, but also to the alternating light-dark cycle of spring and autumn. Thus, flexibility must be seen as of positive survival value in an environment where variation in day and night can be extreme. During summer the light-adapted condition of the retina of the Arctic charr ensures that it is capable of utilizing the abundant summer food resources at high northern latitudes optimally in addition to escaping predators.

This is a preliminary study conducted during the summer. Further studies are being carried out to investigate changes that occur throughout the year under natural and enforced conditions in an attempt to better understand the adaptations that the Arctic charr has made to ensure its success as a candidate for ecosystems of high northern latitudes.

REFERENCES

Ali, M.A. 1965. Retinal structure in the Arctic char (*Salvelinus alpinus* L.). J. Fish. Res. Board Can. 22:221-223.

Ali, M.A. and J.H. Wagner. 1980. Vision in charrs: a review and perspectives, p. 391-422. *In* E.K. Balon [ed.] Charrs: salmonid fishes of the genus *Salvelinus*. Dr. W. Junk Publishers, The Hague.

Basinger, S., R. Hoffman, and M. Matthes. 1977. Photoreceptor shedding is initiated by light in the frog retina. Science 194:1074-1076.

Besharse, J.C., J.G. Hollyfield, and M.E. Rayborn. 1977. Turnover of rod photoreceptor outer segments. II. Membrane addition and loss in relationship to light. J. Cell Biol. 75:507-527.

Bigelow, H.B. 1963. Genus *Salvelinus* Richardson 1836. Sears Found. Mar. Res. Mem. 1:503-546.

Braekevelt, C.R. 1980. Wandering phagocytes at the retinal epithelium — photoreceptor interface in the teleost retina. Vision Res. 20:495-499.

Fridricksson, A. 1939. On the Murta in Thingvellir Lake in relation to the char and trout in that lake. Natturufraed. 9:1-36 (In Icelandic).

Frost, W.E. 1965. Breeding habits of Windermere charr, *Salvelinus willughbii* (Gunther) and their bearing on speciation in these fish. Proc. R. Soc. Edinb. B. 163:232-284.

Hollyfield, J.G. and S.F. Basinger. 1978. Photoreceptor shedding can be initiated within the eye. Nature 274:794-796.

Hollyfield, J.G., J.C. Besharse, and M.E. Rayborn. 1977. Turnover of rod photoreceptor outer segments. I. Membrane addition and loss in relationship to temperature. J. Cell Biol. 75:490-506.

Holeton, G.F. 1973. Respiration of Arctic char, *Salvelinus alpinus*, from a high Arctic lake. J. Fish. Res. Board Can. 30:717-723.

Holeton, G.F. 1974. Metabolic cold adaptation of polar fish: fact or artefact. Physiol. Zool. 47:137-152.

Johnson, L. 1980. The Arctic charr, *Salvelinus alpinus*, p. 15-98. *In* E.K. Balon [ed.] Charrs: salmonid fishes of the genus *Salvelinus*. Dr. W. Junk Publishers, The Hague.

Muller, K. 19' [illegible] flexibility of the circadian system of fish at different latitudes, p. 91-104. *In* J.E. Thrope [ed.] Rhythmic Activity of Fishes. Academic Press, London, New York, San Francisco.

O'Day, W.T., and R.W. Young. 1978. Rhythmic daily shedding of outer segment membranes by visual cells in the goldfish. J. Cell. Biol. 76:593-604.

Raynauld, J.P., J.R. Laviolette, and H.J. Wagner. 1979. Goldfish retina: a correlate between cone activity and morphology of the horizontal cell in cone pedicels. Science 204:1436-1438.

Snorrason, S.S. 1981. The Fishes of Lake Thingvallavatn. Ph. D. Thesis, Univ. of Iceland, Reykjavik, Iceland.

Swift, D.R. 1964. The effect of temperature and oxygen on the growth rate of the Windermere charr (*Salvelinus alpinus willughbii*). Comp. Biochem. Physiol. 12:179-183.

Venable, J.H. and R. Coggeshall. 1965. A simplified lead citrate stain for use in electron microscopy. J. Cell Biol. 25:407-408.

Wagner, H.J. 1980. Light-dependent plasticity of the morphology of horizontal cell terminals in cone pedicels of fish retinas. J. Neurocytol. 9:573-590.

Young, R.W. 1977. The daily rhythm of shedding and degradation of cone outer segment membranes in the lizard retina. J. Ultrastruct. Res. 61:172-185.

Young, R.W. 1978. Rhythmic shedding of rod and cone membranes. Invest. Opthalmol. Vis. Sci. 17:105-116.

Energetic Costs Associated with the Production of Gonads in the Anadromous Arctic Charr (*Salvelinus alpinus*) of the Nauyuk Lake Basin, Canada

JEAN-DENIS DUTIL

Department of Fisheries and Oceans, Quebec, P.Q.

DUTIL, J.D. 1984. Energetic costs associated with the production of gonads in the anadromous Arctic charr (*Salvelinus alpinus*) of the Nauyuk Lake basin, Canada. p. 263–276. *In* L. Johnson and B.L. Burns [eds.] Biology of the Arctic charr, Proceedings of the International Symposium on Arctic Charr, Winnipeg, Manitoba, May 1981. Univ. Manitoba Press, Winnipeg.

In the Nauyuk Lake system, anadromous Arctic charr become sexually mature when 600 mm or more in length. Maturing fish do not migrate to the sea the year they spawn. Individual charr in that population have also been shown never to spawn two years in succession. It is demonstrated that the lost opportunity to go to sea in summer prior to spawning, and the production of gonads, drain so much energy as to prevent the fish from spawning two years in succession. The time required for charr to recover is size dependent: larger fish need more time to recover than smaller fish. The 600 mm length-class has the highest energy content per gram: the content decreases with increasing and decreasing length.

INTRODUCTION

The success of Arctic charr throughout the Arctic suggests that they have developed physiological mechanisms to meet the severe climatic conditions that prevail in northern waters. Charr have been shown to exhibit K-strategies. Their growth rate is slow (summarized by Johnson 1980); they are long lived (Johnson 1976) and sexual maturity is deferred (Martin 1955, Hunter 1970). Northern populations turn less energy to the production of eggs than southern populations (Johnson 1980). Northern populations have also been shown to exhibit intermittent spawning, that is maturation taking place once every two or three years (Sprules 1952, Johnson 1980). This phenomenon also occurs in lake trout (*Salvelinus namaycush*) and brook charr (*Salvelinus fontinalis*) (Dutil 1980). This paper investigates the basis for intermittent spawning.

The charr population of the Nauyuk Lake system has been under study since 1974. Nauyuk Lake is located south of Victoria Island on the Kent Peninsula (68°22′N, 107°40′W). Johnson (1976, 1980) has provided a thorough description of the Nauyuk Lake charr population. The movements of these charr are summarized in a diagram (Fig. 1). When 160 to 240 mm in length, charr make their first migration to the sea. They enter the sea in June and July, returning to freshwater

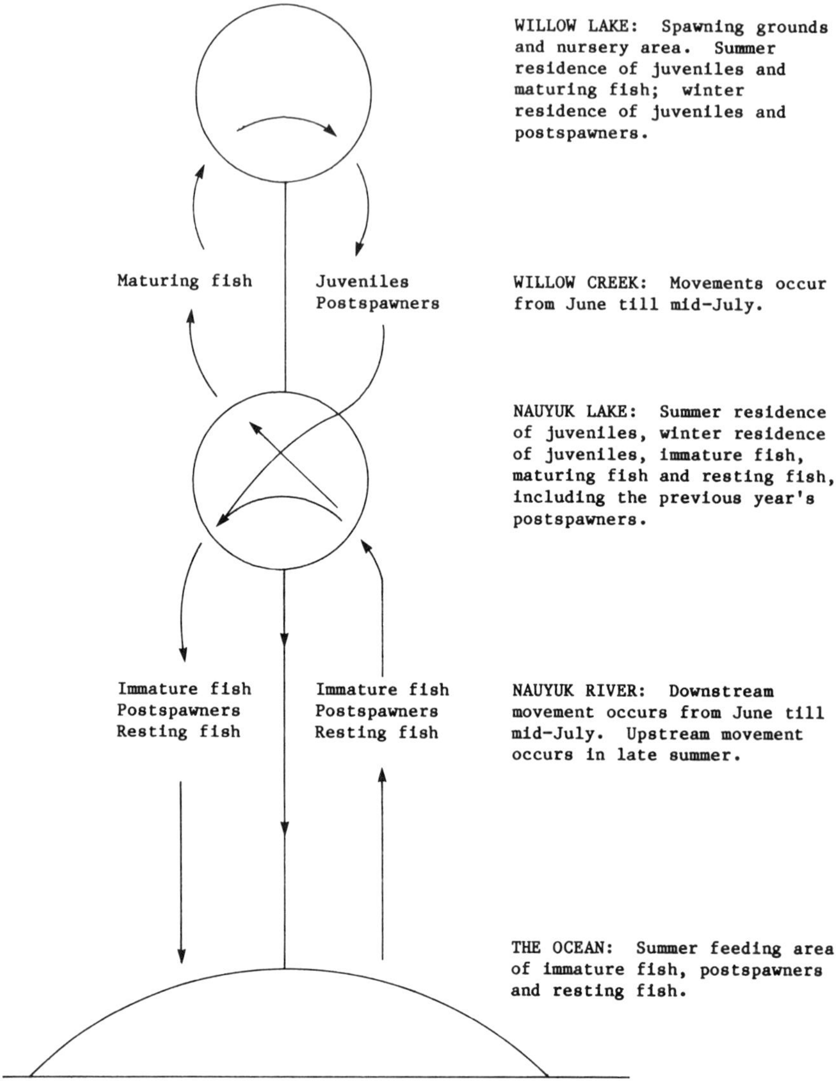

FIG. 1 The movements of Arctic charr in the Nauyuk Lake system.

in the fall. Young fish migrate year after year until they become mature. Maturing fish do not go to sea the year they spawn; they move to Willow Lake where they spend the summer and following winter. Postspawners move down to

Nauyuk Lake in spring, joining immature and resting fish in their migration to the sea.

Throughout the text we use the term maturing fish to indicate those fish that are to spawn in the fall of the current year. Fish that have spawned in the previous fall are postspawners remaining in this category until the following fall. Resting fish are those that have spawned previously, excluding maturing fish and postspawners. Immature fish are those that have not reached sexual maturity on any previous occasion.

MATERIALS AND METHODS

Many northern charr populations have a migratory pattern similar to that in the Nauyuk Lake system, but the situation in Nauyuk Lake made it possible to readily differentiate between maturing fish and postspawners (Willow Creek). Since all fish moving through Willow Creek were tagged at the fence we could distinguish between maturing fish, postspawners, immature fish and resting fish in Nauyuk River. However, to tell immature fish from resting fish was not possible. Therefore, those classes of fish were pooled.

In 1977, two weirs were set up that completely screened Willow Creek and Nauyuk River. From the Willow Creek weir, we collected maturing fish moving upstream and postspawners moving down. From the Nauyuk River weir, we collected immature or resting fish moving down to the sea, and immature or resting fish and postspawners moving upstream from the sea (Table 1). Routine operation of the weirs also required the measurement and weighing of random fish samples; such length and weight data had been recorded annually since 1974.

The fish we collected were placed in plastic bags and shipped frozen to the laboratory where they were measured ($\pm$ 1 mm) and weighed ($\pm$ 1 g). Fish were processed individually through a meat grinder. Duplicate samples were taken for proximate analysis. Lipid content was determined by the Bligh and Dyer method. Nitrogen content was determined by a micro-Kjeldahl digestion method, followed by the Berthelot reaction and from the data obtained the protein content was calculated. The energy content was calculated from proximate analysis data (Brody 1945). The methodology is outlined in more detail in Dutil (1982).

RESULTS

The Cost of Reproduction

Basically, the procedure consisted of determining the energy content of a standard fish. This required the proximate analysis of fish collected in the field. Condition index was calculated as

$$100\ W\ /\ L^3$$

Table 1. Classes of fish to be investigated in the present study.

Maturing fish moving up Willow Creek in 1976	I
Maturing fish moving up Willow Creek in 1977	II
Postspawners moving down Willow Creek in 1977	III
Postspawners moving up Nauyuk River in 1977	IV
Immature fish and resting fish moving down Nauyuk River in 1977	V
Immature fish and resting fish moving up Nauyuk River in 1977	VI

where W = total weight of the fish and L = fork length. Data are presented in Dutil (1982). Then linear regressions were fitted to proximate analysis data and condition index (Tables 2 and 3). Six length-classes were selected to reveal potential length-correlated phenomena (Fig. 2). Notice that the fish were pooled based on their length-class and not on the degree of maturity.

The next step consisted in calculating the length-weight relationships needed (Table 4). They were calculated from the length and weight data collected randomly from fish at the weirs. The mean weights of fish at exact length were calculated from these relationships (Table 5). These weights were then used to compute the condition index of the standard fish. From the linear regressions of Table 2 and Table 3 and the condition index, the standard proximate analysis values were derived and converted to energy values (Table 6).

The cost of production of the gonads (including both the cost of maturation and the lost opportunity for feeding at sea that summer) and starvation over winter, can be assessed from Table 6. Table 6 shows the standard condition of the maturing fish caught in Willow Creek in 1976, and the standard condition of the postspawners caught in Willow Creek the next year. The change in condition was assessed (Table 7). Net loss in energy content was 52 percent in the 12-month period. Indeed, the metabolic condition of postspawners caught in Willow Creek that year was poor (Table 6). They contained between 35-46 percent less energy than the rest of the fish (the immature and resting fish) moving down to the sea that year. Yet, maturing fish caught in Willow Creek in

Table 2. Linear regressions of lipid content (Y (g)) on transformed condition index (X′). Condition index (X) is transformed to $\sin^{-1} (X - 0.4)^{0.5}$.

Length class	Equation	Number of pairs of data	Coefficient of determination
550	Y = -196.11 + 390.28 X'	5	0.91
600	Y = -251.66 + 531.49 X'	24	0.82
650	Y = -408.04 + 733.91 X'	8	0.96
700	Y = -476.26 + 881.41 X'	23	0.95
750	Y = -455.34 + 899.69 X'	9	0.81
800	Y = -368.90 + 980.91 X'	6	0.77

1976 contained 06-15 percent more energy (the percentage decreasing with increasing length) than the immature and resting fish moving down to the sea that year. Let us assume that these percentages were the same in 1976; then production of the gonads in 1976, including the lost feeding opportunity, cost from 6000 KJ for 550 mm fish to nearly 16000 KJ for 800 mm fish. Therefore, postspawners moving to the sea in 1977 had a 35 to 46 percent handicap compared to the immature and resting fish.

How much time will it take for postspawners to recover? From Table 6, the summer increase in the energy content of postspawners was assessed (Table 8). The results are contrasted with those for the immature fish and resting fish (Fig. 3). Postspawners that were 650 mm or more in length stored less reserve than did the immature fish and resting fish. Postspawners that were 600 mm in length or less stored more, so that by fall these fish had made up for some of the initial difference. Nevertheless, late in the summer of 1977, postspawners still contained less in the way of reserve than the immature and resting fish. This difference is contrasted with the summer storage of energy of the immature and resting fish (Fig. 4). Notice that in this study postspawners are not considered as resting fish until a whole year has passed since spawning. Figure 4 shows that 650 mm fish require yet another full summer such as that of 1977 to recover. Smaller fish would need less and larger fish would need more, so that the time needed for postspawners in general to catch up to the immature fish and resting fish invariably is more than one year such as 1977.

Table 3. Linear regressions of protein content (Y (g)) on transformed condition index (X'). Condition index (X) is transformed to $\sin^{-1} (X - 0.4)^{0.5}$.

Length class	Equation	Number of pairs of data	Coefficient of determination
550	Y = -0.93 + 400.79 X'	5	0.87
600	Y = -13.22 + 521.28 X'	27	0.82
650	Y = -101.62 + 781.05 X'	12	0.94
700	Y = -53.22 + 853.53 X'	26	0.84
750	Y = -40.63 + 1063.53 X'	10	0.83
800	Y = 60.54 + 1080.43 X'	6	0.93

The Energy Contained per Gram of Charr

This section comments on length-related variations in the relative energy content. Consider fish with a condition index of 1.00. For a number of length-classes, the total weight of charr with a condition index of 1.00 was calculated. Then, total lipid content and total protein content were derived from the regressions of Table 2 and Table 3 (refer to Dutil 1982 for missing length-classes) and converted to energy contents. Total energy content was then calculated by summation (Table 9). Lowest contents occurred in smaller fish due to their minimal reserves of fat in summer. The 600 mm length-class had the highest energy content per gram.

DISCUSSION AND CONCLUSIONS

The charr population of the Nauyuk Lake system is similar in its characteristics to most northern charr populations (Johnson 1980). That maturing fish do not have access to the spawning beds past late July has no consequence, since charr do not migrate to the sea the year they spawn (Moore 1975). Partitioning fish on the basis of their sex was not possible: the sex ratio being so high in favour of females in the migration of Willow Creek, that to get the required numbers of fish of both sexes would not have been possible without killing a large number of fish. Other research programs in Nauyuk Lake precluded this. Most workers have concluded that females tend to become more depleted (Lien 1978, Craig 1977, Love 1970), although the direct measurement of individually marked pre-

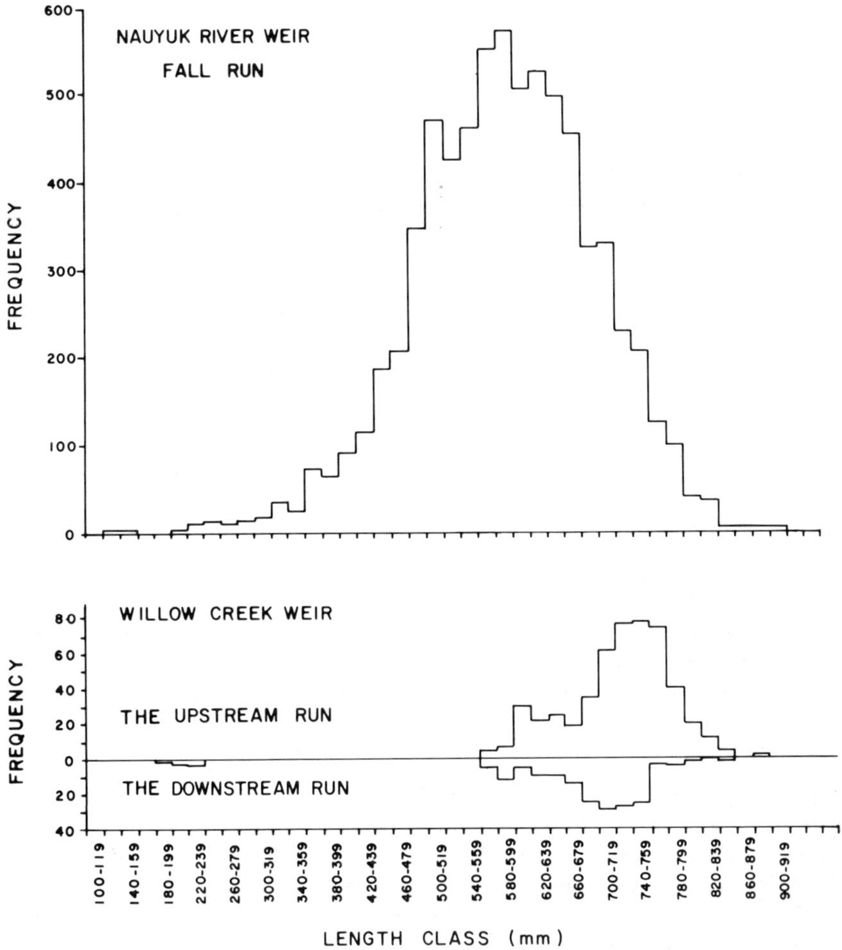

FIG. 2 Catch curves for length of charr caught in Willow Creek and Nauyuk River (fall run only) in 1976.

and postspawning males and females shows little difference in weight loss in the Nauyuk Lake population (Johnson, pers. comm.).

The population was divided into categories based on maturity: maturing fish, postspawners, immature fish and resting fish. Since the immature fish could not be distinguished from resting fish, these two categories were pooled. Furthermore, the border-line between postspawners and resting fish is to a certain extent arbitrary: postspawners are considered as resting fish when more than a year has

Table 4. Length-weight relationships of fish sampled at the weirs, where a and b are the constants in the power curve and r^2 the coefficient of determination. The class numbers refer to Table 1.

Class	Number of fish	a	b	r^2
I	531	7.198×10^{-5}	2.6974	.90
II	225	1.765×10^{-5}	2.9044	.91
III	472	1.790×10^{-5}	2.8581	.89
IV	112	5.387×10^{-4}	2.3686	.78
V	1805	7.218×10^{-6}	3.0326	.94
VI	1080	8.017×10^{-6}	3.0485	.99

passed since they last matured. In any event, partitioning fish into more discrete categories tends to mask the differences we wish to stress and not to magnify them.

The body composition predicted from the condition index is assumed to be the same regardless of the month, the year or the maturity stage. The mean condition index was calculated in this study from random samples of fish routinely collected at the weirs. Caulton and Bursell (1977) found a good correlation between condition index and body composition.

Nitrogen content was used to calculate the protein content. Non-protein nitrogen represents 38 percent of total nitrogen in rays (Shewan 1951), but this fraction is much less (2 percent) in salmonids (Brett et al. 1969). This fraction has not been taken into consideration in this study. Therefore, the protein content and hence the energy content may be slightly too high. Calculating the energy content from proximate composition has been shown to result in higher values (Craig et al. 1978). In this study, results from proximate analysis and calorimetry differed by 4 percent.

The energy content of mature gonads could not be assessed. Nevertheless, this study has shown that postspawners contain 35 to 46 percent less energy in spring than immature and resting fish though maturing fish were in better condition the year before. Postspawners require more than a year (a year like

Table 5. Mean weight (g) of fish at exact length (mm), calculated from the length-weight relationships of Table 4. The class numbers refer to Table 1.

Length (mm)	Class					
	I	II	III	IV	V	VI
550	1775	1606	1216	1677	1476	1811
600	2245	2068	1560	2058	1921	2361
650	2785	2609	1961	2484	2449	3014
700	3402	3236	2424	2956	3066	3777
750	4098	3954	2952	3475	3780	4662
800	4877	4769	3550	4044	4597	5675

1977) to fully recover. Smaller fish (550 mm) require a little more than such a summer whereas larger fish (800 mm) require more than two such summers. Therefore, the conclusion is reached that the production of gonads generally takes so much energy as to prevent the fish from spawning two years in succession.

The mechanism for intermittent maturation has never been examined. This study shows that production of gonads in Nauyuk Lake costs so much to charr as to limit their frequency of maturation. Scott (1962) on *Salmo gairdneri*, Bagenal (1969) on *Salmo trutta*, and Hislop and collaborators (1978) on *Melanogrammus aeglefinus*, concluded that restricted diet in the pre-reproduction period resulted in a lower maturation rate and possibly in no maturation. However, Leggett and Carscadden (1978) concluded that the reproductive characteristics of shad could not be explained in these terms. That individual charr in Nauyuk Lake rarely mature 2 years in succession is not as expected from models of life-history strategies (Stearn 1976). Starvation throughout the cold months and reproduction are the main drain on the reserves charr build up in summer. Since the 2-month migration to sea does not permit the full recovery of postspawners, the conclusion is reached that limitations to maturation rate must be physiological in nature.

The decision whether to breed or not is thus critical. Arctic charr probably have developed mechanisms that permit them to make a balanced decision on whether to reproduce. This could be dependent on the energy reserves stored in summer: maturing fish in spring are in a better condition than the immature and resting fish. Furthermore, maturity is deferred when reserves are low.

Table 6. Total energy content (kJ) of the standard charr as calculated from their lipid and protein contents. The class numbers refer to Table 1.

Length (mm)	Class					
	I	II	III	IV	V	VI
550	16155	13415	7429	14386	11393	16621
600	20766	17757	9741	17757	15736	22409
650	24172	21321	9941	18975	18494	28499
700	28323	25567	12378	21192	22834	34381
750	33235	30795	16226	24115	28379	41739
800	40874	39563	23493	29921	37005	51172

The Nauyuk Lake charr population has a bimodal length distribution (Johnson 1976). Smolts make up the small mode (220 mm). The bulk of the population is composed of larger fish with a modal length of 600 mm. Johnson observed that this structure persists through time. He hypothesized that natural forces push individual fish to reach and to retain a certain size. This study found the quantity of energy stored per gram to be maximum for 600 mm fish. Since the energy reserves are a key factor, fish of modal length certainly are best fitted in this respect.

ACKNOWLEDGEMENTS

I thank Dr. Lionel Johnson, Freshwater Institute, Fisheries and Oceans, for his helpful suggestions in the design of the project and review of the manuscript, A. Lutz for his assistance in the selection of methods for chemical analyses, L. Allard and E. Woods for technical assistance in the laboratory, and my wife Judith for her assistance in data analysis. The author received scholarships from the National Research Council of Canada and from the Ministère de l'Education du Québec.

REFERENCES

Bagenal, T.B. 1969. The relationship between food supply and fecundity in brown trout *Salmo trutta*. J. Fish Biol. 1:167-182.

Brett, J.R., J.E. Shelbourn, and C.T. Shoop. 1969. Growth rate and body composition of fingerling sockeye salmon (*Oncorhynchus nerka*) in relation to temperature and ration size. J. Fish. Res. Board Can. 26:2363-2394.

Table 7. Mean losses (△E) incurred by maturing fish both from the production of gonads and starvation.

Length (mm)	ΔE (kJ)
550	8727
600	11025
650	14231
700	15945
750	17009
800	17381

Brody, S. 1945. Bioenergetics and growth. Reinhold Publishing Co., New York.

Caulton, M.S., and E. Bursell. 1977. The relationship between changes in condition and body composition in young *Tilapia rendalli*. J. Fish Biol. 11:143-150.

Craig, J.F. 1977. The body composition of adult perch, *Perca fluviatilis*, in Windermere, with reference to seasonal changes and reproduction. J. Anim. Ecol. 46:617-632.

Craig, J.F., M.J. Kenley, and J.F. Talling. 1978. Comparative estimations of the energy content of fish tissue from bomb calorimetry, wet oxydation and proximate analysis. Freshwater Biology 8:585-590.

Dutil, J.D. and G. Power. 1980. Coastal populations of brook trout, *Salvelinus fontinalis*, in Lac Guillaume-Delisle (Richmond Gulf) Quebec. Can. J. Zool. 58:1828-1835.

Dutil, J.D. 1982. Periodic changes in the condition of the Arctic charr (*Salvelinus alpinus*) of the Nauyuk Lake system. Ph.D. Thesis. The University of Manitoba, Canada. 150p.

Hislop, J.R.G., A.P. Robb and J.A. Gauld. 1978. Observations on effects of feeding level on growth and reproduction in haddock, *Melanogrammus aeglefinus*, in captivity. J. Fish Biol. 13:85-98.

Hunter, J.G. 1970. Production of Arctic char (*Salvelinus alpinus*) in a small Arctic lake. Fish. Res. Board Can. Tech. Rep. 321:190 p.

Johnson, L. 1976. Ecology of Arctic populations of lake trout, *Salvelinus namaycush*, lake whitefish, *Coregonus clupeaformis*, Arctic charr, *Salvelinus alpinus*, and associated species in unexploited lakes of the Canadian Northwest Territories. J. Fish. Res. Board Can. 33:2459-2488.

Johnson, L. 1980. The Arctic charr, *Salvelinus alpinus*, p. 15-98. *In* E.K. Balon [ed.] Charrs: salmonid fishes of the genus *Salvelinus*. Dr. W. Junk Publishers, The Hague.

Leggett, W.C. and J.E. Carscadden. 1978. Latitudinal variation in reproductive characteristics of American shad (*Alosa sapidissima*): evidence for population specific life-history strategies in fish. J. Fish. Res. Board Can. 35: 1469-1478.

Lien, L. 1978. The energy budget of the brown trout population of Ovre Heimdalsvatn. Holarctic Ecology 1:279-300.

Love, R.M. 1970. The chemical biology of fishes. Academic Press, London and New York. 547 p.

Martin, N.V. 1955. Limnological and biological observations in the region of the Ungava or Chubb Crater, Province of Quebec. J. Fish. Res. Board Can. 12:487-498.

Moore, J.W. 1975. Distribution movements and mortality of anadromous Arctic char, *Salvelinus alpinus*, in the Cumberland Sound of Baffin Island. J. Fish Biol. 7:339-348.

Table 8. Summer increase (△E) in total energy content in postspawners.

Length (mm)	ΔE (kJ)
550	6957
600	8016
650	9034
700	8814
750	7889
800	6428

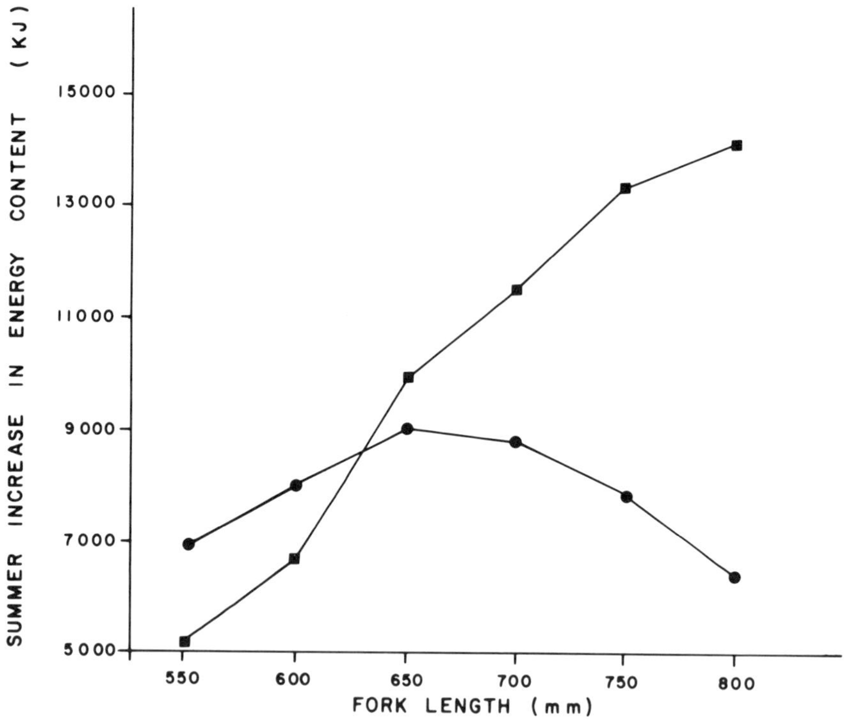

FIG. 3 Summer increase in energy content in the postspawners (—●—) and in the immature and resting fish (—■—), in relation to fork length.

Table 9. Calculation of the quantity of energy stored per gram. The condition index is assumed to be 1.00. Thus the transformed condition index is 0.89 (Table 2 and Table 3).

Length class	Weight (g)	Lipids (g)	Proteins (g)	Lipids ($kJ\text{-}g^{-1}$)	Proteins ($kJ\text{-}g^{-1}$)	Total energy ($kJ\text{-}g^{-1}$)
220	106	4	30	1.49	6.69	8.18
400	640	44	132	2.72	4.88	7.60
500	1250	91	261	2.88	4.94	7.82
550	1664	151	356	3.59	5.06	8.65
600	2160	221	451	4.05	4.94	8.99
650	2746	245	594	3.53	5.11	8.64
700	3430	308	706	3.55	4.87	8.42
750	4219	345	906	3.23	5.08	8.31
800	5120	504	1022	3.89	4.72	8.61

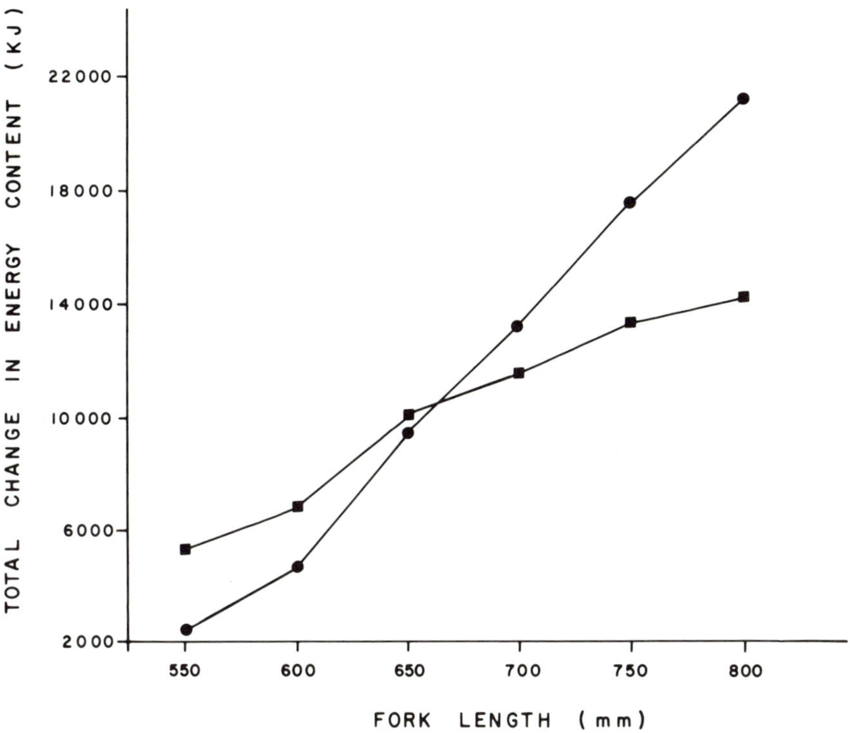

FIG. 4 The difference in energy content between the postspawners and the immature and resting fish in fall (—●—) contrasted to the summer increase in energy content of the immature and resting fish (—■—), in relation to fork length.

Scott, D.P. 1962. Effect of food quantity on fecundity of rainbow trout *Salmo gairdneri*. J. Fish. Res. Board Can. 19:715-731.

Shewan, J.M. 1951. The chemistry and metabolism of the nitrogenous extractives in fish. Biochem. Soc. Symp. 6:28-48.

Sprules, W.M. 1952. The Arctic char of the west coast of Hudson Bay. J. Fish. Res. Board Can. 9:1-15.

Wilkins, N.P. 1967. Starvation of the herring, *Clupea harengus*: survival and some gross biochemical changes. Comp. Biochem. Physiol. 23:503-518.

Management and Utilization of Arctic Charr in Norwegian Hydroelectric Reservoirs

PER AASS

Directorate for Wildlife and Freshwater Fish, As, Norway

AASS P. 1984. Management and utilization of Arctic charr in Norwegian hydroelectric reservoirs, p. 277–291. *In* L. Johnson and B.L. Burns [eds.] Biology of the Arctic charr, Proceedings of the International Symposium on Arctic Charr, Winnipeg, Manitoba, May 1981. Univ. Manitoba Press, Winnipeg.

Next to brown trout, the non-migratory Arctic charr is the most widespread freshwater fish in Norway. It is found from sea level to about 1400 m a.s.l. Crowded populations may be established where there are good natural spawning grounds, limited production of benthos and a lack of predators. The impoundment of lakes increases this predisposition towards crowding and dwarfing, which in turn gives rise to problems in fisheries management. Pronounced year class fluctuations may also arise from variations in the rate and extent of winter drawdown. Impoundment makes the use of the traditional bottom gillnet difficult, and therefore the fisheries in most reservoirs are not fully exploited; surface gillnets and bag nets are used in some of the larger reservoirs. During winter drawdown charr congregate in the currents near the hydroelectric intake tunnels and in narrow straits; popular ice fisheries are found in these locations. In winter charr only leave the reservoirs occasionally, but in spring a considerable part of the adult population may migrate through the dam gates to occupy the rapids below the dams where they are heavily fished. Although sport fishing may be heavy, total yields of charr are moderate and only exceptionally reach 3-5 $kg \cdot ha^{-1} yr^{-1}$. The individual size and age of the fish caught is not usually influenced by fishing. Under poor food conditions some charr become cannibalistic, attaining a weight of 2-3 kg, but this is of little significance in population control. The Tunhovdfjord strain of brown trout may feed mainly on dwarf charr. The consumption of charr may attain 2-3 $kg \cdot ha^{-1} yr^{-1}$. About two million fingerlings of this strain are now stocked in charr reservoirs each year. If fishing and predation do not exploit a stock sufficiently to increase the individual size of charr, the introduction of food animals may be effective. Experimental introductions of *Mysis relicta* seem to have increased the size and quality of the charr.

INTRODUCTION

Next to brown trout, the non-migratory Arctic charr *Salvelius alpinus*, is the most widespread and one of the most valuable freshwater fish in Norway. It occurs naturally along the coast and in some inland districts originally having reached its present locations through rivers emptying into the Baltic. More recently, charr has been widely distributed and is now living in thousands of lakes and tarns, and running water, in Northern Norway (Fig. 1). Due to acid precipitation its distribution in the southern-most part of the country is shrinking (Sevaldrud and Muniz 1980).

CHARR AND LAKE IMPOUNDMENTS

The natural conditions in most Norwegian watercourses favour the charr. Natural waters are cold, clear and oligotrophic with predominantly stony bottoms supporting only a few fish species, mainly brown trout *Salmo trutta*. The charr has the ability to exploit both bottom organisms and zooplankton; therefore the species has a higher production potential than the brown trout. Spawning and early life in still water facilitate production of juveniles, but this may lead to crowded populations of stunted adults. If the food situation is extremely poor, cannibalism may develop and the populations become divided in two distinct parts: a few large fast-growing individuals and the dwarfed majority. In some lakes there exist both dwarf and fast-growing strains not the result of food shortage.

Most of Norway's major watercourses are regulated for hydroelectricity, and about half of the area occupied by lakes is impounded. Due to its wide distribution, charr occur in hundreds of reservoirs. Generally, the short term impoundment effect, causing increased plankton production, favours the charr. The growth increases and in a few years the charr double or quadruple their weight, depending on the area submerged. At this stage of reservoir development the charr may turn to fish-eating if opportunity presents itself. The fast-moving predatory charr is easily caught with hooks or nets. After a few years the food conditions deteriorate and charr growth becomes less rapid than before impoundment. But there are exceptions to this rule of the positive interlude. The impoundment of several lakes and rivers in the upper part of River Usta put large quantities of deposited glacial silt into suspension. Charr react more strongly than other fish species to water turbidity caused by erosion and construction work in the watershed. Therefore catches fell drastically and the effects were discernible, even in parts of the system not actually impounded. In Lake Ustedalsfjord, 6 km below the last dam, the catch of charr per fishing effort decreased to about 1 percent of the pre-impoundment value, and the mean weight of the charr caught was also reduced. The silting lasted only two years, but 15 years later the charr population still had not stabilized (Fig. 2). Adequate measures to improve the fishing have not yet been found.

The long term effect of impoundment is an impoverishment, or increased oligotrophication, of the lakes. The littoral is transformed into a barren zone of sand, gravel and boulders, thus extending the spawning areas of the charr considerably. At the same time, the quantity and quality of bottom organisms

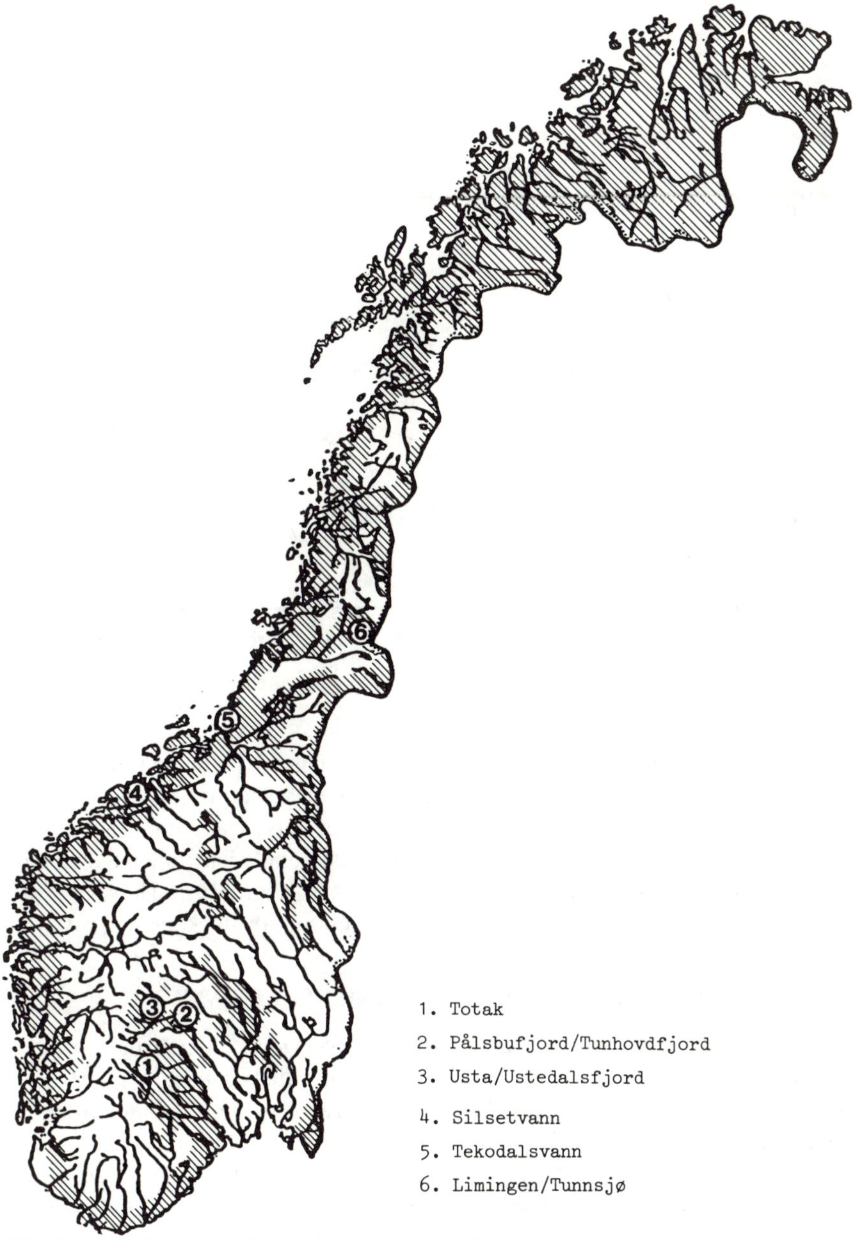

FIG. 1 Distribution of charr in Norway shaded. Reservoirs mentioned in the text are numbered.

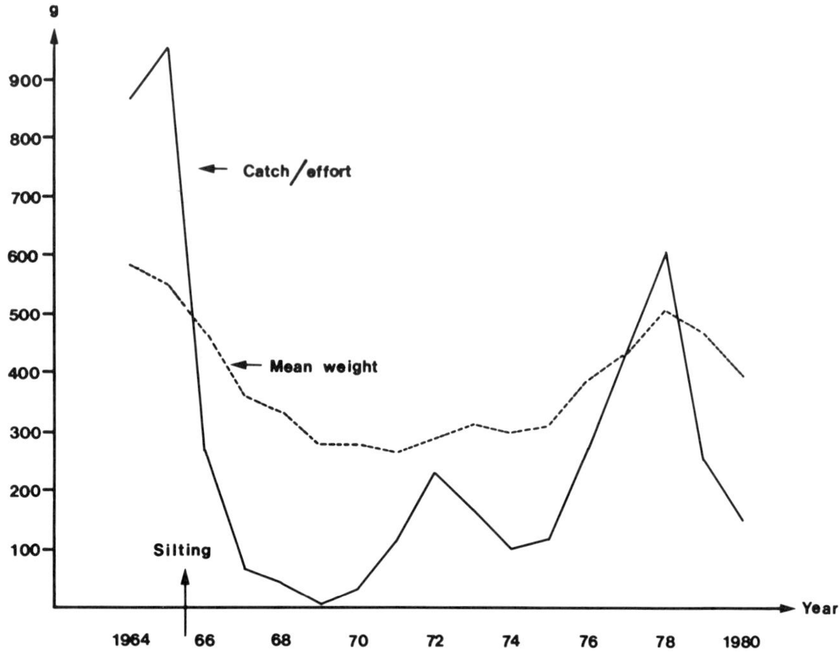

FIG. 2 Lake Ustedalsfjord. Catch/effort and mean weight of charr after silting in 1965-66. Mesh size 40 mm.

available to the charr is drastically reduced, although the plankton production is less affected. Together, these trends lead to increased crowding and eventual dwarfing of the charr populations. In many reservoirs, the mature fish do not achieve a weight of 100 g. The utilization of such populations, often existing in mountainous, far away places, has always been a problem. However, the management of the reservoirs may in some cases control stock and improve the harvest.

IMPOUNDMENTS AND CHARR RECRUITMENT

Typical of reservoirs is the fluctuating water level, which may have a pronounced effect on charr recruitment. Although tied to their old spawning sites, the charr gradually invade the eroded littoral to spawn. The proportion of eggs and newly hatched fry left high and dry by winter drawdown depends on the rate and extent of the usage; the strength of the year classes varies accordingly. The lakes Pålsbufjord and Tunhovdfjord in the Numedalslågen's watercourse are good examples. The lakes are situated close to each other with maximum amplitudes of 24.5 and 18 metres respectively (for details, see Aass 1970).

Formerly, the water level in Pålsbuford was lowered early in winter; fully drawndown its surface area was reduced by 75 percent. The egg mortality was high and year classes became small. The size of the mature fish was fairly good and the stock was reasonably well exploited. Five winters during the period 1948-70 the reservoir was lowered only 9.3-17.9 metres. This resulted in rich year classes which formed the basis of improved fishing for some years (Fig. 3). The draining of Tunhovdfjord did not follow a fixed pattern, but in winters with rapid drawdown a large proportion of eggs and fry were exposed and frozen, and the ensuing year classes were small (Aass 1964). In the last decade the pattern of lowering has been more consistent. To obtain maximum power production Tunhovdfjord has been lowered as slowly and as late as possible, while Pålsbufjord has been emptied first. This has changed the population densities in both lakes. The reduced egg and fry mortality in Tunhovdfjord has led to a series of large year classes and decreasing fish size, although in spite of this trend the yearly catch has increased. The charr stock of Pålsbufjord is now so small that the fishing effort is greatly reduced notwithstanding satisfactory fish size. The yearly fluctuations are thus reflected in the total yield, catch per effort, mean weight and growth. Through the manipulation of reservoir levels the size of charr populations may be controlled, but the economics of such a measure cannot compete with the efficient use of water for electricity.

CHARR MIGRATIONS

In summer and autumn charr is mainly fished using bottom nets. Damming renders this harvesting difficult because of drowned vegetation and drifting debris. This situation may last for more than half a century (Jensen and Aass 1958) and the traditional fishing is often brought to an end. However, turning the lake into a reservoir may start charr movement on a large scale, leading to concentrations that facilitate exploitation. Migrations were first reported from Lake Aursunden (Schmidtnielsen 1925) but charr movements, also within the lake, have been extensively studied in Lakes Pålsbufjord and Tunhovdfjord (Aass 1970, 1973).

The movement of most individuals taking place in winter and spring are nearly always downstream. It is assumed that these migrations start as a search for food, but the later phases of the migration are controlled by hydrological factors. In winter the active search turns into a passive movement under the ice. Movement is interrupted by currents which form upstream from the draining tunnels or dam gates or in narrow straits that may be created when the lake level is lowered. When the water velocity increases to a level perceptible to the charr, they respond by halting and keeping to the same place for the rest of the winter. The charr in the currents are exploited using baited spoons (Fig. 4). By regulation

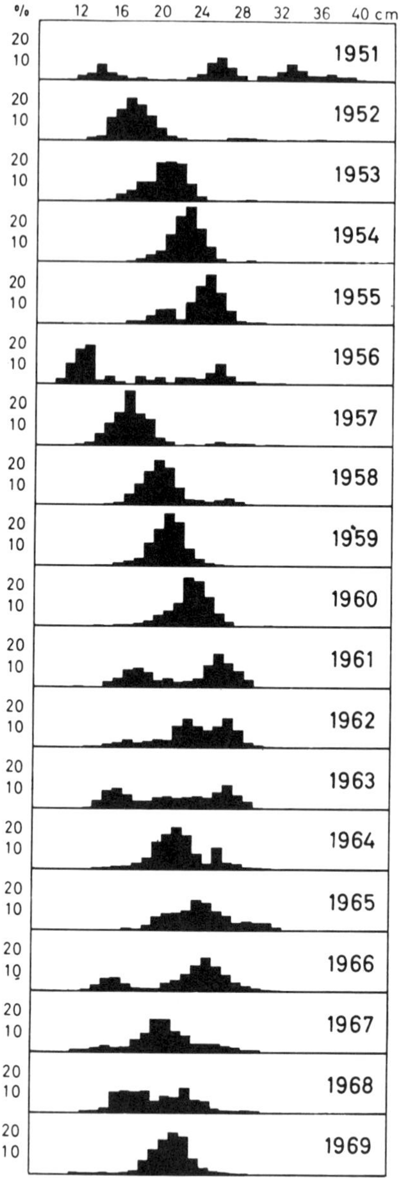

FIG. 3 Lake Pålsbufjord. Length distribution of charr caught during winter angling 1951-69. Rich year classes were born in 1949, 1954, 1959, 1961 and 1964. The large charr caught in 1951 were the result of the damming in 1946-47.

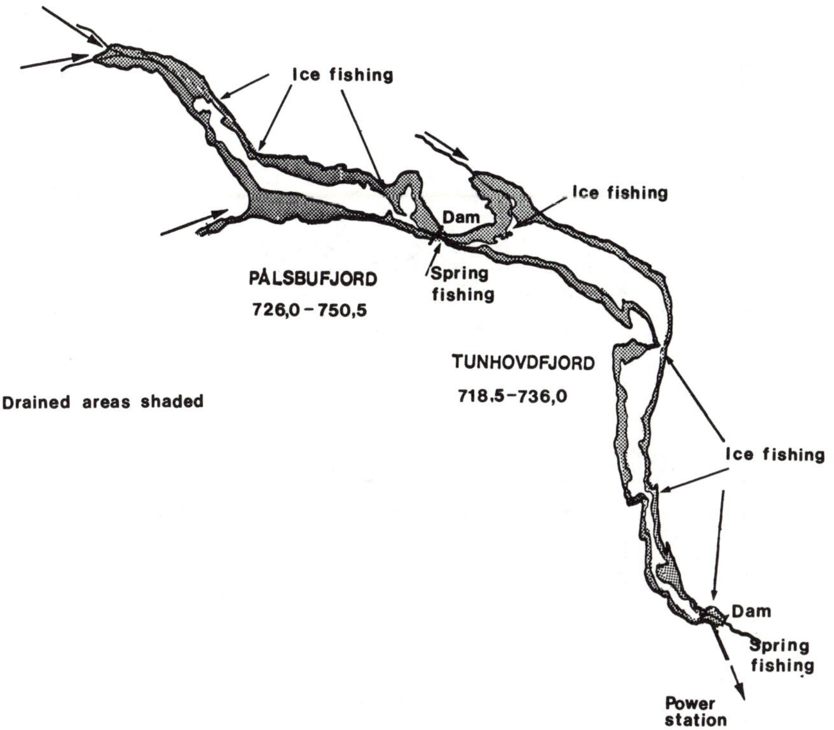

FIG. 4 Lakes Pålsbufjord and Tunhovdfjord. Straits and currents form the main angling places.

of the water level it is possible to influence the length of the fishing season and distribution of the charr as well as the number moving. Rapid draining early in winter will get the migrations off to an early start and a heavy drawdown may create greater fish concentrations than slight ones. Generally the winter migrations involve a great part of the total stock. It is estimated that in Tunhovdfjord more than 50 percent including even the remotest spawning population, moves downstream in winter (Aass 1970).

YIELD OF CHARR

Table 1 shows the estimated yearly catch in Tunhovdfjord, by number and weight during the 20 year period 1960-80. The number varies between 47,100 and 158,250. These figures are based on daily creek censuses at the main fishing places and the statements of anglers and fishermen. On average, 65 percent of the yearly catch is taken in the currents, while the remaining 35 percent is evenly spread between ice fishing along the shores and net fishing on the spawning sites

Table 1. Lake Tunhovdfjord. Estimated yield of charr 1961-1980.

Year	Number				Weight	
	Winter		Summer/autumn	Total	Kg	$Kg \cdot ha^{-1}$
	currents	still water				
1961	59 900	7 300	12 500	79 700	7 930	3.13
1962	44 650	6 100	10 000	60 750	6 000	2.37
1963	34 050	5 050	8 000	47 100	4 410	1.74
1964	36 900	7 600	6 900	51 400	4 550	1.79
1965	47 700	10 600	9 300	67 600	5 140	2.03
1966	73 800	12 500	7 500	93 800	6 320	2.49
1967	71 100	14 600	5 800	91 500	6 280	2.48
1968	59 250	22 800	10 900	92 950	6 550	2.58
1969	65 950	19 100	12 600	97 650	7 020	2.77
1970	51 950	11 200	22 600	85 750	7 360	2.90
1971	65 300	11 700	21 500	98 500	8 770	3.46
1972	37 450	10 300	27 100	74 850	7 345	2.90
1973	32 800	13 500	17 100	63 400	6 440	2.54
1974	38 950	21 100	9 700	69 750	4 340	1.71
1975	45 850	27 200	16 200	89 250	5 455	2.15
1976	58 150	27 500	20 350	106 000	7 265	2.87
1977	72 800	18 200	26 100	117 100	8 445	3.33
1978	101 850	18 400	38 000	158 250	11 690	4.61
1979	71 500	14 550	16 100	102 150	7 220	2.85
1980	81 750	20 850	29 000	131 600	7 885	3.11

in the autumn. In recent years net fishing has increased its share of the catch at the expense of angling in the currents.

The yearly variations in number are partly due to climatic conditions, number of anglers and the water level, but the year class strength has a dominating influence. Figure 5 shows the yearly catch in number and the rate and extent of drawdown. It is assumed that the improved yield subsequent to 1976 is due to slow or incomplete utilization of the reservoir food reserves in previous years. The altered management of the reservoir favours the hatching and survival of charr fry. As a result the mean size of charr caught has decreased. A rich year class is easily observed in the ice fishery where 2-3 year old charr (140-180 mm) constitute the most significant part of the catch. Due to net selection it is more difficult to distinguish between the various year classes in the autumn fishing. Nor does a single year class dominate the net catches over a number of years because few charr are older than 8 years. A reduction of the age is typical of most reservoir populations and must be due to the altered food conditions. This should have consequences for the exploitation of the populations.

The relation between reservoir management and charr numbers, movements and exploitation is especially conspicuous in Lake Tunhovdfjord. Similar conditions are found in many reservoirs, for instance the neighbouring Lake Pålsbufjord. Migrations within the lake give rise to a winter fishery upstream from the outlet, but the outward migration is even more striking. The charr leave the lake through the bottom gate in spring while the water level is rising. For the years 1962-68 between 11 and 33 percent of the mature charr were assumed to leave Pålsbufjord. The manipulation of the gates was thought to be responsible for most of the yearly variations (Aass 1973). Thus the management of the reservoir may further influence the exploitation of the charr stock. After entering the short rapids below Pålsbufjord the charr stop in a vain attempt to re-enter the lake. This gives rise to intensive angling; a large proportion of the Pålsbufjord yield is thus harvested outside the lake. Table 2 shows the charr catches 1961-71 in the lake and rapids; the latter constitute, on average, 28 percent of the total yearly catch. The charr often leave the reservoirs through the tunnels leading to the power stations, thus the rate of emigration is thus impossible to estimate. To judge from all the dead charr covering the trash racks in Tunhovdfjord, this mortality is certainly considerable.

The estimated yearly charr catch in kilos and kilos $\cdot$ ha^{-1} in Tunhovdfjord, is given in Table 1. The yearly average for the last two decades is 2.69 kg $\cdot$ ha^{-1}. The figure is based on the area of the reservoir at its highest regulated level. If the area at the lowest regulated level is made the basis, the figure should be multiplied by 1.8. In comparison the impounded Lake Limingen in Namdalen yields only 0.075 kg $\cdot$ $ha^{-1}yr^{-1}$. Differences in yield between reservoirs are as

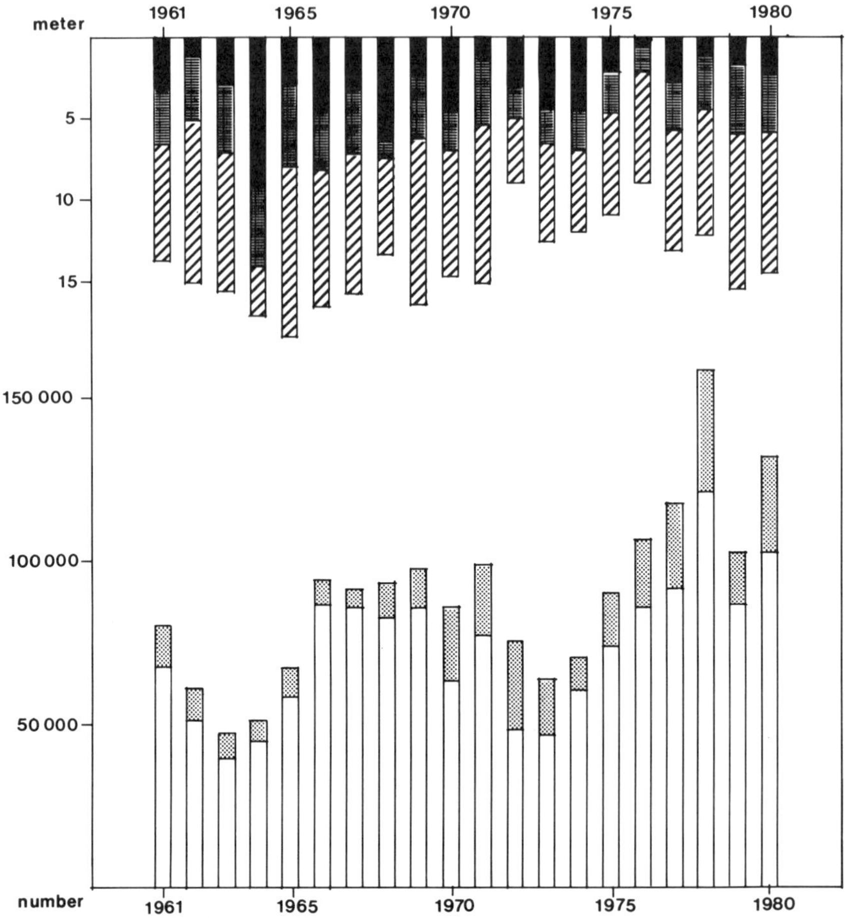

FIG. 5 Lake Tunhovdfjord. Bottom: yearly yield of charr in number — ice angling, net fishing. Top: yearly lowering in metres — before 1 February, 1 February - 1 March, after March 1.

large as those between unimpounded lakes. This is due partly to natural conditions and partly to the regulation height and management of the reservoir. The original conditions in Tunhovdfjord and Pålsbufjord were very much the same, yet the average yearly yield of the latter, including the emigrants, was only 1.83 $kg \cdot ha^{-1}$ for the years 1961-70. The extensive winter draining, reducing the area from 19.50 to 5.25 km^2, is assumed to be responsible for the difference. Both recruitment and food production are affected.

The estimated fishing mortality of the two lakes is based on very extensive investigations. Sources of natural mortality include charr escaping through drainage gates and tunnels and fish eaten by predatory brown trout. The loss may be roughly calculated on the basis of tagging and estimated yield. If emigrations from Pålsbufjord and Tunhovdfjord are proportional, the total weight of adult charr removed from Lake Tunhovdfjord as a result of emigration, fishing and predation is about 10 $kg \cdot ha^{-1}$ each year.

IMPROVEMENT OF CHARR GROWTH

Because of its small size, the market value of the reservoir charr is often low. Nor do the time-consuming conventional fishing methods, angling and bottom net fishing, encourage the utilization of the reservoir stocks. However, more efficient gears are starting to be used; for example, floating monofilament nets have been successfully introduced in many reservoirs. In Tunhovdfjord, the catch per effort during summer may be compared with that of the autumn fishery for spawners. The increased length of the fishing season has the additional advantage that the pelagic charr is superior in quality to the spawners.

As an experiment, floating bag nets have been introduced into other reservoirs. Korsen and Møkkelgjerd (1980) report that one bag net in the Teksdalsvatn reservoir of 400 ha caught 28.5 $charr \cdot ha^{-1}$ (2.9 $kg \cdot ha^{-1}$) in the 1979 season. But neither floating nets, nor bag nets, are effective in long periods of calm, warm weather because charr avoid high surface temperatures.

Reducing the stock by increased fishing effort may improve the quality, but large quantities must be removed, and because plankton form the basis of the food, growth will increase only marginally. In the Totak reservoir preliminary experiments with rearing charr in floating pens has been tried (Paal Wendelbo, pers. comm.), but data on the economics are not yet available.

INTRODUCTION OF FOOD ORGANISMS

Swedish attempts to improve the fish food situation in reservoirs have been more comprehensive (Fürst 1972), but introductions of *Pallasea quadrispinosa* Sars and *Mysis relicta* Lovén have also been made in Norway (Aass 1973, Garnås et al. 1980). In the autumn of 1969 *Mysis* was transferred to Limingen from the adjacent Swedish Lake Blåsjön, where the species had been introduced in 1964 (Fürst 1965). The highly oligotrophic Lake Limingen covers an area of 95 km^2 having a maximum depth of more than 200 m. When the lake is lowered 8.7 m the narrow littoral shelf is completely exposed; this results in both charr size and number being reduced. The general effects of the Scandinavian *Mysis* introductions on charr populations are not yet known, but the development in

Table 2. Lake Pålsbufjord and River Numedalslågen (Torkelsbustryket). Estimated yield of charr 1961-1971.

Year	Number		Weight kg		Total weight
	Lake	River	Lake	River	$kg \cdot ha^{-1}$
1961	20 100	11 100	2 820	1 500	2.22
1962	17 700	6 900	2 310	1 000	1.70
1963	16 800	3 800	2 040	490	1.30
1964	25 200	8 000	2 780	820	1.85
1965	17 000	13 600	2 180	1 780	2.03
1966	21 750	13 400	2 560	1 630	2.15
1967	22 600	12 150	1 950	1 030	1.53
1968	22 200	7 950	1 600	640	1.15
1969	41 500	7 050	4 180	570	2.44
1970	24 750	10 500	3 125	920	2.07
1971	22 300	3 300	3 030	340	1.73

Limingen has so far been positive. During the experimental fishing in 1972, no *Mysis* was found in charr stomachs; in 1974 *Mysis* constituted 8 and 10 percent of the food weight in summer and autumn respectively. The amount increased to 31 and 33 per cent in 1978; simultaneously the mean weight of charr in the experimental catches doubled, at the same time showing an improvement in flesh quality. (Fig. 6, Table 3). The turning point coincided with the introduction of *Mysis*, which is possibly responsible for the change. A corresponding development has not yet taken place in the adjacent Lake Tunnsjø where *Mysis* was introduced some years later. So far the number of charr in Limingen seems unaltered. Mysis may have a positive effect only on the food situation of the adult charr, but the potential of the spawning sites was severely reduced by drawdown and power plant construction.

CHARR AS BROWN TROUT FOOD

The improvement and utilization of a charr stock takes time. Until then, the stocks may be exploited by brown trout strains consuming charr. In most larger lakes with mixed populations of charr and brown trout, some of the latter become charr-eaters leaving the littoral zones. The change to a fish diet is accelerated by the food situation after impoundment. Tunhovdfjord offers perhaps the best example of a brown trout population living on charr. After a youth spent in a lean environment along the shorelines, nearly all trout start eating charr when about 25 cm long, their growth rate increasing accordingly. The maximum weight, earlier about 2-3 kg, has now increased to 14-15 kg. The change from bottom organisms to charr was completed in 20-30 years.

Assuming a food conversion factor of 10, a yearly brown trout production of 1500 kg, of which ⅔ is due to charr consumption, indicates that about 10,000 kilos or 4 $kg{\cdot}ha^{-1}yr^{-1}$ of charr is consumed. A possible error is that the production estimate is too low because of missing recapture reports. If this is so, the charr consumption is even greater. Possibly the brown trout exploits the charr population more heavily than the thousands of anglers. However, the effect on charr number and growth may only be marginal when rich year classes occur.

Experiments with tagged or fin-clipped juvenile of brown trout show that the Tunhovdfjord strain keeps its charr-eating disposition even when transferred to other reservoirs. The small Silset reservoir (1 km^2) offers a good example, where the average size of the spawning charr is 170-180 mm and 40-50 g. Some 4 to 6 years after their introduction, Tunhovdfjord trout were recaptured weighing 2-4 kilos. The Tunhovdfjord strain of brown trout is now widely used in the stocking of charr-brown trout reservoirs. Every year nearly 2 million eggs are hatched and

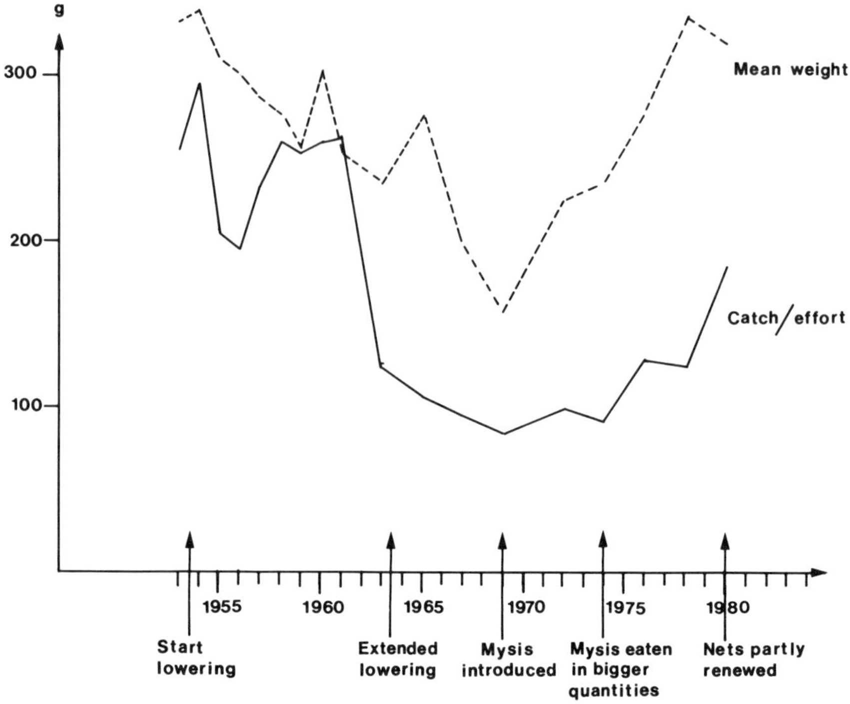

FIG. 6 Lake Limingen. Catch/effort and mean weight of charr in experimental fishing before and after the introduction of *Mysis*.

the resulting young are released after one or two summers. However, a successful change to a charr diet seems not only to be dependent on a numerous charr population, the adult charr should not exceed a length of 200-250 mm.

REFERENCES

Aass, P. 1964. Reguleringer og røyerekruttering. N.J. and F.F. 93:378-381. (In Norwegian).

Aass, P. 1970. The winter migrations of char, *Salvelinus alpinus* L., in the hydroelectric reservoirs Tunhovdfjord and Pålsbufjord, Norway. Inst. Freshw. Res. Drottningholm Rep. 50:5-44.

Aass, P. 1973. Some effects of lake impoundments on Salmonids in Norwegian hydroelectric reservoirs. Acta Univ. Upsaliensis. Abstr. Uppsala Diss. Sci. 234: 1-14.

Aass, P. 1973. Fiskeforholdene i Limingen og Tunnsjø efter reguleringene. Stortingsmelding nr. 70 (1973-74): 32-41. (In Norwegian).

Fürst, M. 1965. Experiments on the transplantation of *Mysis relicta* Lovén into Swedish lakes. Inst. Freshw. Res. Drottningholm Rep. 46: 79-89.

Fürst, M. 1972. On the biology of the opposum shrimp *Mysis relicta* Lovén and its introduction in impounded lakes in Scandinavia. Acta Univ. Upsaliensis. Abstr. Uppsala Diss. Sci. 207: 1-7.

Table 3. Lake Limingen. Yield and mean weight of charr in experimental fishing 1952-1980.

	Year	g/net night		Mean weight	
		summer	autumn	summer	autumn
	1952				284
Lowering →	1953	256	652	332	289
	1954	294	658	339	245
	1955	205	617	310	253
	1956	192	457	300	244
	1957	233	724	285	230
	1958	260	666	277	241
	1959	254	760	260	228
	1960	264	822	301	225
	1961	269	749	255	217
	1962				
Extended lowering →	1963	123	345	236	170
	1964				
	1965	107	268	274	202
	1966				
	1967	92	209	198	202
	1968				
Introduction of *Mysis* in autumn →	1969	85	205	159	193
	1970				
	1971				
	1972	99	184	224	217
	1973		179		208
First year *Mysis* eaten in bigger quantities →	1974	88	167	235	214
	1975				
	1976	129	118	280	195
	1977				
	1978	124	172	336	195
	1979				
Nets partly renewed →	1980	185	202	318	224

Garnås, E., T. Hesthagen and T.B. Gunnerød. 1980. Fiskeribiologiske undersøkelser fra 1973-79 i 3 sjøer med utsatt *Mysis relicta* i Sør-Trøndelag. DVF, Reguleringsundersøkelsene, Rapp. nr. 11-1980, 31 p. (In Norwegian).

Jensen, K.W. and P. Aass. 1958. Bunnforholdene i regulerte sjøer. N.J. and F.F. 87: 44-55. (In Norwegian).

Korsen, I. and P. Møkkelgjerd. 1980. Forsøk med kilenot i ferskvann. TOFA 1979/80: 61-67. (In Norwegian).

Schmidtnielsen, L. 1925. Aursundsjøen. Vassdragsregulering og Fiskeri. Røros. 25 p. (In Norwegian).

Sevaldrud, I.H. and I.P. Muniz. 1980. Sure vatn og innlandsfisket i Norge. Resultater fra intervjuundersøkelsene 1974-1979. SNSF-prosjektet IR 77/80 95 p. (Summary in English).

The Introduction of Arctic Charr (*Salvelinus alpinus salvelinus*) in Nordrhein Westfalen (Federal Republic of Germany)

TOMÁS BRENNER

Landesanstalt für Fischerei NW,
5942 Kirchhundem, Albaum, Federal Republic of Germany

BRENNER, T. 1984. The introduction of Arctic charr (*Salvelinus alpinus salvelinus*) in Nordrhein Westfalen (Federal Republic of Germany), p. 293–301. *In* L. Johnson and B.L. Burns [eds.] Biology of the Arctic charr, Proceedings of the International Symposium on Arctic Charr, Winnipeg, Manitoba, May 1981. Univ. Manitoba Press, Winnipeg.

Increasing demand for water for industry and domestic purposes, combined with the lack of natural lakes has necessitated the construction of many reservoirs in Nordrhein Westfalen. Reservoirs at altitudes between 300 and 400 m have been found suitable for stocking with salmonid fishes because of their good water quality. Initially, brown trout, *Salmo trutta*, from the affluents, colonized these artificial lakes, but to guarantee sportfishing, stocking is necessary. As an alternative, Arctic charr, *Salvelinus alpinus*, has been introduced to some standing waters of Nordrhein Westfalen.

In 1961 the Sorpe-reservoir (surface area 3.25 km^2) was stocked with fry and fingerlings of charr. The eggs came from the Mondsee (Austria) and were reared in the Landesanstalt für Fischerei in Albaum. It was found that natural reproduction of charr takes place at specific spawning sites in the Sorpe-reservoir. Another introduction of charr was made in 1980 into the Rur-reservoir (surface area 7.8 km^2). This paper reports the present distribution of charr in Nordrhein Westfalen and describes the biology of the charr in the Sorpe-reservoir.

INTRODUCTION

In the past many attempts have been made to introduce fish species in Europe: Vooren (1972) reports a total of 39 introductions of fish into European waters since the 19th century; not all were successful. The failure of several introductions has been attributed to physical, chemical and biological habitat characteristics. Introductions of Arctic charr have been reported by Hubault (1955) in France, by Grimaldi (1972) for the Lago Maggiore in Italy and by Nilsson (1972) for Scandinavia. Introductions of charr in Nordrhein Westfalen started in 1961 (Tack 1968).

As there are few natural lakes in Nordrhein Westfalen, the standing waters are mainly man-made. A total of 64 reservoirs has been constructed in the past;

32 of them serve as drinking water storages. Except for the Möhne-reservoir, which is managed by a professional fisherman, they are used for sportfishing only. The largest in surface area are: the Möhne-reservoir (10.3 km^2), the Rur-reservoir (7.8 km^2) and the Bigge-reservoir (7.1 km^2). Most of the reservoirs consist of one or more basins. Besides the Sorpe-reservoir, the Rur-reservoir was also stocked with Arctic charr in 1980 and it is planned to introduce this species into the Wehebach-reservoir (1.6 km^2). The introduction of charr into the Breitenbach-reservoir (0.28 km^2) has not been successful (Schmidt 1975). This paper is part of a study concerned with the documentation of all fish species in Nordrhein Westfalen (Brenner 1980): it deals with the biology of Arctic charr in the Sorpe-reservoir.

DESCRIPTION OF THE RESERVOIR AND THE PRESENT FISHERIES SITUATION

The Sorpe-reservoir was constructed in 1935; after being damaged during World War II and reconstructed in 1959, it consists at present of two basins covering 3.25 km^2 (Table 1). The catchment area is 65 percent forested; 30 percent is used for agricultural purposes. Detailed limnological studies of this man-made lake have been carried out by Nusch (1975).

In the deep and eutrophic Sorpe-reservoir the hypolimnetic oxygen reserve is sufficient to allow the aerobic decomposition of all organic matter produced; even at the end of a period of stagnation there is still 50 percent oxygen saturation in the bottom layer. The Sorpe-reservoir, similar to other man-made lakes, was colonized first by the brown trout of the region; other local species are: minnow (*Phoxinus phoxinus*), bullhead (*Cottus gobio*), gudgeon (*Gobio gobio*), grayling (*Thymallus thymallus*), pike (*Esox lucius*) and perch (*Perca fluviatilis*) (Tack 1964). In addition to these autochthonous fishes additional species have been introduced in the last 20 years. For instance whitefish (*Coregonus lavaretus*), the lake-adapted brown trout (*Salmo trutta* f. *lacustris*) and charr (*Salvelinus alpinus salvelinus*) (Tack 1964, 1972). In 1961 the Sorpe-reservoir was stocked with fry and fingerlings of charr. The eggs (30,000) had been obtained from the Mondsee (Austria) and were reared in the Landesanstalt für Fischerei in Albaum. Since that time no restocking has taken place. The first specimens of charr were caught in 1966 (Tack 1968).

Presently the reservoir is populated by 19 species of fish. The ichthyofauna consists of salmonids, coregonids, percids, cyprinids, pike and eel. Because of a steep littoral zone and an oscillating water level, natural reproduction of fish is possible only in a few cases. Similar to natural lakes, these man-made lakes are stocked annually with a large quantity of fish for the purpose of sportfishing.

Table 1. Sorpe-reservoir data

Latitude: 51° 17′	Volume: 70 X 10^6 m^3
Longitude: 7° 56′	Max. surface area: 3.25 km^2
Altitude: 280 m a.s.l.	Catchment area: 99.2 km^2
Max. depth: 60 m	

With the exception of the results given by Tack (1964, 1968, 1972, 1979) not much is known about the fish in the Sorpe-reservoir.

Because of a lack of detailed statistics, the total sportfishing yield can only be estimated to lie somewhere between 50 and 100 $kg \cdot ha^{-1} yr^{-1}$. Only a few specimens of charr have been caught by sportfishermen in the past, amounting to less than 0.01 percent of the yield. The percentage weight composition of salmonids, percids, cyprinids, pike and eel in the yields from 1976 to 1979 are given in Figure 1. The reservoir is characterized by a high proportion of cyprinids.

MATERIALS AND METHODS

This paper is based on investigations done by Tack (1979) from November 1966 to December 1970 and my own studies from May 1979 to February 1981. Charr were caught using bottom gillnets of various mesh-sizes (10 mm to 40 mm stretched mesh) in different parts of the main basin. The catches were obtained in the main tributaries from 1966 to 1970 and in the deeper parts of the reservoir from 1979 to 1981. The following data were recorded: shape of the mouth, skin colour, total length, total weight, number of gillrakers on the first gill arch of the right hand side and sex. Eggs and intestines were preserved for further investigation. Age was estimated from otoliths (Brenner 1980).

Age was back calculated down to age group I, but because of the small number of fishes available, correction for allometric growth could not be made.

RESULTS

In the reservoir, charr inhabit the deeper parts of the lake; therefore the best catches were obtained in depths of about 40 m. Only during the spawning time from November to January were charr observed in the shallow parts and the tributaries of the reservoir. Following Haempel (1924, 1930) three different forms of charr can be distinguished: the "Wildfangsaibling", the "Normalsaibling" and the "Tiefseesaibling"; according to the observations and investigations of

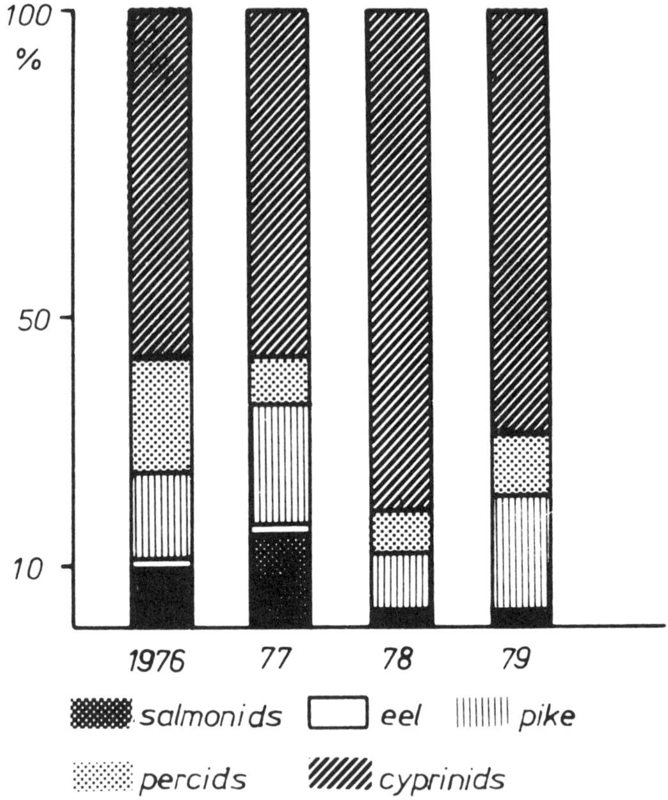

FIG. 1 Percentage weight composition of salmonids, percids, cyprinids, pike and eel in the yields from 1976 to 1979.

Tack (1979) charr in the Sorpe-reservoir should be classified as the so-called "Wildfangsaibling". As mentioned above the eggs originated from the Mondsee progeny of the "Normalsaibling" which normally has a maximum weight of 200 g and feeds on zooplankton. But due, apparently, to the environmental conditions in the Sorpe-reservoir, the charr transformed to the so-called "Wildfangsaibling", which grows better and is mainly piscivorous.

During my own investigations more females than males were caught, contrary to the results of Tack (1979). Males and females were coloured and 82 percent showed the mouth inferior. The number of gillrakers is close to that given for the "Normalsaibling" of other lakes (Table 2). The slight difference is probably due to the limited number of fish investigated. Tack (1979) describes

Table 2. Range of gillraker counts for different charr populations ("Normalsaibling").

Lake	n	Range	Mean	S.D.	Author
Achensee	298	16-30	25.03	2.17	Schulz (1977)
Lake Constance	40	25-31	27.70	1.18	Dörfel (1974)
Sorpe-reservoir	25	18-27	24.00	1.95	(1981)

sexual dimorphism related to the total weight, but this could not be found in the following years (Table 3). The heaviest charr ever caught in the Sorpe-reservoir had a weight of 2345 g and a total body length of 630 mm.

Spawning Time and Egg Counts

Spawning time is between November and February. Tack (1979) observed spawning males and females in the mouth of the Setmecke tunnel in the eastern part of the main basin. Normally, during this time of the year, the water current is strong enough to attract spawners. The water depth at this spawning ground is 2 m and the bottom is covered with stones of about 20 mm in diameter. According to my observations it can be concluded that spawning also takes place in the deeper parts of the reservoir, at approximately 30 m, as running and ripe females and males have been caught at this depth.

The number of eggs per kilogram of body weight is 3240 for a female of average body length of 440 mm. The diameter of the Sorpe charr eggs is similar to that of other populations (Schulz (1977) and Steiner (1972)) being within the range 4.3-5.2 mm.

Food

The summer diet consists mainly of chironomids although the larger charr are piscivorous. Generally in winter the intestines were empty: the few exceptions had the same diet as found in summer.

Growth and Age

Charr caught in the period from 1979 to 1981 were 6 and 7 years old (Fig. 2); only one specimen, having a body length of 600 mm, was older; it was estimated to be about 10 years old. Exact aging was not possible in this specific case. Comparison with earlier years was not possible since no data were available.

Table 3. Comparative lengths, weights and condition factors of Arctic charr caught in the Sorpe-reservoir during the period from 1966 to 1970 (A) (Tack 1979) and from 1979 to 1981 (B).

		n	Length (mm)		Weight (g)		condition factor	
			range	mean	range	mean	mean	S.D.
A	♀	18	365-520	441.6	400-1250	780.00	0.89	0.07
	♂	30	340-630	492.0	330-2345	1248.14	0.93	0.12
B	♀	18	354-600	416.0	395-2160	643.60	0.85	0.12
	♂	9	390-480	435.0	440- 910	675.10	0.80	0.07

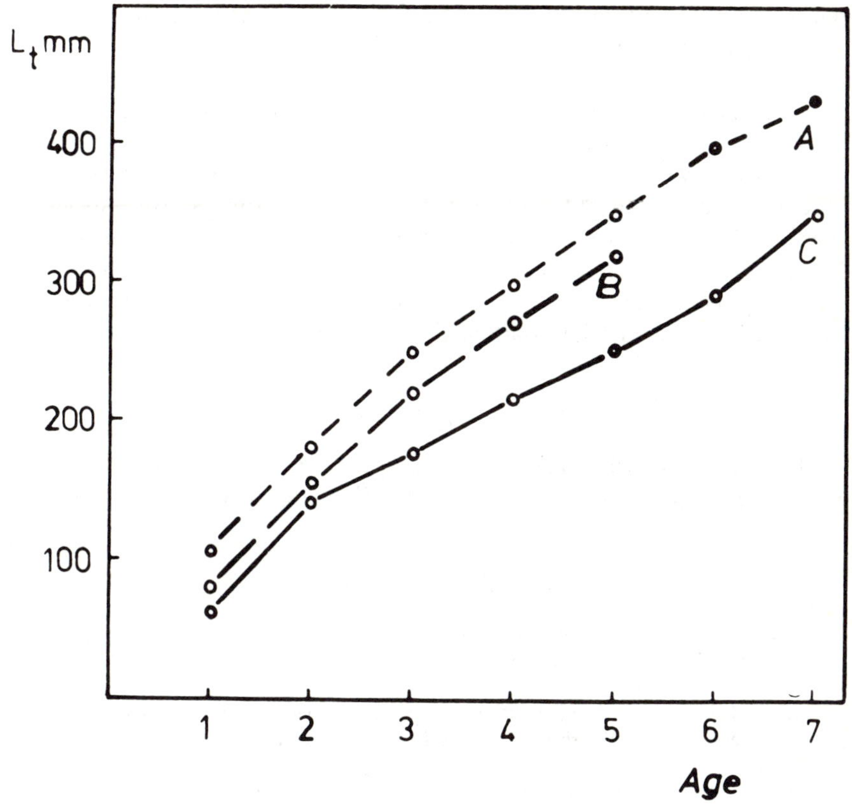

FIG. 2 Growth of Arctic charr in the eutrophic Sorpe-reservoir (A), the mesotrophic Lake of Constance, Überlinger See (B) (Dörfel 1974) and the oligotrophic Achensee (C) (Schulz 1977).

Compared with other charr populations e.g. the oligotrophic Achensee (Schulz 1977) and the mesotrophic Lake of Constance (Dörfel 1974) the growth rate of charr in the eutrophic Sorpe-reservoir appears to be better.

DISCUSSION

In recent years the intention has been to restock the Sorpe-reservoir with charr but it has been impossible to get stocking material. Natural reproduction seems to result in a small stock as no small charr could be caught. This is probably due to environmental factors that prevent annual reproduction.

The failure of several introductions has been attributed to various biotic and abiotic factors (Vooren 1972). In the Sorpe-reservoir the abiotic factors, e.g.

temperature and oxygen conditions, are favourable; even during the stagnation period there is no dangerous oxygen depletion. It can be further concluded that eutrophication appears to accelerate growth.

From the point of view of nature conservation, a successful fish introduction can be risky in that it may be unfavourable to endemic fish populations destroying the diversity of natural ecosystems. Therefore to protect the endemic species indiscriminate stocking of fish of various origin must be stopped. Nevertheless in the case of man-made lakes, which are suitable for Arctic charr, stocking may be legitimate. A successful introduction would be a contribution to the protection of an endangered fish species, which in natural lakes has little chance to survive due to increasing eutrophication (Hartmann, this volume).

ACKNOWLEDGEMENTS

I wish to thank Ulrich Wolff for collecting the samples and Cristoph Hebbeker who processed part of the samples and assisted in preparation of the figures.

REFERENCES

Brenner, T. 1980. The Arctic charr, *Salvelinus alpinus salvelinus*, in the prealpine Attersee, Austria, p. 765-772. *In* E.K. Balon [ed.] Charrs: salmonid fishes of the genus *Salvelinus*. Dr. W. Junk Publishers, The Hague.

Brenner, T. 1980. Ein Beitrag zur Fischfauna einiger stehender Gewässer Nordrhein-Westfalens und der derzeitige Stand der Arbeiten über die Erstellung eines Fischereikatasters. Fischwirt 30: 25-26.

Dörfel, H.J. 1974. Untersuchungen zur Problematik der Saiblingspopulationen (*Salvelinus alpinus* L.) in Überlinger See (Bodensee). Arch. Hydrobiol. Suppl. 47: 80-105.

Grimaldi, E. 1972. Lago Maggiore: effects of exploitation and introductions on the salmonid community. J. Fish. Res. Board Can. 29: 777-785.

Haempel, O. 1924. Studien am Seesaibling mehrerer Alpenseen. Verh. Int. Ver. Limnol. 2:129-134.

Haempel, O. 1930. Fischereibiologie der Alpenseen, 259 p. *In* A. Thienemann [ed.] Die Binnengewässer, Vol. 10, Schweizerbart'sche Verlagsbuchhandlung, Stuttgart.

Hartmann, J. 1983. The charrs (*Salvelinus alpinus*) in the Lake of Constance, a lake undergoing cultural eutrophication. Paper presented at Int. Symp. on Arctic charr, Winnipeg, 1981 (This volume).

Hubault, E. 1955. Introduction of species into lakes in the eastern part of France. Verh. Int. Ver. Limnol. 12:515-519.

Nilsson, N.-A. 1972. Effects of introductions of salmonids into barren lakes. J. Fish. Res. Board Can. 29:693-697.

Nusch, E.A. 1975. Comparative investigations on extent, causes and effects of eutrophication in Western German reservoirs. Ver. Int. Ver. Limnol. 19: 1871-1879.

Schmidt, G.W. 1975. Fischerträge einer westdeutschen Trinkwassertalsperre (Breitenbachtalsperre, Kreis Siegen) nach mehrjährigen Fangberichten von Sportanglern. Gewässer und Abwässer 57/58: 79-116.

Schulz, N. 1977. Untersuchungen zum Problem der Seesaiblinge des Achensees (Tirol, Österreich). Veröffentlichungen des Museum Ferdinandeum Innsbruck 57: 79-102.

Steiner, V. 1972. Die Temperaturtoleranz des Seesailblings (*Salvelinus alpinus* L.). Ph. D. Thesis, University of Innsbruck. 146 p.

Tack, E. 1964. Fischarten und Fischdichten in westfälischen Talsperren. Fischwirt 14: 280-283.

Tack, E. 1968. Seesaiblinge eingebürgert in Talsperren Westfalens. Fischwaid 7.

Tack, E. 1972. Die Fische des südwestfälischen Berglandes mit Einschlub von Möhnetalsperre und Ruhr. Decheniana 125: 63-77.

Tack, E. 1979. Biometrische Untersuchungen an Fischbeständen aus rheinisch-westfälischen Gewässern. Gewässer und Abwässer 65: 107 p.

Vooren, C.M. 1972. Ecological aspects of the introduction of fish species into natural habitats in Europe, with special reference to the Netherlands. J. Fish Biol. 4: 565-583.

Effects of Water Level Fluctuations on the Recruitment of Arctic Charr

MAGNUS FÜRST and JOHAN HAMMAR

Institute of Freshwater Research, S-170 11 Drottningholm, Sweden

FÜRST, M. AND J. HAMMAR. 1984. Effects of water level fluctuations on the recruitment of Arctic charr, p. 303–311. *In* L. Johnson and B.L. Burns [eds.] Biology of the Arctic charr, Proceedings of the International Symposium on Arctic Charr, Winnipeg, Manitoba, May 1981. Univ. Manitoba Press, Winnipeg.

Most charr lakes in the high mountains of Sweden have been impounded; in extreme cases the lake level is permitted to fluctuate over a range of 35 m. In lakes with extreme fluctuations the recruitment of charr is severely reduced; recruitment is also affected by summer temperature and interspecific interactions. The influence of a declining water level in spring, at the time of hatching was studied and a negative correlation between low water level and the relative strength of the current year class was established; this was also found to exist in lakes with only a moderate amplitude in the fluctuation of lake level (12-13 m). The influence of lake level is so pronounced that it over-rides temperature effects. In hatcheries, eggs hatch in December-January but in their natural environment it is a few months later, in fact, at the time when the water level is close to the minimum permitted. After hatching, free-swimming fry move very close to the shoreline and often become trapped on stones or other objects by surface tension. This shore-seeking behaviour is probably an adaptation to the natural rise of water level at this time of the year.

INTRODUCTION

Most of the Swedish high mountain charr lakes are impounded and function as hydroelectric reservoirs. The water is stored during summer and autumn and largely drawn down in winter. The highest and lowest permitted water levels differ between lakes. There are amplitudes of 35 metres. The drawdown limit is often reached in early April. (More information is given by Aass, this volume.)

The littoral region within the limits of regulation is eroded and consists of stony or solid bottoms (Fig. 1). The organic material has settled below the drawdown limit and often forms an extremely soft layer (Grimås 1961).

In all Swedish charr lakes spawning occurs in shallow water. If there are two or more sympatric species of charr, it is usual that the larger and more valuable one spawns in shallow water (Hammar, this volume).

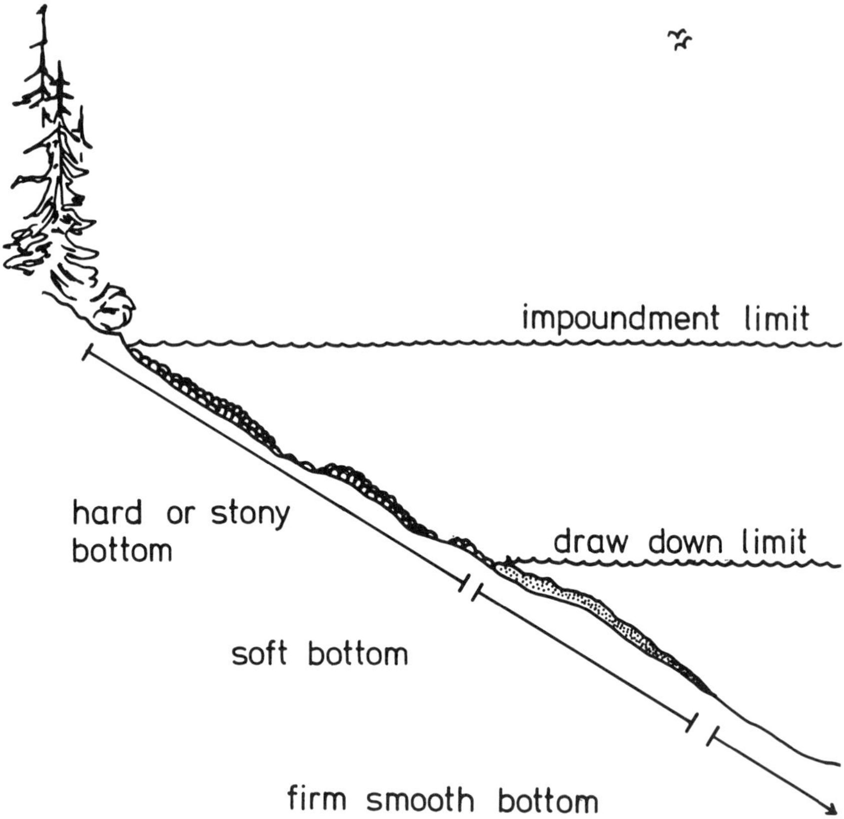

FIG. 1 Bottom structure in the zone effected by regulation. The figure is schematic and shows the new spawning habitat for Arctic charr in relation to the water level.

Large areas of the bottom in the lake reservoirs within the zone seem to be ideal for spawning; in spite of this there has, by this time, been a shift in spawning area to the deeper parts of the zone (Runnström 1951, Fürst et al. 1978).

There are a number of examples of damage to charr recruitment caused by rapid drawdown (18 and 24.5 m) of the water level in lake reservoirs (Aass 1964, 1970). In other cases, as for example Lake Torrön, with a smaller water level amplitude (12.85 m), it has not been possible up to the present to document any damage (Runnström 1946, 1951). The reason given was that alevins developed early so that they succeeded in escaping before the water level reached a critical stage.

Mysis relicta Lovén was introduced into several reservoirs such as Lake Blåsjön and Lake Torrön (Fürst et al. 1978, 1981, Fürst 1981) and became a readily available food item during the entire year and for all life stages of the charr. The charr grew faster but did not increase in number. On the contrary, the species of charr which spawned in shallow water decreased in number. There are several possible reasons for this phenomenon. The fishing intensity increased as the quality of the charr induced by *Mysis* feeding increased. Earlier, for instance there had been five year classes of females spawning at the same time in Lake Blåsjön but after the increased fishing there were only two (Fürst et al. 1978). The recovery of the population of brown trout (*Salmo trutta* L.) — possibly an effect of *Mysis* consumption — also probably depressed the charr population.

In spite of an abundant and stable *Mysis* population, available as food during the entire year, there was a fluctuation in year class strength that did not seem to depend on climatic factors.

These findings made it mandatory to carry out a new analysis of the relationship between year class strength, the lowest water level, and the summer temperature, in lake reservoirs of moderate amplitude.

MATERIALS AND METHODS

The water level of Lake Torrön has been increased 8.25 m and decreased 4.60 m from the natural yearly mean level. The mean natural amplitude was about 1.90 m before impoundment in 1937, and the new amplitude is 12.85 m.

The lake is ice-covered between December 22 and May 24, although the mean daily temperature exceeds zero after April 25; melting of the snow starts earlier because of the difference in temperature between day and night.

The relative year class strength of charr in this lake was studied during the years 1966-79. Fishing was carried out each year at the same time, occupying the same stations, using the same gangs of gillnets. Nine different mesh sizes were used ranging from 10 to 50 mm. Sampling was carried out twice during August (total 18 sets).

Aging of the fish was done using otoliths. To quantify the number of surviving charr from each year's spawning the relative strength of each year class was calculated by a method described by Svärdson (1961). The percentage of charr in each year class, each year, was related to a mean survival curve calculated on the whole material.

The water level is slowly drawn down in lake reservoirs during late winter and spring. To study the behaviour of the latest stages of the alevins close to the shoreline some experiments were made in a tank.

The tank was 3 m long, sloping 0.3 m from one end to the other. The bottom was covered with gravel and stones closely resembling the natural bottom. In two successive experiments the tank was stocked with 100 alevins two weeks and three weeks after hatching respectively. The yolk-sac was not yet resorbed. The water temperature varied between 4 and 10°C. The water level was allowed to fall slowly, resembling the drawdown in Lake Torrön. The water was not allowed to stream within the tank.

RESULTS

Incubation Time

Some data on the time of spawning, time to hatching and the time to the point when alevins become free-swimming (having almost totally or totally resorbed their yolk-sac are available for Swedish charr populations. (Table 1).

Observations on hatching and free-swimming stages were made on only a few occasions, but the spawning period is well documented for these lakes. It never lasts more than three weeks but is, in most cases, shorter. Runnström (1946, 1951) observed hatching in Lake Torrön from mid-December to the beginning of January. In his investigations he used boxes containing fertilized eggs kept in the lake. Other details about the method are not described.

There are also some data on incubation time from hatcheries. The temperature in hatcheries is low compared with that of the lakes during the fall.

The cumulative temperature unit (CTU) values at very low temperatures in hatcheries correspond well with Swift's (1965) experimental findings in constant but higher temperature regimes.

Figure 2 shows the temperature changes in lakes Torrön and Vättern during the period between spawning and observation of free swimming alevins. In the same figure there is another curve showing the development of fertilized eggs in a hatchery.

The results of several direct observations indicate that charr in reservoirs normally hatch later than would be expected from the experimental findings or CTU values obtained in hatcheries. In spite of the fact that the temperature in lakes is higher than in hatcheries, development is slower. It might be possible to speculate on the causes for this difference but in this case it is sufficient to state that slow development means that the eggs, alevins or fry risk being left behind on terra firma when the water level is drawn down.

The lowest water level of the year is reached between February 17 and April 30 but the actual level varies between the years (Fig. 3).

The Behaviour of the Alevins

If the alevins become free-swimming and can escape from being caught on land, how do they then behave if the water level still continues to sink?

Table 1. Data concerning spawning, incubation and juvenile stages in three lakes and one small river.

Lake	Charr species (Hammar 1983)	Spawning	Hatching records	Other observations
Lake Torrön	S	Sep 20-25	-	-
Lake Juveln	S	Sep 20-25	-	Caught in outlet stream April 3-6 yolk almost completely resorbed.
Lake Näckten (Gönczi)	N	Oct 18	Mar 5-10	Direct observation yolk resorbed April 1
River Storbäcken (Hanell)	F?	Sep 10-15	April 10	Catch of drifting charr. May 15 yolk almost completely resorbed.
Lake Vättern (Essvik)	N	Nov 12	May 1	Boxes on spawning grounds. May 15-20 yolk resorbed.
Lake Aursunden (K. Dahl, Norway)	?	-	Mar 15-	Direct observation
Lake Tunnhovdfjord (P. Aass, Norway)	F	-	Feb 26-	Direct observation

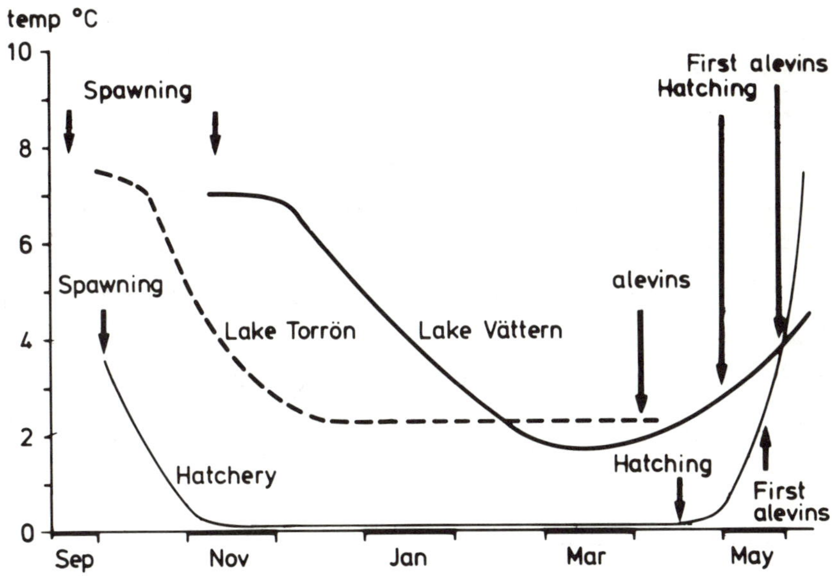

FIG. 2 Temperatures in two lakes and in a hatchery related to spawning, incubation time and juvenile stages.

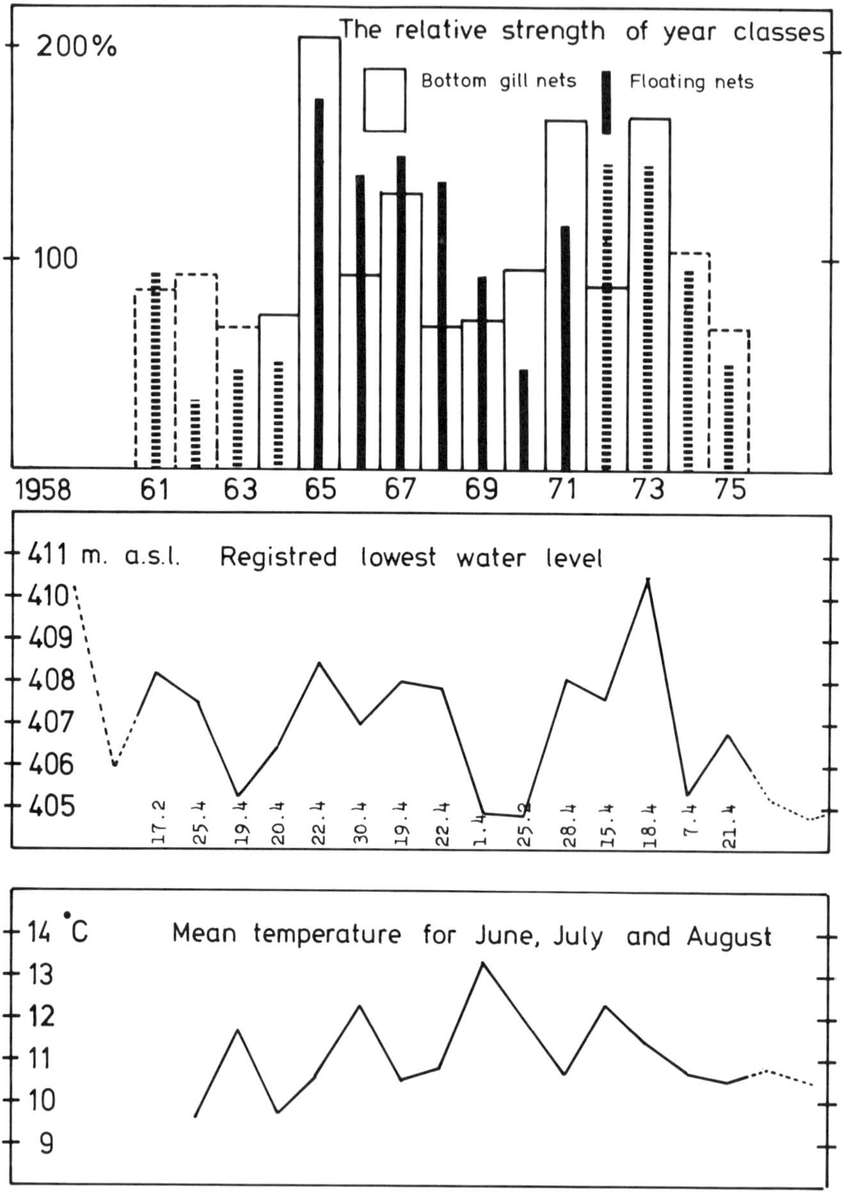

FIG. 3 The relative strength of year classes of Arctic charr in Lake Torrön in benthic and pelagic catches (top). The lowest recorded water levels (centre) and the mean temperature in June, July and August in the same years (bottom) are also shown.

The tank experiment was designed to model a reservoir with a falling water level. The two-week old alevins released in the tank behaved in quite a distinctive manner. They congregated close to the shoreline of the tank; sometimes they hid under stones and became trapped but frequently they swam so close to the shore that they got stuck to the sides of stones, trapped by the surface film.

If water was allowed to flow slowly through the tank the alevins either behaved as in the previous experiment or they swam against the current. A second experiment was made one week later. This showed that as the alevins grew older, they gradually changed behaviour and succeeded in escaping by following the sinking water. Aass (pers. comm.) found only alevins having remnants of yolk in the vicinity of the shoreline, never fry without yolk.

Water Level and Temperature

There is a well recognized relationship between strong year classes and summer temperature in unimpounded charr lakes (Runnström 1951, Aass pers. comm.). This relationship seems to be a common phenomenon in various fish species (Svärdson and Molin 1981). In Lake Torrön the relationship between year class strength, summer temperature and lowest water level was studied.

Figure 3 shows these relationships: in 1963, 1966, 1969 and 1972 the total mean temperature of June, July and August was higher than normal; in spite of this the year classes were weak, but the water levels were lower than ''normal'' in these same years. The summer of 1969 was the warmest during the entire period but the corresponding year class was weaker than all others during the same period; the drawdown was also exceptionally low. In 1965, 1971 and 1973 rich year classes were formed; the temperature was about normal but the water level was unusually high in springtime. Statistical analysis shows a significant negative correlation between the strength of the charr year class and the lowest water level during springtime ($p =\leqslant 0.001$), but there is no correlation between year classes and summer temperatures over the same period.

There is however, a significant correlation between year class strength of charr spawning in shallow waters and water level during spring. This relationship totally outweighs the normal relation to summer temperature which of course still has an influence.

DISCUSSION

Several lines of evidence indicate that hatching and the following stages of development occur later in lakes than would be expected on the basis of experiments, or from hatchery experience.

Not even the first free-swimming stages seem to be able to escape the falling water level. The behaviour of these alevins is probably an adaptation to a rising water level and the sinking level now acts as a death-trap. The mean daily temperature exceeds 0°C during late April in the Swedish high mountain region but the day temperature is often above freezing several weeks earlier. The water formed by melting snow along the lake shores contains fertilizing salts and probably also organic drift; both these items could form an important food supply for the alevins.

The reservoir is, in winter, characterized by an increasing area of dry bottom covered with ice. This area is totally eroded and sterile and it is possible to believe that the melting water penetrates these bottom areas as they are not frozen. The normal food supply is thus probably partially excluded from reaching the lake.

Yoshihara (1974) found alevins still in the redds as late as about two weeks after the total resorption of the yolk. This behaviour does not correspond with the charr studied in our tank experiments. There are several populations of stream spawning charr in Sweden. The alevins migrate or drift downstream when they have almost resorbed the yolk (Hanell pers. comm.). This seems to be about the same stage of development at which alevins in the tank experiments left the vicinity of the shoreline.

In conclusion it is possible to discern several different elements connected with the changes of water level of the reservoirs, which may cause initial mortality to the year classes. Such elements are late embryonic development compared to the rate of drawdown (even early free-swimming stages run the risk of being trapped on land) and the change of spawning site to deeper layers which seems to be an adaptation but does not compensate for the losses.

The consequences of this are that it might be necessary to compensate for the losses either by regulation of the water level to meet the biological needs or by stocking.

REFERENCES

Aass, P. 1964. Reguleringer och rojerekruttering. N.J. and F.F. 93:378-381.

Aass, P. 1970. The winter migrations of char *Salvelinus alpinus* L., in the hydroelectric reservoirs Tunhovdfjord and Palsbufjord, Norway. Inst. Freshw. Res., Drottningholm Rep. 50:5-44.

Aass, P. 1983. Management and utilization of Arctic charr in Norwegian hydro-electric reservoirs. Paper presented at Int. Symp. on Arctic charr, Winnipeg, 1981. (This volume).

Fürst, M. 1981. Results of introductions of new fish-food organisms into Swedish lakes. Inst. Freshw. Res. Drottningholm Rep. 59: 33-47.

Fürst, M., U. Boström and J. Hammar. 1978. Effects of new fish-food organisms in Lake Blåsjön. Inform. Inst. Freshw. Res. Drottningholm (15). 94 p. (In Swedish with English summary).

Fürst, M., U. Boström and J. Hammar. 1981. Effects of new fish-food organisms in Lake Torrön. Inform. Inst. Freshw. Res. Drottningholm (1). 48 p. (In Swedish with English summary).

Grimås, U. 1961. The bottom fauna of natural and impounded lakes in northern Sweden. Inst. Freshw. Res. Drottningholm Rep. 42: 183-237.

Hammar, J. 1983. Ecological characters of different combinations of sympatric populations of Arctic charr in Sweden. Paper presented at Int. Symp. on Arctic charr, Winnipeg, 1981 (This volume).

Runnström, S. 1946. Sjöregleringar och fisket. (Föredrag vid Svenska fiskevardsforbundet.) Lantbruksveckans Handl.: 141-163. (In Swedish).

Runnström, S. 1951. The population of char *Salmo alpinus* Linne, in a regulated lake. Inst. Freshw. Res., Drottningholm Rep. 32: 66-78.

Swift, D.R. 1965. The effect of temperature and oxygen on the growth rate of the Windermere char (*Salvelinus alpinus willughbii*). Comp. Biochem. Physiol. 12: 179-183.

Svärdson, G. 1961. Ingen effekt av sikodlingen i Kalmarsund. Svensk Fisk. Tidskr. 70(2): 23-26. (In Swedish).

Svärdson, G. and G. Molin. 1981. The impact of eutrophication and climate on a warmwater fish community. Inst. Freshw. Res. Drottningholm Rep. 59: 142-151.

Yoshihara, H.T. 1974. Some life history aspects of Arctic char. Federal Aid in Fish Restoration. Project F-9-5, Job F-III-A, Sport Fish Invest. Alaska 14:1-63.

Solution to the "Charr Problem" Based on Arctic Charr (*Salvelinus alpinus*) in Norway

HANS NORDENG

Zoological Institute, University of Oslo, P.O. Box 1050,
Blindern, Oslo 3. Norway

NORDENG, H. 1983. Solution to the ''charr problem'' based on Arctic charr (*Salvelinus alpinus*) in Norway. Can. J. Fish. Aquat. Sci. 40: 1372-1387.

Rearing and transplantation experiments demonstrate that three coexisting forms of Arctic charr *Salvelinus alpinus* (anadromous, small and large freshwater residents) belong to the same gene pool. The parr of each form segregate into all three forms. Single individuals may manifest all three forms during their lifetime, successively attaining the appearance and spawning colour of each form. Males mature at a younger age and smaller size than females. Hence, in charr populations featuring two or more forms males dominate in the early maturing form and females in the late maturing form. Charr populations that feature only one form (sex ratio 1:1) seem to be established through natural selection for age at sexual maturity. Segregation during the young stage depends upon their genetic constitution and access to food. Offspring of the small resident parents produced more resident individuals and fewer smolts than did offspring of anadromous parents. An increased amount of food increases the resident fraction and reduces the fraction of anadromous charr. It is demonstrated that the potential for anadromy exists in populations of charr in the southern nonanadromous area, and that one reason for resident behaviour in the smolt may be infection of the ureter fluke *Phyllodistomum conostomum* Olsson.

As this paper has been published in the Canadian Journal of Fisheries and Aquatic Sciences only the abstract is reproduced in this volume in order to avoid duplicate publication.

Reproductive Strategy of Dwarf and Normal Arctic Charr (*Salvelinus alpinus*) from Vangsvatnet Lake, Western Norway

BROR JONSSON and KJETIL HINDAR

Zoological Institute, University of Oslo, P.O. Box 1050, Blindern, Oslo 3. Norway

JONSSON, B., AND K. HINDAR. 1982. Reproductive strategy of dwarf and normal Arctic charr (*Salvelinus alpinus*) from Vangsvatnet Lake, western Norway. Can. J. Fish. Aquat. Sci. 39:1404-1413.

In this paper, we investigate the reproductive strategy maintaining dwarf and normal phenotypes of an Arctic charr (*Salvelinus alpinus*) population. Both morphs occurred together in the spawning area, but the peak abundance of dwarf spawners was significantly deeper, and later in the spawning period than normal spawners. Both phenotypes attained sexual maturity at the same age and did not differ in length-specific fecundity or egg diameter. Our data support the hypothesis that the females matured sexually at the age that maximizes their fitness within the constraints imposed by growth and survival. The two female morphs appeared not to have equal fitness; these differences may be due to random events. The males were more heterogeneous by size than the females when maturing sexually. This may be due to intrasexual competition between males making large males the principal spawners. The dwarf males may circumvent the effects of intrasexual competition by mimicking immature fish, sneaking at the redds of normal females, and participating in the spawning acts of normal charr pairs. Also, they may fertilize the eggs of dwarf females spawning in deeper areas.

The two charr morphs developed from a uniform young phenotype (parr). The dwarf spawners were precocious individuals maturing sexually in their parr stage, whereas normal spawners had a colour shift, analogous to smoltification in anadromous charr, before they attained sexual maturity. The different spawning colouration of the two morphs may be due to sexual selection.

As this paper has been published in the Canadian Journal of Fisheries and Aquatic Sciences only the abstract is reproduced in this volume in order to avoid duplicate publication.

Habitat and Food Segregation of Dwarf and Normal Arctic Charr (*Salvelinus alpinus*) from Vangsvatnet Lake, Western Norway

KJETIL HINDAR and BROR JONSSON

Zoological Institute, University of Oslo, P.O. Box 1050,
Blindern, Oslo 3. Norway

HINDAR, K. AND B. JONSSON. 1982. Habitat and food segregation of dwarf and normal Arctic charr (*Salvelinus alpinus*) from Vangsvatnet Lake, western Norway. Can. J. Fish. Aquat. Sci. 39:1030-1045.

The aim of the present study was to test possible habitat and food segregation between two charr morphs in Vangsvatnet Lake. The charr population of Vangsvatnet consists of one pale phenotype with parr marks along the flanks and poorly developed spawning colouration (dwarf charr), and another phenotype with silvery flanks in nonbreeding condition and bright red spawning colours (normal charr). The habitus of the two morphs indicated that they were adapted to different niches during the growth season. The analyses of distribution and food showed that there was a marked habitat and food segregation between the two morphs during summer, when dwarf charr dwelled deep-benthic and normal charr roamed in the limnetic zone. The segregation broke down in periods of food abundance indicating that the habitat segregation was due to intraspecific competition for food. A comparison of charr morphs of Vangsvatnet with the nearby Lonavatnet Lake suggested that the number of charr morphs is adapted to local environmental conditions, among which lake morphology, food availability and other fish species are important. We hypothesize that the number of charr morphs within a locality depends on the number of available niches during the growth season.

As this paper has been published in the Canadian Journal of Fisheries and Aquatic Sciences only the abstract is reproduced in this volume in order to avoid duplicate publication.

Dwarf Populations of Arctic Charr in High-Mountain Lakes of the Alps Resulting from Under-Exploitation

ROLAND PECHLANER

Limnology Division, Department of Zoology, University of Innsbruck, Austria

PECHLANER, R. 1984. Dwarf populations of Arctic charr in high-mountain lakes of the Alps resulting from under-exploitation, p. 319–327. *In* L. Johnson and B.L. Burns [eds.] Biology of the Arctic charr, Proceedings of the International Symposium on Arctic Charr, Winnipeg, Manitoba, May 1981. Univ. Manitoba Press, Winnipeg.

Fish biologists generally consider high-mountain lakes to be extremely unproductive, but new understanding of energy flow in such ecosystems compels considerable modification of this view. The scarcity of autochthonous animal food for fish in alpine lakes is largely due to the fish populations themselves; the considerable potential for food production by zooplankton and zoobenthos is strongly reduced by the feeding pressure imposed by overcrowded fish stocks. Heavy grazing impedes the flow of energy from benthic and planktonic primary production (and bacterial transformation of allochthonous detritus) to higher trophic levels.

Exploitation by angling and commercial fishing methods is seen to be a particularly appropriate method of ensuring good growth of charr in alpine lakes by the regulation of population density to conform to the food production potential.

INTRODUCTION

The origin of Arctic charr, *Salvelinus alpinus* populations in high-mountain lakes of the Austrian Tyrol has been discussed elsewhere (Pechlaner 1966, 1969, 1979, 1984). For the present purposes it is sufficient to state that these charr populations become overcrowded, producing large numbers of small, adult fish of low condition factor, having little interest for anglers or commercial fishermen.

In order to explore the potential for improvement of these lakes through an understanding of their limnology, a comprehensive ecosystem study of Vorderer Finstertaler See was carried out between 1968 and 1974 as part of the International Biological Programme, but it had to be terminated in 1975 when the lake was emptied during the construction of a hydroelectric reservoir. However, good data on the factors leading to the stunting of charr were obtained.

As stressed elsewhere (Pechlaner 1984) the small resident charr of high-mountain lakes are not hereditary dwarfs; they are morphs: stunted specimens of a stock capable of producing large fish. Heckel and Knerr (1858) were probably the first to observe that when stunted fish of alpine lakes were transplanted to lakes with better feeding conditions, growth rates improved. Similar results were observed by Heller (1969) in Oberer or Mittlerer Plenderlesee, Steinböck (1950 p. 2) in Schwarzsee ob Sölden and by Balon and Penczak (1980) in Dösener See (2218 m a.s.l.). A stunted population of brook charr, *Salvelinus fontinalis*, has been described by Reimers (1958, 1979) for the alpine Bunny Lake (3322 m a.s.l.) in the Sierra Nevada, California; in this case stunting was also caused by too dense a stocking.

Overcrowding in high-mountain lakes of the Alps is generally due to the early maturity of resident Arctic charr, spawning as a rule taking place in the third, sometimes even in the second year of age; this is combined with favourable conditions for reproduction in these stony, oxygen rich lakes. Predatory fishes are missing in most cases and, as there is no planned fishery to regulate numbers, overpopulation of the lake and malnutrition of the charr population is the result.

Inadequate food supply is the result of two different causes:

a) There are too many fish living on a limited food supply: each fish needs a relatively high proportion of its energy uptake for covering the costs of searching for food.

b) Production of autochthonous food for charr is strongly reduced by the feeding pressure of the overcrowded fish stock. Utilization of energy from benthic and planktonic primary production (and bacterial transformation of allochthonous detritus) by intermediate producers (such as macrozoobenthos and planktonic Crustacea) is impeded by the heavy grazing of charr on such food organisms.

RESULTS

Typically, Arctic charr in high-mountain lakes of the Alps has small size and a low condition factor. Figures 1 and 2 show examples of length frequencies of charrs caught in alpine lakes. Maximum total lengths of 220 mm (Vorderer Finstertaler See) or 190 mm (Schwarzsee ob Sölden) can be considered as representative of the respective lakes, whereas the peak of the length frequency histogram (between 140 and 160 mm) is caused by an underestimation of smaller fish in catches with gillnets and fishing rod.

As estimated by Pechlaner (Pechlaner et al. 1972, Steiner and Pechlaner 1975), there were about 10,000 charrs of age two years or older in Vorderer Finstertaler See, their biomass (fresh weight) being 6.7 $kg \cdot ha^{-1}$, with annual production being 3.8 $kg \cdot ha^{-1}$. Under the assumption of a mean stomach evacuation rate of 48 hours, food consumption of the charr was estimated to amount to 76.4 $kg \cdot ha^{-1} yr^{-1}$, 57 percent being autochthonous zoobenthos, 36

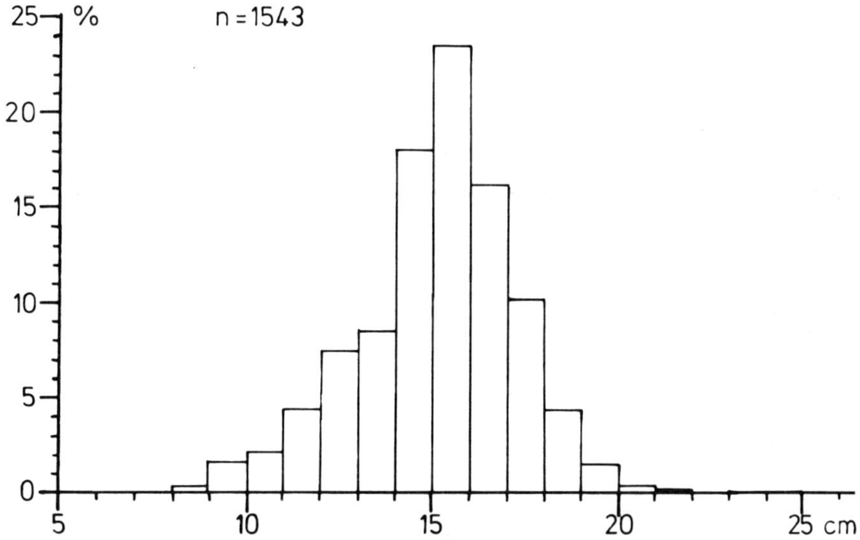

FIG. 1 Frequency distribution of total lengths of 1543 Arctic charr of Vorderer Finstertaler See caught between 1962 and 1972.

percent surface food, 5 percent drift of the inflowing stream, and only 1 percent zooplankton and fish (mainly consumed eggs).

Food consumption of the brown trout population (about 1700 individuals of age 2 years or older, fresh weight 6.2 kg·ha^{-1}, production 1.6 kg.·ha^{-1}yr^{-1}) has not been quantified adequately for the period of ice cover, but reached 175 kg·ha^{-1} for the seven month period of the ice-free season: 80 percent surface food, 12 percent drift, the rest autochthonous food organisms (mainly zoobenthos).

Standing crop and production rate of macro zoobenthos in Vorderer Finstertaler See has been quantified by Bretschko (1974, and pers. comm.; chironomids, chydorids and ostracods) and Wagner (1975; oligochaets) as summarized in Table 1. The figures for chydorids and ostracods in Table 1 refer to "minimum production", calculated from observed biomass differences.

The production of the dominant Chironomidae *Lauterbornia coracina* and *Heterotricocladius marcidus*, for which elimination rates could be calculated, was above "minimum production" by the factors 2.7 and 3.8 respectively (Bretschko, pers. comm.). Even if these factors are considerably higher for the benthic crustaceans, the actual consumption of benthic food by charr and trout is very close to the zoobenthos production. Oligochaets did not occur as food items in any of the stomachs studied, and midwinter food of brown trout and annual food consumption of all young-of-the-year salmonids are not included in the

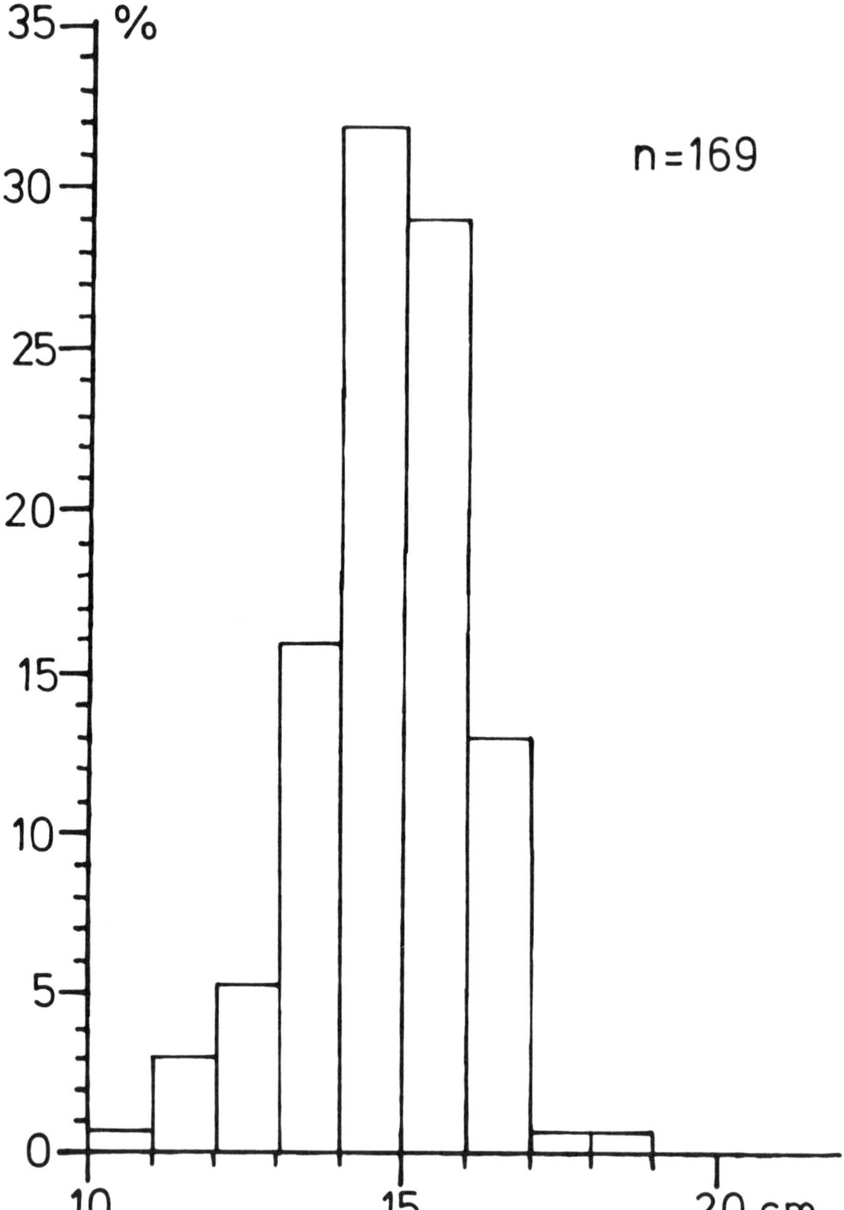

FIG. 2 Frequency distribution of total lengths of 169 Arctic charr of Schwarzsee ob Sölden, caught in 1969 and 1970.

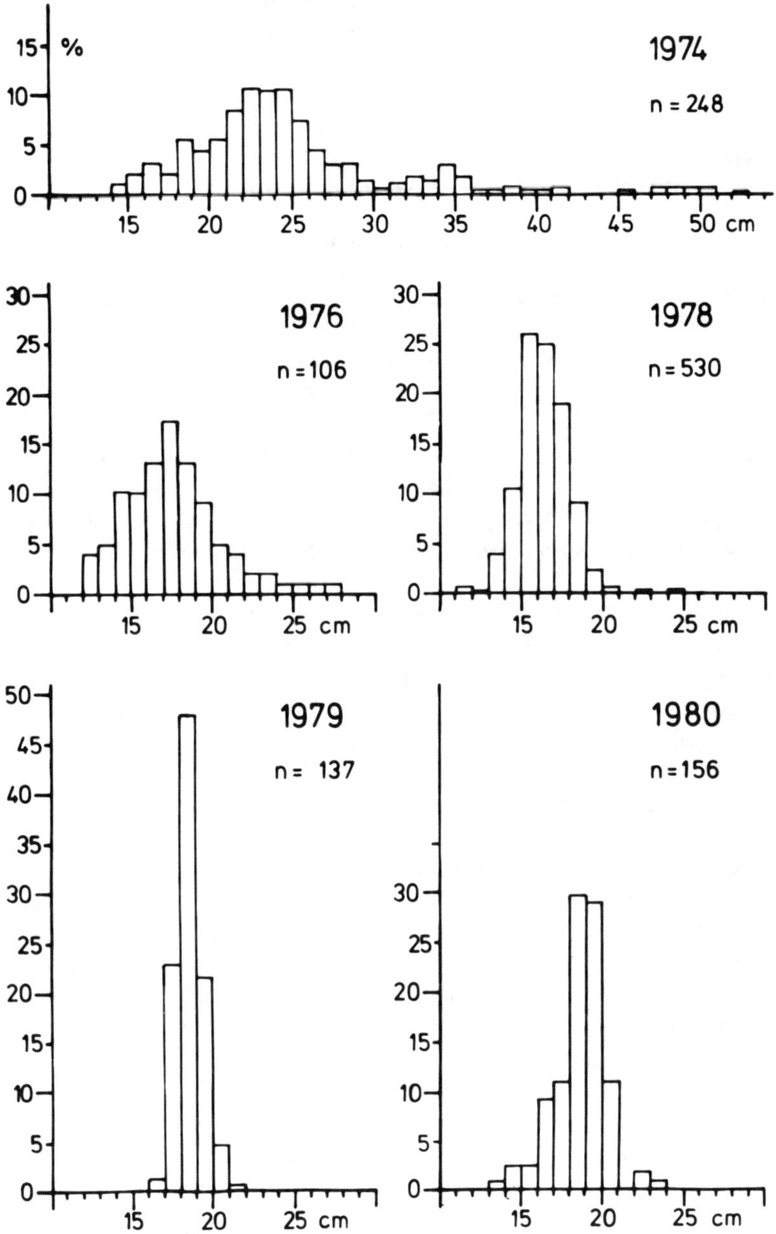

FIG. 3 Length frequencies of Arctic charr from Drachensee, 1974 through 1980.

consumption figure of 5.8 $kg \cdot ha^{-1}yr^{-1}$. Molluscs (*Pisidium* sp.) are occasionally found in charr intestines but occurred too infrequently in benthos samples to be quantified.

This close correspondence between production and consumption points to strong pressure being exerted by charr and trout on the food resources. Moreover, over-grazing is evident when macrozoobenthos production is compared with biomass and production of micro-organisms in Vorderer Finstertaler See.

DISCUSSION

Bretschko (1975) has shown that the bacterial breakdown of allochthonous organic material is a very important energy source for zoobenthos. Tautermann (1977) has quantified biomass and production rate of benthic bacteria in Vorderer Finstertaler See. According to his results the biomass of bacteria in the topmost 5 cm of sediment (annual mean 1523 $kg \cdot ha^{-1}$) is 2 orders of magnitude higher than macrozoobenthos biomass, and the production rate (504,000 $kg \cdot ha^{-1}yr^{-1}$) exceeds the production of benthic fish food organisms by 5 orders of magnitude. Primary production and standing crop of phytobenthos and phytoplankton together correspond to less than one percent of production rate and biomass of benthic bacteria (Pfeifer 1974, Tilzer 1972).

If the energy demand of meio- and micro-zoobenthos is assumed to be as high as that of macro-zoobenthos (Table 1), there still remains a considerable potential for additional production of benthic fish food.

It is difficult to say, whether charr and trout in Vorderer Finstertaler See have only reduced the abundance of the zoobenthos forms still present, or whether they have actually eliminated certain species. Although Bretschko (1976) found ephippia of large daphnids in the sediment of Vorderer Finstertaler See below a depth of 15 cm, since such crustaceans do not occur in the lake at present or even in its catchment basin; an annual sedimentation rate of 0.4 mm would be in good agreement with both total sediment thickness and lake age, as well as with the time of postulated salmonid introductions in late medieval times. The eradication of a pelagic daphnid by charr in Vorderer Finstertaler See seems highly probable.

From Switzerland, Schmassmann (1919) and Suchlandt and Schmassmann (1935) report that *Diaptomus* and *Daphnia* disappeared in Lunersee (1943 m a.s.l.) after the introduction of Arctic charr and other salmonids in 1887. Rainbow trout eradicated *Gammarus* in Grünsee (2140 m a.s.l.), and the same occurred with *Rivulogammarus lacustris* in Sünser See (Bregenzer Wald, Austria, 1810 m a.s.l.) after the introduction of charr (Amann 1980). Reimers (1958, 1979) found that due to the overstocking of Bunny Lake all but 3 benthic invertebrates, as well as *Daphnia* and *Diaptomus*, were eliminated within the first year.

Table 1. Annual mean biomass (in kg fresh-weight per hectare) and production rate (in kg FW per hectare and year) of macrozoobenthos in Vorderer Finstertaler See.

chironomids	14.3	45.0
oligochaets	3 - 13	16 - 40
chydorids	0.8	2.8
ostracods	0.4	1.4
	28.5	89.2

Gliwicz (1967) found a drastic reduction in the species diversity and abundance of crustaceans in two alpine lakes of the Polish Tatra Mountains; Anderson (1980) gives further references and presents results from lakes of the Canadian Rocky Mountains.

Results from a stocking experiment carried out in the Tyrolean lake Drachensee (1880 m a.s.l., area 6 ha, maximum depth 25 m), in the area were the northern calcareous Alps rise above the timber-line, are of interest in the present context. The lake had been stocked with "trout" in 1504 (Diem 1964), but Heller (1869) found it populated with Arctic charr. The lake was undoubtedly fishless in 1959, and was stocked with 1,000 Arctic charr fingerlings in 1969. In 1974, 248 Arctic charr were caught having lengths from 140 to 530 mm (Fig. 3), and weights up to 1 kg, considerably more than charr ever reach in Grundlsee (Neresheimer 1937, Steiner and Pechlaner 1975), where the planted stock originated. But, as is evident from Figure 4, both the size range of charr and the length of the dominant size class decreased considerably between 1974 and 1980. Arctic charr larger than 250 mm were caught in all years from 1973 to 1977 by anglers, but were not taken either in the gillnet catches of 1978 through 1980, or during intensive angling in 1981.

Age readings from otoliths, population estimates and other studies are underway and will perhaps explain why the size classes dominating the catch changed from lengths between 150 and 160 mm in 1978 to 180 and 190 mm in 1979 and 1980. The harvesting of 20 kg of charr in 1978 and the planting of 200 two-year old *Salvelinus namaycush* in 1976 may have stopped or even reversed the over-population which led, within a few years, to the significant dwarfing of Arctic charr in Drachensee.

The mayfly *Siphlonurus* sp., extremely abundant in 1959, disappeared from Drachensee; *Daphnia middendorffiana* is no longer found in the plankton. When phytoplankton biomass and production rate in this lake were investigated for the first time in 1959, the lake turned out to be much more productive than alpine soft water lakes in the Tyrol and Swedish Lappland (Rodhe et al. 1966); but the impoverishment of the zooplankton and zoobenthos leaves much microplankton and microbenthos unutilized.

In Zürser See (Vorarlberg, Austria, 2150 m a.s.l., 5.5 ha, maximum depth 15 m), which contained no fish until 1959, 600 charr fingerlings from Grundlsee (among other salmonids) were planted in 1960. These attained lengths of 280 to 320 mm within a few years, but in 1974 only a large number of stunted charr with a maximum length of 165 mm were observed (Schurig 1967, Amann 1980).

There are a few instances of high-altitude lakes where "nature" regulates population densities of charr adequately for normal growth. In Sünser See oxygen depletion occasionally kills charr recruits (Amann 1980); in Zireiner See (Tyrol, 1750 m a.s.l.), a good charr lake even in the early 16th century, egg and/or fish predation by burbot (*Lota lota*) and bullheads (*Cottus gobio*) apparently ensures that the fish develop in harmony with the available food (Pechlaner 1979).

However, as a rule, high-mountain lakes stocked by man need to be exploited adequately to ensure the efficient functioning of the food chain. This obviously was done by Kaiser Maximilian I and his fishermen; fishermen at that time did not restrict themselves to angling, but were advised, with respect to Drachensee, "to put in gillnets by night and lift them again at dawn" (Diem 1964, p. 94).

Why not reverse the fish movement of late medieval times, harvesting valuable charr fry from high-mountain lakes for the stocking of lowland lakes with deteriorating spawning sites, at the same time regulating population densities of the charr in the alpine lakes so as to ensure a supply of normal and trophy fish for recreational angling?

REFERENCES

Amann, E. 1980. Limnologische und fischereiliche Untersuchungen am Sünser See. Österreichs Fischerei 33: 205-215.

Anderson, R.St. 1980. Relationships between trout and invertebrate species as predators and the structure of the crustacean and rotiferan plankton in mountain lakes, p. 635-641. *In* W.C. Kerfoot [ed.] Evolution and ecology of zooplankton communities. Spec. Symp. Vol. 3, Amer. Soc. Limnol. Oceanogr., Univ. Press of New England, Hanover, New Hampshire, U.S.A.

Balon, E.K. and T. Penczak. 1980. The dwarfed charr of Dösener See, an alpine lake in Austria, p. 773-794. *In* E.K. Balon [ed.] Charrs: salmonid fishes of the genus *Salvelinus*. Dr. W. Junk Publishers, The Hague.

Bretschko, G. 1974. The chironomid fauna of a high-mountain lake (Vorderer Finstertaler See, Tyrol, Austria, 2237 m a.s.l.) — Ent. Tidskr. 95, Suppl.: 22-33.

Bretschko, G. 1975. Annual benthic biomass distribution in a high-mountain lake (Vorderer Finstertaler See, Tyrol, Austria, 2237 m a.s.l.). Verh. Int. Ver. Limnol. 19: 1279-1285.

Bretschko, G. 1976. Ökosystemstudie Vorderer Finstertaler See: 3.2 Subfossile Daphniden-Ephippien. Jber. Abt. Limnol. Innsbruck 2: 162-163.

Diem, H. 1964. Beiträge zur Fischerei Nordtirols. Veröff. Museum Ferdinandeum 3: 5-132.

Gliwicz, Z.M. 1967. Zooplankton and temperature-oxygen conditions of two alpine lakes of the Tatra Mountains. Pol. Arch. Hydrobiol. 14: 53-72.

Heckel, J. and R. Kner. 1858. Die Süsswasserfische der Östreichischen Monarchie. Leipzig.

Heller, C. 1869. Die Seen Tirols und ihre Fischfauna. Festschrift der Naturforscher-Versammlung in Innsbruck: 51-56.

Neresheimer, E. 1937. Die Lachsartigen (Salmonidae) I.Teil. Handb. d. Binnenfischerei Mitteleuropas, III. 219-370.

Pechlaner, R. 1966. Salmonideneinsätze in Hochgebirgsseen und — tumpel der Ostalphen. — Verh. Int. Ver. Limnol. 16: 1182-1191.

Pechlaner, R. 1969. Hochgebirgsseen als Lebensraum für Salmoniden. Zool. Anz., Suppl. 32: 750-757.

Pechlaner, R. 1979. High-mountain lakes in Tyrol. — "Tirol immer einen Urlaub wert" (Innsbruck) 14:3-24.

Pechlaner, R. 1984. Historical evidence for the introduction of Arctic charr into high-mountain lakes of the Alps by man. Paper presented at Int. Symp. on Arctic charr, Winnipeg, 1981. (This volume).

Pechlaner, R., G. Bretschko, P. Gollmann, H. Pfeifer, M. Tilzer und H.P. Weissenbach. 1972. Ein Hochgebirgssee (Vorderer Finstertaler See, Kühtai, Tirol) als Modell des Energietransportes durch ein limnisches Ökosystem. Verhandl. Zool. Ges. 65: 47-56.

Pfeifer, H. 1974. Das Phytobenthos des Vorderen Finstertaler Sees. — Diss. Innsbruck: 1-119.

Reimers. N. 1958. Conditions of existence, growth and longevity of brook trout in a small, high-altitude lake of the Eastern Sierra Nevada. Calif. Fish and Game 44: 319-333.

Reimers, N. 1979. A history of a stunted brook trout population in an alpine lake: A life span of 24 years. Calif. Fish and Game 64: 196-215.

Rodhe, W.J., J.E. Hobbie and R.T. Wright. 1966. Phototrophy and heterotrophy in high-mountain lakes, — Verh. Int. Ver. Limnol. 16: 302-313.

Schmassmann, W. 1919. Zur Besetzung alpiner Seen. Schweiz. Fischereizeitung 3: 36-38.

Schurig, H. 1967. Fischereiwirtschaftliche Erfahrungen an einem Hochgebirgssee. Österreichs Fischerei 20: 49-55.

Steinböck, O. 1950. Probleme der Ernährung und des Wachstums bei Salmoniden. Schweiz. Fischereizeitung 1950: 1-7.

Steiner V. and R. Pechlaner. 1975. Mark-recapture experiments with Arctic char (*Salvelinus alpinus*) in Austrian lakes. EIFAC Technical paper No. 23 (suppl. 1), Vol. II: 672-688.

Suchlandt, O. und W. Schmassmann. 1935. Limnologische Beobachtungen an acht Hochgebirgsseen der Landschaft Davos. Schweiz. Z. Hydrobl. 7: 1-201.

Tautermann, G. 1977. Benthische Bakterien im Vorderen Finstertaler See (2237 m, Kühtai, Tirol), Diss. Abt. Limnol. Innsbruck 10: 1-103.

Tilzer, M. 1972. Dynamik und Produktivität von Phytoplankton und pelagischen Bakterien en einem Hochgebirgssee (Vorderer Finstertaler See, Österreich). Arch. Hydrobiol. Suppl. 40: 201-273.

Wagner, B. 1975. Populationsdynamik der Oligochaeten im Vorderen Finstertaler See (2237 m, Kühtai, Tirol). Jber. Abt. Limnol. Innsbruck. 2:1-102.

The Biology and Bioenergetics of Arctic Charr in Char Lake, N.W.T., Canada

WAYNE R. MacCALLUM

Ontario Ministry of Natural Resources, Thunder Bay, Ontario

HENRY A. REGIER

Institute for Environmental Studies, University of Toronto
Toronto, Ontario

MACCALLUM, W.R. AND H.A. REGIER. 1984. The biology and bioenergetics of Arctic charr in Char Lake, N.W.T., Canada p. 329–340. *In* L. Johnson and B.L. Burns [eds.] Biology of the Arctic charr, Proceedings of the International Symposium on Arctic Charr, Winnipeg, Manitoba, May 1981. Univ. Manitoba Press, Winnipeg.

The biology, population dynamics and bioenergetics of the landlocked population of Arctic charr in a small Arctic lake (52.6 ha) was investigated. This research was part of the Canadian I.B.P. Char Lake Project whose principal objective was to study energy flow through all the major components in the lake.

The adult population of charr was estimated by the Jolly-Seber and the Delury methods to be 18,384 and 19,034 respectively. The standing stock of the adult population was 9.23 $kg \cdot ha^{-1}$; the length-frequency distribution was unimodal with the mode at 140-150 mm.

Growth, estimated by mark-recapture and by apparent growth from an age-length relationship was extremely slow and variable. The mean age of a 150 mm fish was 13.7 years. Males and females grew at the same rate.

There was no significant difference in the mean age of maturity between males and females (5.8 years). Some males and some females spawned for the first time at the age of four and three years respectively; all were mature by age nine. Not all the charr spawn annually. Gonad maturation began in April and was completed by late August; spawning occurred in September. Eggs hatched in late March and early April.

The energy content of the somatic tissue of females initiating egg maturation in the spring was significantly greater than that of those that were not (6,403 vs 6,165 $kcal \cdot g^{-1}$ ash-free).

INTRODUCTION

The International Biological Program (I.B.P.) had among its objectives the initiation and encouragement of comprehensive studies needed to test and utilize the production-energy flow approach to studying ecosystems.

The Char Lake Project, of which this study was part, was one of the Canadian contributions to the I.B.P. (Rigler 1975). The study of a high Arctic lake would be a valuable contribution to the I.B.P. because of the relative biotic simplicity and also because of the paucity of information that exists on such lakes. A prime criterion for choosing a lake to study was that it would have a landlocked population of Arctic charr, *Salvelinus alpinus*, and no other species of fish.

The primary objective of this study was to provide data concerning energy flow and production of the Arctic charr population. Another major objective was to contribute a second detailed study to that of Hunter (1970) on a fish species that is widely distributed in Canada but not well known. A third objective was to compare the bioenergetics of the population with those of rainbow trout, *Salmo gairdneri*, and kokanee salmon *Oncorhynchus nerka* in Marion Lake, British Columbia described by Sandercock (1969) whose work was one of the contributions to the I.B.P. study of that lake.

This paper focuses on abundance, size structure, growth and reproductive aspects of bioenergetics.

METHODS AND MATERIALS

Char Lake is a sub-circular, single basin lake with a surface area of 52.6 ha, a maximum depth of 27.5 m, and a mean depth of 10.2 m (Schindler et al. 1974).

It was one of the objectives of the Char Lake project to study an undisturbed system, so care was taken by all participants to disturb the lake and its biota as little as possible. The major implication of this was that an absolute minimum number of fish were removed from the lake: less than 600 during the five years of the study.

Six methods were used to capture fish: angling through the ice with number 10 and 12 barbless hooks mainly in the spring, trapnetting with small nylon nets (Beamish 1972) during the fall spawning period, gillnetting in the summer of 1970, electric shocking (McCrimmon and Berst 1963) and seining for juveniles along the shoreline, and by hand, by SCUBA diving. All fish were anaesthetized with MS-222, weighed to 0.1 g, fork length (mm) recorded, and either uniquely marked by hot branding, or batch marked with partial fin clips.

Fish required for energy content and aging studies were frozen for later examination. The sex, state of maturity and stomach contents were recorded and gonads weighed to 0.1 g, eggs preserved in Gilson's solution for subsequent counts and size determination, and saccular otoliths removed and stored dry in envelopes.

Ages were determined using split otoliths immersed in a 3:1 solution of methyl salicylate and benzyl benzoate. The determination of the ages from otoliths was consistent with the interpretations of Grainger (1953), Nordeng (1961), Hunter (1970), and Power (1978). The age of maturity was estimated from an examination of the growth patterns of the otoliths, following the interpretation of Nordeng (1961). Briefly, a zone of wide juvenile growth bands, is surrounded by a zone of narrow bands laid down after maturity.

The population size was estimated using the methods of Delury (1958) and Jolly-Seber (Jolly 1965, Seber 1965, 1973). The samples were stratified into one-cm length groups and estimates made for each stratum to reduce the effects of gear selectivity and to permit size dependent population energetics estimates to be made. The Jolly-Seber method requires a minimum of three sampling periods. Each year was considered to be a sampling period because most fish were captured in a short, six week period in the spring. The estimates presented are those calculated for 1970. Fish captured more than once during a sampling period were treated as single captures.

The standing stock was calculated by combining the length stratified population estimates, a length-weight regression and then summing the estimates of the weights.

Growth was estimated by two methods, the growth of fish marked and later recaptured, and the apparent growth from an age-length relationship. The average yearly growth for each recaptured fish was calculated by dividing the growth increment between capture and recapture by the time interval in years. Fish that were recaptured the same year they were marked were not included in the estimates to reduce the effects of capture and branding. The average yearly growth was summarized as a function of length, because the ages of fish in the mark-recapture study were not known. A length-weight relationship was used to calculate the weights of each length category of fish at the beginning and the end of a year. The difference was then multiplied by the average number of fish in each category during the year, and the results summed to estimate the total yearly somatic growth.

The energy content of somatic and gonadal tissue was determined using a Gallenkamp Ballistic Bomb Calorimeter (Miller and Payne 1959). Because of their small size, whole fish, from which stomach contents and gonads had been removed, were homogenized in a blender, dried for 24 hours at 110°C, powdered with a mortar and pestle or a rotary coffee grinder, and further dried for 12 hours. Pelletized 0.5 to 1.0 g samples were then ignited in the calorimeter and the mean of two samples for each fish was used in subsequent calculations.

To investigate the relationship between the energy content of fish that were initiating egg maturation and those that were not, bomb calorimetry samples of the somatic tissues of the two groups of fish were compared using covariance analysis. Females having eggs larger than 1.6 mm in April were placed in the maturing category, and those with eggs less than 1.6 mm were placed in the non-maturing one. April was chosen because it was the earliest that fish that were going to spawn in the fall could be differentiated from those that were not. The terminology used in discussions of gonad development is that of Vladykov (1956).

The factor 1.06 Kcal per wet gram was used to convert wet weights to energy equivalents (Winberg 1971).

RESULTS AND DISCUSSION

Population Estimates

Some 500 fish were captured by angling in 1968, the first year of the project and by the fall of 1971 the total was about 4,200. Approximately 5,000 fish were captured on the spawning grounds by trapnets in 1971 and 1972. Less than 100 fish were captured by gillnets.

Length-frequency catch curves were unimodal at 140 to 150 mm and positively skewed for both angling and trapnetting, although the distribution from the trapnet was more leptokurtic (Fig. 1). Angled recaptures exhibited a shift in the mode towards larger fish, possibly reflecting the influence of gear selectivity, size dependent mortality, and growth (Fig. 2). It is likely that selectivity is the major cause of the shift because fish marked and recaptured within the main six week sampling periods in all years showed similar shifts.

The Delury and Jolly-Seber estimates for the population were similar, 19,034 and 18,384 respectively (Table 1, Fig. 3). The calculation of standard errors and confidence limits for the estimates was abandoned as a fruitless exercise when upper limits for several estimates were negative numbers. Arnason and Baniuk (1978) discuss this type of anamoly. The standing stock of fish 130 mm and greater was 9.23 kg•ha^{-1}, (9.79•10**3 Kcal•ha^{-1}).

Growth

Growth was extremely slow and variable. Growth rates of the Char Lake charr are among the slowest reported in the literature (Carlander 1969). Figure 4 gives the average yearly growth of fish in the range of 100 to 400 mm, from the mark-recapture study. It was possible to test for differences in growth rates between sexes using branded fish recaptured on the spawning grounds in 1971 and 1972. A two way analysis of variance showed no significant difference ($p < .05$, $n = 138$).

The relationship between age and length for fish sampled throughout the study are given in Table 3 and Figure 5. The significant features are the extremely slow growth and the large variation in the size at any given age. A semi-log regression of length on age was highly significant:

$$\ln(\text{length}) = 4.47 + 0.45*\text{Age} \qquad n = 366, r = 0.82$$

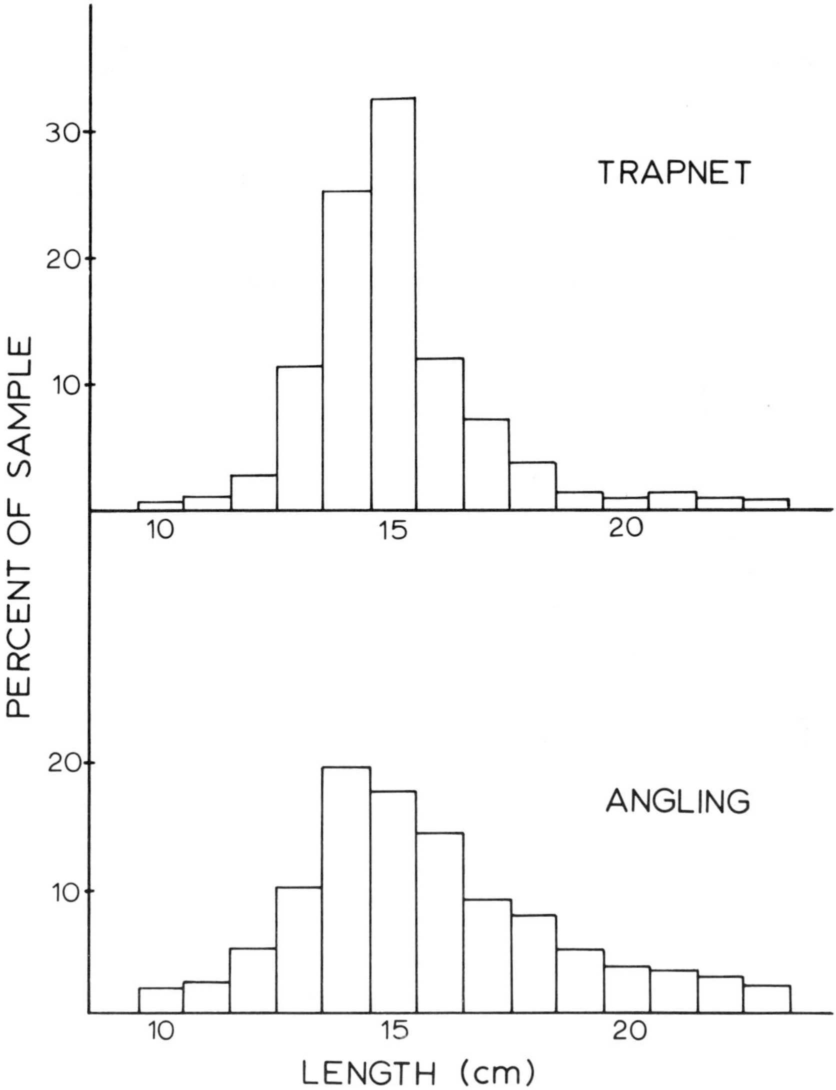

FIG. 1 A comparison of the length frequency distributions of fish caught by angling in the spring and trapnetting on the spawning grounds in the fall.

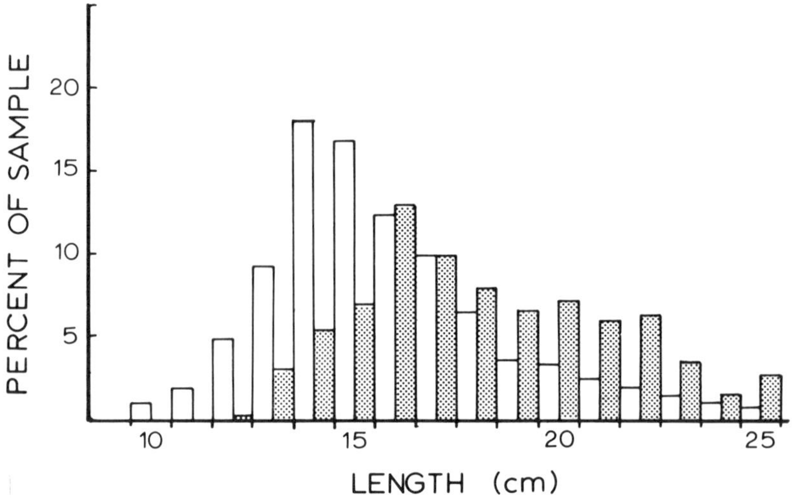

FIG. 2 A comparison of the length frequency distributions of newly captured fish caught by angling. The shaded portions are recaptured fish.

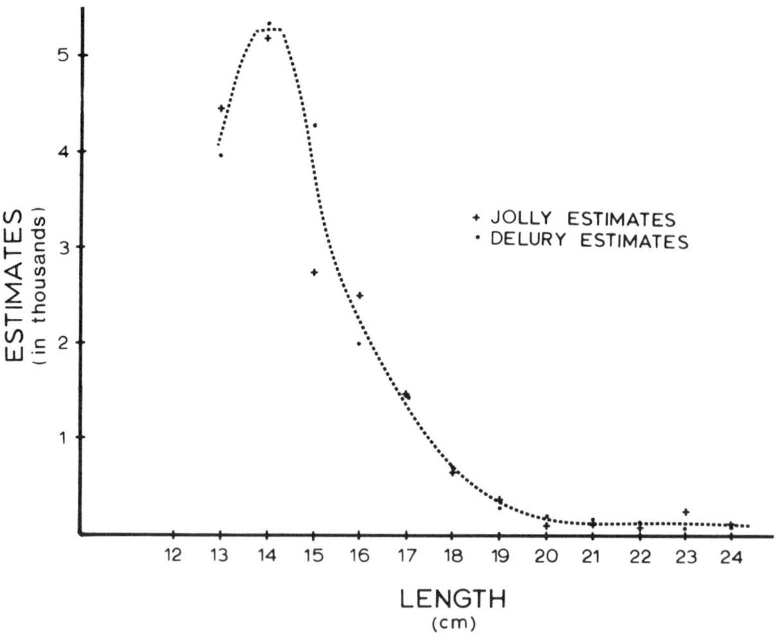

FIG. 3 The length frequency distribution of the Char Lake population from Jolly-Seber and Delury population estimates.

Although most of the fish were small, the lake supported several charr among the largest reported in the literature (Johnson 1976, 1980) for landlocked populations. Five charr greater than 500 mm were caught, one of which was a recapture. The oldest age assigned to a fish was 35 years for a 274 mm female.

The total yearly somatic growth for fish 130 mm and greater was 0.547 kg $\times$ ha^{-1}, (0.580 $Kcal \cdot ha^{-1}$).

Age of Maturity

The mean age of maturity for males and females combined was 5.8 years (Table 2). There was no significant difference between the sexes ($p < 0.5$, $n = 319$).

Maturation of Gonads

Recruitment stock eggs, 1.2 to 1.6 mm diameter, were the largest eggs found in the ovaries from October to April. From April until spawning time in September, progressively larger eggs were found in some fish, while other fish during this time had eggs no larger than 1.6 mm. A similar relationship was exhibited when gonad weight was expressed as a percentage of body weight. A maximum of about three percent during the October to April period was increased to a maximum of 18 percent in September. The maturation of the ovaries occurred one to two months earlier than those in the populations reviewed by Johnson (1980).

Energy Content and Egg Production

Not all charr spawn annually (Grainger 1953, Hunter 1970, Johnson 1980). It was expected that the initiation of egg maturation would depend on the physiological well-being of the fish. Bagenal (1969) demonstrated a relationship in brown trout (*Salmo trutta*) between ration level and the percentage of fish maturing eggs. Brett (1971) found a relationship between ration level and the energy content of young sockeye salmon (*Oncorhynchus nerka*). Sandercock (1969) observed a seasonal pattern in the energy content of rainbow trout and kokanee. Several weeks before the spring spawning period, the energy content of mature rainbow trout somatic tissue started dropping until it reached a low at spawning time. During the summer feeding period, the energy content increased until it reached a high in the fall. Kokanee had a similar pattern except that the drop in energy content began several weeks before the fall spawning, after which they die.

Charr that had just begun to mature a set of eggs had a significantly higher energy content than those which were not maturing eggs (6,403 vs 6,165 calories

Table 1. Jolly-Seber (J-S) and Delury (D) population estimates for Char Lake.

	Length Categories in millimetres													
	130	140	150	160	170	180	190	200	210	220	230	240	250	Total
J-S	4482	5188	2715	2525	1454	728	381	99	158	92	251	57	254	18384
D	3989	5335	4293	1999	1460	706	268	212	172	147	84	82	287	19034

Table 2. Age of maturity of male and female charr interpreted from otoliths.

	Males		Females	
Age at Maturity	Percent of Sample	Cumulative Percentage	Percent of Sample	Cumulative Percentage
3	0	0	0.5	0.5
4	4.6	4.6	12.7	13.3
5	27.7	32.3	29.6	42.8
6	46.9	79.2	31.2	74.0
7	16.9	96.1	21.2	95.2
8	3.1	99.2	3.7	98.9
9	0.8	100.	1.1	100.

$\cdot g^{-1}$ ash-free) (Cov., $p < .05$, df = 1,22). The energy content of the somatic tissue of ripe females on the spawning grounds was significantly lower than those which were initiating egg maturation in the spring (6,163 vs 6,403 cal $\cdot g^{-1}$ ash-free (Cov., $p < 05$, df = 1,16). These observations are consistent with non-annular spawning of the female charr and the suggestion that they may have to achieve a threshold energy level before egg maturation can begin. After having spawned, the females in a low energy state, would be facing a winter in which food intake would be low, resulting in their reaching spring in a condition below the threshold necessary for spawning in the current year. One or two summers might pass before a fish was able to begin egg maturation again.

REFERENCES

Arnason, A.N., and L. Baniuk. 1978. POPAN-2. A data maintenance and analysis system for mark-recapture data. The Charles Babbage Research Centre. St. Pierre, Manitoba. 269 p.

Bagenal, T.B. 1969. The relationship between food supply and fecundity in brown trout *Salmo trutta* L. J. Fish Biol. 1: 167-182.

Beamish, R.J. 1972. Design of a trapnet for sampling shallow-water habitats. Fish. Res. Board Can. Tech. Rep. 305: 14 pp.

Brett, J.R. 1971. Growth responses of young sockeye salmon (*Oncorhynchus nerka*) to different diets and planes of nutrition. J. Fish. Res. Board Can. 28(10): 1635-1643.

Carlander, K.D. 1969. Handbook of freshwater fishery biology. Volume one. Iowa State University Press. Ames, Iowa.

Delury, D.B. 1958. The estimation of population size by marking and recapture procedure. J. Fish Res. Board Can. 15: 19-25.

Table 3. The mean lengths of fish from young-of-the-year to 23 years old.

Age	Mean length (mm)	Standard deviation	Sample size
0	27	1.93	50
1	48	3.10	174
2	65	3.58	141
3	77	4.51	16
4	89	15.04	3
5	110	4.04	3
6	107	4.88	7
7	106	5.77	3
8	126	17.1	10
9	128	6.89	13
10	129	11.3	15
11	137	11.0	23
12	139	33.4	41
13	152	19.0	82
14	171	26.1	51
15	183	29.2	23
16	173	26.5	9
17	195	35.3	13
18	190	30.8	13
19	217	33.2	11
20	223	38.8	13
21	215	35.9	8
22	231	43.0	7
23	215	23.5	6

Grainger, E.H. 1953. One the age, growth, migration, reproductive potential and feeding habits of the Arctic char (*Salvelinus alpinus*) of Frobisher Bay, Baffin Island. J. Fish. Res. Board Can. 10(6): 326-370.

Hunter, J.G. 1970. Production of arctic char (*Salvelinus alpinus*) in a small arctic lake. Fish. Res. Board Can. Tech. Rep. 231: 190 pp.

Johnson, L. 1976. Ecology of arctic populations of lake trout, *Salvelinus namaycush*, lake whitefish, *Coregonus clupeaformis*, Arctic charr, *S. alpinus*, and associated species in unexploited lakes of the Canadian Northwest Territories. J. Fish. Res. Board Can. 33(11): 2459-2488.

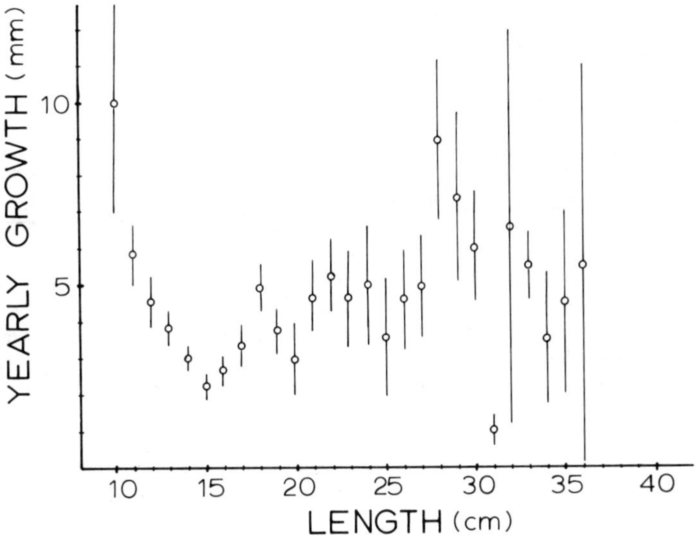

FIG. 4 Estimates of annual growth for charr in the range 10 to 36 cm estimated from mark-recapture data. The bars represent the standard deviations. n = 856

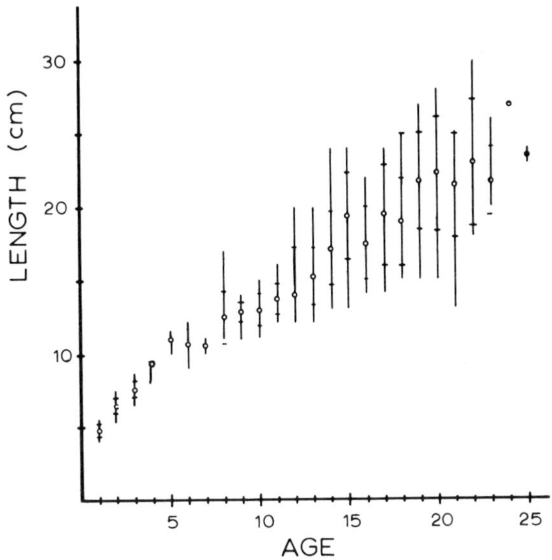

FIG. 5 The relationshp between length and age. The circles are the means; the horizontal bars, the standard deviations; and the vertical lines, the range of the observations. Sample sizes are given in Table 3.

Johnson, L. 1980. The Arctic charr, *Salvelinus alpinus*, p. 15-98. *In* E.K. Balon [ed.] Charrs: salmonid fishes of the genus *Salvelinus*. Dr. W. Junk Publishers, The Hague.

Jolly, G.M. 1965. Explicit estimates from capture-recapture data with both death and immigration-stochastic model. Biometrika 52: 225-247.

McCrimmon, H.R. and A.H. Berst. 1963. A portable A.C.-D.C. backpack fish shocker designed for operation in Ontario streams. Progr. Fish-Cult. 24: 159-162.

Miller, D.S., and P.R. Payne. 1959. A ballistic bomb calorimeter. Brit. J. Nutr. 12: 501-508.

Nordeng, H. 1961. On the biology of char (*Salmo alpinus* L.) in Salangen, North Norway. 1. Age and spawning frequency determined from scales and otoliths. Nytt Mag. Zool. 10: 67-123.

Power, G. 1978. Fish population structure in arctic lakes. J. Fish. Res. Board Can. 35: 53-59.

Rigler, F.H. 1975. The Char Lake Project. An introduction to limnology in the Canadian Arctic. p. 171-198. *In* T.W.M. Cameron and L.W. Billingsley [eds.] Energy flow — its biological dimensions. Royal Society of Canada, Ottawa, for Canadian Committee for the International Biological Program (IBP).

Sandercock, K.F. 1969. Bioenergetics of the rainbow trout (*Salmo gairdneri*) and the kokanee (*Oncorhynchus nerka*) populations of Marion Lake, British Columbia. Ph.D. Thesis, University of British Columbia.

Schindler, D.W., H.E. Welch, J. Kalff, G.J. Brunskill, and N. Kritsch. 1974. The physical and chemical limnology of Char Lake, Cornwallis Island (75 N lat.) J. Fish. Res. Board Can. 31: 585-607.

Seber, G.A.F. 1965. A note on the multiple-recapture census. Biometrika 52: 249-259.

Seber, G.A.F. 1973. The estimation of animal abundance. Griffin. London 506p.

Vladykov, V.D. 1956. Fecundity of wild speckled trout (*Salvelinus fontinalis*) in Quebec Lakes. J. Fish. Res. Board Can. 13: 799-84.

Winberg, G.G. [ed.] 1971. Symbols, units and conversion factors in studies of fresh water productivity. IBP London. 23 p.

The Ecology and Management of the Arctic Charr in Lake Mývatn, North Iceland

JÓN KRISTJÁNSSON

Institute of Freshwater Fisheries, P.O. Box 754, Reykjavik, Iceland

HÁKON ADALSTEINSSON

National Energy Authority, Grensásveg 9,108 Reykjavik, Iceland

KRISTJÁNSSON, J. AND H. ADALSTEINSSON. 1984. The ecology and management of the Arctic charr in Lake Mývatn, North Iceland, p. 341–347. *In* L. Johnson and B.L. Burns [eds.] Biology of the Arctic charr, Proceedings of the International Symposium on Arctic Charr, Winnipeg, Manitoba, May 1981. Univ. Manitoba Press, Winnipeg.

Lake Mývatn is a shallow eutrophic lake. It is ice-free on the average for 190 days between May and October. The maximum summer temperature of 15-17°C is comparatively high for Icelandic lakes. Total primary production has been estimated to be 800-1000 g dry weight$\cdot$m^{-2}yr^{-1} and secondary production 40-50 g dry weight$\cdot$m^{-2} yr^{-1}.(approximately 300 g wet weight). As a result, the lake supports a large duck population. Documentation on extensive charr fishing is available. The Arctic charr population which has been declining of late, largely as a result of overfishing, does not fully utilize the available food resources. A total of 150 gillnets (25 m long, 45 mm knot to knot) are fished daily for eight months of each year; recently 85 percent of the stock has been removed annually. The size of the 3-5 year old charr is 400-1200 g.

INTRODUCTION

Arctic charr, *Salvelinus alpinus,* is widespread throughout Iceland, from the seashore to high mountain lakes. Sea-run charr is found in the freshwaters around the whole coastline, often together with lake resident forms.

Shallow lakes, 1-3 m deep are usually inhabited by only one form of charr, but in deeper lakes with a pronounced pelagic zone, two or more different forms, or ecotypes, are found; in Thingvallavatn (surface area 84 km^2, maximum depth: 114 m) at least four forms of charr have been identified. Thingvallavatn yields 3-5 tons of the benthic form of charr and 60-70 tons of the pelagic form annually (total 8.9 kg$\cdot$ha^{-1}). On average, charr lakes in Iceland yield 10-15 kg$\cdot$ha^{-1}yr^{-1}, with better lakes yielding up to 45 kg$\cdot$ha^{-1}yr^{-1}.

An exploitation experiment was started in East Fridmundarvatn in 1975 to attempt to 'fish out' the stock completely (Kristjánsson 1978). The lake, a typical shallow mountain charr lake, is located 345 m a.s.l., with a surface area of 2.36 km^2. Maximum depth is 1.15 m and average depth 0.8 m. The bottom of the lake is soft mud covered mostly with *Myriophyllum alterniflorum;* the blue green algae *Nostoc* sp. is also common. Annual primary production has been estimated to be in the range of 200-400 $g \cdot C\ m^{-2}\ yr^{-1}$. The dominant invertebrate species, *Eurycercus lamellatus,* is also the most important food of the charr.

In 1975 the gillnet catch increased from a previous level of 1-2 $kg \cdot ha^{-1} yr^{-1}$ to an average value of 7.5 $kg \cdot ha^{-1}$; the highest catch was 15.2 kg in 1977. In that year the standing crop of charr was estimated to be 65 $kg \cdot ha^{-1}$. At the end of the period of intensive fishing the composition of the stock had changed; both mean and maximum weight were reduced. The population number has probably increased. It is proposed that recruitment has increased due to removal of fully grown competitors. This experiment has shown that in order to bring the population down to suitable level in such lakes gillnets are not effective.

As most lakes are underexploited, work is being undertaken to find more effective fishing methods. Of special interest is a modified type of the Danish seine used in the Icelandic coastal fisheries.

In Mývatn three species of fish are found: Arctic charr, brown trout, *Salmo trutta,* and threespine stickleback, *Gasterosteus aculeatus.* Arctic charr is the dominant species and it is found in two distinct forms, a situation commonly encountered in Scandinavia (Nilsson and Filipsson 1971, Nyman 1972, Klemetsen and Grotnes 1975). Data on fishing, fish growth and fish feeding is available from the beginning of the present century, due to the enthusiasm of the first fishery biologist in Iceland (Saemundsson 1915). Much is also known about fishing methods and the gear used from the middle of the 19th Century to recent times (reviewed by Adalsteinsson 1975). Lamby (1941) studied both the growth and feeding of char in Iceland in the early 1930's.

During the Mývatn-Laxá research 1971-74 (Jonasson et al. 1979) feeding, growth and several other aspects of the charr biology were examined (Adalsteinsson 1976 to 1979). Until about the middle of this century, beach seines and hooks were used for fishing. Then came the outboard motors and nylon gillnets, and now monofilament gillnets are used exclusively. During 1973-76, the Institute of Freshwater Fisheries collected material from Mývatn that suggested the lake was being heavily overfished. In that period up to 700 nets (25 m long) were used each day throughout the fishing season (230 days). In April 1977 the fishermen agreed not to use more than about 150 nets each day. This opportunity was used to obtain reliable information on catch and effort.

Mesh size used was 45 mm (knot to knot); the age of the charr was 3+ and 4+ with the mean weight around 700 grams.

RESULTS AND DISCUSSION

The fishing season can be divided into two periods. Winter fishing under the ice from February 1 to about mid-April depending on ice conditions, and summer fishing when the lake is ice-free. From September 27 to February 1 no fishing is allowed. Spawning takes place from late September to December.

Basic data on the charr catch and effort are shown in Table 1; it can be seen that both catch and catch per unit effort decrease from February to May. In the winter months the nets are set under the ice and conditions are thought to be stable when the lake is ice-covered (temperature, fish activity, fish growth, and recruitment are all low). The hypothesis is that the decrease in catch per unit effort is a result of depletion of the catchable stock by fishing. If this is the case then Leslie's method (Ricker 1975) can be used to estimate the size of the catchable stock prior to the fishing season. Based on the catch and effort in February, March and April (three data pairs) the values shown in Table 2 were obtained. In 1980 each month was divided into two periods in order to obtain more precise estimates.

Figure 1 shows the plots of catch per unit effort against cumulative catch in the winter fishing period.

The calculated catchability of the gillnets does not change much from one year to another based on data from the winter season only.

In order to construct a population model the catchability in the summer months has to be known. For various reasons this has not been possible so the winter catchability has been used throughout the year. Mean stock each month can be computed from Baranov's catch equation:

$$C = F \cdot N = \frac{N \cdot F \cdot A}{Z}$$

where:

C = catch
F = annual fishing mortality
N = catchable stock
A = annual mortality
Z = instantaneous mortality

The stock size at the beginning of each month can be computed if M (natural mortalty) is known ($Z = F + M$). As the natural mortality is not known a value of $M = 0.15$ has been used until more information is available.

Table 3 shows the population model thus obtained. The stock has been divided into veteran stock, which was present on February 1, and recruits which

Table 1. Catch of charr and fishing effort in Lake Mývatn 1977-1980.

Month	1977 C_t[a]	1977 f_t[b]	1978 C_t	1978 f_t	1979 C_t	1979 f_t	1980 C_t	1980 f_t
February	2217		5407	2378	1954	1942	3728	3440
March	1457		3467	2624	1349	2432	2269	3939
April	872		2239	1965	797	1513	988	2212
May	707		1651	1009	828	1108	929	1120
June	3979	1720	3982	1744	1119	1052	1857	1811
July	7113	2700	5180	2167	1903	1763	1787	1825
August	12403	2646	432	238	1893	866	3212	2515
September	4096	1120	361	194	1383	477	2428	1610
TOTAL	32653		22713	12319	11325	11452	17198	18472

[a] C_t = Catch (number of fish) taken during time interval t (one month).
[b] f_t = Fishing effort (gillnet nights) during time interval t.

Table 2. Catchable charr stock prior to fishing season in Lake Mývatn 1978-1980.

Year	No	q	Fi
1978	16431	1.61×10^{-4}	1.12
1979	6211	1.84×10^{-4}	1.08
1980	9164	1.49×10^{-4}	1.43

No = Catchable stock prior to fishing season.

q = Catchablility - the fraction of the population taken by one unit of fishing effort.

Fi = Instantaneous rate of fishing mortality in the period 1 February to 30 April.

Table 3. Population model of the charr stock in Lake Mýyvatn 1980.

Date	Nvet	Cvet	NDvet	F	M	Nrecr	Crecr	NDrecr
1 Feb	9164							
		2261	73.0	.26	.008			
15 Feb	6830							
		1707	54.0	.25	-			
1 Mar	5269							
		1088	42.0	.28	-			
15 Mar	4139							
		1181	33.0	.31	-			
1 Apr	2925							
		796	23.0	.25	-			
14 Apr	2106							
		192	17.0	.08	-			
1 May	1897					3915		
		284	28.0	.17	.015		645	58
1 June	1585					6118		
		363	24.0	.27	-		1494	91
1 July	1198					6191		
		275	18.0	.27	-		1512	92
1 Aug	905					9296		
		273	13.0	.37	-		2939	138
1 Sept	619					10303		
		129	9.0	.24	-		2299	155
1 Oct	481					7939		
Sum		8343	334	2.75	.12		8889	534

Estimated population size on October 1, 1980: 481 veterans + 7939 recruits = 8420

Nvet = Catchable stock of veterans in the beginning of each period.

Nrecr = Catchable stock of recruits in the beginning of each period.

NDvet = NDrecr = Number of fish naturally dead during the period.

F = Estimated instantaneous fishing mortality.

M = Estimated instantaneous natural mortality.

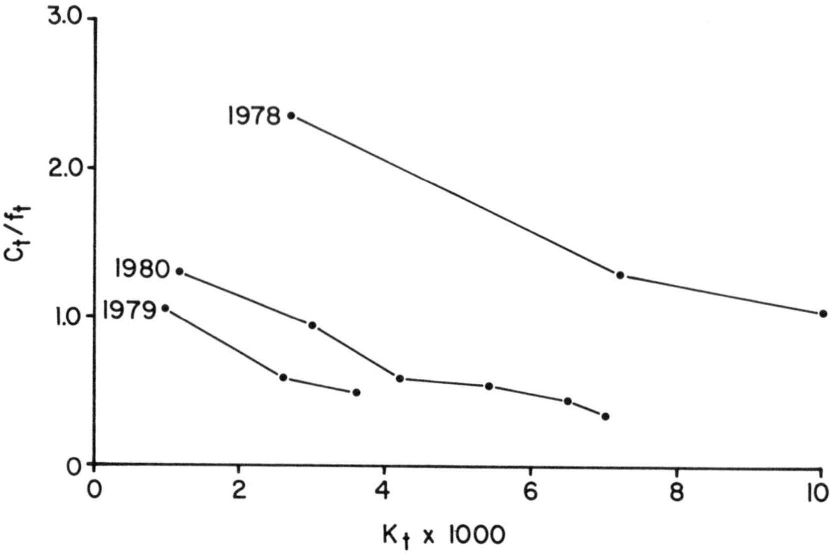

FIG. 1 Catch per unit effort against cumulative catch in the period February 1 to April 30, 1978-1980.

start to appear in May. It can be seen that the yearly fishing mortality, F, is 2.75 which means that about 94 percent of catchable fish is harvested each year. This is supported by the fact that fish of age 5+ are very rare in the catch and older fish almost absent. The oldest charr which has been aged in recent years was 7+ and some 600 mm in size. The standing stock on October 1, 1980 was calculated to be 8420 fish.

With so few years to compare it is difficult to estimate the survival of the calculated stock in the period from October 1 to February 1 the following year. For 1978-79 the survival was 63 percent and 77 percent the year after.

Fish caught by 19.5-26.0 mm nets in June each year are to be the recruits in the following year. By knowing the catchability of these nets it should be possible to estimate their numbers. It has not been possible to use this method properly yet because of unusually low recruitment since 1977. Only the general effect of extreme low recruitment has been documented.

Valuable information on stock size and catchability of the fishing gear has been gathered since 1977. As the years pass these methods will be refined, especially if a strong year-class appears. The relative population density has been so low since 1977 that live-tagging of fish in the immediate pre-recruit stage has been impossible. Therefore, no reliable information on catchability in the summer months exists. One might expect that catchability would be higher in the summer due to increased activity of the fish, but there are other factors that would decrease catchability, such as *Cladophora* and detritus blocking the nets, a very pronounced factor in Lake Mývatn.

Until 1981 the fishing pressure was regulated by the number of nets. This has not been satisfactory and the fishermen have now agreed upon a pre-determined quota with the aim at decreasing the fishing mortality to the 50-60 percent level over the next few years.

REFERENCES

Adalsteinsson, H. 1976. Fiskstofnar Mývatns. Náttúrufr. 45: 154-177. (In Icelandic).

Adalsteinsson, H. 1979. Size and food of Arctic char *Salvelinus alpinus* and stickleback *Gasterosteus aculeatus* in Lake Mývatn. Oikos 32:228-231.

Jónasson, P.M. 1979. The Lake Mývatn ecosystem, Iceland. Oikos 32: 289-305.

Klemetsen. A. and P.E. Grotnes. 1975. Food and habitat segregation by two sympatric Arctic char populations. Verh. Int. Ver. Limnol. 19: 2521-2528.

Kristjánsson, J. 1978. Growth rates of brown trout and Arctic char in Iceland. Isl. Landbún. (J. Agricult. Res. Iceland) 10: 125-134.

Lamby, K. 1941. Sur Fischereibiologie des Mývatn, Nord Island. Z. Fisch. 39: 749-805.

Nilsson, N.-A. and O. Filipsson 1971. Characteristics of two discrete populations of Arctic char (*Salvelinus alpinus* L.) in a North Swedish lake. Inst. Freshw. Res. Drottningholm Rep. 51: 90-108.

Nyman, L. 1972. A new approach to the taxonomy of the Arctic char "*Salvelinus alpinus* species complex". Inst. Freshw. Res. Drottningholm Rep. 52:103-131.

Ricker, W.E. 1975. Computation and interpretation of biological statistics of fish populations. Fish. Res. Board Canada. Bull. 191.

Saemundson, B. 1915. Fiskirannsóknir 1913 og 1914. Andvari 40.

The Arctic Charr Sport Fishery at Tree River, Northwest Territories, Canada, 1964-78

R.W. MOSHENKO, R.F. PEET, L.W. DAHLKE and D.H. DOWLER

Department of Fisheries and Oceans, Winnipeg, Manitoba, Canada

MOSHENKO, R.W., R.F. PEET, L.W. DAHLKE AND D.H. DOWLER. 1984. The Arctic charr sport fishery at Tree River, Northwest Territories, Canada, 1964-78, p. 349–364. *In* L. Johnson and B.L. Burns [eds.] Biology of the Arctic charr, Proceedings of the International Symposium on Arctic Charr, Winnipeg, Manitoba, May 1981. Univ. Manitoba Press, Winnipeg.

Today the Tree River provides the best sports fishing for Arctic charr in North America, but this is a relatively recent development. Prior to 1964 the only organized fisheries were based on Coppermine: the Inuit for domestic purposes and the Royal Canadian Mounted Police for dog food. Since that time various sport fishing lodges, mostly those based on Great Bear Lake, have utilized the river. In 1964, 745 anglers harvested 1029 charr; in addition, about 4700 fish were taken by the domestic fishery. A possession limit of two charr per angler and an overall quota of 500 fish were introduced in 1965. For the next ten years fishing rights were granted to one lodge established on Great Bear Lake. In 1976, the overall quota was increased to 700 charr and three other lodges were granted permits to bring in anglers for a specified number of days.

Studies indicate that since 1965 an average of 394 anglers per year each spent 10.0 hours fishing with a success rate of 0.4 charr/angler hour. The mean annual harvest over this period, by angling and domestic fishing was 445 and 210 charr respectively. Anglers retained about 32 percent of their catch. Angled fish held in cages for a period up to seven days prior to release experienced one percent mortality. Sampled charr had a mean age of 10.6 years, were 759 mm in length and weighed 4.5 kg. Age, growth, maturity and fecundity are briefly discussed.

During the 15 year period the fishery experienced a cycle of decline and recovery. The future management of this trophy sport fishery is discussed.

INTRODUCTION

The Arctic charr, *Salvelinus alpinus* (Linnaeus), in all parts of its circumpolar range has been of considerable economic importance, particularly to indigenous people (Johnson 1980). In many coastal streams of Canada's Arctic, Arctic charr is an important sport fish species. The Tree River, flowing into the Coronation Gulf (Fig. 1), is perhaps the best known trophy fishery for Arctic charr in North America.

Since the onset of an organized sport fishery in 1964, a creel census and biological sampling program has been carried out by the Federal Department of Fisheries. The major objectives of this study were to determine the harvest by sport and domestic fishermen, and to accumulate information on the life history of Arctic charr. Certain aspects of this fishery were reported by Robertson and Dowler (1973) Falk and Dahlke (1979).

This paper describes the history of the sport fishery from 1964 to 1978, and the basic life history information generated during this period. Recommendations are made for future management and research.

MATERIALS AND METHODS

Study Area

The Tree River is located on the Canadian Shield east of the town of Coppermine, Northwest Territories, at 67° 38′N and 111° 53′W (Fig. 1). Starting at Inulik Lake, the river flows 170 km north into the Coronation Gulf, draining an area of 5,772 km^2 (Douglas 1970). Due to the rather rugged terrain, numerous rapids are found along its course. All upstream fish movement is prevented at a point 11 km from the river mouth by falls 7m high. It is only in the last 8 km of its length in the coastal lowland that boat navigation is possible. Ice breakup occurs between the last week of June and the first week of July.

The Fishery

Traditionally, charr were used by the Inuit to provide food for themselves and their dogs. This apparently was done on a cyclic basis; fishing continued until a low catch-per-unit-of-effort necessitated a transfer to other rivers until the original one had recovered. The Hudson Bay Company maintained a trading post at Tree River between 1917 and 1927 and the Royal Canadian Mounted Police (RCMP) had a detachment there from 1919 to 1926; both agencies moved to another location, following the Inuit, who apparently left due to a scarcity of charr. Local residents from Coppermine indicate that a scarcity of charr again occurred about 1938 and 1950. During the late 1950s and early 1960s an estimated 22,000 kg (about 7,000 charr) were harvested by the Inuit domestic and RCMP fisheries.

The sport fishery at Tree River is unique due to the availability to anglers of large and red-coloured pre-spawning charr. Since anadromous Arctic charr in the Canadian Arctic frequently remain in freshwater in the year they spawn, the 11 km of stream between the ocean and the falls is inhabited by prespawners throughout the open-water months (July-August). In fact, these charr are heavily concentrated in several major pools below fast rapids. Since 1964 anglers from

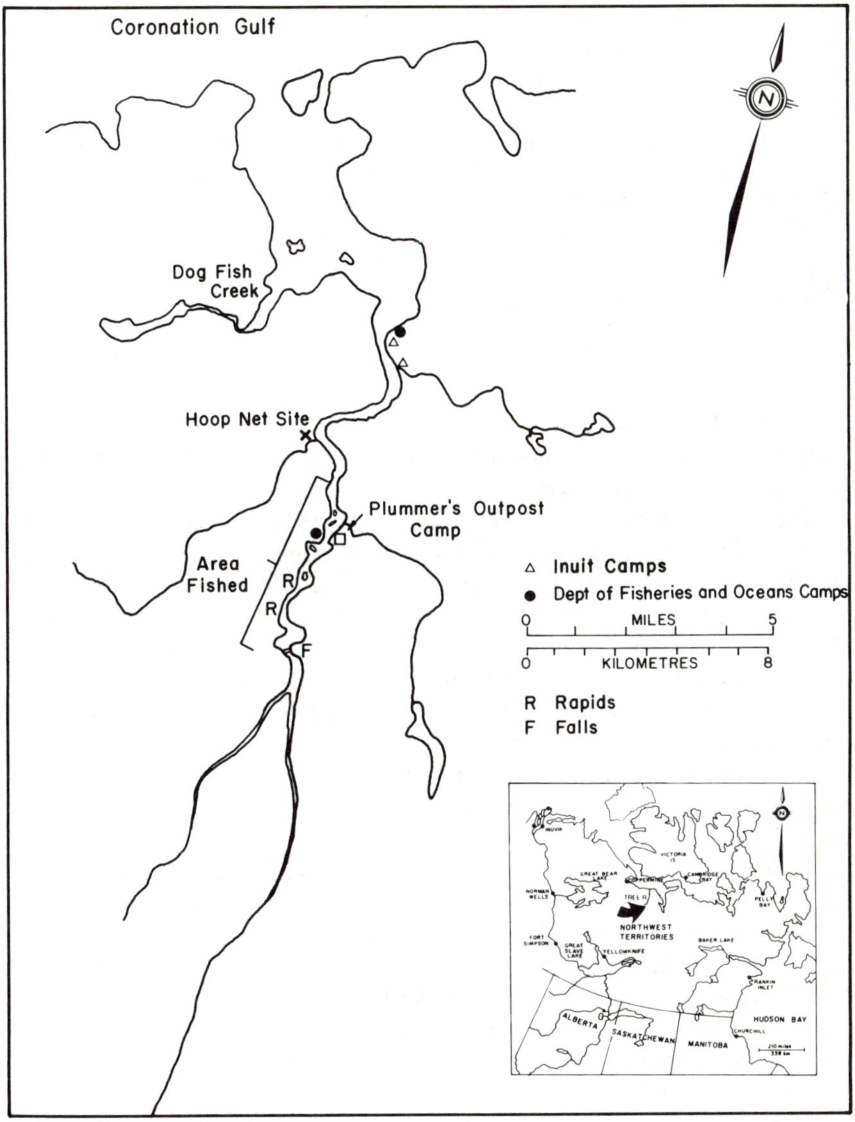

FIG. 1 Map of Tree River showing the location of fishing camps, areas fished, and sampling locations.

various sport fishing lodges, mostly based on Great Bear Lake, have utilized the river. A possession limit of two charr per angler and an overall quota of 500 fish per year were introduced in 1965 and enforced since 1966. For the following 10 years fishing rights were granted to one lodge established on Great Bear Lake. In 1976, the overall quota was increased to 700 charr and three other lodges were granted permits to bring anglers for a specified number of days. Anglers generally purchase a side trip from the lodge and stay one day and one night at Tree River. An Inuit guide accompanies every two or three fishermen. Angling is primarily spin-casting from shore. The angling season ranges from the first week in July to the end of August.

A small domestic fishery is carried out by the Inuit who are also employed as guides. Usually 5-8 families are present, harvesting 100-300 charr per season (Fig. 2). These are either dried or eaten fresh.

Methods

Since 1964, either fishery officers, guardians or locally hired Inuit collected creel census data. Anglers were interviewed for number of fish caught, kept and released, and numbers of hours spent fishing.

During the monitoring period from 1964 to 1978 all charr retained were measured for fork length, weighed and were, if possible, sexed. In 1964 only weights were obtained; in 1978 maturity data and otoliths were also obtained from a portion of the angled catch and additional specimens caught in a hoop net. Otoliths (for aging) were obtained from gillnetted Arctic charr, and some of the angled fish from other years. In the period 1975-78 small charr (less than 200 mm) were captured by hooks, minnow traps and seines, then processed for detailed examination in the laboratory. In 1976 skulls of trophy charr were obtained from a local taxidermist for their otoliths along with data on sex and fork length. All otoliths were stored dry and aged as described by Falk and Dahlke (1979).

In addition to opportunistic observation on mortality following hooking and release during 1965-69 an intensive hooking mortality program was undertaken in 1978 using the procedure decribed in Falk et al. (1974). Barbed and barbless lures and jigs of the type used by the anglers were used. Once a fish was hooked, landing time, hook removal time and the time taken to place the fish in the holding pen were recorded. Hook placement, damage and the number of embedded barbs were noted. An internal and external autopsy was performed on all fish that died in the holding pen.

Maturity was determined by examining the gonads and coded as described in Falk and Dahlke (1979).

Ovaries were obtained from 28 charr to determine egg diameter and fecundity.

RESULTS AND DISCUSSION

The Sport Fishery

The first intensive sport fishing effort occurred in 1964. Sport lodges from Great Bear Lake brought in over 745 anglers which often resulted in their being crowded in the prime 2 km angling area (up to 40 anglers/0.5 km).

Short-term variations in angling success can be attributed to weather and number of anglers. Since 1965 a general trend was established in which overall catches and catch/unit of effort (CPE) increased as number of anglers decreased. A high density of anglers at the few favourite pools resulted in some foul-hooking of charr (usually 5-10 percent but as high as 35 percent in 1972 during one 8 day period). Since that time snagging or foul-hooking has been minimized by enforcement officers temporarily closing some of these pools — to give the charr a ''rest''.

Since 1965 the mean number of fish harvested annually by angling and domestic fishing was 445 and 210 charr respectively; 394 anglers per year visited Tree River spending 10.0 hours fishing per angler and have a success rate of 0.4 charr/angler hr. These mean values are slightly higher if 1964 data is used (Fig. 2, Table 1). Other charr sport fisheries in the NWT had success rates as follows: Arctic Outpost Camp (Victoria Island) 0.3 charr/angler hr (Kristofferson and Dahlke In prep.), Hadley River (Victoria Island) 0.2-0.3 charr/angler hr (G. Carder, pers. comm.), and Robertson River (Baffin Island) 1.4 charr/angler hr (Moshenko 1981).

The high exploitation rates of charr prior to 1965 probably contributed to the gradual decline in mean size of angled fish up to 1970 (Fig. 3 and Table 2). However, management measures (limits, restricted number of anglers etc.) imposed since 1965 resulted in a general trend of increasing size from 1970 to 1976. From 1977 onwards mean size of angled charr declined slightly (Fig. 3). This may be a short-term variation, however, as there is some concern that angler effort and domestic harvest have also increased since 1976 (Fig. 2). The annual mean weight of angled charr varied from 4.5-5.9 kg over the 15 year period (Table 1). A world record charr (by angling) of 14.8 kg was caught in 1981 (A. Kristofferson, pers. comm.). Angled charr from Tree River had a mean fork length of 759 mm and a mean age of 10.6 yr. These fish are considerably faster growing and larger than charr in other sport fisheries in the NWT (Table 3). At Tree River anglers released about 68 percent of all charr caught while sport fisheries at Robertson River and Arctic Outpost Camp released 63 percent and 40 percent respectively (Table 3).

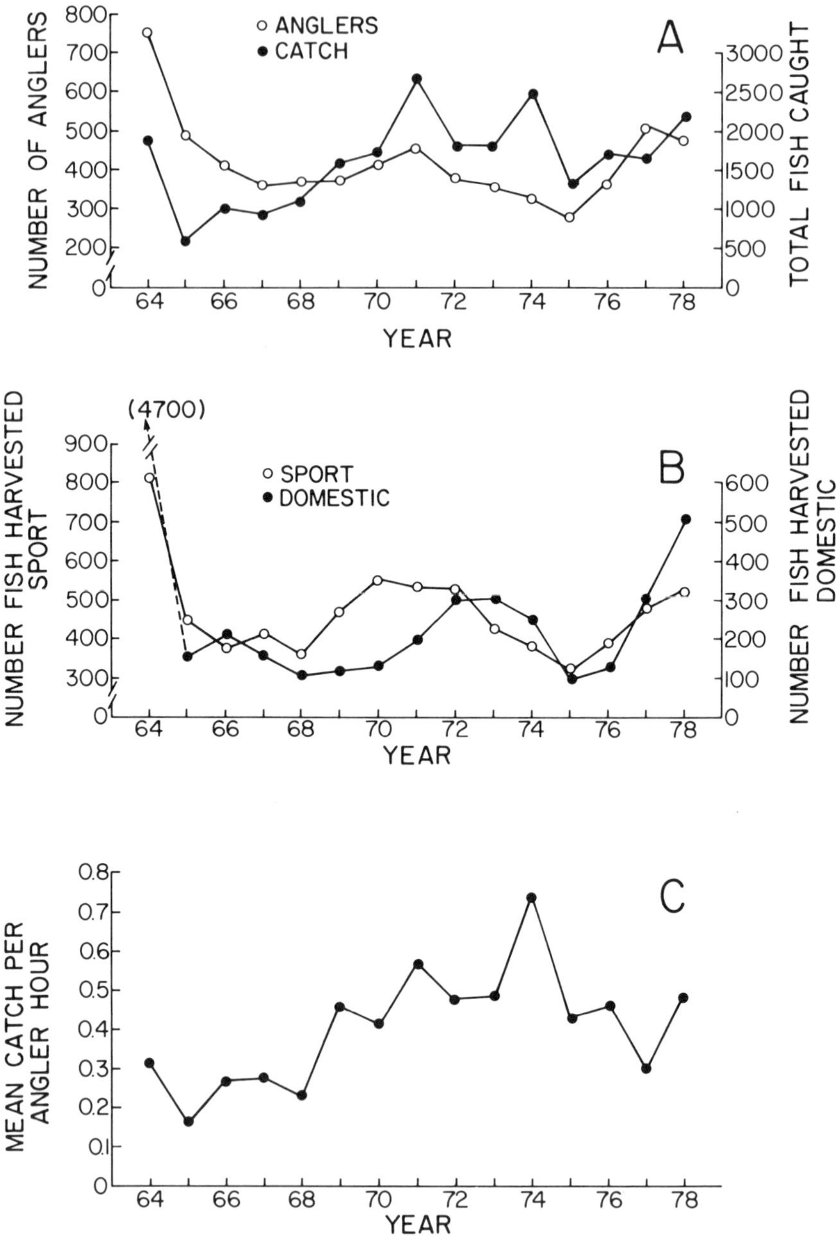

FIG. 2 Catch and effort data for the sport and domestic fisheries on the Tree River, 1964-78.

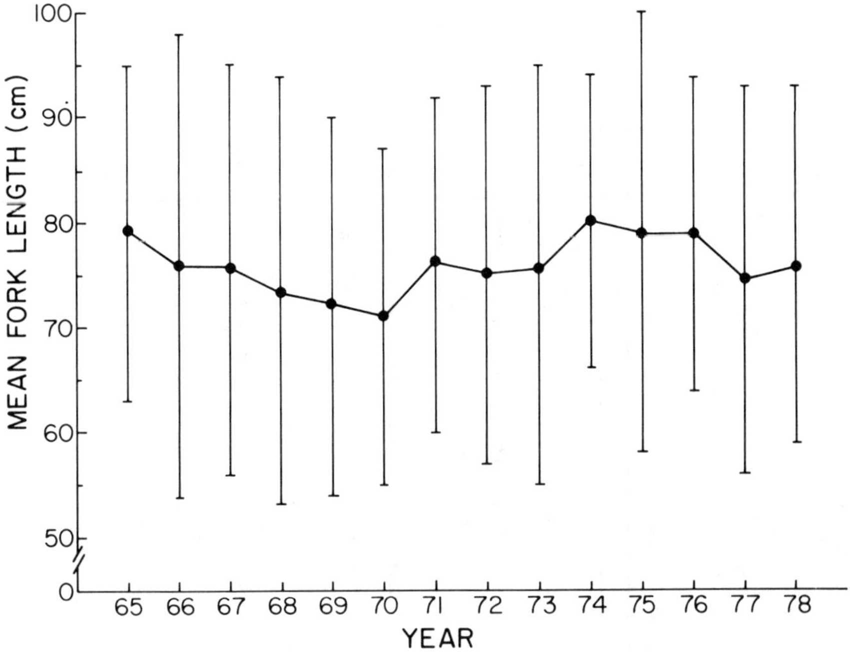

FIG. 3 Annual mean fork length for Arctic charr from the Tree River sport fishery, 1965-78.

Biological Information

Growth in length of the charr of Tree River is rapid from age 5 to about age 10 or 11 (Fig. 4 and Fig. 7). Fish of ages 1-4 were abundant and readily caught in freshwater; it is assumed the charr four years of age or older make their first migration to sea (similar to those of the Nauyuk Lake population (Johnson 1980)) and that this is the reason for the rapid increase in growth at this time. Figure 4 indicates a slightly faster growth rate for males compared with females beyond age 7. Growth of charr in Tree River is much more rapid than in many charr populations in the central and eastern Arctic (Fig. 5), possibly due to the early age of first migration to the sea. The weight-length relationship is shown in Figure 6; however, it should be cautioned that most of the data are taken from pre-spawners (that is, those fish available to anglers) and as Johnson (1980) points out it is important to state this when developing equations relating length and weight (see also Dutil, this volume).

Sexual maturity (50 percent) is reached at age 8-9 yr at a size of about 500 mm; all fish are mature by about age 11 and 700 mm in fork length (Fig. 7). Maturity is reached at a younger age and size than that reported for Nauyuk Lake

Table 1. Summary of creel census data from the sport fishery at Tree River, 1964-78.

	1964	1965	1966	1967	1968	1969	1970	1971	1972	1973	1974	1975	1976	1977	1978	Mean
No. of anglers	745	448	402	361	371	376	408	451	386	358	326	277	369	505	472	417
No. of charr caught	1853	595	1011	927	1083	1595	1685	2501	1831	1796	2492	1295	1709	1762	2186	1628
Charr retained, no:	1029	458	355	434	364	459	561	539	524	427	392	317	397	484	524	486
%:	56	77	35	47	34	29	33	21	29	24	16	24	23	27	24	32
Mean hours/ angling trip	8.3	8.1	9.3	9.7	12.4	9.2	10.0	10.1	9.6	10.4	10.4	11.0	10.2	10.8	9.5	10.0
Mean no. fish caught/trip	2.5	1.3	2.5	2.6	2.9	4.2	4.1	5.8	4.7	5.0	7.6	4.7	4.6	3.5	4.6	3.9
Mean no. fish caught/angler-hr	0.3	0.2	0.3	0.3	0.2	0.5	0.4	0.6	0.5	0.5	0.7	0.4	0.5	0.3	0.5	0.4
Mean weight of fish (kg)	5.0	5.4	4.9	5.0	4.2	4.5	4.2	5.2	4.8	4.6	5.9	5.5	5.4	4.8	4.9	4.5

Table 2. Fork length distribution by number (sexes combined) for Arctic charr from the sport fishery on Tree River, 1965-78. Data for 1978 from Falk and Dahlke (1979).

Length Interval (cm)	1965	1966	1967	1968	1969	1970	1971	1972	1973	1974	1975	1976	1977	1978
25.0 - 29.9	-	-	-	-	-	-	-	-	-	-	1	-	-	-
30.0 - 34.9	-	-	-	-	-	-	-	-	3	-	-	-	-	-
35.0 - 39.9	-	1	-	-	-	-	-	-	2	-	4	-	-	-
40.0 - 44.9	-	3	-	-	-	-	-	-	6	-	1	-	-	1
45.0 - 49.9	-	3	-	-	-	2	-	-	4	-	1	-	-	7
50.0 - 54.9	4	11	2	5	2	2	1	7	3	-	1	-	-	4
55.0 - 59.9	4	18	9	11	23	9	8	5	5	-	3	5	11	11
60.0 - 64.9	10	27	29	68	74	89	23	45	29	9	18	14	44	36
65.0 - 69.9	20	29	69	62	125	174	91	102	60	29	40	32	84	69
70.0 - 74.9	76	46	59	58	119	128	112	115	84	46	48	67	99	98
75.0 - 79.9	101	54	45	56	66	58	86	90	85	92	47	63	60	160
80.0 - 84.9	101	64	50	42	63	64	108	81	88	102	72	88	57	113
85.0 - 89.9	74	46	52	19	42	25	67	59	61	86	88	60	53	52
90.0 - 94.9	37	24	38	21	16	10	29	24	18	13	32	19	14	20
95.0 - 99.9	8	8	6	11	5	2	2	1	3	5	5	2	3	2
100.0 - 104.9	-	-	1	2	1	-	-	-	-	-	-	-	-	-
105.0 - 109.9	-	-	1	-	-	-	-	-	-	-	-	-	-	-
110.0 - 114.9	-	-	-	-	-	-	-	-	-	-	-	-	-	-
115.0 - 119.9	-	-	-	-	1	-	-	-	-	-	-	-	-	-
Total number of fish	436	334	361	355	537	563	527	529	451	382	361	350	425	573
Mean length (cm)	79.7	75.9	76.5	73.4	72.7	71.2	76.3	74.9	75.4	79.9	78.6	78.5	74.5	75.6
SD	8.15	11.50	10.37	10.56	9.12	7.85	8.33	8.82	10.56	6.96	10.58	7.77	8.83	8.61

Table 3. Comparison of various types of sport fisheries for Arctic charr in the Northwest Territories.

	Trophy (catch limit: 2 daily, 2 maximum)	High/Medium Quality (catch limit: 4 daily, 7 maximum)	
	Tree River (Coronation Gulf)	Robertson River[b] (Baffin Island)	Arctic Outpost Camp[c] (Victoria Island)
No. fish caught/ angler-hour	0.4 (0.1 in 1978)	1.4	0.3
Percentage fish released	68	63	40
Type of hooks used	barbed/barbless	barbless	barbed
Mean no. of fish harvested/angler	1.1	5.3	> 4.5
Type of water	Freshwater-river	mostly estuary	mostly freshwater-lake
Reproductive condition of charr (for present yr)	almost entirely pre-spawners	non-spawners	> 75% pre-spawners
Charr caught:			
Mean weight (kg)	4.5[a]	3.0	2.9
Mean length (cm)	75.9	65.9	64.1
Mean age (yr)	10.6	17.9	15.2

[a]World angling record of 13.3 kg (> 29 lb) in 1970.

[b]Preliminary study (Moshenko 1981).

[c]Kristofferson and Dahlke (In prep.).

(Kent Peninsula) and other charr stocks in the central and eastern Arctic (Johnson 1980).

Fecundity studies showed a mean number of 4,647 (range 2457-7849) ova per female and a mean number of 146 (range 82-247, SD = 34) ova per 100 g body weight. Fecundity-length relationship is shown in Figure 8. When compared to data in Johnson (1980), charr fecundity from Tree River is similar to Nauyuk Lake but higher than that for most charr stocks in the central and eastern Arctic. Ova diameters were as follows: 3.4 mm (1-15 July), 3.5 mm (16-30 July), 4.1 mm (1-15 August) and 4.4 mm (15-31 August). Since, ova are ripe at 5.1-5.5 mm in Nauyuk Lake during September (Johnson 1980), charr in Tree River may not spawn until late September or well into October.

Management Recommendation

The sport fishery at Tree River, by nature of the availability of large fish, low harvest levels, low catch limits and high release (Table 3) is designated as a high quality (trophy) fishery. Other criteria used for this designation include: no

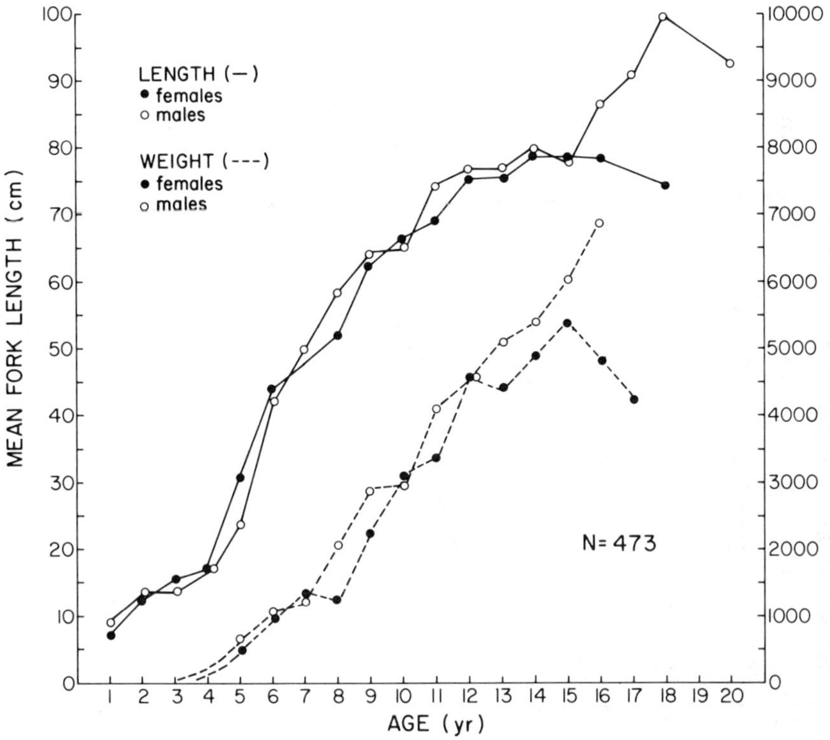

FIG. 4 Mean fork length and weight by age for Arctic charr from the Tree River.

commercial fishing and restricted road or boat access. Since this fishery exploits the spawning segment of the population, it is important that good conservation and management practices be employed to maintain the fishery in the healthy state at which it now appears to be. The following recommendations should be implemented to ensure the above status:

1) Sport fishing quota to remain at 700 charr with catch and possession limit to remain at 2 fish. A quota should be set for the domestic harvest — based on consultation with the local people based on "reasonable needs."
2) Travel permits providing for a maximum of 70 anglers per week should remain in force. This results in a 5 day angling-week with approximately 14 anglers/day. The spacing of anglers over the season plus the 2 days each week of "no angling" minimizes the angling stress on the Arctic charr, minimizes crowding, anxiety and frustration of anglers and produces greater angling success.

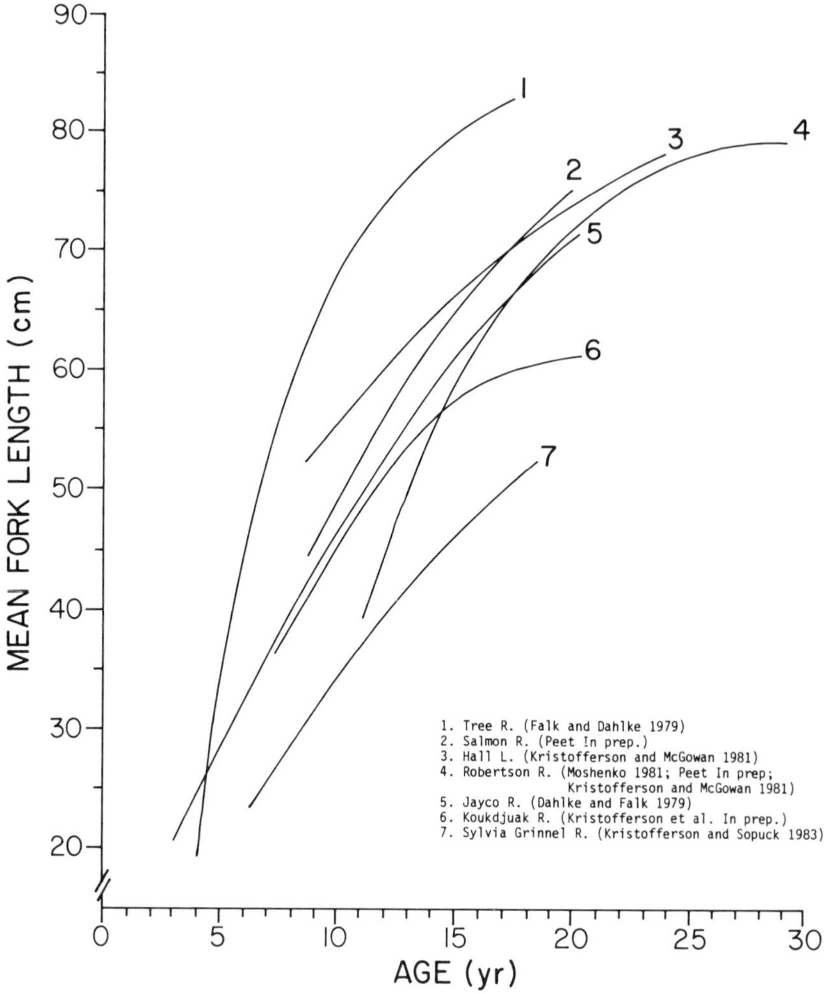

FIG. 5 Comparison of mean fork length by age for Arctic charr from various locations in the Northwest Territories.

3) The use of barbless hooks and the proper handling and release of fish should be emphasized continually. This is a high release fishery (68 percent) and mortality studies have shown that there is about a one percent mortality in lure-caught fish being held in captivity for 7 days (Falk and Dahlke 1979). The use of barbless hooks and the quick and proper release of fish minimize mortality.

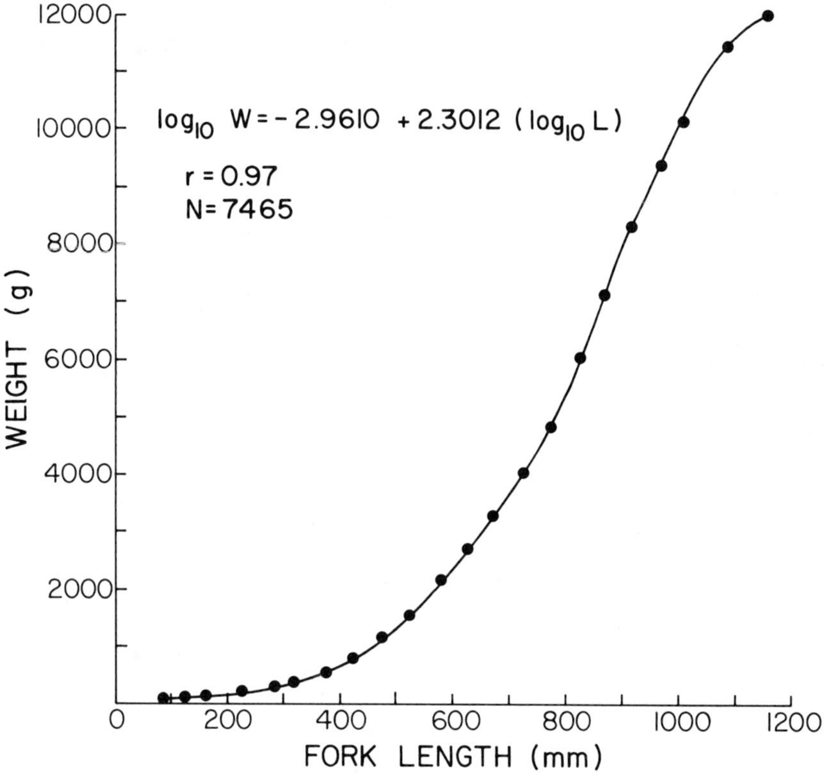

FIG. 6 Weight-length relationship for Arctic charr from Tree River.

4) There is a need to determine the standing crop and the optimum yield (for trophy fishery). Prior to 1965 harvests of up to 22,000 kg resulted in a crash of the population. The present harvest of 3,000 kg (15 percent of 22,000) annually appears to be quite safe — but an optimum should be calculated.
5) The sport and domestic fisheries should be monitored at least once every three years to determine catch, catch/effort, harvest and biological status of the population. Note: this fishery was monitored in 1981 and 1983.

ACKNOWLEDGMENTS

The authors wish to thank M. Angel, J. Delancey, J. Fleming, D. Green, J. Hunter, R. shepard and especially F. Taptuna of Coppermine for their assistance in the studies. G. Carder aged the Arctic charr by the otolith method. The studies

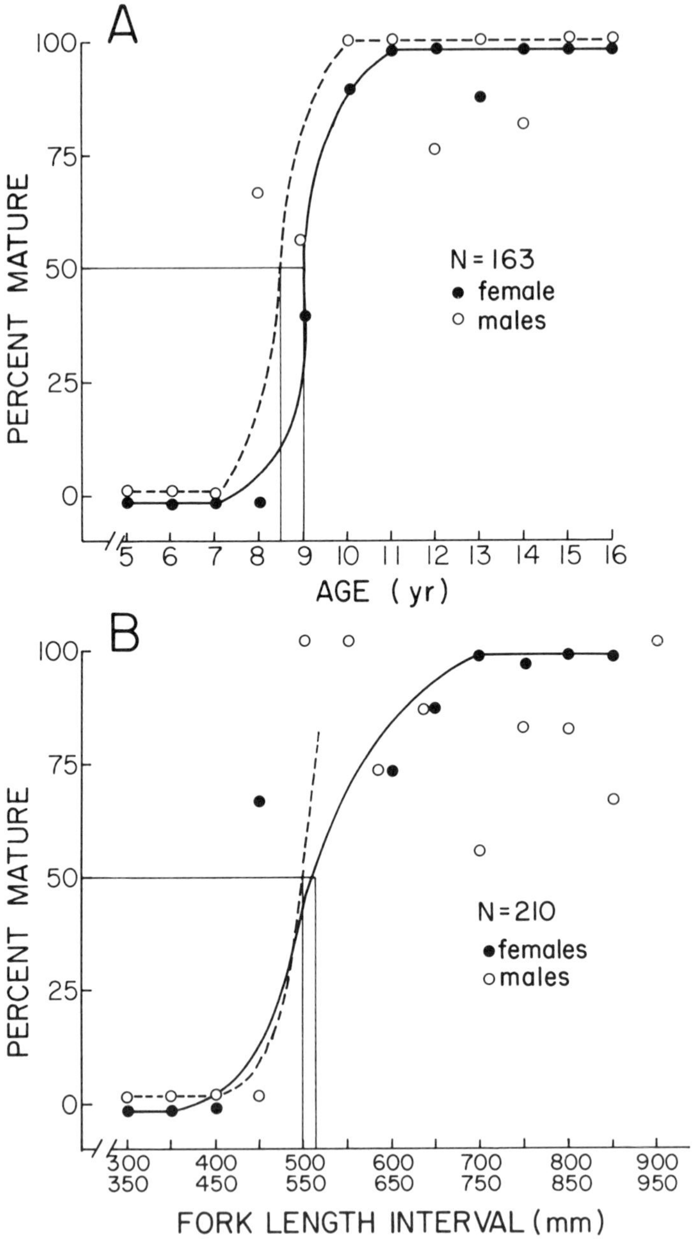

FIG. 7 Age (A) and length (B) at maturity for Arctic charr from the Tree River.

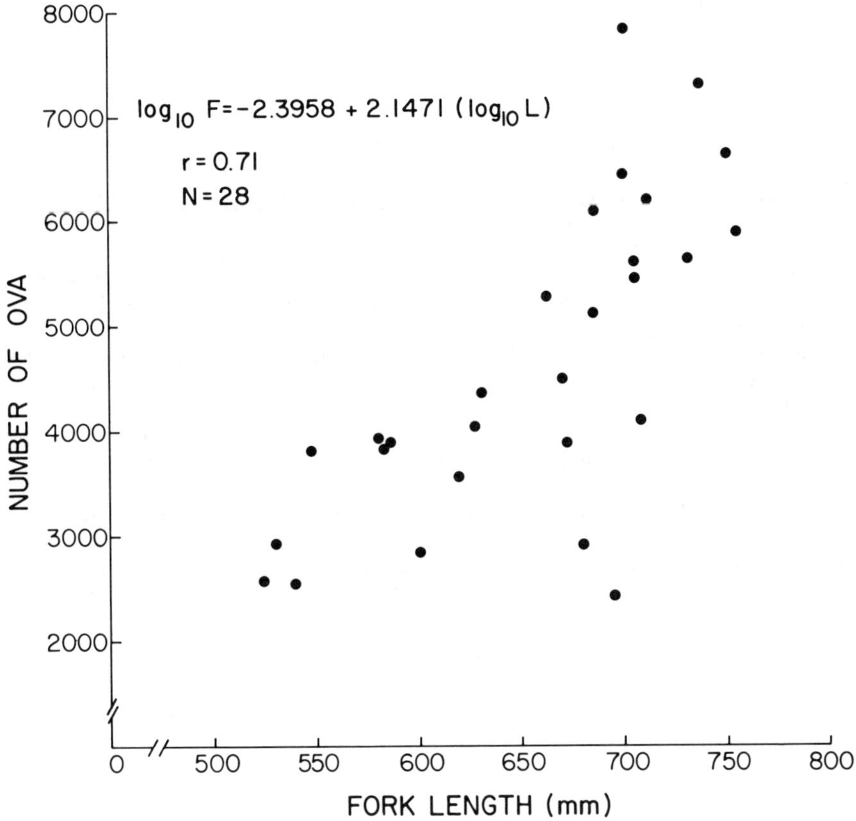

FIG. 8 Fecundity-length relationship for Arctic charr from Tree River.

would not have been successful without the assistance and cooperation of W. Plummer and C. Plummer of Great Bear Lake Lodge and the Inuit from Coppermine employed by the lodge. Data on some trophy fish were made available through Hawkins Taxidermists, Winnipeg.

Figures were prepared by Graphic Services (DFO). Thanks to D. Holland and M. Smith who typed the manuscript.

REFERENCES

Dahlke, L.W., and M.R. Falk. 1979. Data on Arctic char from the Jayco River, Northwest Territories, 1975. Can. Fish. Mar. Serv. Data Rep. 156: iv + 11 p.

Douglas, R.J.W. [ed.] 1970. Geology and economic minerals of Canada. 5th ed. (Vol. 1 — text). Geological Survey of Canada, Ottawa, 838 p.

Falk, M.R., and L.W. Dahlke. 1970. Data on the Arctic char sport fishery at Tree River, Northwest Territories, 1978. Can. Fish. Mar. Serv. Data Rep. 153: iv + 32 p.

Falk, M.R., D.V. Gillman and L.W. Dahlke. 1974. Comparison of mortality between barbed and barbless hooked lake trout. Can. Fish. Mar. Serv. Tech. Rep. Ser. CEN/T-74-1: 28 p.

Johnson, L. 1980. The Arctic charr, *Salvelinus alpinus*, p. 15-98. *In* E.K. Balon [ed.] Charrs: salmonid fishes of the genus *Salvelinus*. Dr. W. Junk Publishers, The Hague.

Kristofferson, A.H., and L.W. Dahlke, In prep. The sport fishery at Arctic Outpost Camp, Victoria Island, Northwest Territories, 1975. Can. MS. Rep. Fish. Aquat. Sci. (Title-provisional).

Kristofferson, A.H., and D.K. McGowan. 1981. Data from test fisheries on Baffin Island, Northwest Territories, 1975-79. Can. Data Rep. Fish. Aquat. Sci. 255. iv + 43 p.

Kristofferson, A.H., and R.D. Sopuck. 1983. The effects of exploitation on the Arctic charr population of the Sylvia Grinnell River, Northwest Territories. Can. Manuscr. Rep. Fish. Aquat. Sci. 1721: v + 34 p.

Kristofferson, A.H., R. Sopuck, and D.K. McGowan. In prep. The Arctic charr fishery on the Koukdjuak River, Baffin Island, 1975-77. Can. MS. Rep. Fish. Aquat. Sci. (Title and Authors, provisional).

Moshenko, R.W. 1981. A preliminary assessment of the Arctic charr sport fishery on the Robertson River (Koluktoo Bay), Northwest Territories, 1979. Can. Data Rep. Fish. Aquat. Sci. 306: iv + 9 p.

Peet, R.F. In prep. Data on the bilogy of Arctic charr from the eastern Arctic, Northwest Territories. Can. Data Rep. Fish. Aquat. Sci. (Title provisional).

Robertson, M.R., and D.H. Dowler. 1973. The history of exploitation of Arctic char of Tree River, N.W.T. Paper presented at the Con. Cttee for Fish. Res. Halifax, N.S. Jan 1973 (Mimeographed).

Commercial Processing of Charr in Canada's Eastern Arctic

D.G. IREDALE

Department of Fisheries and Oceans, Winnipeg, Manitoba, Canada

IREDALE, D.G. 1984. Commercial processing of charr in Canada's eastern Arctic, p. 365–369. *In* L. Johnson and B.L. Burns [eds.] Biology of the Arctic charr, Proceedings of the International Symposium on Arctic Charr, Winnipeg, Manitoba, May 1981. Univ. Manitoba Press, Winnipeg.

To generate local employment and enhance the value of the charr resource to the native fisherman, a fish canning operation was initiated in 1963 on the west coast of Hudson Bay and eventually established at Rankin Inlet, N.W.T. in 1966. Economic pressures as well as operational problems associated with the remoteness of the cannery location caused its termination in 1976. During that period charr and other species indigenous to the region were canned in various forms. This operation and the resulting products are discussed as well as proposed future initiatives for charr processing.

HISTORY

Since the mid-1950's Arctic charr (*Salvelinus alpinus* L.) has become well recognized as a food delicacy in North America. This reputation may well be founded not only on the intrinsic gustatory qualities of the species but perhaps also on its relative scarcity and high cost in the larger markets south of production sources. The historical economic importance of charr to the Inuit has declined to some degree due to the availability of alternative protein sources, although it still remains an important food source.

Charr for export has been produced commercially in small quantities in Greenland since 1930 (Dunbar and Hildebrand 1952). In 1932 permission was granted by the Minister of Fisheries to a Mr. Ingebrigtsen of Churchill, Manitoba, to fish charr on a small commercial scale along the west coast of Hudson Bay. He used a sailing vessel which was not equipped with auxilliary motors, being unable to finance other equipment for a more extensive fishery, so that his range was limited to approximately 250 km north of Churchill. However, 1000 kg of lightly salted "sea trout" was marketed locally on his return to Churchill (Sprules 1952). Commercial exploitation began on the Labrador coast in the early

1940's. Further records indicate that a small commercial fishery was undertaken at Frobisher Bay at the Shaw Steamship Co. during the summer of 1948. Some of the production from this fishery was salted and apparently, some was canned for export shipment (Grainger 1948). This appears to be the first documented instance of the commercial canning of charr in Canada's Arctic, although a sample of charr was canned experimentally in 1945 by a Mr. Martin, a trader in the Churchill district, and submitted to the Pacific Experimental Station of the Fisheries Research Board of Canada for evaluation (Sprules 1952).

Interest in the potential for canning Arctic marine mammals and fish was considered in the early 1960's by the Department of Indian Affairs and Northern Development's Industrial Division, with the experimental operation of a so-called portable pilot cannery (Hofmann, pers. comm.). Such equipment could be flown into appropriate fishing locations and set up under tents with other tents provided to house the processing labour. With an investment of approximately $4,000.00 for the canning equipment, the facility was capable of producing 250 $kg{\cdot}day^{-1}$ of canned charr.

In 1964 a semi-permanent facility was established at Winchester Inlet, approximately 700 km north of Churchill, Manitoba, to can charr caught in nearby Daly Bay (Lantz and Iredale 1972). The pre-fabricated cannery building was constructed of galvanized sections on a tubular steel frame with opaque fibreglass panels to admit light. As in the case of the portable pilot cannery, tents were again provided to house processing labour. The operation relied on propane fired retorts to process the canned charr and gasoline generators to power other equipment. This operation was only marginally successful due to the less than anticipated supply of charr from Daly Bay, limitations in water supply and power and lack of housing facilities.

RECENT DEVELOPMENTS

Thus in 1966 the operation was dismantled and re-established permanently at Rankin Inlet, approximately 200 km to the south; it was registered as the Issatik Food Plant. The advantages here included good transportation and communication, with housing, education and health services being available in an established settlement. Greater access to fish resources also meant that a year round processing operation could be considered by freezing seasonally caught fish for off season processing. Another consideration was that the economy of this settlement had formerly been based largely on a mining operation that had since ceased to function; the establishment of a cannery, it was hoped, would provide needed jobs and help to sustain a population of approximately 350 Inuit.

The Winchester Inlet cannery building was incorporated into an enlarged structure which included a 7.5 m x 7.5 m blast freezer and cold storage room.

The size of the cannery building was increased to permit the installation of two commercial can closing machines for round and rectangular cans, and a commercial-size retort. A room adjoining the cannery building housed an oil fired steam boiler, a smoking kiln, labelling equipment and provided temporary storage space for canned products.

Up until this time (1966) only a steak style charr product had been processed, packed in 220 g (7¾ oz) round cans under the ''Ookpik'' brand label. Access to a smoking kiln and other processing equipment allowed some product diversification. This included smoked slices of charr in oil, packed in a 106 g (3¾ oz) flat rectangular aluminum can as well as similar packs using lake trout (*Salvelinus namaycush*) caught from inland lakes.

In addition to these canned products the cannery became a multi-use processing plant during this period, with seal and whale meat also being can processed. Following the seasonal processing of charr and lake trout, the plant was ''converted'' to a meat-packing facility following a strictly detailed wash-down and sanitizing procedure under the supervision of the Department of Agriculture, Health of Animals Branch, in order to process seal and whale meat. A variety of products was packed, including muktuk (the skin and subcutaneous layer of fat from whale (beluga)), whale meat and seal meat in various sauces. A special pack was processed to fill the needs of the local market with the excess being canned under a different ''Ookpik'' brand label for export to southern Canadian and U.S. markets. However, in 1967, analyses of the whale and seal meat being processed revealed the presence of abnormally high levels of mercury, hence mammal processing was discontinued.

Further product diversification of charr was initiated in 1967 and a new ''Arctic Circle'' brand label introduced. Also, lake whitefish (*Coregonus clupeaformis*) caught from inland lakes was added to the species processed by the cannery. Efforts were made to maximize the utilization of charr which included a grading scheme according to intensity of flesh colour. Only charr with the most intense orange pink colour was used for the steak style and smoked sliced canned products. The lighter coloured fillets and all fillet trimmings were frozen and reserved for subsequent processing as a canned charr chowder packed in a 400 g (14 oz) tall round can. This product was eventually discontinued because of the high cost associated with export shipping. Even though the cannery had negotiated a favourable ''back-haul'' rate for shipping products by air to southern markets, the cost, particularly in the case of the canned chowder, which contained approximately 80 percent water, was still prohibitively high. Consistent with maximizing utilization, the lighter coloured charr fillets and fillet trimmings were processed alternatively as a canned charr cocktail spread packed in a 106 g (3¾ oz) round can, which contained greater than 68 percent charr. A fourth charr product was also introduced: this was a charr fillet portion in aspic jelly, packed in the 106 g (3¾ oz) flat rectangular aluminum can.

The incorporation of freezing and cold storage facilities in the Issatik Food Plant cannery meant that continuity in processing could be planned. Regular staffing requirements could be pre-determined as they could now draw from a frozen charr inventory and refreeze secondary use material for the added cocktail spread product line as required, rather than be dependent on the day to day variability of charr production. At peak production the cannery employed twenty-two Inuit processors plus a non-Inuit manager experienced in canned food processing. Can processing lasted approximately four to five months each year, commencing at the beginning of August and continuing through until December. However, the cannery was plagued continually with operational and economic problems. What had not been recognized was the inherent difficulty of conducting a food processing operation of this nature in an Arctic location, where limited access to equipment, equipment spare parts, material suppliers, technical expertise and services prevails, a condition that does not exist in a southern industrial location. An additional complicating factor was that although there had been plans, once the canning concept was proven, to establish the operation as an independent Inuit co-op, the cannery remained the administrative responsibility of the Economic Development Branch of the Territorial Government, headquartered in Yellowknife. Yellowknife is over 1400 km to the west, although the management division responsible for Keewatin activities was located in Churchill, 500 km south. These problems were compounded by less than anticipated volumes of all species, the high production cost of whitefish and lake trout from inland lakes and cannery waste disposal problems resulting in a prohibitively higher per unit can production cost. These factors contributed to the decision to terminate the canning operation in 1976.

The Issatik Food Plant continues today as a co-op producing frozen dressed charred. Charr in excess of local food requirements is shipped to the Freshwater Fish Marketing Corporation in Winnipeg for marketing.

A consideration at this time is to increase the value of charr to the co-op plant, by processing cold smoked charr fillets. A process has been developed that results in a product comparable to cold smoked salmon fillets. The product was test marketed during the past year through brokers in some major Canadian cities and received an encouraging response. If production costs can be rationalized, consumers may again see a limited amount of a value added gourmet charr product in the marketplace.

REFERENCES

Dunbar, M.J. and H.H. Hildebrand 1952. Contribution to the study of the fishes of Ungava Bay. J. Fish. Res. Board Can. 9(2): 83-128.

Grainger, E.H. 1948. Report on biological observations on Arctic char (*Salvelinus alpinus*). Unpublished MS. cited by Sprules, W.M., 1952. The Arctic char of the west coast of Hudson Bay. J. Fish. Res. Board Can. 9(1): 1-15.

Hofman, E. 1966. Personal communication.

Lantz, A.W. and D.G. Iredale, 1972. Canadian Northwest Territories fish and mammal processing plant. Can. Inst. Food Science and Technol. J. Vol. 5, no. 1.

Sprules, W.M. 1952. The Arctic char of the west coast of Hudson Bay. J. Fish. Res. Board Can. 9(1): 1-15.

Parasites And Arctic Charr Management — An Academic Curiosity Or Practical Reality?

TERRY A. DICK

Department of Zoology, University of Manitoba, Winnipeg R3T 2N2

DICK, T.A. 1984. Parasites and Arctic charr management — an academic curiosity or practical reality? p. 371–394. *In* L. Johnson and B.L. Burns [eds.] Biology of the Arctic charr, Proceedings of the International Symposium on Arctic Charr, Winnipeg, Manitoba, May 1981. Univ. Manitoba Press, Winnipeg.

One hundred and seven species of parasite have been reported from charr throughout its range. Sixty-six species have been identified in charr from North America, 18 from Europe (Norway) and 69 from the U.S.S.R. Differences exist between the three regions of the world, most notable is the absence of acanthocephalans and protozoa from Norway with only one species of protozoan reported from North America. Most parasites found in Arctic charr utilize copepods and amphipods as intermediate hosts. Differences in the distribution of parasite species are pronounced in North America. *Crepidostomum farionis* has not been found in Arctic charr north of 70° north latitude and *Diphyllobothrium* prevalence and intensity decrease westward from Baffin Island; it is absent from Alaska and Labrador. *Cystidicola* is found in Arctic charr in the central and western part of North America only. Its absence in Baffin Island and Labrador is probably related to absence of lake trout and/or the intermediate host, *Mysis*. Parasite species of Arctic charr fall into two categories. One group is holarctic and includes *C. farionis*, *Phyllodistomum* sp., *D. dendriticum*, *D. ditremum*, *Cystidicola* sp., *Philonema* sp. and *S. edwardsi*. The second group contains those parasites that are at the most northern part of their hosts' distribution ranges, generally in water bodies close to the Arctic Ocean or its connecting seas. These include *Bonodera*, *Dipiostomum*, *Pelichnibothrium*, *Capillaria*, *Contracaecum*, *Cucullanus*, *Rhabdochona*, *Acanthobdella*, *Metechinorhynchus* and *Neoechinorhynchus*. Trematodes in the Arctic are widely distributed in marine environments and are probably not holarctic in distribution; the one exception is *B. crenatus* which may utilize charr as one of its intermediate hosts. Charr are probably accidental hosts for the other species. Charr from neighbouring lakes in the Arctic may have quite different parasitofaunas. These differences appear to be related to the absence of the intermediate host or perhaps some lakes were initially populated by charr lacking certain parasites. The three likely candidates which could affect growth and mortality in Arctic charr are *D. ditremum*, *D. dendriticum* and *Metechinorhynchus*. Parasites as biological tags are useful in separating sea-run and non-migrating charr under certain situations, such as occur in the Nettilling Lake-Koukdjuak River. Good indicator species for this system were: *Diphyllobothrium* spp., *E. salvelini* and *P. longicollis* for non-migrating charr and *B. crenatus*, *P. squamatus*, and *B. sturionis* for sea-run charr. Two methods were used to separate the charr: presence or absence of six indicator species and a qualitative index for *Diphyllobothrium* cyst levels. Approximately 800 charr from the commercial catch were evaluated and found to contain approximately 20 percent non-migrating charr and 80 percent sea-run. The use of parasites as biological tags in other parts of the Arctic is difficult to predict because distribution of parasites is patchy. Each system must be evaluated separately.

Parasites affect men directly through infection with *Diphyllobothrium*, presumably *D. dendriticum*. Indirectly, parasites such as *D. ditremum* and *D. dendriticum* are known to affect charr growth and mortality. *Metechinorhynchus* and *Cystidicola* are also likely candidates but more work needs to be done. It is assumed that the relationship between charr and parasites is relatively stable for a given system but changes such as stocking, the connecting of several distinct aquatic systems or the introduction of intermediate hosts could drastically alter this stability by allowing the introduction of a ''new'' parasite(s).

Exploitation of charr tends to concentrate on sea-run charr for a number of reasons. This genetic selection process may be detrimental in the long run to a given population of charr.

INTRODUCTION

Publications by Margolis and Arthur (1979) and Bykhovskaya-Pavlovskaya et al. (1964) have produced valuable catalogues of parasites of fishes. However, studies by Beverley-Burton (1978), Butorina (1975), Curtis (1979), Dick and Belosevic (1981), Eddy and Lankester (1978), Jamieson and Freeman (1975), Mudry and McCart (1976), and a series of studies by Kennedy (1977, 1978 a, b) have not only added to our knowledge of parasite species in Arctic charr *Salvelinus alpinus* but also contributed substantially to our understanding of the biology of these parasites. Knowledge on the zoogeographical characteristics of charr parasites is limited to the work of Kennedy (1977). Arctic charr parasites as biological tags have recently received attention from Dick and Belosevic (1981) and Eddy and Lankester (1978) but their application needs to be tested further. There are still considerable lacunae in our knowledge of charr parasites. For example the importance of parasites on the reproductive capacity of charr is now being investigated and will be discussed in this symposium; further, the relationship between human parasitism and species of parasite involved is not well understood (Jamieson and Freeman 1975) and only recently have efforts been initiated to evaluate the population dynamics of charr and their parasites (Henricson 1978).

This review brings up to date our current knowledge of parasites from all charr species. In addition zoogeographical characteristics of parasites recovered from *Salvelinus alpinus* in North America, Europe (Norway) and the U.S.S.R. will be compared and the situation in North America discussed. Differences in the parasites of charr from neighbouring lakes will be evaluated. The harmful effect of parasites on charr and the usefulness of parasites as biological tags will be developed. Finally interactions between charr, parasites and man will be discussed.

CHARR PARASITES AND ZOOGEOGRAPHY

A total of 107 species of parasite has been reported from charr throughout its range. Sixty-six species have been identified in charr from North America, 18 from Europe (Norway) and 69 from the U.S.S.R. Appendices I, II, III list the type, genus and species reported from charr. The most striking difference when comparisons are made (Appendices I, II, III and Table 1) is the absence, or very low numbers of, protozoan species from Norway and North America. While it is probable that most studies on parasites of Arctic charr from North America did not search for protozoa this is not the case in Norway where Kennedy (1977) looked for, but did not find, any species of protozoa. The presence of acanthocephalans in Arctic charr from North America and the U.S.S.R. is in contrast with their absence from Arctic charr in Norway. This is unusual considering Kennedy (1977) reported the presence of amphipod intermediate hosts and recovered acanthocephalans from *Esox lucius* in some lakes. It is conceivable that the charr which populated these lakes as the glaciers receded did not have acanthocephalans. Clearly further studies are needed to clarify this question.

A common denominator when one compares parasites of *S. alpinus* from North America, Norway and the U.S.S.R. is the presence of parasites which utilize copepods and amphipods as intermediate hosts. Parasites which have copepods as the intermediate host include *Diphyllobothrium, Eubothrium, Proteocephalus* and *Philonema*. The only parasite common to all three regions which utilizes amphipods as the intermediate host is *Cystidicola*. It is interesting that *Cyathocephalus* also requiring an amphipod intermediate host is absent from North America (Table 4) but common in both Norway and the U.S.S.R. *Crepidostomum* is the only parasite utilizing a mulluscan host (sphaerid clams) throughout the circumpolar range of *S. alpinus*. Parasites with direct life cycles such as *Tetraonchus* and *Salmincola* do not seem to have any advantage under Arctic conditions, either by numbers or overall distribution.

Information on parasites from anadromous charr is scanty but there appear to be differences between North America, Europe and the U.S.S.R. (Table 3). The dominance of the parasitofauna by digenetic trematodes (*Brachyphallus, Derogenes, Lecithaster,* and *Prosorhynchus*) indicates the importance of marine molluscs and amphipods in the life cycle of these parasites and in the diet of sea-run charr. These parasites are much more common in North America and the U.S.S.R. than in Europe. The estuarian trematode, *Brachyphallus crenatus*, is widely distributed but it is surprising to find it absent in Norway. Its distribution must be related to the absence of the intermediate molluscan host. The cestode, *Bothrimonus* is the only tapeworm common to sea-run charr throughout its range; it requires a marine amphipod intermediate host. Although *Bothrimonus* and *Diplocotyle* are

Table 1. Number of parasite species reported from *Salvelinus alpinus*.

Parasite type	North America	Europe	USSR
Protozoa	1	-	10
Monogenea	1	1	2
Digenea	11	3	4
Cestoda	14	6	9
Nematoda	17	6	8
Acanthocephala	5	-	9
Crustacea	3	1	3
Mollusca	-	-	1
Total	52	17	47

still reported from charr, *Diplocotyle* was synonymized with *Bothrimonus* by Burt and Sandeman (1969). *Corynosoma* and *Echinorhynchus* are acanthocephalans found in marine environments but only *Echinorhynchus*[1] is common to North America, Norway and the U.S.S.R. *Corynosoma* occurs as an immature form in the intestine of charr; this requires seals as a definitive host. It should be noted that while it appears that all anadromous charr from North America have a relatively rich marine parasitofauna this is misleading; the most extensive records were found in those charr feeding in the Foxe Basin (Dick and Belosevic 1981); but studies done on migrating charr in the Nauyuk Lake System, Kent Peninsula and on the Ekaluk River in Victoria Island, found fewer marine species of parasite, primarily *Brachyphallus* and *Bothrimonus* (Dick unpublished data). Mudry and McCart (1976) working on the north slope of the Yukon and Alaska also reported small numbers of *Brachyphallus* and *Bothrimonus* from migrant charr. Differences in numbers and kinds of marine parasites probably relate to feeding patterns. Charr penetrating further out to sea and feeding more extensively upon truly marine fauna would be expected to have higher numbers of marine parasites.

[1] There is increasing evidence that *Echinorhynchus* from charr in North America is a mixture of *E. godi* (marine) and *E. leidyi* (freshwater) (Shostak, pers. comm.).

Table 2. Parasites of resident *Salvelinus alpinus*.

Parasite	Region of the world		
	North America	Europe (Norway)	USSR
Tetraonchus	X	-	X
Discocotyle	-	X	-
Bunodera	X	-	-
Crepidostomum	X	X	X
Diplostomum	-	-	X
Phyllodistomum	X	X	X
Cyathocephalus	-	X	X
Diphyllobothrium	X	X	X
Eubothrium	X	X	X
Pelichnibothrium	-	-	X
Proteocephalus	X	X	X
Cystidicola	X	X	X
Capillaria	-	X	X
Contracaecum	-	-	X
Cucullanus	-	-	X
Philonema	X	X	X
Rhabdochona	-	X	X
Acanthobdella	-	-	X
Metechinorhynchus	X	-	X
Neoechinorhynchus	-	-	X
Salmincola	X	X	X

A comparison of the distribution of Arctic charr parasites within North America (Table 4) shows some interesting trends. *Crepidostomum farionis* disap-

Table 3. Parasites in anadromous *Salvelinus alpinus*.

Parasite	Region of the world		
	North America	Europe (Norway)	USSR
Brachyphallus	X	-	X
Derogenes	X	-	X
Lecithaster	X	X	X
Prosorhynchus	X	-	X
Bothrimonus	X	-	-
Diplocotyle	-	X	X
Corynosoma	X	-	X
Echinorhynchus	X	X	X

pears from Arctic charr as one moves northwards; perhaps this is related to the distribution of sphaerid clams. *Bunodera* and *Phyllodistomum* are absent from the high Arctic but are present towards the Atlantic and Pacific Oceans. *Diphyllobothrium* is common in the high Arctic but its prevalence and intensity decrease westward from Baffin Island and it is absent in Alaska and Labrador. Both *Eubothrium* and *Proteocephalus* are found throughout the area surveyed in North America. *Cyathocephalus* is found only in Alaska while *Cystidicola* is found only in Arctic charr in the central and western part of North America (Table 4). Its absence from the southern part of Baffin Island is understandable as both lake trout and the intermediate host *Mysis* are absent from this region. Its absence from Labrador is difficult to explain as lake trout are found throughout this region (Scott and Crossman 1973) and lake trout are known to harbour the same species found in the high Arctic. *Philonema* and *Metechinorhynchus* are present in most localities but their distribution is somewhat spotty. The crustacean *Salmincola* is distributed throughout North America and appears to be an important component of Arctic charr parasitofauna.

The parasite species of resident or non-migrating Arctic charr fall into two categories. The first group is comprised of those parasites that have a wide holarctic distribution and includes *Crepidostomum farionis, Phyllodistomum,*

Table 4. Parasites of resident *Salvelinus alpinus* (North America).

Parasite	Region				
	Alaska	Victoria Island Kent Peninsula	Cornwallis Island	Baffin Island	Labrador
Tetraonchus	X	-	X	-	-
Crepidostomum	X	X	-	X	X
Bunodera	X	-	-	-	X
Phyllodistomum	X	-	-	-	X
Diplostomum	-	-	-	-	X
Diphyllobothrium	-	X	X	X	-
Eubothrium	X	X	X	X	X
Proteocephalus	X	X	X	X	X
Cyathocephalus	X	-	-	-	-
Cystidicola	X	X	X	-	-
Bulbodacnitis	X	-	-	-	-
Metrabronema	-	-	-	-	X
Philonema	X	X	-	X	-
Metechinorhynchus	X	X	-	-	X
Neoechinorhynchus	-	-	-	X	-
Salmincola	X	X	X	X	X

Diphyllobothrium dendriticum, D. ditremum, Cystidicola cristivomeri and/or *C. farionis, Philonema agubernaculum* and/or *P. oncorhynchi,* and *Salmincola edwardsi.*[2] The second group contains those parasites that are at the most northern part of their hosts' distributional range, generally in water bodies close to the Arctic Ocean or its connecting seas: *Bunodera, Diplostomum, Cyathocephalus, Pelichnibothrium, Capillaria, Contracaecum, Cucullanus, Rhabdochona, Acanthobdella, Metechinorhynchus,* and *Neoechinorhynchus.*

It is difficult to determine if marine parasites in Arctic charr have a truly holarctic distribution. Probably *Bothrimonus, Ecinorhynchus* and *Corynosoma* are holarctic but the marine trematodes are not. Marine trematodes are widely distributed in marine environments but Arctic charr are largely accidental hosts and therefore play little or no role in maintaining the life cycle of these parasites. The one exception may be the estuarian trematode, *B. crenatus* which appears to utilize charr as one of its definitive hosts.

VARIATION IN PARASITOFAUNA OF ARCTIC CHARR FROM CLOSELY RELATED LAKES

It is evident from the previous section that the parasite fauna of Arctic charr vary markedly and that there are few predictable indicator species for a continent or region of a continent. This is reinforced when charr populations are compared from neighbouring lakes. Kennedy (1977) found that parasite prevalences and intensities varied between lakes from widely diverging geographical areas and attributed these differences to local invertebrate fauna and feeding habits of char. By contrast Kennedy (1978a) compared three lakes on Bjornoya and found a high degree of similarity in parasites between lakes. Studies in the Nauyuk system revealed differences in the parasitofauna of charr from different lakes (Tables 5, 6, 7, 8). *Cystidicola* was the dominant parasite in Little Nauyuk and Gavia lakes, low in Willow Lake and absent from Proven Lake. This is probably related to the presence of *Mysis* in some lakes and its absence from others. *Diphyllobothrium* and *Eubothrium* are present in charr from all four lakes, but *Proteocephalus* while low in charr from Gavia, Proven and Willow lakes is absent from Little Nauyuk Lake. The environment supports some copepods since *Diphyllobothrium* and *Eubothrium* are present but the species required as an intermediate host for *Proteocephalus* is missing.

It has been suggested that high levels of *Diphyllobothrium* in Arctic charr are related to the absence of lake trout as the main piscivorous fish in a given system; in other words if charr are the top of the aquatic food chain one should expect

[2] Evidence is accumulating that *S. carpionis* is more common on charr than previously thought and that many identifications of *S. edwardsi* are probably incorrect (Bernier, pers. comm.).

Table 5. Parasites of Little Nauyuk Lake charr (Kent Peninsula).

Parasite	Prevalence %	Intensity	Abundance	Dominance
Cystidicola	91.07	138.92	126.51	89.99
Diphyllobothrium	38.71	19.04	7.37	5.24
Crepidostomum	37.09	15.87	5.89	4.20
Eubothrium	27.42	3.00	0.82	0.58

Table 6. Parasites of Willow Lake charr (Kent Peninsula).

Parasite	Prevalence %	Intensity	Abundance	Dominance
Metechinorhynchus	74.07	94.28	69.83	68.22
Crepidostomum	64.82	18.66	12.10	11.82
Diphyllobothrium	51.85	13.79	7.15	6.99
Cystidicola	50.00	14.19	7.10	6.94
Eubothrium	48.12	11.92	5.74	5.61
Proteocephalus	14.81	3.00	0.44	0.43

them to have higher levels of *Diphyllobothrium*. Data from these four lakes clearly does not show any trend as Willow Lake has lake trout and Arctic charr. There is no question that charr in the eastern Canadian Arctic have high levels of *Diphyllobothrium* but it appears that factors other than absence of lake trout are important in maintaining such high levels. Perhaps certain species of copepod are better intermediate hosts or different species exist in different lakes. The presence of breeding pairs of aquatic birds, such as the red-throated loon, may be essential for maintenance of the life cycle. Also Henricson and Nyman (1976) and Butorina (1975) clearly indicated that the parasitofauna of charr differed depending on whether they were benthic or pelagic feeders: charr feeding on benthos such as *Gammarus* were more heavily infected with *Cyathocephalus*

Table 7. Parasites of Proven Lake charr (Kent Peninsula).

Parasite	Prevalence %	Intensity	Abundance	Dominance
Metechinorhynchus	70.00	24.71	17.30	29.25
Diphyllobothrium	65.00	29.20	18.98	32.09
Crepidostomum	56.00	36.00	20.16	34.08
Eubothrium	25.00	10.67	2.67	4.51
Proteocephalus	2.00	2.00	0.04	0.68

Table 8. Parasites of Gavia Lake charr (Kent Peninsula).

Parasite	Prevalence %	Intensity	Abundance	Dominance
Cystidicola	94.00	235.85	221.69	92.85
Diphyllobothrium	38.00	19.57	7.44	3.12
Crepidostomum	41.00	8.93	3.66	1.53
Proteocephalus	25.00	3.67	0.92	0.39
Eubothrium	13.00	38.89	5.06	2.12

while pelagic feeders were more heavily infected with *D. dendriticum, D. ditremum* and *Proteocephalus*. Clearly more work is needed to assess reasons for different levels of parasitism of Arctic charr and considerable emphasis should be placed on the composition of the invertebrate fauna.

EFFECTS OF PARASITISM ON ARCTIC CHARR

It is only recently that parasites have been implicated in increased mortality of charr (Henricson 1978), reduction of growth and fecundity (Curtis, this volume) and, perhaps, prevention of their migration. Henricson (1978) concluded that the loss of *Diphyllobothrium* from a system was due to death of heavily infected hosts as determined by a lowering of intensity of infection and overdispersion. Henricson (1978), also stated that *D. dendriticum* was more widely distributed in different organs than *D. ditremum* and may damage charr and render them more susceptible to predation. Low numbers of *Diphyllobothrium* (10-20 cysts), in the study by Henricson (1978), were sufficient to cause mortality, yet numbers of 1000 or more have been found in Arctic charr in the eastern Canadian Arctic. The relationship between size and lethal levels in charr has still to be established; perhaps the *Diphyllobothrium* from one system is more pathogenic than that from another. Some of the host-parasite systems in the Arctic may be of a relatively recent origin and therefore exhibit a propensity towards greater pathogenicity, expecially if heavy concentrations of parasites occur.

The second major candidate as a pathogen of Arctic charr is the acanthocephalan, *Metechinorhynchus*; it is found in very large numbers (up to 1300/fish) and on several occasions found to have penetrated the intestinal wall and entered the body cavity. Acanthocephalans are known to cause mortality and reduce growth in a wide range of freshwater fish species.

PRACTICAL USE OF PARASITES OF CHARR

Parasites as biological tags have been used in distribution and migration studies of marine fish (Sindermann 1961 a,b; Kabata 1963; Scott 1969; Gibson 1972; Lubieniecki 1977) and anadromous (Margolis 1963; Pippy 1969 a,b; Hare and Burt 1976; Eddy and Lankester 1978; Dick and Belosevic 1981). Some advantages to the use of parasites as biological tags are: (1) fish do not have to be caught, tagged and then recaught for information, (2) biological tags do not introduce arbitrary and artificial external influences which may upset the natural balance thereby producing misleading results, (3) a relatively low cost once parasite tags are established. Major disadvantages are (1): amount of information needed on life cycles, (2) time spent on the initial survey, (3) no guarantee that every system will produce useful parasites as tags.

Criteria for parasites as biological tags have been established by Sindermann (1961 a,b) and Kabata (1963). Information on use of parasites as biological tags for separating ecological groups of charr is limited. Kennedy (1977) examined resident and migratory charr in northern Norway and found differences in their parasites. Eddy and Lankester (1978) suggested that high levels of *C. cristivomeri* were found in charr that did not migrate to sea and suggested this was due to heavy feeding on *Mysis* in the lake. Dick and Belosevic (1981) reported on the use of parasites to separate sea-run and non-migrating charr in a commercial fishery at the outlet of the Koukdjuak River from Nettilling Lake, Baffin Island. The approach in this system was to take large fish (500-600 mm in length) from lake and river sites prior to the commercial fishing period. Large fish were chosen so that comparisons could be made with the commercial catch. Parasites recovered from the non-migrating and sea-run charr are given in Tables 9 and 10.

Table 9. Parasites of charr from Nettilling Lake (Baffin Island).

Parasite	Prevalence	Intensity
Diphyllobothrium spp.	95.92	433.62
Eubothrium salvelini	96.94	138.37
Proteocephalus longicollis	86.73	148.14
Crepidostomum farionis	21.43	45.52
Echinorhynchus gadi	23.47	34.74
Neoechinorhynchus rutili	8.16	3.88
Salmincola edwardsii	26.53	2.96
Philonema agubernaculum	3.0	2.66

The parasites chosen as good indicator species for non-migrating charr were *Diphyllobothrium* spp., *Eubothrium salvelini* and *Proteocephalus longicollis* and for sea-run charr were *Brachyphallus crenatus, Bothrimonus sturionis* and *Prosorhynchus squamatus*. These parasites were chosen because prevalance and intensity values were high and they constituted the majority of parasites in charr from their respective freshwater or marine environments. Levels of *Diphyllobothrium* spp. differed substantially between non-migrating and sea-run charr, with abundance values of 411.98 and 4.59, respectively.

Two hundred and ten Arctic charr were examined for the six indicator parasites and it was determined that 82 percent of the commercial catch was sea-run and 18 percent was non-migrating charr. A qualitative index for *Diphyllobothrium* spp. cyst levels (Table 11) was used to assess 600 Arctic charr from the commercial fishery: the catch consisted of 78 percent sea-run and 22 percent non-migrating charr. In this particular system the cyst assessment method was quicker than the more accurate, but more time consuming, identification of the indicator parasites. Clearly then, in this system, parasites could be used to separate two ecological groups of charr but it also raised some interesting but unresolved questions. Morphometric measurements taken from the migratory and non-migratory groups of charr showed minor differences but some were

Table 10. Parasites of charr collected from Koukdjuak River (Baffin Island).

Parasite	Prevalence	Intensity
Brachyphallus crenatus	97.09	1146
Bothrimonus sturionis	97.89	352
Prosorhynchus squamatus	61.05	80.57
Cystidicoloides teniussima	45.59	11.44
Salmincola edwardsii	35.79	4.47
Diphyllobothrium spp.	33.68	13.91
Crepidostomum farionis	31.58	83.45
Echinorhynchus gadi	30.00	3.03
Eubothrium salvelini	24.41	25.24
Neoechinorhynchus rutili	30.53	2.69
Contracaecum sp.	13.57	9.08
Lecithaster gibbosus	6.31	240.83
Philonema agubernaculum	6.32	4.33
Derogenes varicus	4.21	8.75

statistically significant (Table 12) (Dick and Belosevic 1981). In contrast, condition factors for the two groups of charr were similar i.e., 1.005 for sea-run and 1.015 for non-migratory charr. Also strontium levels were similar between the two groups of charr (Curtis, pers. comm.). If it is the same population of charr why did sea-run fish have less *Diphyllobothrium*? Several explanations are suggested: (1) charr going to sea purge themselves of *Diphyllobothrium*, (2) heavily parasitized charr do not go to sea (3) there is a turnover of the population at or near the mouth of the Koukdjuak River, some going to sea in a given year and some remaining behind. There is evidence suggesting that Arctic charr returning from the sea have less *Diphyllobothrium*; preliminary work done at Nauyuk Lake showed that landlocked charr from Proven and Gavia lakes transplanted to the Nauyuk River estuary had a major decrease or complete loss of *Diphyllobothrium* cysts at the end of a two week period. This is not equivocal evidence as charr transplanted to freshwater also showed a decline in numbers of

Table 11. *Diphyllobothrium* spp. cyst levels in body cavity and on the digestive tract.

Categories	No. of cysts
+1	0- 25
+2	26 - 50
+3	51 - 100
+4	101 - 200
+5	201 - 500
+6	500+

cysts. There is no evidence to date which proves that heavily parasitized charr do not migrate to sea. Studies designed to prove this would be most interesting. A turnover of one population of charr at, or near, the mouth of the Koukdjuak River is possible but why then are the two groups of charr growing at the same rate? We know from tagging data that fish tagged the previous year returned to sea. Perhaps the large volume of water flowing from Nettilling Lake may be concentrating insect larvae in the region of the outlet, thus making it a fairly productive site for non-migrating charr.

Use of parasites in other Arctic charr systems needs to be evaluated but it is known that *Bothrimonus* and *Brachyphallus* are useful tags at Nauyuk Lake and the Ekaluk River on Victoria Island. Since we see such marked differences in level and kinds of parasites between closely related lakes it is possible that the freshwater origin of sea-run charr may be determined in the catch from estuarian or marine charr fisheries. *Cystidicola* is a likely candidate as are *Eubothrium* and *Crepidostomum*. Studies on charr transplanted to the Nauyuk River estuary showed these parasites were reduced in numbers but were not lost. The patchy nature of Arctic charr parasite distribution precludes a generalized approach to parasites as biological tags. It appears more likely that each system will have to be evaluated separately.

Table 12. Morphic measurements of Arctic charr.

Measurement	Nettilling Lake	Koukdjuak River	F-value*
Fork length	54.57 ± 4.34	54.49 ± 4.54	0.01
Standard length	50.12 ± 4.15	51.11 ± 4.29	2.70
Total length	57.93 ± 4.45	56.86 ± 4.75	2.63
Depth of caudal peduncle	3.29 ± 0.41	3.13 ± 0.36	9.17
Pectoral fin length	7.29 ± 0.90	7.37 ± 1.00	0.30
Head length	11.07 ± 1.17	11.49 ± 1.34	0.20
Interorbital width	3.94 ± 0.53	3.79 ± 0.49	3.94
Upper gillrakers	9.65 ± 1.25	9.44 ± 1.22	1.43
Lower gillrakers	13.91 ± 1.42	13.65 ± 1.71	1.33
Total gillrakers	23.65 ± 2.20	23.09 ± 1.64	2.88
Weight	1621.58 ± 390.40	1659.53 ± 430.80	0.42
Girth	26.27 ± 2.61	26.32 ± 2.71	0.01
Age	14.38 ± 1.75	15.24 ± 2.22	6.85

*F-value > 3.84, significant at the 0.05 level.

CHARR — PARASITES — MAN

The interrelationships of these three components directly affects man's health and has an impact on environmental quality. Parasites may detrimentally affect charr production on the one hand, but, on the other, parasites may have value in distinguishing charr populations or stocks for management purposes.

Two parasite species found in charr can infect man: *Diphyllobothrium* and *Anisakis*. Only *Diphyllobothrium* is known to infect humans throughout the Arctic, presumably from eating raw or undercooked charr. *Anisakis* infections, while common to man consuming marine fish, have not been directly associated with the consumption of Arctic charr. Although the fish tapeworm, *Diphyllobothrium latum* is reputed to occur in Arctic charr, most researchers agree that the most frequently encountered species are *D. dendriticum* and *D. ditremum*. Even though these two parasites are readily identified (Andersen 1975 a,b) researchers persist in referring to them as *Diphyllobothrium* sp. plerocercoids and continue to convey the uncertainty of identification. Laird and Meerovitch (1961) suggested that human infections of "the northern *Diphyllobothrium*" in the Canadian Arctic were obtained from *S. alpinus* but the source of infection to charr was somewhat obscure. Beverley-Burton (1978) suggested that the definitive hosts may include *Gavia stellata* (red-throated loon), *Larus argentatus* (herring gull), *Xema sabini* (Sabine's gull), and *Sterna paradisea* (Arctic tern). On the other hand, Jamieson and Freeman (1975) examined herring gulls, dogs, red-throated loons, and ravens; only worms, tentatively identified as *D. ditremum*, were recovered. Jamieson and Freeman (1975) state "it is reasonably certain the worms recovered are *D. dendriticum*, which elsewhere in the world is considered to be primarily a parasite of gulls." Human infections are thought to occur from eating contaminated ripe roe and from plerocercoids adhering to the wall of the abdominal cavity, but Freeman and Jamieson (1976) suggest it is probably of little consequence to health. The nature of Arctic diphyllobothriasis needs further work and clearly the species present in charr from freshwater and marine environments need to be clarified. A detailed study, including experimental infections in the definitive host and better methods to identify marine forms are required.

Diphyllobothrium through its effects on growth and mortality of charr also affects man. *Diphyllobothrium* as a natural pathogen of Arctic charr needs to be assessed, particularly in the eastern Canadian Arctic. Furthermore, since the lethal levels of parasitism that Henricson (1978) determined experimentally are much lower than actual levels found in charr from several parts of the Canadian Arctic, we need to know whether there are strains of *Diphyllobothrium* which are more virulent or if the mixture of *D. ditremum* and *D. dendriticum* is important in death of charr. The effects of *Metechinorhynchus* and *Cystidicola* on charr growth and mortality are unknown and we have no information on the effects of *Cystidicola* on swimming behaviour of charr.

It is well known that a recent association of parasite and host is more likely to result in mortality. Consequently any change in the Arctic that results in parasites entering a system where they have not previously existed (i.e., artificial stocking, changes in waterways through hydroelectric development, etc.) could have detrimental effects on charr populations. For example, the introduction of an invertebrate host that is missing from one system could ultimately have detrimental effects on another aquatic system: heavy levels of parasitism are particularly pronounced in Arctic environments as species diversity is low and invertebrate production and feeding by charr are concentrated in time.

Overexploitation of charr populations has been in the past, and continues to be, a serious problem in the Arctic. Estuarine or sea fisheries may be removing one genetic component of a given aquatic system disproportionately, as it removes only sea-run charr. Perhaps fisheries managers should be looking more closely at exploiting a proportion of non-migrating charr as well. Certainly we need to know more about systems like the Nettilling Lake-Koukdjuak River system which seems to be producing charr of equal quality from the lake and the marine environment. There is no doubt that parasites as biological tags are one useful tool in assessing this type of fishery. Other information is urgently needed such as: What is the charr productivity from the outlets of these large lakes? Do these large river systems support permanent charr populations? Is there a turnover of individuals from a single population on a one, two or three year basis? Strontium studies may help us, but tagging and parasite identification will also be required. Studies on charr populations from small aquatic systems have given us invaluable information on the biology of charr but I do not believe this information is directly applicable to charr populations from large systems. This is somewhat disconcerting.

In conclusion we can provide a qualified answer to the question posed in the title: Are parasitological studies of Arctic charr of practical value or do they merely represent another facet of academic curiosity? There is no doubt that a better understanding of the biology of Arctic charr, in all its aspects, will foster the improvement of management techniques, in particular studies of the parasites will contribute to understanding the make-up of different, sympatric stocks; ultimately there may be benefit to be derived from fishing programmes designed to reduce parasitic infestation; but, above all, the immediate benefits may be that it can prevent us from making disastrous and irreversible mistakes, through the deliberate or inadvertant transfer of parasites or their intermediate hosts from one waterbody to another.

APPENDIX 1 Parasites of charr from North America

Parasite species	*S. alpinus*	*S. malma*
Protozoa		
Haemogregarina irkalukpiki	X	
Myxosporida gen. sp.		X
Monogenea		
Tetraonchus alaskensis	X	
Discocotyle sagittata		X
Gyrodactyloidea gen. sp.		X
Digenea		
Brachyphallus crenatus	X	X
Bunodera luciopercae	X	X

Parasite species	S. alpinus	S. malma
Crepidostomum cooperi	X	
Crepidostomum farionis	X	X
Crepidostomum transmarinum	X	
Derogenes varicus	X	
Diplostomum sp. metacercariae		X
Diplostomum spathaceum	X	
Lecithaster gibbosus	X	X
Neascus sp.		X
Phyllodistomum sp.	X	
Phyllodistomum limnosa		X
Prosorhynchoides ozakii	X	
Prosorhynchus squamatus	X	
Tetracotyle sp.		X
Tubulovesicula lindbergi		X
Cestoda		
Bothrimonus sturionis	X	X
Bothriocephalus cuspidatus		X
Cyathocephalus truncatus	X	X
Dibothrium fasteni	X	
Diphyllobothrium sp.	X	X
Diphyllobothrium cordiceps		X
Diphyllobothrium dendriticum	X	
Diphyllobothrium ditremum	X	
Eubothrium crassum	X	
Eubothrium salvelini	X	
Nybelina surmenicola	X	
Phyllobothrium sp.		X
Pelichnibothrium caudatum	X	
Proteocephalus articus	X	
Proteocephalus longicollis	X	
Proteocephalus salmonidicola	X	
Proteocephalus tumidocollus	X	
Proteocephalus sp.	X	
Nematoda		
Anisakis sp.	X	
Anisakis laevis	X	
Bulbodacnitis alpinus	X	
Capillaria sp.		X

Parasite species	*S. alpinus*	*S. malma*
Capillaria salvelini	X	
Contracaecum aduncum	X	
Contracaecum sp.	X	X
Cystidicola sp.		X
Cystidicola cristivomeri	X	
Cystidicola farionis	X	
Cystidicola stigmatura	X	
Cystidicoloides tenuissima	X	X
Dacnitis laevis	X	
Eustrongyloides sp.		X
Metrabronema salvelini	X	
Philonema agubernaculum	X	
Philonema oncorhynchi	X	X
Philonema salvelini	X	
Philonema sp.	X	
Rhabdochona milleri		X
Salvelinema salmonicola		X
Skrjabinocapillaria bakeri		X
Thynnascaris adunca	X	
Truttaedacnitis truttae	X	X
Acanthocephala		
Corynosoma strumosum	X	
Echinorhynchus gadi	X	X
Metechinorhynchus lateralis	X	
Metechinorhynchus leidyi		X
Metechinorhynchus salmonis	X	
Neoechinorhynchus rutili	X	X
Crustacea		
Lepeophtheirus salmonis	X	
Salmincola californiensis		X
Salmincola carpionis	X	X
Salmincola edwardsii	X	
Salmincola siscowet		X

APPENDIX II Parasites of charr from Europe (Norway)

Monogenea
Discocotyle sagittata

Digenea
  Crepidostomum farionis
  Lecithaster gibbosus
  Phyllodistomum conostomum

Cestoda
  Cyathocephalus truncatus
  Diphyllobothrium ditremum
  Diphyllobothrium dendriticum
  Diplocotyle olrikii
  Eubothrium salvelini
  Proteocephalus exiguus

Nematoda
  Anisakis sp.
  Capillaria salvelini
  Cystidicola farionis
  Philonema oncorhynchi
  Phocanema decipiens
  Rhabdochona denudata

Crustacea
  Salmincola edwardsii

APPENDIX III Parasites of charr from the USSR

	Fish Species			
Parasite species	*S. alpinus*	*S. lepechini*	*S. malma*	*S. leucomaenia*
Protozoa				
Chloromyxum coregoni	X			
Dermocystidium salmonis	X			
Henneguya zschokkei	X			
Leptotheca krogiusi	X			
Myxidium oviforme	X			
Myxidium salvelini	X			
Myobolus neurobius	X			
Trichodina nigra kamchatika	X			
Trichodina truttae			X	X
Trichophyra piscium	X			
Tripartiella californica			X	X
Zschokkella orientalis	X			

Parasite species	Fish Species			
	S. alpinus	*S. lepechini*	*S. malma*	*S. leucomaenia*
Monogenea				
Gyrodactylus birami	X			
Tetraonchus alaskensis	X		X	
Digenea				
Bucephalus gracilescens			X	
Brachyphallus crenatus			X	X
Crepidostomum farionis	X		X	X
Derogenes varicus			X	
Diplostomum sp.	X			
Genarches mulleri			X	
Genolinea anura				X
Hemiurus levinseni			X	
Lecithaster gibbosus	X			
Phyllodistomum conostamum	X			
Prosorhynchus crucibulum			X	
Tubulovesicula lindbergi				X
Cestoda				
Cyathocephalus truncatus	X		X	X
Diphyllobothrium norvegicum	X	X		
Diphyllobothrium sp. (L)	X			
Diplocotyle olrikii	X			X
Eubothrium crassum	X	X	X	X
Eubothrium salvelini	X			
Nybelina surminicola			X	
Pelichnibothrium speciosum	X		X	X
Proteocephalus exiguus	X			X
Scolex pleuronectis	X		X	X
Nematoda				
Anisakis sp. (1)	X			
Ascarophis malmae			X	
Capillaria salvelini				X
Contracaecum aduncum	X		X	
Contracaecum sp.	X			
Cucullanus laevis			X	
Cucullanus truttae	X			
Cystidicola farionis	X		X	X

Parasite species	Fish Species			
	S. alpinus	*S. lepechini*	*S. malma*	*S. leucomaenia*
Ichthyobronema gnedini		X		
Metrabronema canadense			X	X
Metrabronema salvelini		X	X	X
Philonema oncorhynchi	X		X	X
Rhaphidascaris acus	X	X		
Rhabdochona denudata	X			
Rhabdochona amago			X	
Rhabdochona salvelini			X	
Acanthocephala				
Acanthobdella livanowi	X			
Acanthobdella peledina	X	X		
Bolbosoma caenoforme	X			
Corynosoma strumosum	X			
Echinorhynchus gadi	X			
Metechinorhynchus salmonis	X			
Metechinorhynchus truttae	X			
Neoechinorhynchus rutili	X			
Pomphorhynchus laevis	X			
Crustacea				
Salmincola biaculiculata			X	
Salmincola edwardsii	X	X		
Salmincola faculata			X	
Salmincola gibber			X	
Salmincola salmonea	X			
Salmincola smirnovi			X	
Salmincola thymalli	X			
Mollusca				
Anodonta yukonensis	X			

REFERENCES

Andersen, K. 1975a. Ultrastructural studies on *Diphyllobothrium ditremum* and *D. dendriticum* (Cestoda, Pseudophyllidea), with emphasis on the scolex tegument and the tegument in the area around the genital atrium. Z. Parasitenk. 46:253-264.

Andersen, K. 1975b. Comparison of surface topography of three species of *Diphyllobothrium* (Cestoda, Pseudophyllidea) by scanning electron microscopy. Int. J. Parasitol 5:293-300.

Beverley-Burton, M. 1978. Metazoan parasites of Arctic char (*Salvelinus alpinus* L.) in a high Arctic, landlocked lake in Canada. Can. J. Zool. 56:365-368.

Burt, M.D.B. and I.M. Sandeman. 1969. Biology of *Bothrimonus (= Diplocotyle)* (Pseudophyllidea: Cestoda) Part 1. History, Description, Synonymy, and Systematics. J. Fish. Res. Board Can. 26:975-995.

Butorina, T.E. 1975. Dynamics of the parasite fauna of different forms of *Salvelinus alpinus* from the basin of Lake Azabachje U.S.S.R. Parazitologiya 9:237-246.

Bykhovskaya-Pavlovskaya, I.E., A.V. Guseu, M.N. Dubinina, N.A. Izyumova, T.S. Smirnova, I.L. Sokolovskaya, G.A. Shtein, S.S. Shul'man, and U.M. Epshtein. 1962. Key to parasites of freshwater fish of the U.S.S.R. 1964 Isr. Progr. Sci. Trans., Jerusalem. (Translated from Russian). 917 p.

Curtis, M.A. 1979. Metazoan parasites of resident Arctic char (*Salvelinus alpinus*) from a small lake on southern Baffin Island. Le Natural Can. 106:337-338.

Dick, T.A. and M. Belosevic. 1981. Parasites of Arctic charr *Salvelinus alpinus* (Linnaeus) and their use in separating sea-run and non-migrating charr. J. Fish Biol. 18:339-347.

Eddy, S.B. and M.W. Lankester. 1978. Feeding and migratory habits of Arctic char, *Savelinus alpinus*, indicated by the presence of the swimbladder nematode *Cystidicola cristivomeri* White. J. Fish. Res. Board Can. 35:1488-1491.

Freeman, R.S. and J. Jamieson. 1976. Parasites of eskimos at Igloolik and Hall Beach, p. 306-315. *In* R.J. Shepard and S. Itoh [eds.] Circumpolar health, Proceedings of the 3rd International Symposium, Yellowknife, N.W.T., University of Toronto Press, Toronto.

Gibson, D.I. 1972. Flounder parasites as biological tags. J. Fish Biol. 4:1-9.

Hare, G.M. and M.D.B. Burt. 1976. Parasites as potential biological tags of Atlantic salmon (*Salmo salar*) smolts in the Miramichi River System, New Brunswick. J. Fish. Res. Board Can. 33:1139-1143.

Henricson, J. and L. Nyman. 1976. Parasites as indicators of two sibling species of char (*Salvelinus alpinus* L.) Norw. J. Zool. 24:465-467.

Henricson, J. 1978. The dynamics of infection of *Diphyllobothrium dendriticum* (Nitzsch) and *D. ditremum* (Creplin) in the char *Salvelinus alpinus* (L) in Sweden. J. Fish Biol. 13:51-71.

Jamieson, J. and R.S. Freeman. 1975. Parasitology: Parasites of man and of Arctic char at Iglooik, p. 290-293. *In* T.W.M. Cameron and L.W. Billingsley [eds.] Energy Flow: its biological dimension. Summary of the I.B.P. in Canada.

Kabata, Z. 1963. Parasites as biological tags. Int. Commn. N.W. Atlantic. Fish., Special Publ. 4:31-37.

Kennedy, C.R. 1977. Distribution and zoogeographical characteristics of the parasite fauna of char *Salvelinus alpinus* in Arctic Norway, including Spitsbergen and Jan Mayon islands. Astarte 10:49-55.

Kennedy, C.R. 1978a. The parasite fauna of resident char *Salvelinus alpinus* from Arctic islands, with special reference to Bear Island. J. Fish Biol. 13:457-466.

Kennedy, C.R. 1978b. The biology, specificity and habitat of the species of *Eubothrium* (Cestoda: Pseudophyllidea), with reference to their use as biological tags: a review. J. Fish Biol. 12:393-410.

Laird, M. and E. Meerovitch. 1961. Parasites from northern Canada. I. Entozoa of Fort Chimo Eskimos. Can. J. Zool. 39:63-67.

Lubieniecki, B. 1977. The plerocerus of *Grillotia erinaceus* as a biological tag for haddock *Melanogrammus aeglefinus* in the North Sea and the north-east Atlantic. J. Fish Biol. 11:555-565.

Margolis, L. 1963. Parasites as indicators of geographical origin of sockeye salmon, *Oncorhynchus nerka* (Walbaum), occurring in the north Pacific Ocean and adjacent seas. Bull. Int. N. Pacif. Fish. Commn. 11:101-156.

Margolis, L., and J.R. Arthur. 1979. Synopsis of the parasites of fishes of Canada. Bull. Fish. Res. Board Can. 199: 269.

Mudry, D.R. and P. McCart. 1976. Metazoan parasites of Arctic char (*Salvelinus alpinus*) from the north slope of Canada and Alaska. J. Fish. Res. Board Can. 33:271-275.

Pippy, J.H.C. 1969a. Preliminary report on parasites as biological tags in Atlantic salmon (*Salmo salar*). I. Investigations 1966-1968. Fish. Res. Board Can. Tech. Rep. 134: 44 p.

Pippy, J.H.C. 1969b. *Pomphorhynchus laevis* (Zoega) Muller, 1976. (Acanthoce-phala) in Atlantic salmon (*Salmo salar*) and its use as a biological tag. J. Fish. Res. Board Can. 26:909-918.

Scott, J.S. 1969. Trematode populations in the Atlantic Argentine, *Argentina silus* and their use as biological indicators. J. Fish. Res. Board Can. 26:879-891.

Scott, W.B. and E.J. Crossman. 1973. Freshwater Fishes of Canada. Fish. Res. Board Can. Bull. 184: 955 p.

Sindermann, C.J. 1961a. Parasite tags for marine fish. J. Wildl. Manage. 25:41-47.

Sindermann, C.J. 1961b. Parasitological tags for redfish of the western North Atlantic. Int. Commn. N.W. Atlantic Fish, Spec. Publ. 3:111-117.

Diphyllobothrium spp. and the Arctic Charr: Parasite Acquisition and its Effects on a Lake-Resident Population

MARK A. CURTIS

Institute of Parasitology, Box 231, Macdonald College, Ste. Anne de Bellevue, Quebec, Canada H9X 1C0

CURTIS, M.A. 1984. *Diphyllobothrium* spp. and the Arctic charr: parasite acquisition and its effects on a lake-resident population, p. 395–411. *In* L. Johnson and B.L. Burns [eds.] Biology of the Arctic charr, Proceedings of the International Symposium on Arctic Charr, Winnipeg, Manitoba, May 1981. Univ. Manitoba Press, Winnipeg.

Certain lake-resident populations of Arctic charr in northern Canada are moderately to heavily infected by plerocercoid larvae of the cestodes *Diphyllobothrium ditremum* and *D. dendriticum*. Research has been directed towards determining why fish in some lakes are so heavily infected and quantifying the effect of these parasites on their host populations.

During the summers of 1977 and 1978 a small lake system near Frobisher Bay, N.W.T. was sampled intensively. Virtually all fish in the lakes acquire plerocercoids by a length of about 200 mm and an age of 6-8 years. *D. ditremum* generally encysts on the stomach wall of charr while the larger and more mobile *D. dendriticum* also encysts in other tissues, including the liver, gonads and, very rarely, the flesh. Charr feeding on *Pungitius pungitius* acquire heavy worm burdens; the transfer of plerocercoids between the two fish species has been confirmed experimentally for both *Diphyllobothrium* species. Massive infection appears to inhibit gonadal development in both sexes, particularly in females, and there is evidence of increased mortality in charr harbouring more than approximately 100 plerocercoids.

Food web interactions in lakes inhabited by charr may account for some observed geographical differences in the level of *Diphyllobothrium* parasitism. Charr appear to be less infected where they are not piscivorous. Competition with other piscivorous fish in northern lakes may thus sometimes serve as a natural control on *Diphyllobothrium* infection.

INTRODUCTION

Throughout its circumpolar range *Salvelinus alpinus*, the Arctic charr, is known to harbour parasitic cestodes of the genus *Diphyllobothrium*. These worms are larval forms found encysted in and on the visceral organs of the charr, occasionally occuring in the flesh. During the past decade or so our knowledge of the taxonomy and biology of the *Diphyllobothrium* species occurring in charr has been advanced by a number of researchers, particularly those in Scandinavia (Andersen 1975; Halvorsen 1970; Henricson 1977, 1978; Vik 1964). This body of work provides a firm basis from which we can now attempt to examine and quantify the actual effects of parasitism by *Diphyllobothrium* on its fish host, the Arctic charr. Research of this nature is essentially a synthesis of fish biology and parasitology, and it is representative of a growing body of current ecological work dealing with the effects of pathogenic organisms on host populations.

In the Canadian eastern Arctic, non-migratory lake-resident charr, particularly those of large size, are often heavily parasitized by *Diphyllobothrium*. These infections are characterized by masses of cysts and by worms free in the viscera. Native people and local sports fishermen from northern communities are often aware that various lakes in the region tend to harbour "wormy" fish. One such location is Shona Lake (Ikalugajukuluit), a small lake situated about 8 km southeast of the Frobisher Bay townsite at 63° 42′ N, 68° 23′ W. When first sampled by the Arctic Biological Station of Fisheries and Oceans, Canada, in 1958 and 1959, the high levels of *Diphyllobothrium* occurrence in the charr were noted. For the purposes of the present study, the lake was visited again in 1977 and 1978 when over 1000 charr collected from it were utilized for parasitological work. The initial goal of the research was to discover how charr in the lake acquired *Diphyllobothrium* infections and why such a sizeable proportion of fish in the population should be so heavily infected. A further objective was to determine the extent of any effects the plerocercoids might have upon the fecundity and mortality of charr.

Two species of *Diphyllobothrium, D. ditremum* and *D. dendriticum*, occur in the charr of Shona Lake. Both parasite species utilize copepods as first intermediate hosts and freshwater fish as second intermediate hosts. The usual definitive hosts for *D. ditremum* are ducks and loons, predominately at Shona Lake the red-throated loon (*Gavia stellata*). Gulls (*Larus* spp.) normally serve as definitive hosts for *D. dendriticum*, but adult forms of the tapeworm can also occur in various mammals, including sled dogs and man. A diagram of the life cycles for *D. ditremum* and *D. dentriticum* in Shona Lake is presented in Figure 1. The transfer of plerocercoids from fish to fish, indicated in the life cycle diagram, is an important feature of the lake and will be discussed later in this presentation.

The plerocercoid of *D. ditremum*, (Fig. 2a) the smaller of the two species (20-30 mm), is most commonly found encysted on the stomach of its fish host, although also occurring in other visceral organs. *D. dendriticum* (Fig. 2b), a much larger worm of up to 100 mm in length, is distributed among various organs, particularly the liver and stomach; occasionally it is found in the flesh of its host. Heavily infected charr harbour up to several thousand plerocercoids of the two species and present an abhorrent sight when cut open; the viscera is bound together by adhesions and scarr tissue formed in response to wandering plerocercoids, making it difficult to dissect the organs from one another and from

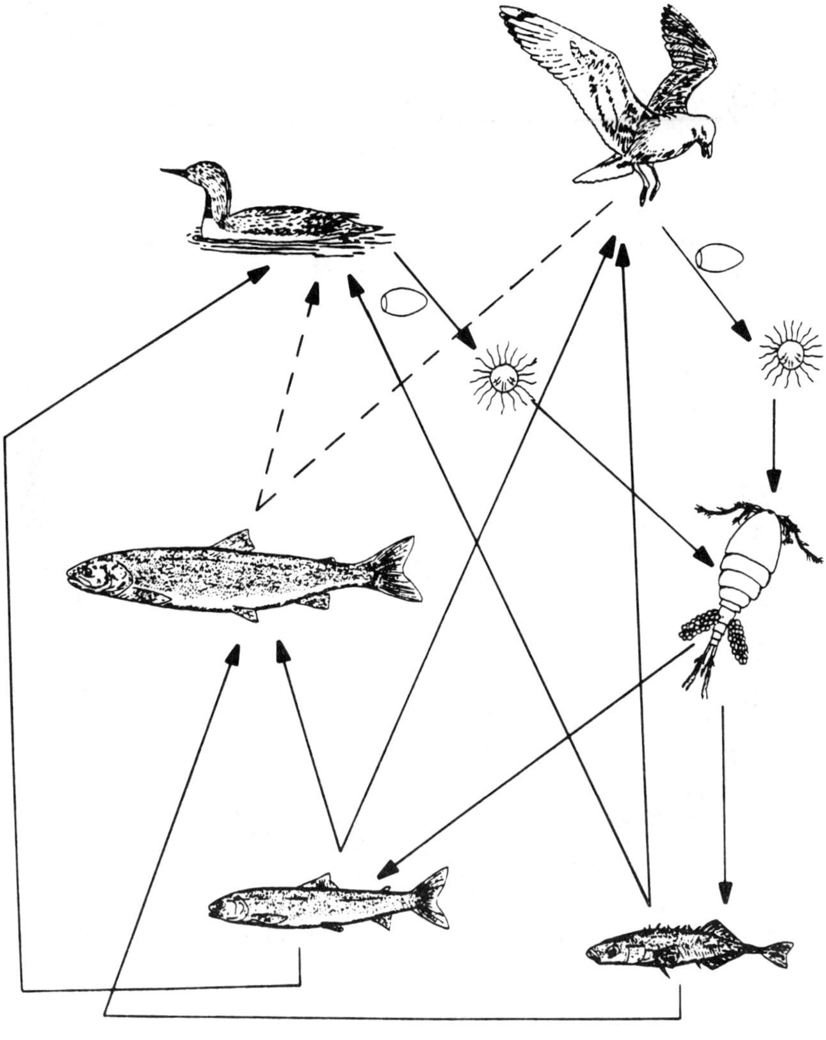

FIG. 1 Life cycles of *D. ditremum* and *D. dendriticum* at Shona Lake near Frobisher Bay, N.W.T. Definitive hosts are the red-throated loon *(Gavia stellata)* for *D. ditremum*, and the herring gull *(Larus argentatus)* for *D. dendriticum*. Life cycle stages include free-living ciliated coracidia, procercoids in copepods *(Cyclops* or *Diaptomus* spp.) and plerocercoids in ninespine sticklebacks or small Arctic charr. Plerocercoids of both *Diphyllobothrium* species can be transmitted from smaller fish to piscivorous charr.

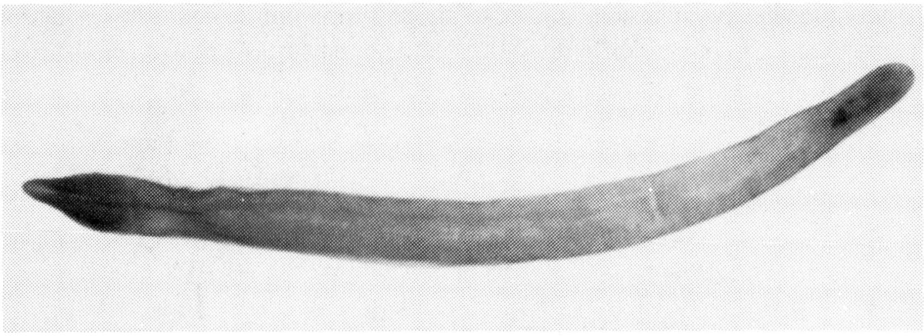

FIG. 2a Plerocercoid of *D. ditremum* from the stomach tissue of an infected charr. Length of plerocercoid = 25 mm. Scolex at left side of the photograph.

FIG. 2b Plerocercoid of *D. dendriticum* from the liver of an infected charr. Length of plerocercoid = 55 mm. Scolex at left side of photograph.

the body wall of the fish (Fig. 3a). Much internal hemorrhaging is evident in such infections and the stomach sometimes appears to be little more than a thickened mass of cysts (Fig. 3b).

For the purpose of this presentation, I will speak in terms of total worm counts for *Diphyllobothrium*, combining both *D. ditremum* and *D. dendtriticum*. Although the two species differ in their locations within the viscera of the fish

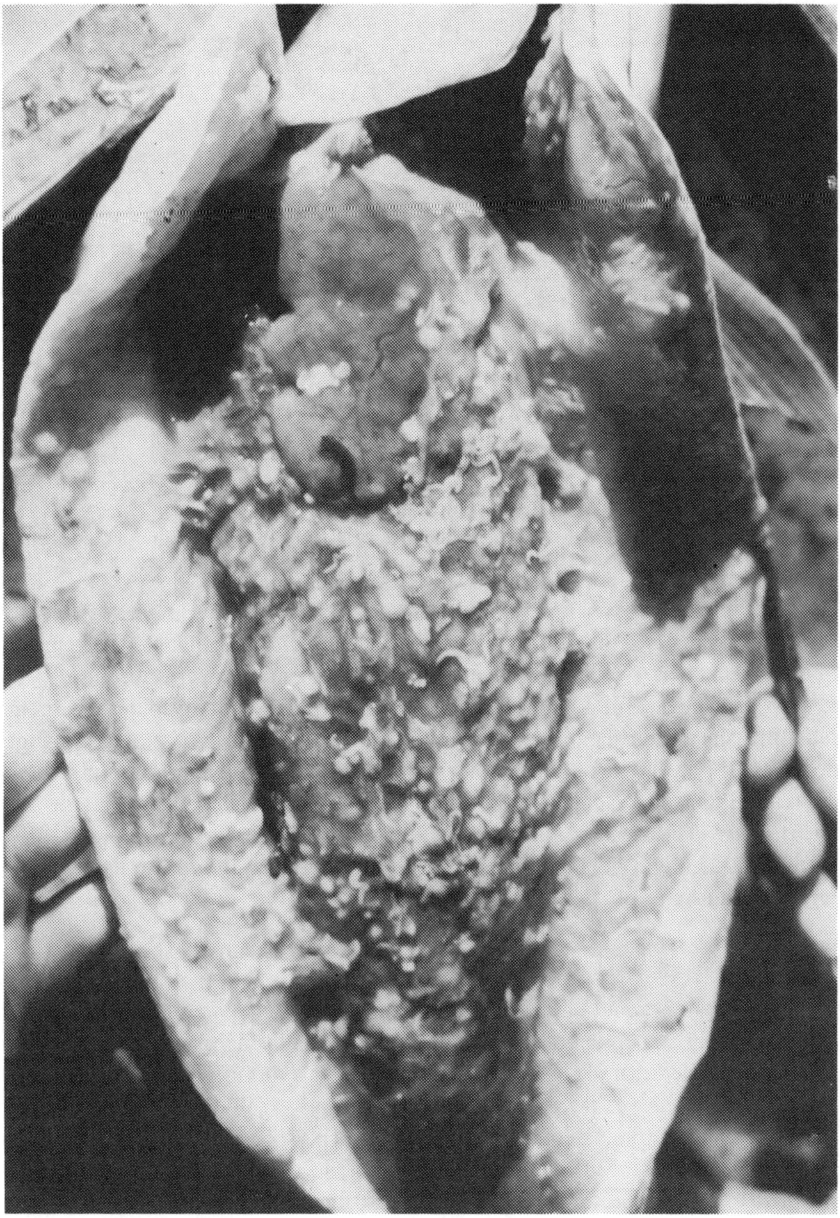

FIG. 3a Ventral view of an Arctic charr heavily infected by plerocercoids of *D. ditremum* and *D. dendriticum*.

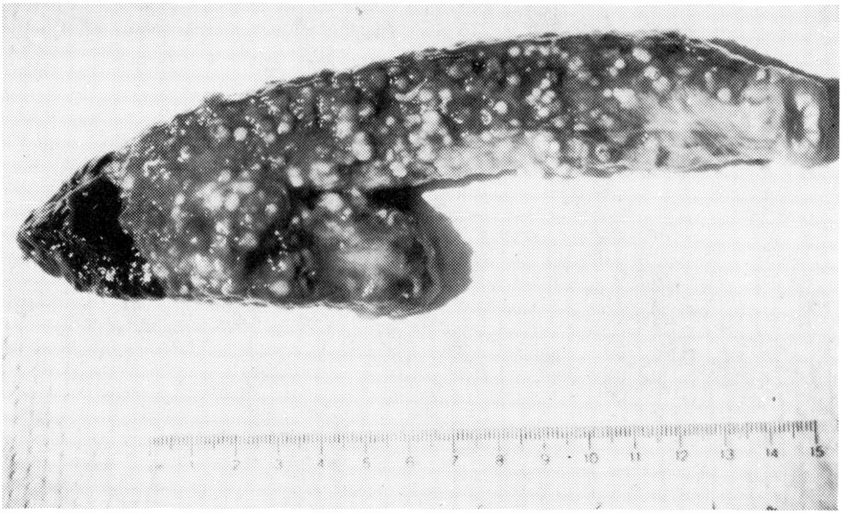

FIG. 3b Lateral view of an Arctic charr stomach with numerous cysts, primarily of *D. ditremum.*

host and probably exert varying degrees of pathogenicity, it is possible to illustrate some major features of the charr/*Diphyllobothrium* relationship in terms of total counts only. With deference to the management perspective of this paper, it may also be useful to report results as total plerocercoid counts, without requiring any further taxonomic separation of the species involved. In passing, however, it is well to note that 90-95 percent of the *Diphyllobothrium* plerocercoids in Shona Lake charr generally are *D. ditremum* and the remainder *D. dendriticum*. The relative numbers of plerocercoids of each species are closely correlated, as might be anticipated from the similarities in their life cycles with regard to transmission.

METHODS

Shona Lake is about one kilometer in length with a maximum depth of 18 m. Approximately 50 percent of the lake bottom is at depths of only 3-5 m. Lying in a valley separated from Frobisher Bay by a low range of hills, Shona is the most elevated of a series of three lakes, connected by a small stream and separated from one another by ancient glacial moraines. The seçond of the three lakes is about one-half the size of Shona and also supports a charr population, which was sampled in 1978. The third lake, that nearest the sea, is shallow and probably without a permanent charr population. Shona Lake and the lake immediately downstream from it appear to harbour only charr and ninespine sticklebacks

(*Pungitius pungitius*); threespine sticklebacks (*Gasterosteus aculeatus*) occur in a lake at a higher elevation one km distant, but were never found in the Shona Lake system. On the basis of the parasitological evidence, and based also on strontium analyses of the charr, only non-migratory lake-residents occur in Shona Lake.

Charr were collected from Shona Lake from June to August 1977, and in August 1978, using floating and sunken gillnets (mesh sizes (stretched measure) 1″ (25 mm), 1½″ (18 mm), 2½″ (64 mm), 3½″ (89 mm)) set for 24 hour periods in all areas of the lake. Some additional specimens were obtained by angling and electrofishing. Handling of charr for parasitological work was carried out in three ways: 1) Fish were examined on site at the lake while fresh; 2) Fish were transported to Frobisher Bay, frozen and shipped to McGill's Institute of Parasitology for later examination; 3) Fish were eviscerated soon after capture and all viscera from the coelomic cavity were secured in "whirlpak" bags with 10 percent formalin. An injection of formalin into the stomach served to preserve its contents for later analysis.

Data obtained for the fish included fork length, total weight, eviscerated weight, sex, state of gonadal development (oocyte diameter in females and testis width in males), gonad weight, and identification of stomach contents. Saccular otoliths were taken from each specimen for age determination by annular counts (Grainger 1953). *Diphyllobothrium* plerocercoids were dissected from their cysts and enumerated; counts of *D. ditremum* and *D. dentriticum* were made separately. For charr harbouring more than about 500 plerocercoids, total numbers were estimated after counting and identifying approximately 50 percent of the worms present. In testing this method against total counts for individual fish, it was found to be accurate within 20 percent of actual *Diphyllobothrium* burden.

RESULTS AND DISCUSSION

Some Quantitative Characteristics of the Charr Population in Shona Lake

Size- and age-frequency histograms for the charr collected in Shona Lake during 1977 and 1978 are presented in Figure 4. Net selectivity and fishing effort played some role in ultimately determining the shape of these; however, by August 1977 catches of charr exceeding 350 mm fork length were very small, indicating that many of the larger charr in the lake had already been collected. This observation was further confirmed by low yields of large charr in 1978. Charr of 300 mm and less remained abundant in the lake and were not fished to depletion as sufficient numbers for parasitological work were readily obtained. The size-frequency histogram is thus somewhat biased towards underestimating the numbers of small charr in the lake, and this is further accentuated by the obvious paucity of fish smaller than 100 mm, which were below the size range

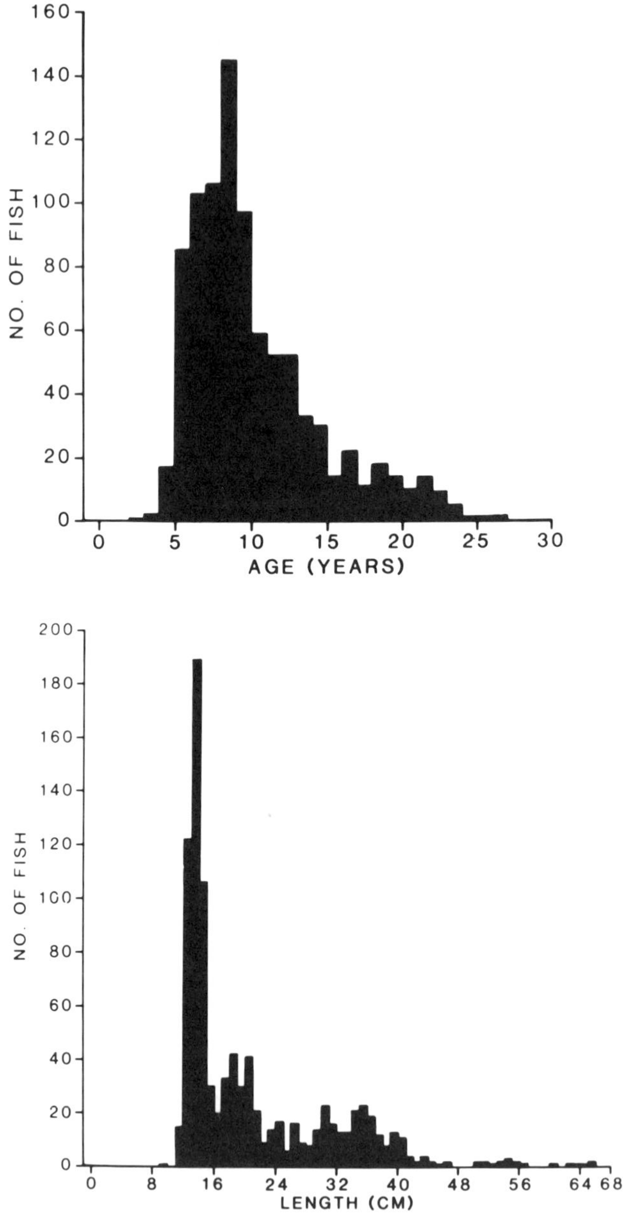

FIG. 4 Size-frequency and age-frequency distributions of Arctic charr from Shona Lake collected during 1977 and 1978 for *Diphyllobothrium* counts.

sampled by the nets. Nonetheless, the number of fish exceeding 300 mm in length and 10 years of age may be considered to reflect, adequately, the real size- and age-class structure of this important segment of the population.

Using as a criterion for gonadal development an oocyte diameter greater than 3.0 mm or testis width greater than 5.0 mm, the percentage of each age class developing towards spawning condition in August 1977 is given in Figure 5. The onset of sexual maturation occurs at an age of 3 to 4 years as is typical for other lake-resident charr populations in northern Canada (Hunter 1970, Johnson 1980). It is shown in Figure 5 that there is a trend towards a decline in the population of maturing fish among the older age classes such that only about 10-15 percent of males and females, age 10 or older, are reproductively active.

Diphyllobothrium in the Charr: Acquisition and Its Effects

It is informative to examine the pattern of plerocercoid acquisition with regard to the size and age of charr at Shona Lake. Figure 6a illustrates the prevalence of *Diphyllobothrium* in charr of all age classes sampled, and Figure 6b gives similar information for fish of various size classes. Prevalence reaches approximately 100 percent for fish exceeding 14 years of age and over 250 mm fork length.

Mean parasite burdens, expressed as the $\log_{10}$ of the total number of *Diphyllobothrium* per fish, steadily increase with age until about the tenth year, then appear to level off at an asymptotic value of roughly 100 plerocercoids per fish (Fig. 7). This leveling off of plerocercoid numbers could result from either a death rate of plerocercoids that compensates for recruitment of larvae into the fish, forming a dynamic equilibrium, or it could be the result of higher mortality among the more heavily parasitized fish. Data lending support to the latter explanation is presented in Figure 8, where the variance-to-mean (s^2/x) ratio of plerocercoid counts is indicated for each age class of charr in June, July, and August 1977. The marked rise in s^2/x during the summer months may indicate one segment of a typical seasonal cycle: heavily infected fish appear in mid and late summer as *Diphyllobothrium* transmission occurs, followed by winter mortalities of these heavily infected fish and a reduction in s^2/x values. Given the stable nature of mean plerocercoid burden among the older age-classes in the lake, as shown in Figure 7, the seasonal changes in s^2/x offer some evidence that parasite-induced host mortality is a regular feature of the char/*Diphyllobothrium* system. Henricson (1978) also gives similar information on charr infected with *Diphyllobothrium* in Swedish lakes, where again the asymptotic value of mean plerocercoid burden is approximately 100 plerocercoids per fish.

The trend mentioned previously towards a decline in the percentage of maturing charr among the older cohorts represented at Shona Lake may also be

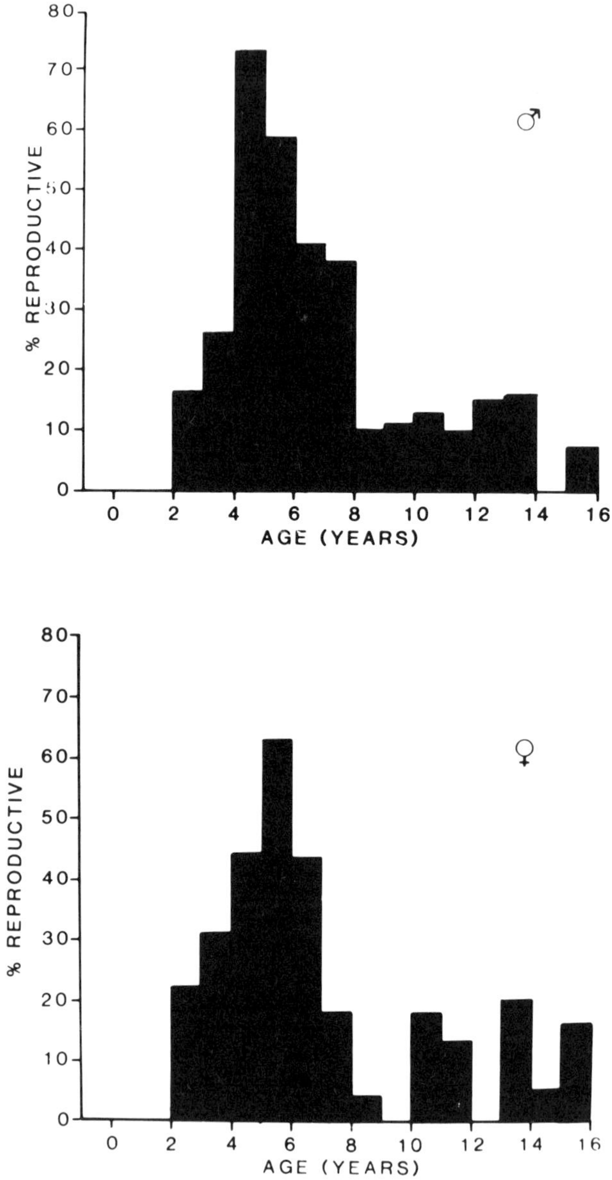

FIG. 5 Age-maturation histograms for male and female Arctic charr at Shona Lake in August 1977-78. Female charr were categorized as mature if oocyte diameters exceeded 3.0 mm. Males were categorized as mature if maximal testes width exceeded 5.0 mm.

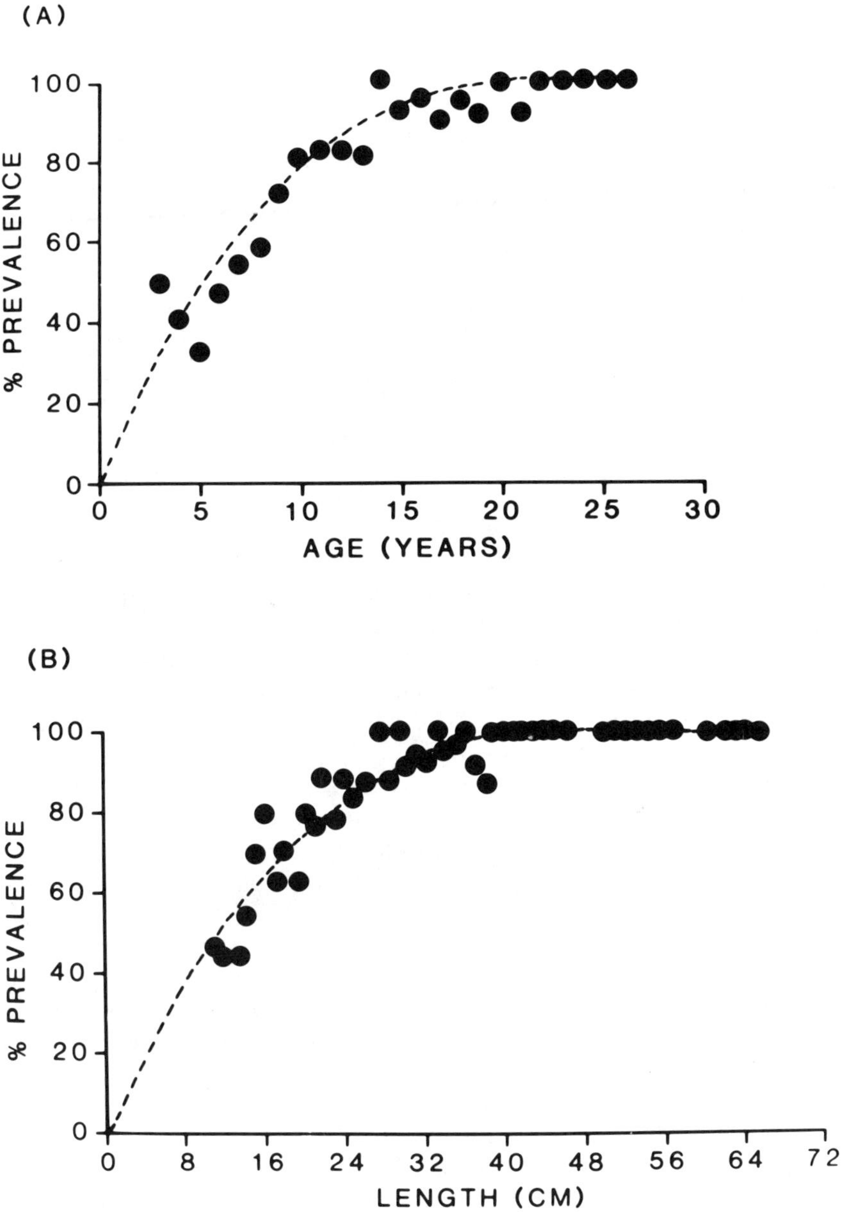

FIG. 6 Age-prevalence and size-prevalence curves for infections of *Diphyllobothrium* spp. in Arctic charr at Shona Lake.

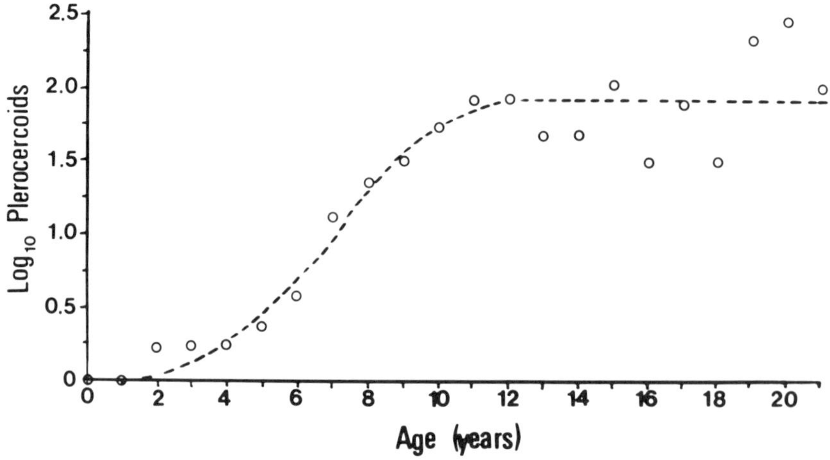

FIG. 7 Trend line for age-specific mean abundance of *Diphyllobothrium* spp. in Arctic charr from Shona Lake.

related to plerocercoid burden (Fig. 9). There is clearly a relationship between *Diphyllobothrium* intensity of infection and the likelihood of fish maturing. Spawning inhibition of charr with plerocercoid burdens between 5 and 25 plerocercoids per 100 g is not absolute; presumably it depends upon other factors such as general condition of the fish. Females appear more significantly affected than males. No reproductively developing charr were found with plerocercoid burdens over 30 plerocercoids per 100 g.

In first examining the charr of Shona Lake it soon became evident that fish feeding upon sticklebacks usually harboured heavy infections of *Diphyllobothrium* plerocercoids. This observation is confirmed by data in Figure 10 where frequency distributions for *Diphyllobothrium* occurrence in piscivorous and non-piscivorous charr are compared. The data included are for charr greater than 250 mm to ensure that only charr of a size large enough to be able to adapt to piscivorous feeding are used for comparison. Piscivorous charr were those whose stomachs contained recognizable remains of sticklebacks or small charr. Charr were considered as non-piscivorous when their stomachs contained food items which did not consist of fish. The smallest charr found with sticklebacks in its stomach measured 200 mm in length. Any charr of that size or greater may be considered capable of adopting a piscivorous habit.

It is already known that fish can acquire *D. dendriticum* by direct transmission of plerocercoids, but on the basis of experimental work on *Salmo trutta* (Halvorsen and Wissler 1973), it has been considered unlikely that *D. ditremum* plerocercoids could penetrate the stomach tissue and encyst. Therefore in 1978

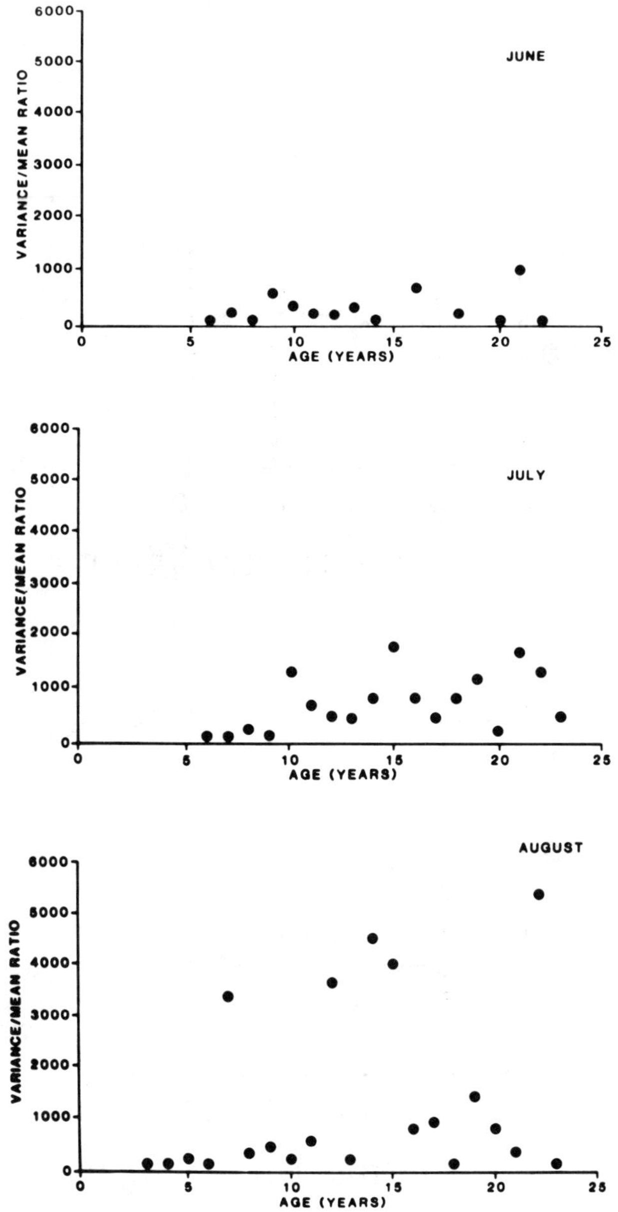

FIG. 8 Age-specific variance-to-mean (s^2/x) values for *Diphyllobothrium* spp. in Shona Lake charr during June, July and August 1977.

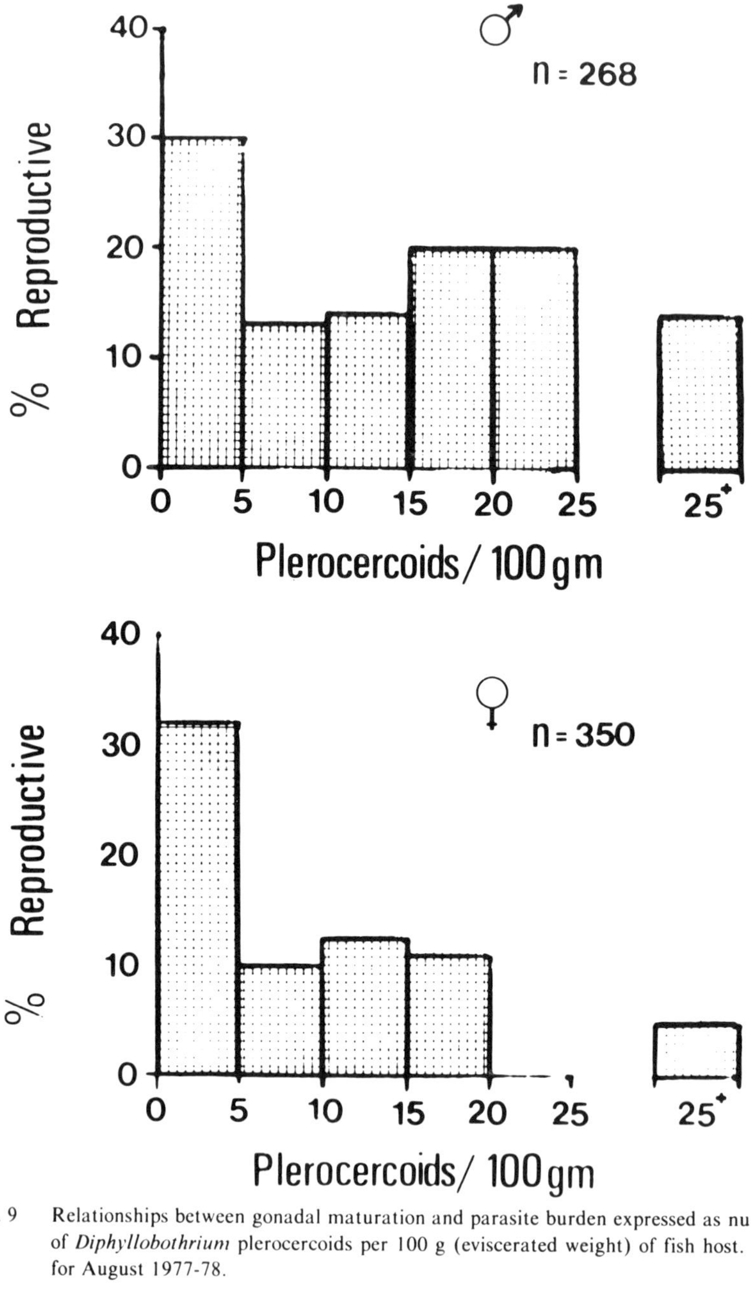

FIG. 9 Relationships between gonadal maturation and parasite burden expressed as number of *Diphyllobothrium* plerocercoids per 100 g (eviscerated weight) of fish host. Data for August 1977-78.

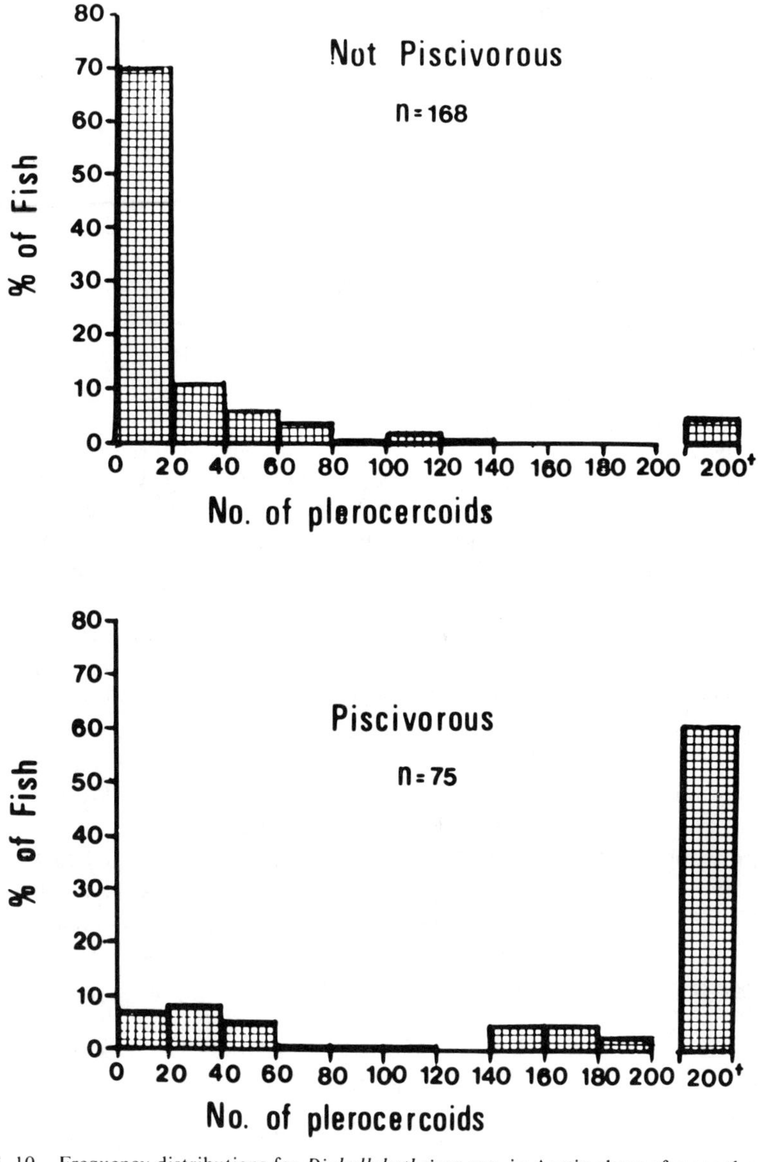

FIG. 10 Frequency distributions for *Diphyllobothrium* spp. in Arctic charr of more than 250 mm in length. Charr categorized as piscivorous had remainders of ninespine sticklebacks or small charr in their stomachs. The stomach contents of charr categorized as non-piscivorous consisted primarily of larval and adult insects. Fish with empty stomachs were not included in the non-piscivorous category.

experiments were carried out on site at Shona Lake to test whether or not *D. ditremum* plerocercoids could infect charr. The results indicated that *D. ditremum* plerocercoids are capable of infecting charr, with as many as 40 of the 100 force-fed plerocercoids encysting on the stomachs of the small (100-120 mm) charr used in the experiments.

CONCLUSION

The finding that parasitism by *Diphyllobothrium* spp. exerts deleterious effects upon a host such as Arctic charr is not surprising, particularly in view of the gross pathology observed in heavy infections. The evidence presented here for inhibition of reproductive development and for increased mortality among parasitized fish is based upon field data rather than experimental work, but it seems reasonably sound. In view of this, perhaps there should be a heightened awareness among fish biologists as to the possibility that parasites may seriously affect the essential characteristics of charr populations. The significance of *Diphyllobothrium* infections may be especially acute for lake-resident charr which, in eastern Arctic Canada, tend to be much more heavily parasitized than anadromous charr.

Recent theoretical work in modelling host-parasite systems has involved the premise that parasites are inherently capable of reducing the growth rate (r) of host populations (Anderson 1979). A reduction in host population growth rate can be accomplished either by decreasing the birth rate or increasing the death rate or, as likely for the charr/*Diphyllobothrium* system, by simultaneously affecting both factors. While the actual calculations for the decrease in r for Shona Lake charr depend upon more detailed information than can be presented here, it is appropriate to note that there is indeed evidence for a significant effect upon the population. It remains to be seen how generally these results may be applied to other charr stocks; however, it does at least appear that in future work on Arctic charr dynamics the potential impact of *Diphyllobothrium* and other pathogenic parasites should be duly considered.

REFERENCES

Andersen, K. 1975. The morphological stability of *Diphyllobothrium* Cobbold. A comparison of adult *D. dendriticum* (Nitzsch), *D. latum* (L.) and *D. ditremum* (Creplin) developed in different hosts. Norw. J. Zool. 23: 45-53.

Anderson, R.M. 1979 The influence of parasitic infection on the dynamics of host population growth, p. 245-281. *In* R.M. Anderson, B.D. Turner and L.R. Taylor [eds.] Population Dynamics. Blackwell Scientific, Oxford.

Grainger, E.H. 1953. On the age, growth, migration, reproductive potential and feeding habits of the Arctic char (*Salvelinus alpinus*) of Frobisher Bay, Baffin Island. J. Fish. Res. Board Can. 10:326-370.

Halvorsen, O. 1970. On the taxonomy and biology of plerocercoids of *Diphyllobothrium* Cobbold, 1858 (Cestoda, Pseudophyllidea) from northwestern Europe. Nytt Mag. Zool. 18:113-174.

Halvorsen, O., S.K. Wissler. 1973. An experimental study of the ability of *Diphyllobothrium latum* (L.), *D. dendriticum* (Nitzsch) and *D. ditremum* (Creplin) (Cestoda, Pseudophyllidea) to infect paratenic hosts. Norw. J. Zool. 21:201-210.

Henricson, J. 1977. The abundance and distribution of *Diphyllobothrium ditremum* in the char *Salvelinus alpinus* in Sweden. J. Fish Biol. 11:231-248.

Henricson, J. 1978. The dynamics of infection of *Diphyllobothrium dendriticum* (Nitzsch) and *D. ditremum* (Creplin) in the char *Salvelinus alpinus* (L.) in Sweden. J. Fish Biol. 13:51-71.

Hunter, J.G. 1970. Production of Arctic char (*Salvelinus alpinus* L.) in a small Arctic lake/Keyhole Lake, Victoria Island, N.W.T. Fish. Res. Board Can. Tech. Rep. 321:190 p.

Johnson, L. 1980. The Arctic charr, *Salvelinus alpinus*, p. 15-98. *In* E.K. Balon [ed.] Charrs: salmonid fishes of the genus *Salvelinus*. Dr. W. Junk Publishers, The Hague.

Vik, R. 1964. Plerocercoids of *Diphyllobothrium* spp. from the Rossaga water system, Nordland County. Nytt Mag. Zool. 12:1-9.

Distribution and Biology of Swimbladder Nematodes, *Cystidicola* spp. (Habronematoidea), in Charr, *Salvelinus* spp.

GEOFF A. BLACK

Department of Zoology, University of Guelph, Guelph, Ontario N1G 2W1

MURRAY W. LANKESTER

Department of Biology, Lakehead University, Thunder Bay, Ontario P7B 5El

BLACK, G.A. AND M.W. LANKESTER. 1984. Distribution and biology of swimbladder nematodes, *Cystidicola* spp. (Habronematoidea), in charr, *Salvelinus* spp. p. 413–429. *In* L. Johnson and B.L. Burns [eds.] Biology of the Arctic charr, Proceedings of the International Symposium on Arctic Charr, Winnipeg, Manitoba, May 1981. Univ. Manitoba Press, Winnipeg.

The distribution of *Cystidicola cristivomeri* White (1941) in North America and of *C. farionis* Fischer (1798) in North America and Eurasia in fish hosts of the genus *Salvelinus* is reviewed. The systematic relationship of these swimbladder nematodes is discussed.

A field study of the biology of *C. cristivomeri* in Arctic charr, *S. alpinus*, in Gavia Lake, Kent Peninsula, N.W.T., and in *S. namaycush* in three lakes in northwestern Ontario illustrates that this nematode has a life span of at least 10 years. Growth of female worms and the rate at which they release eggs is regulated in a density-dependent way.

INTRODUCTION

Nematodes of the genus *Cystidicola* are conspicuous parasites inhabiting the lumen of the swimbladder of salmonid fishes; two species are recognized: *Cystidicola farionis* Fischer (1798) which is circumpolar and has been reported from a variety of salmonids, and *Cystidicola cristivomeri* White (1941) found only in North America, and maturing only in *Salvelinus* spp. The two species are most readily distinguished by the morphology of their eggs. The eggs of *C. farionis* have polar, or polar and lateral, filaments while those of *C. cristivomeri* have only lateral lobes.

Cystidicola farionis develops to the infective stage in various amphipods including *Pontoporeia affinis, Gammarus fasciatus*, and *Hyalella azteca* (see

Baylis 1931; Smith and Lankester 1979). *Mysis relicta* is the only known intermediate host of *C. cristivomeri* (see Smith and Lankester 1979; Black and Lankester 1980). Upon ingestion, third-stage larvae migrate via the pneumatic duct to the swimbladder of the fish host (Black and Lankester 1980). Fish paratenic hosts are not believed to play a role in the transmission of swimbladder nematodes to piscivorous salmonids (Black and Lankester 1981).

In this paper we review the distribution of *Cystidicola* spp. and comment on their systematic relationship. The biology of *C. cristivomeri* in *Salvelinus* spp. is also discussed.

Distribution and Systematic Relationships of Cystidicola spp. In Fishes of the Genus Salvelinus

Cystidicola cristivomeri matures only in fish hosts of the genus *Salvelinus* and is found only in North America (Fig. 1). It has been reported in landlocked and anadromous Arctic charr, *S. alpinus*, from the northern Foxe Basin, Northwest Territories (N.W.T.) (Jamieson 1972); from Nauyuk, Willow (authors' unpublished data) and Gavia lakes, Kent Peninsula, N.W.T. (Black and Lankester 1981); from Stanwell-Fletcher Lake, Somerset Island, N.W.T. (Eddy and Lankester 1978); and from Char Lake, Cornwallis Island, N.W.T. (Beverly-Burton 1978). The nematode has been reported in lake trout, *S. namaycush*, from several inland lakes in central (White 1941; Ko and Anderson 1969) and northwestern Ontario (Dechtiar 1972; Lankester and Smith 1980), from La Martre Lake, N.W.T. (Ko and Anderson 1969), and from Nauyuk Lake (authors' unpublished data). Brook trout, *S. fontinalis*, has been reported only once as a host of *C. cristivomeri* in White Partridge Lake, Algonquin Provincial Park, Ontario (Lankester and Smith 1980). *Cystidicola cristivomeri* probably has a more continuous distribution in the north-central Canadian waters occupied by Salvelinus spp. and *Mysis relicta* than is indicated by reports in the literature. The nematode has not been found in western waters of the Arctic coast of Alaska and Yukon Territory (Mudry and McCart 1976; Arthur et al. 1976) nor in eastern Canadian waters of Quebec (Choquette 1948; Hanek and Molnar 1974), coastal Labrador (Hicks and Threlfall 1973) and Newfoundland (Threlfall and Hanek 1970). Curiously, *C. cristivomeri* is absent in *S. namaycush* and *S. fontinalis* in the upper Great Lakes although *M. relicta* is present (Lankester and Smith 1980).

Cystidicola cristivomeri probably speciated as a parasite of *S. namaycush* rather than *S. alpinus*, possibly during the last glaciation in an upper Mississippi refugium. The known distribution of this nematode coincides with that of *S. namaycush* thought to have dispersed from such a refugium (Khan and Qadri

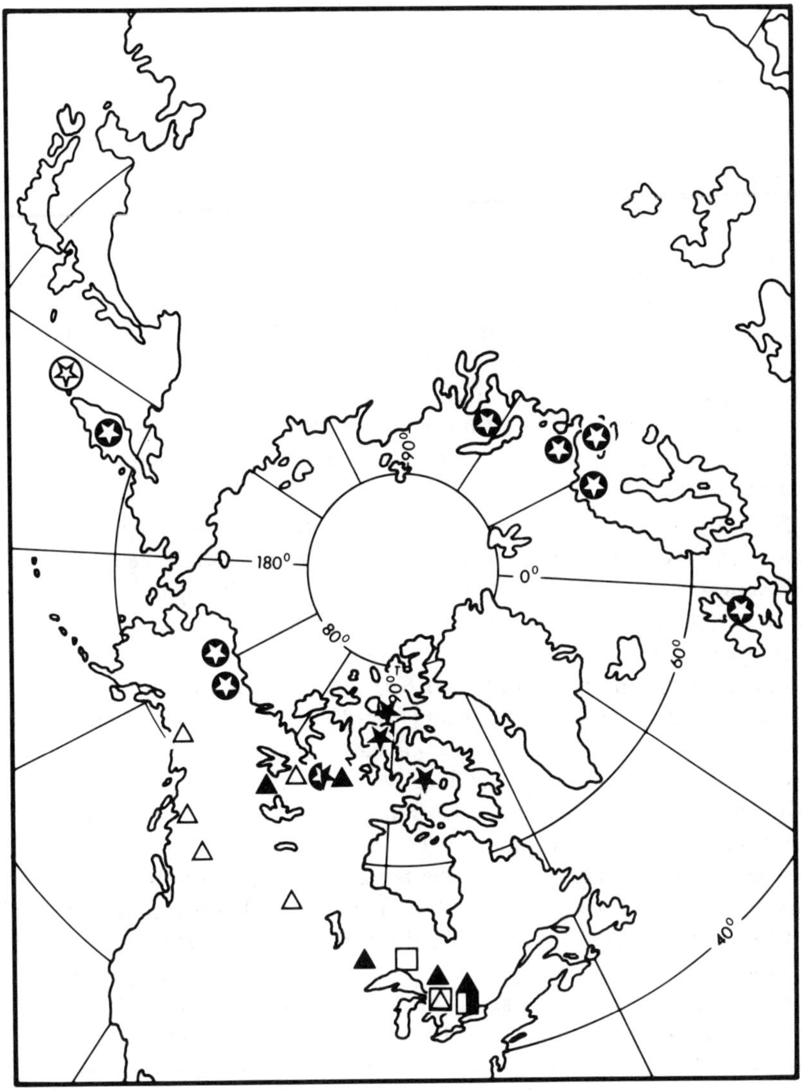

FIG. 1 Distribution of *Cystidicola cristivomeri* and *Cystidicola farionis* in *Salvelinus* spp.
stars = *S. alpinus;* closed, *C. cristivomeri;* open, *C. farionis.*
triangle = *S. namaycush;* closed, *C. cristivomeri;* open, *C. farionis.*
square = *S. fontinalis;* closed, *C. cristivomeri;* open, *C. farionis.*
(circled star) = *S. malma, C. farionis.*
(square with triangle) = *S. fontinalis* X *S. namaycush, C. farionis*

1971). If *C. cristivomeri* arose as a parasite of *S. alpinus*, presumably it would have dispersed with this host from the Bering and other glacial refugia (McPhail and Lindsey 1970) to the Arctic coast of Asia as well as to North America, but it did not. Probably, *C. cristivomeri* evolved with *S. namaycush* and later, with the postglacial movement of this host into Arctic Canada, infected *S. alpinus*.

Cystidicola farionis (= *C. stigmatura* according to Campana-Rouget 1955; Moravec and Ergens 1970; Arthur et al. 1976) occurs widely in both North America and Eurasia in fish hosts of the genera *Brachymystax, Coregonus, Hucho, Oncorhynchus, Osmerus, Prosopium, Salmo, Salvelinus, Stenodus, Thymallus* (see Bykhovskaya-Pavlovskaya et al. 1962; Skrjabin et al. 1967; Moravec and Ergens 1970; Margolis and Arthur 1979). Other species in which *C. farionis* has been reported (*Esox*, Walton 1928; *Couesius*, Richardson 1942; *Gasterosteus, Leuciscus*, and *Lota*, Bykhovskaya-Pavolvskaya et al. 1962; *Apeltes*, Hanek and Threlfall 1970; and *Catastomus*, Arthur et al. 1976) probably represent accidental hosts and, in some cases, the nematode identification needs confirmation.

In North America, members of the genus *Salvelinus* reported as hosts of *C. farionis* include *S. alpinus, S. namaycush, S. fontinalis*, and splake, *S. fontinalis X S. namaycush* (Fig. 1). The extent of the distribution of this parasite throughout the ranges of these fishes undoubtedly is incompletely known. The worm has been reported in *S. alpinus* from the Shaviovik River, Alaska and Joe Creek, Yukon Territory (Mudry and McCart 1976), and from Nauyuk and Willow Lakes (authors' unpublished data). One charr from Nauyuk Lake had both *C. farionis* and *C. cristivomeri*. *Cystidicola farionis* has not been found in anadromous or landlocked charr of the Arctic islands (Beverly-Burton 1978; Eddy and Lankester 1978). Some early reports of *C. farionis* in *S. namaycush* and *S. fontinalis* cannot be interpreted clearly as *C. farionis* and *C. cristivomeri* both mature in these hosts and authors did not mention the presence or absence of filaments on the eggs of worms recovered (Wright 1879; Leidy 1886; Ward and Magath 1917; Richardson 1942; MacLulich 1943; Miller and Kennedy 1948; Rawson 1951; Dickson 1964). *Cystidicola farionis* with filamented eggs is reliably known in *S. namaycush* from Redrock Lake, N.W.T. (Ko and Anderson 1969), Aishihik Lake, Yukon Territory (Arthur et al. 1976), and Lake Nipigon, Ontario (Lankester and Smith 1980). *Cystidicola farionis* with filamented eggs is known in *S. fontinalis* from Hogan Lake, Algonquin Provincial Park, Ontario (Ko and Anderson 1969) and from Lake Nipigon and Wolf Camp Lake, northwestern Ontario (Lankester and Smith 1980). Splake, *S. fontinalis X S. namaycush*, from South Bay, Lake Huron, are also infected with *C. farionis* (Collins and Dechtiar 1974).

Fishes of the genus *Salvelinus* are apparently unsuitable hosts for *C. farionis* found in some Canadian waters. In Lakes Superior and Nipigon, for example, few, if any worms reach sexual maturity in *Salvelinus* spp. while large numbers of *C. farionis* mature in sympatric salmonids with the exception of lake whitefish, *Coregonus clupeaformis* (Lankester and Smith 1980). The swimbladder nematode with filamented eggs that matures in large numbers in *S. namaycush* from Aishihik Lake, for example (Arthur et al. 1976), may represent a distinct, as yet unrecognized, species transmitted from co-habiting *C. clupeaformis* (Lankester and Smith 1980; Black and Lankester 1980).

The systematic relationships of swimbladder nematodes in coregonids of North America are not entirely clear. *Cystidicola cristivomeri* is quite distinct and probably evolved with *S. namaycush*, presently only maturing in *Salvelinus* spp.. *Cystidicola farionis* and *C. stigmatura* (Leidy 1886, Ward and Magath 1917) were synonymized because morphological differences could not be detected. Apparent biological differences between swimbladder nematodes with filamented eggs (Lankester and Smith 1980; Black and Lankester 1980) now encourage speculation that more than one species exist. Hypothetically, one species may have evolved with lake herring (subgenus *Leucichthys*) or *Prosopium* spp. now maturing in these fishes in the Great Lakes watershed, as well as in the introduced *Oncorhynchus* spp. and *Salmo gairdneri*. Another species, maturing in *C. clupeaformis may infect Coregonus* spp., *S. namaycush*, and *S. alpinus* in western Canadian and Arctic waters. When *Coregonus* sp. invaded North America in the late Pliocene (Behnke 1972), it may have been accompanied by a swimbladder nematode. Whether this was the first swimbladder nematode in North America or an addition to existing fauna is difficult to answer.

In Eurasia, *S. alpinus* (= *S. leucomaenis, S. boganida*, and *S. jacuticus*, considered eastern races of Arctic charr in accordance with McPhail and Lindsey 1970) and *S. malma* are listed as hosts of *C. farionis* (see Bykhovskaya-Pavlovskaya et al. 1962; Skrjabin et al. 1967). Only a few additional references available to us specify the locality of infected charr (Fig. 1). Ko and Anderson (1969) described museum specimens of *C. farionis* from *S. alpinus* from Ennerdale Water, Cumberland, England. Kennedy (1977) reported *C. farionis* in *S. alpinus* from Fiskelausvann Lake, northern Norway but did not find the nematode in other lakes of this region nor in charr from Spitsbergen or Jan Mayen Islands. *Cystidicola farionis* has been reported in anadromous *S. alpinus* entering the White Sea (Dogiel and Petrushevski 1935) and the Barents Sea from coastal Murmansk (Polyanskii 1966) and from the Arctic island, Novaya Zemlya (Dogiel and Markov 1937). Makhovenko (1972) reported *C. farionis* in *S. alpinus* from Kamchatka and Roytman (1968) suggested that *C. farionis* existed in *S. malma* from the Kurile Archipelago.

RESULTS

Biology of *C. cristivomeri* in *Salvelinus* spp.

The biology of *C. cristivomeri* was studied in *S. alpinus* in Gavia Lake, Kent Peninsula, N.W.T. (68° 20′N 107° 45′ W; area = 17 ha; depth: maximum = 8 m, mean = 2.9 m) and in *S. namaycush* in Greenwater Lake (48° 34′N 90° 26′W; area = 3060 ha; depth: maximum = 55 m, mean = 24 m), Burchell Lake (48° 35′N 90° 38′W; area = 1027 ha; depth: maximum = 75 m, mean = 24 m) and Squeers Lake (48° 31′N 90° 33′W; area = 387 ha; depth: maximum = 34 m, mean = 11 m) of northwestern Ontario (Black and Lankester 1981).

In Gavia Lake, the mean TNW (total number of worms) per infected fish and mean NFEMW (number of sexually mature female worms with eggs) increased steadily in *S. alpinus* up to age 9 and then remained fairly constant (Fig. 2). The mean TNW in four-year-old fish was probably overestimated by a small sample size including one fish with 892 worms.

In Greenwater Lake, the mean TNW in *S. namaycush* increased almost linearly with age up to age five and then remained fairly constant (Fig. 3). Only nine-year-old fish had a significantly higher number of worms than lake trout of other age classes (Duncan's multiple range test $F = 3.1$ $df = 7{,}195$; ages listed in increasing order of magnitude of mean TNW; ages underscored by the same line are not significantly different $p > 0.05$).

$$\underline{6 \quad 10 \quad 8 \quad 5 \quad 7 \quad 11 \quad 12} \quad 9 \qquad [1]$$
$$ \underline{11 \quad 12 \quad 9}$$

The mean NFEMW increased slowly in fish up to age five and then increased more rapidly up to age nine (Fig. 3). Understandably, the mean number of immature worms reached a maximum in *S. namaycush* at age five and then declined steadily in older fish as they became primarily piscivorous. Also, the mean minimum length of immature worms (an inverse measure of the rate of recent infection estimated by measuring the smallest worm in each fish) increased sharply in fish from age five to age six.

In Burchell Lake, the mean TNW in lake trout increased up to age nine (Fig. 4). The mean NFEMW in fish increased slowly up to age five and more rapidly in older fish.

In Squeers Lake, the mean TNW increased in lake trout up to age six and then remained fairly constant (Fig. 5). The mean NFEMW was low in fish up to age five and then increased sharply.

There were considerable differences in the numbers of worms in fishes between the lakes (Fig. 2-6). The mean NFEMW was not significantly different in four-year-olds or in five-year-olds between the lakes despite significant differences in the mean TNW at each fish age.

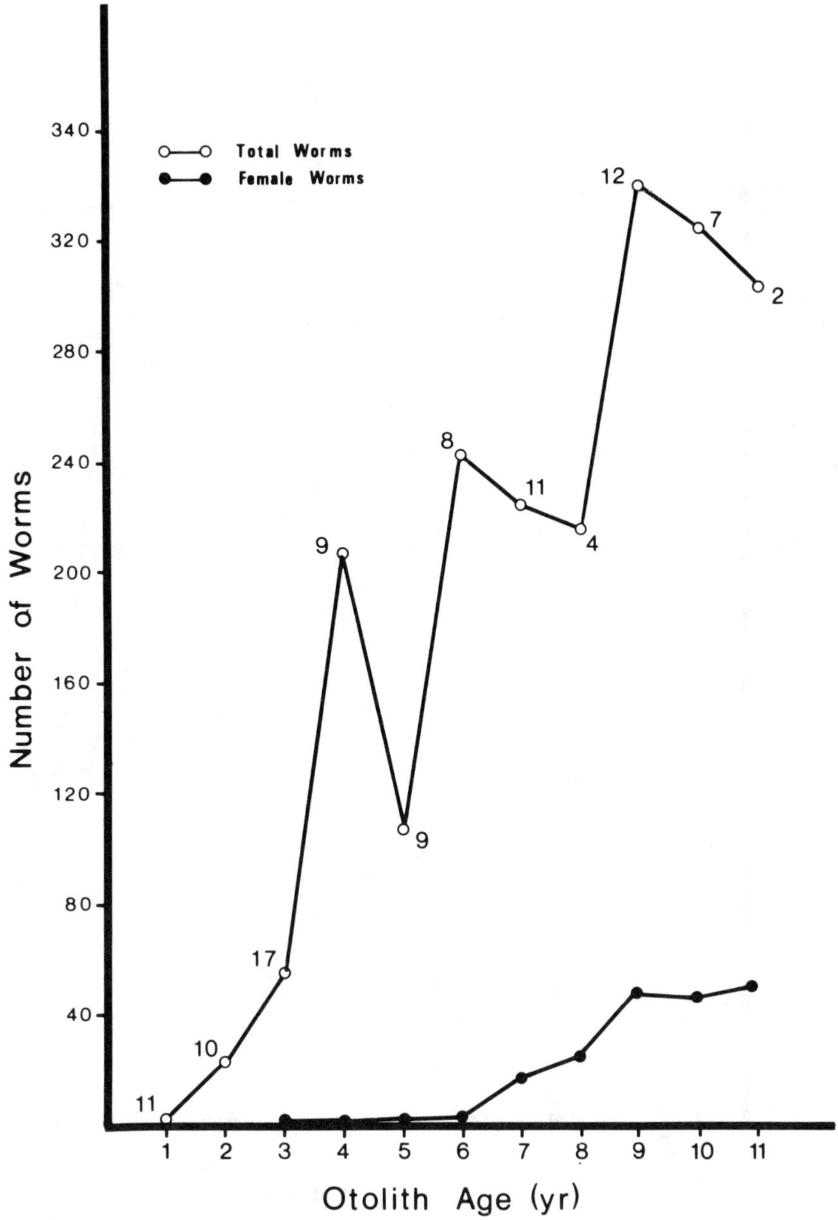

FIG. 2 Mean numbers of *C. cristivomeri* in relation to age of *S. alpinus* from Gavia Lake, Northwest Territories (fish sample sizes indicated).

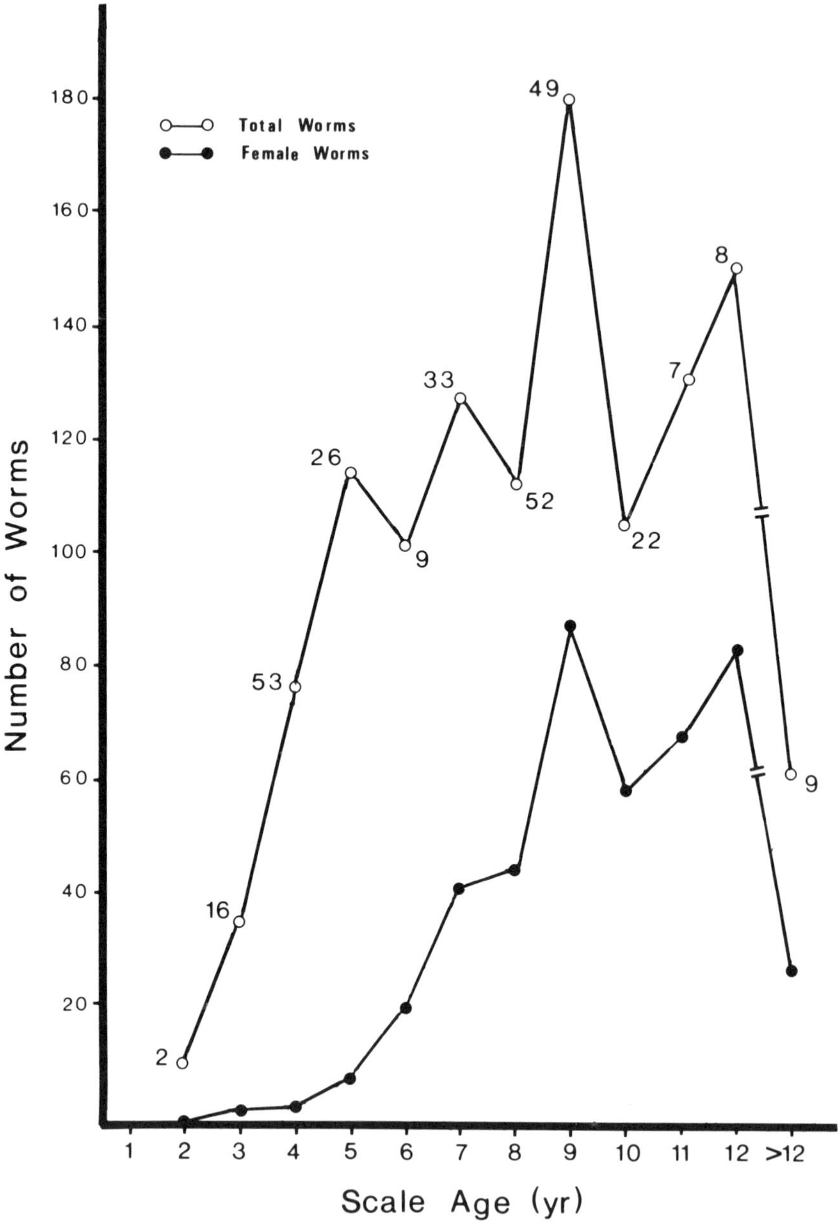

FIG. 3 Mean numbers of *C. cristivomeri* in relation to age of *S. namaycush* from Greenwater Lake, northwestern Ontario (fish sample sizes indicated).

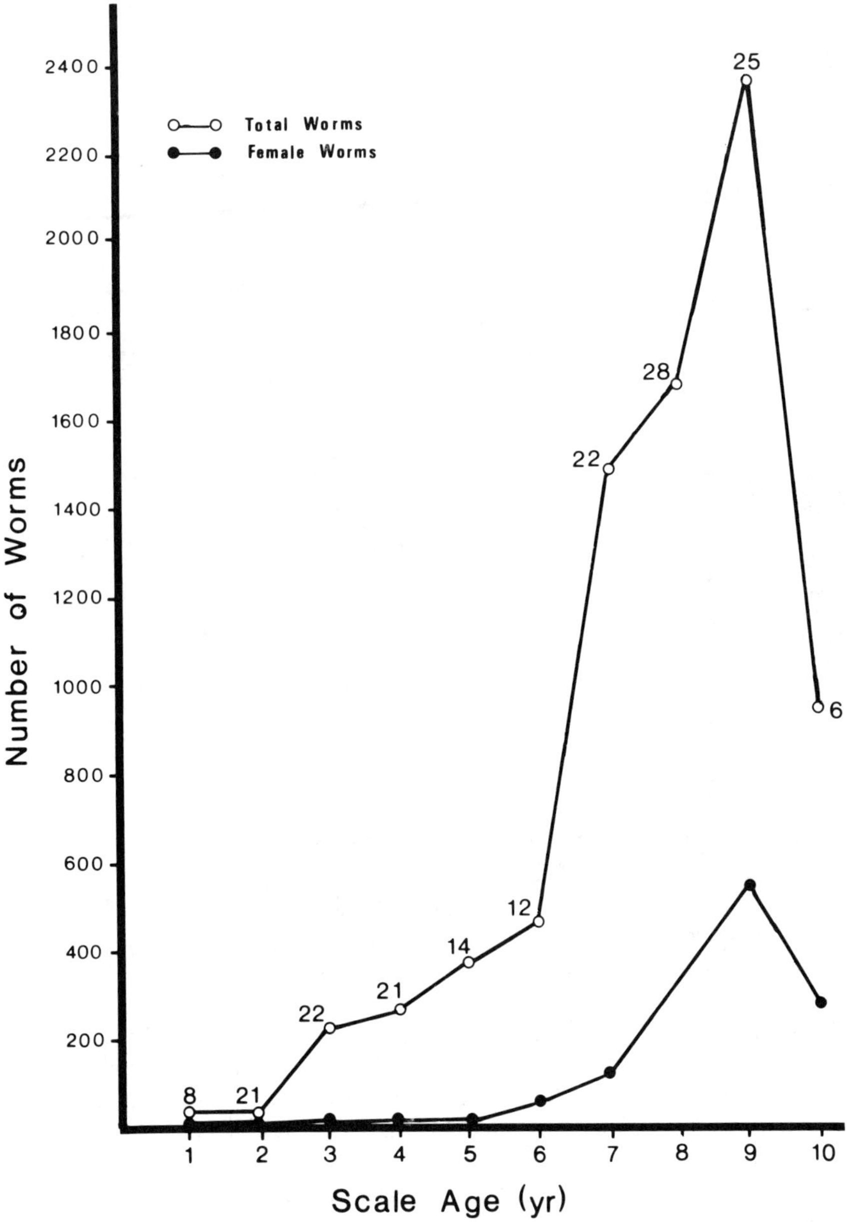

FIG. 4 Mean numbers of *C. cristivomeri* in relation to age of *S. namaycush* from Burchell Lake, northwestern Ontario (fish sample sizes indicated).

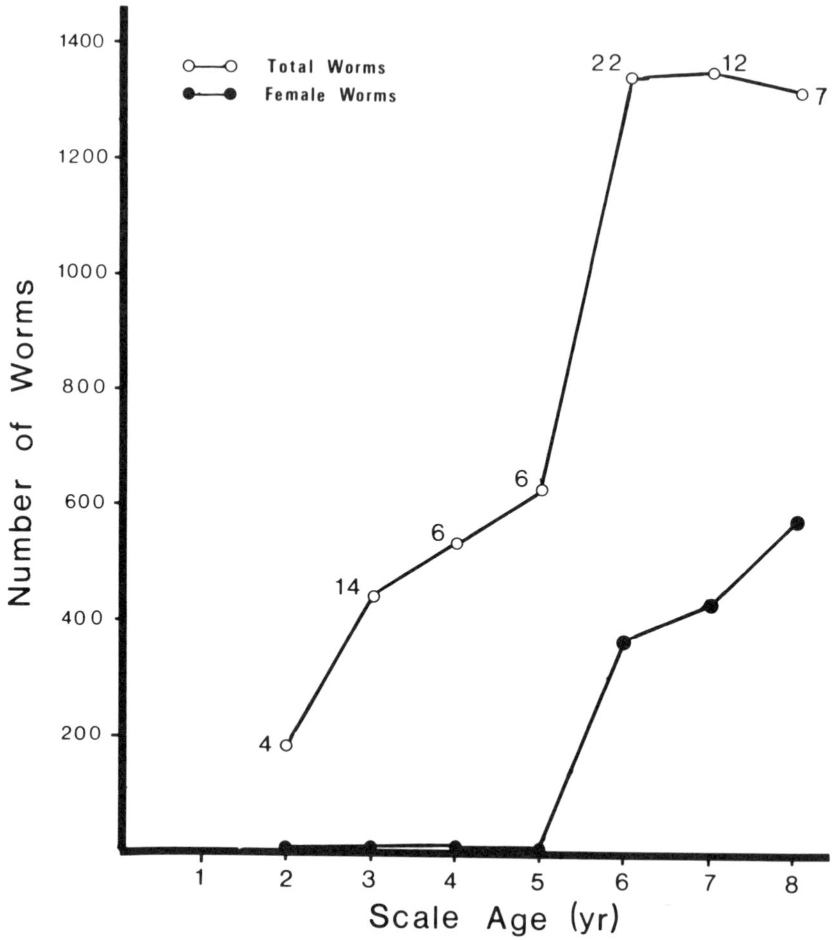

FIG. 5 Mean numbers of *C. cristivomeri* in relation to age of *S. namaycush* from Squeers Lake, northwestern Ontario (fish sample sizes indicated).

In experimentally infected *S. namaycush, Cystidicola cristivomeri* developed to sexual maturity within two years at 4-10°C with a mean of 4.6 worms per fish (Black and Lankester 1980). However, in the wild, *S. namaycush* with large numbers of *C. cristivomeri*, mature female worms were not abundant until fish were older than five years. These observations suggest that the development of sexual maturity in *C. cristivomeri* may be retarded when large numbers of worms are present in the swimbladder. If *C. cristivomeri* develops more slowly under crowded conditions, the area of inner surface of the swimbladder available to

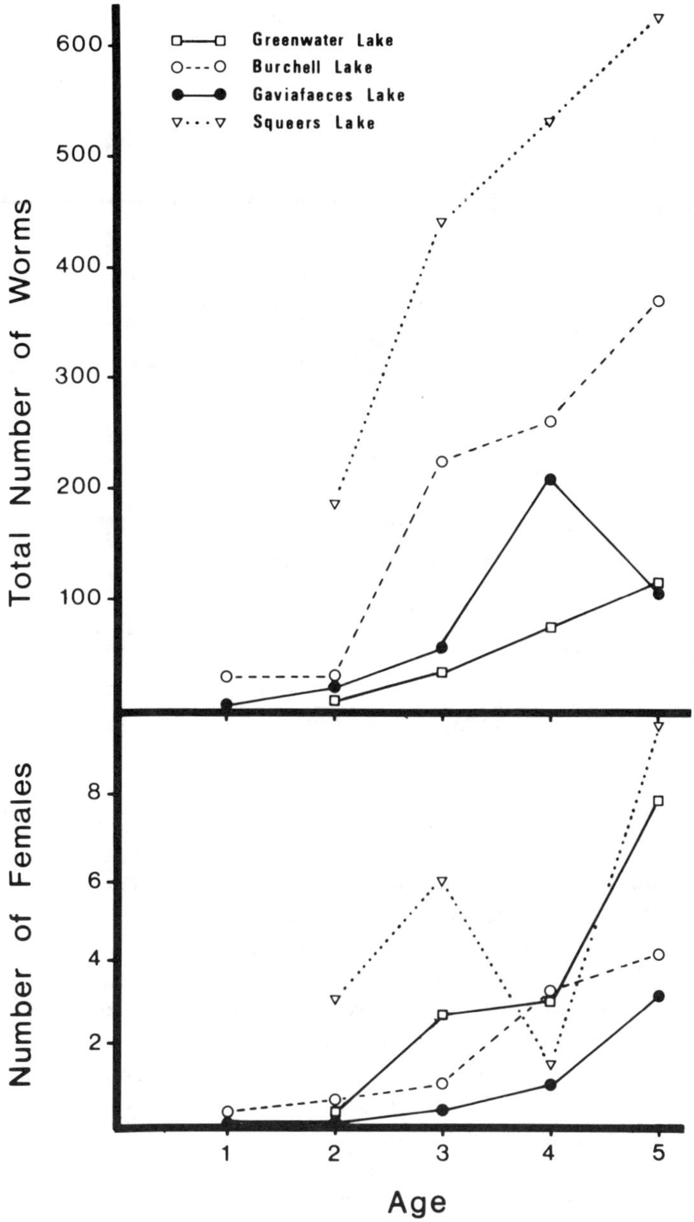

FIG. 6 Mean numbers of *C. cristivomeri* in *S. namaycush* and *S. alpinus* 1-5 yr old from lakes in northwestern Ontario and Gavia Lake, N.W.T., respectively.

individual worms may also be important. These hypotheses were tested by combining data for *S. namaycush* in all lakes in a stepwise multiple regression analysis to determine the effect of total number of worms and length of fish on the percentage of worms present in three-year-old fish that had become mature females in fish at age five ($PFEMW_{3\text{-}5}$). This analysis accounts for differences in recruitment rates of immature worms into fish in all lakes. Length of fish was used as an approximate measure of swim bladder surface area since swim bladder length is linearly related to fish length ($r = 0.99$, $n = 368$). A correction factor (cf) was calculated to estimate the number of worms a five-year-old fish had when it was three years old. This was calculated by assuming, for example, that individual five-year-old fish with more than the mean number of worms in fish at that age class ($\overline{TNW_5}$) would have had proportionately more worms than the mean ($\overline{TNW_3}$) when it was three years old.

Therefore:

$$cf = \frac{\text{TNW in a fish at age five in lake X}}{\overline{TNW_5} \text{ in fish from lake X}} \qquad [2]$$

$$PFEMW_{3\text{-}5} = \frac{\text{NFEMW in a fish at age five in lake X}}{(\overline{TNW_3} \text{ in lake X}) \times cf} \qquad [3]$$

The average number of worms present in a fish from age three to age five ($TNW_{3\text{-}5}$) was:

$$TNW_{3\text{-}5} = \frac{(\overline{TNW_3} + \overline{TNW_4} + \overline{TNW_5}) \times cf}{3} \qquad [4]$$

The $PFEMW_{3\text{-}5}$ was significantly related to the $TNW_{3\text{-}5}$ and fish length. $TNW_{3\text{-}5}$ was the most important independent variable and explained 14 percent of the variance. Fish length was also significant and explained an additional 10 percent of the variance. The multiple regression equation is:

$$PFEMW_{3\text{-}5} = -23.9 - 0.035\, TNW_{3\text{-}5} + 1.5\, \text{Length} \qquad [5]$$

and was significant ($F = 6.7$, $df = 2{,}43$).

Multiple regression analyses were used with data at individual ages (greater than five) of *S. namaycush* and with all ages of *S. alpinus* in Gavia Lake, to examine the effect of TNW and fish length on the PFEMW. Generally, the analyses indicate that the PFEMW is correlated negatively with the TNW and positively with fish length.

The mean maximum length of female *C. cristivomeri* (determined by measuring the three longest worms from each fish examined) was greatest in *S. namaycush* from Greenwater Lake where fish had the least number of worms (Fig. 7). These observations suggest that the maximum length attained by female

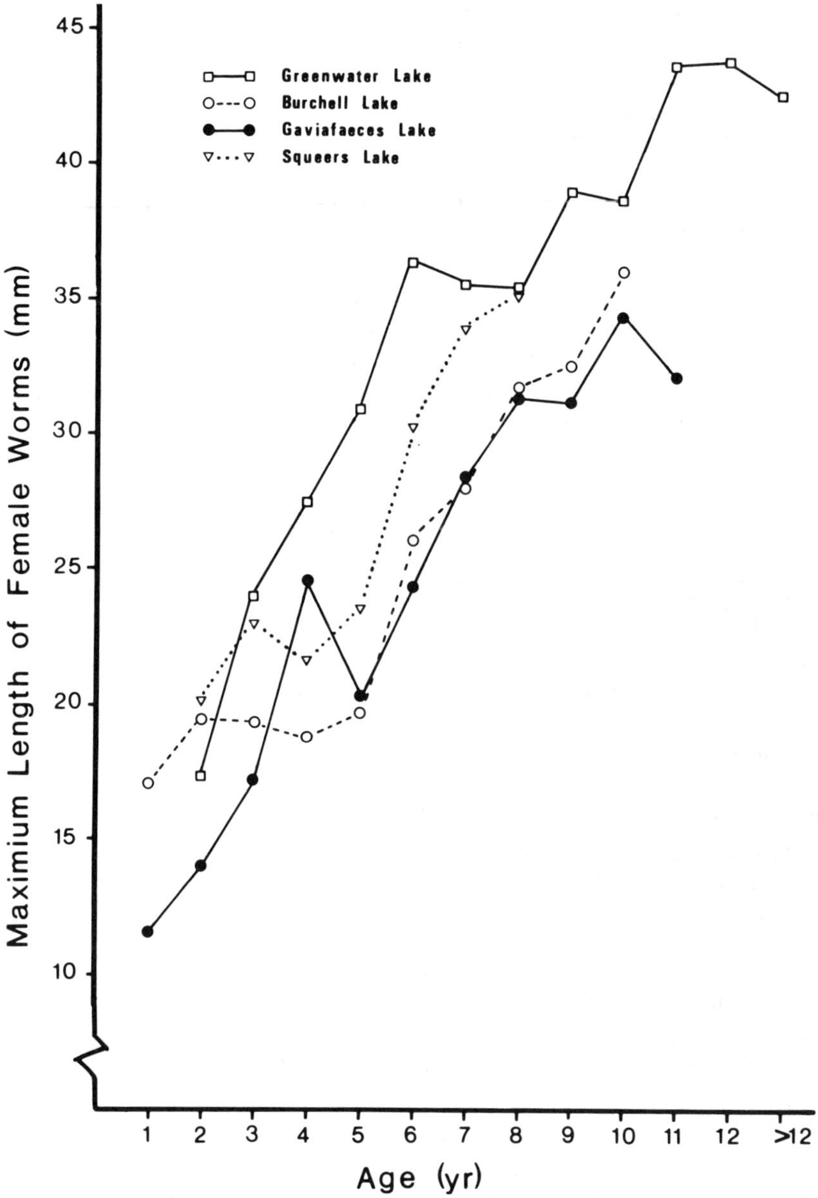

FIG. 7 Mean maximum length of female *C. cristivomeri* in relation to age of *S. namaycush* and *S. alpinus* from lakes in northwestern Ontario and Gavia Lake, N.W.T., respectively.

worms might be determined by the total number of worms present and perhaps by the space available in the swimbladder. Since the maximum length of female worms is positively correlated with fish age (Spearman r = 0.90, 0.83, 0.83 and 0.83 in Gavia, Greenwater, Burchell and Squeers lakes, respectively), subsequent analyses to determine the effect of mean TNW and fish length on the maximum length of females was done within individual age classes for each of the lakes (small sample sizes necessitated combining all ages in Gavia Lake). Generally, fish length was the most important independent variable and was positively correlated with the maximum length of female worms. The length of female worms was significantly and negatively correlated with the total number of worms only in Gavia Lake.

In vitro studies illustrate that the number of eggs released by female worms is positively correlated with worm length (Spearman r = 0.90 for worms taken from lake trout from Squeers Lake).

DISCUSSION

Laboratory and field evidence indicates that *C. cristivomeri* is long-lived. There was no mortality of worms after 600 days in experimentally infected *S. namaycush*; the maximum length of nematode attained after this time was 26.1 mm (Black and Lankester 1980). In Greenwater Lake, the maximum length of mature worms increased with increasing fish age (ages two to twelve) suggesting that worms present in young fish keep growing and are still present when the fish is much older. Also, there was no recruitment to fish older than five years (Black and Lankester 1981), the number of worms remaining essentially constant in fish from age 5 until over age 12 yr. Therefore, most *C. cristivomeri* in *S. namaycush* from Greenwater lake appear to live at least 10 years and some probably longer. Similarly, worms may be long-lived in other lakes where fishes are more heavily infected since the maximum length of mature worms continues to increase with fish age.

Female *C. cristivomeri* appear to mature more slowly in heavily infected fish than in those with fewer worms suggesting a density-dependent regulation of maturation. This is apparent upon casual observation of data for five-year-old fishes where mean numbers of mature females are essentially the same in fishes from each of the four lakes, despite significant differences between the total numbers of worms present. Analysis of data from five-year-old *S. namaycush* indicates that the percentage of mature female *C. cristivomeri* in individual fish is inversely correlated with the total number of worms and positively correlated with fish length. The effect of fish length probably reflects the space available to worms in the swimbladder. The same variables affecting the percentage of mature females were shown to be important, with reasonable consistency, when data for older *S. namaycush*, and data for *S. alpinus* of all ages from Gavia Lake, were analyzed.

The retarded growth of *C. cristivomeri* in heavily infected fishes appears to continue after worms reach maturity. In Greenwater Lake, where the intensity of infection is low, the maximum length of females was consistently higher than in more heavily infected fish from other lakes. The similarity in the maximum length of female worms in older fish from Greenwater and Squeers lakes is probably a result of underestimating the ages of older fish from Squeers Lake. In Gavia Lake, the maximum length of females was negatively correlated with total numbers of worms and positively correlated with fish length. If the maximum length attained by female worms is determined in a density-dependent way then the laboratory observation that short worms lay eggs at a slower rate than larger worms is of obvious significance. As recruitment of nematodes to individual fish increases, the proportion of female worms maturing decreases as does their growth after maturity and the number of eggs released. Density-dependent regulatory mechanisms minimize fluctuations in the size of parasite populations within a system (Anderson 1978) and have been previously identified in only a few fish parasite populations (Holmes et al. 1977; Kennedy 1977; Mills et al. 1979).

ACKNOWLEDGEMENTS

We are indebted to Dr. Lionel Johnson, Freshwater Institute, Winnipeg, for providing facilities, equipment and hospitality at Nauyuk Lake, N.W.T. and for aging Arctic charr. Mrs. Irene Weir, London, Ontario aged lake trout. We also gratefully acknowledge the technical assistance of Mrs. M. Jones and Mrs. L. Ryder. Personnel of the Ontario Ministry of Natural Resources, Thunder Bay, Ontario, including Messrs. T. Chisholm, R. Ryder, D. Sakiyama, C.A. Elsey and R. Hamilton were extremely helpful during this study. Financial support was provided in the form of a Natural Sciences and Engineering Research Council of Canada operating grant to M.W.L.

REFERENCES

Anderson, R.M. 1978. The regulation of host population growth by parasite species. Parasitology 76:119-157.

Arthur, J.R., L. Margolis and H.P. Arai. 1976. Parasites of fishes of Aishihik and Stevens lakes, Yukon Territory, and potential consequences of their interlake transfer through a proposed water diversion for hydroelectrical purposes. J. Fish. Res. Board Can. 33:2489-2499.

Baylis, H.A. 1931. *Gammarus pulex* as an intermediate host of trout-parasites. Ann. Mag. Nat. Hist. 7:431-435.

Behnke, R.J. 1972. The systematics of salmonid fishes of recently glaciated lakes. J. Fish. Res. Board Can. 29:639-671.

Beverly-Burton, M. 1978. Metazoan parasites of Arctic char (*Salvelinus alpinus* L.) in a high Arctic landlocked lake in Canada. Can. J. Zool. 56:365-368.

Black, G.A. and M.W. Lankester. 1980. Migration and development of swim-bladder nematodes, *Cystidicola* spp. (Habronematoidea), in their definitive hosts. Can. J. Zool. 58:1997-2005.

Black, G.A. and M.W. Lankester. 1981. The transmission, life span, and population biology of *Cystidicola cristivomeri* White, 1941 (Nematoda: Habronematoidea) in char, *Salvelinus* spp. Can. J. Zool. 59:498-509.

Bykhovskaya-Pavlovskaya, I.E., et al. 1962. Key to the parasites of freshwater fish of the USSR, Transl. Israel Program for Sci. Transl. 1964.

Campana-Rouget, Y. 1955. Sur deux nouveaux genres de spirurides parasites de poissons; discussion systematique des genres voisins. Ann. Parasitol. Humaine Comp. 30:346-362.

Choquette, L.P.E. 1948. Parasites of freshwater fish. IV. Internal helminths parasitic in speckled trout (*Salvelinus fontinalis* (Mitchill)) in rivers and lakes of the Laurentide Park, Quebec, Canada. Can. J. Res. Sect. D. 26:204-211.

Collins, J.J. and A.O. Dechtiar. 1974. Parasite fauna of kokanee salmon (*Oncorhynchus nerka*) introduced into Lake Huron. J. Fish. Res. Board Can. 31:1818-1821.

Dechtiar, A.O. 1972. Parasites of fish from Lake of the Woods, Ontario. J. Fish. Res. Board Can. 29:275-283.

Dickson, I.W. 1964. Parasites of some fishes in Manitoba. M.Sc. thesis, Univ. of Manitoba, Winnipeg, Manitoba.

Dogiel, V.A. and G.S. Markov. 1937. Age dynamics of parasite fauna of *Salvelinus alpinus* in Novaya Zemlya. Trudy Soc. Nat. Leningr. 64:434-455. (In Russian).

Dogiel, V.A. and G.S. Petrushevski. 1935. An attempt at an ecological study of salmon in the White Sea. Probl. Ecol. Biocen. I. Leningrad. (In Russian).

Eddy, S.B. and M.W. Lankester. 1978. Feeding and migratory habits of Arctic char, *Salvelinus alpinus*, indicated by the presence of the swimbladder nematode *Cystidicola cristivomeri* White. J. Fish. Res. Board Can. 35:1488-1491.

Hanek, G. and K. Molnar. 1974. Parasites of freshwater and anadromous fishes from Matamek River system, Quebec. J. Fish. Res. Board Can. 31:1135-1139.

Hanek, G. and W. Threlfall. 1970. Helminth parasites of the fourspine stickleback (*Apeltes quadracus* (Mitchill)) in Newfoundland. Can. J. Zool. 48:404-406.

Hicks, F.J. and W. Threlfall. 1973. Metazoan parasites of salmonids and coregonids from coastal Labrador. J. Fish Biol. 5:399-415.

Holmes, J.C., R.P. Hobbs and T.S. Leong. 1977. Populations in perspective: Community organization and regulation of parasite populations. In Regulation of parasite populations. Edited by G.W. Esch. Academic Press, New York.

Jamieson, J.L. 1972. Parasites of *Salvelinus alpinus* (Salmonidae) in the northern Foxe Basin, Northwest Territories, with emphasis on those of medical importance. M.Sc. thesis, Univ. of Toronto, Toronto, Ontario.

Kennedy, C.R. 1977. The regulation of fish parasite populations. In Regulation of parasite populations. Edited by G.W. Esch. Academic Press, New York.

Khan, N.Y. and S.U. Qadri. 1971. Intraspecific variations and postglacial distribution of lake char (*Salvelinus namaycush*). J. Fish. Res. Board Can. 28:465-476.

Ko, R.C. and R.C. Anderson. 1969. A revision of the genus *Cystidicola* Fischer, 1798 (Nematoda: Spiruroidea) of the swim bladder of fishes. J. Fish. Res. Board Can. 26:849-864.

Lankester, M.W. and J.D. Smith. 1980. Host specificity and distribution of the swim-bladder nematodes, *Cystidicola farionis* Fischer, 1798 and *C. cristivomeri* White, 1941 (Habronematoidea), in salmonid fishes of Ontario. Can. J. Zool. 58:1298-1305.

Leidy, J. 1886. Notices of nematoid worms. Proc. Acad. Nat. Sci. Philadelphia. 38:308-313.

MacLulich, D.A. 1943. Parasites of trout in Algonquin Provincial Park, Ontario. Can. J. Res. Sect. D. 21:405-412.

Margolis, L. and J.R. Arthur. 1979. Synopsis of the parasites of fishes of Canada. Bull. Fish. Res. Board Can. 199.

Makhovenko, E.T. 1972. On the parasite fauna of different forms of the Arctic charr, *Salvelinus alpinus*, from Kamchatka. Parazitologiya 6:369-375.

McPhail, J.D. and C.C. Lindsey. 1970. Freshwater fishes of northwestern Canada and Alaska. Bull. Fish. Res. Board Can. 173. Ottawa.

Miller, R.B. and W.A. Kennedy. 1948. Observations on the lake trout of Great Bear Lake. J. Fish. Res. Board Can. 7:176-189.

Mills, C.A., R.M. Anderson and J.P. Whitfield. 1979. Density dependent survival and reproduction within populations of the ectoparasitic digenean *Transversotrema patialense* on the fish host. J. Anim. Ecol. 48:383-399.

Moravec, F. and R. Ergens. 1970. Nematodes from fishes and cyclostomes of Mongolia. Folia Parasitol. (Prague), 17:217-232.

Mudry, D.R. and R.S. Anderson. 1977. Helminth and arthropod parasites of fishes in the mountain national parks of Canada. J. Fish Biol. 11:21-33.

Mudry, D.R. and P. McCart. 1976. Metazoan parasites of Arctic char (*Salvelinus alpinus*) from the north slope of Canada and Alaska. J. Fish. Res. Board Can. 33:271-275.

Polyanskii, Y.I. 1966. Parasites of the fish of the Barents Sea. Israel Program of Scientific Translations. Jerusalem.

Rawson, D.S. 1951. Studies of the fish of Great Slave Lake. J. Fish. Res. Board Can. 8:207-240.

Richardson, L.R. 1942. The parasites of the fishes of Lake Wakonichi, central northern Quebec. Trans. Am. Fish. Soc. 71:286-289.

Roytman, V.A. 1968. On certain species of nematodes from the swimbladder of salmonid fishes of the genera *Oncorhynchus* and *Salvelinus* of the far east. Izdatel'stvo ''Nauka'', Moscow, 4:144-150 (In Russian).

Skrjabin, K.I., A.A. Sobolev and V.M. Ivashkin. 1967. Essentials of Nematodology: Spirurata of animals and man and the diseases caused by them. Part 4, Thelazioidea. Israel Program of Scientific Translations, Jerusalem, 1971.

Smith, J.D. and M.W. Lankester. 1979. Development of swimbladder nematodes (*Cystidicola* spp.) in their intermediate hosts. Can. J. Zool. 57:1736-1744.

Threlfall, W. and G. Hanek. 1970. Metazoan parasites of salmonids and coregonids from the Avalon Peninsula, Newfoundland. J. Fish. Res. Board Can. 27:1894-1897.

Walton, A.C. 1928. A revision of the nematodes of the Leidy collection. Proc. Acad. Nat. Sci. Phila. 79:49-163.

Ward, H.B. and T.B. Magath. 1917. Notes on some nematodes from freshwater fishes. J. Parasitol. 3:57-64.

White, F.M. 1941. Studies on the morphology of *Cystidicola cristivomeri* sp. nov. from the swimbladder of lake trout (abstract). Proc. Indiana Acad. Sci. (for 1940) 50:211.

Wright, R.R. 1879. Contributions to American helminthology. Proc. Canadian Inst. 1:54-75.

Cytotaxonomy of North American *Salvelinus*

TED M. CAVENDER

Museum of Zoology, The Ohio State University, Columbus, Ohio 43210

CAVENDER, T.M. 1984. Cytotaxonomy of North American *Salvelinus*, p. 431–445. *In* L. Johnson and B.L. Burns [eds.] Biology of the Arctic charr, Proceedings of the International Symposium on Arctic Charr, Winnipeg, Manitoba, May 1981. Univ. Manitoba Press, Winnipeg.

Chromosome analyses of five species of North American *Salvelinus* yielded the following results: *S. fontinalis* 2n = 84, NF = 100; *S. namaycush* 2n = 84, NF = 100; *S. confluentus* 2n = 78, NF = 102; *S. malma* 2n = 82, NF = 98; and *S. alpinus* 2n = 78, NF = 98.

S. fontinalis and *S. namaycush* were included with the three other forms in this study to provide a direct comparison of karyotypes for all the North American species; the karyotypes being achieved with a uniform set of methods by the same person. *S. fontinalis* and *S. namaycush* together with *S. leucomaenis* possess a primitive karyotype for *Salvelinus*. They have 16 metacentric-submetacentric and 68 acrocentric chromosomes. When arranged according to length from longest to shortest, there is an even gradation through the metacentric and acrocentric series.

S. confluentus, S. malma and *S. alpinus* possess derived karyotypes from the primitive condition in *Salvelinus*. The karyotype of *S. confluentus* is unique, having 24 metacentric-submetacentric chromosomes and an arm number of 102. One submetacentric pair is interpreted as having resulted from a pericentric inversion involving a uni-armed chromosome. The other bi-armed chromosomes probably result from Robertsonian fusion of acrocentric elements. The relatively large degree of difference between the karyotypes of *S. confluentus* and *S. malma* provides another criterion for taxonomic separation of these two species. *S. kronocius* Viktorovsky from the Kamchatka River basin has a karyotype (2n = 78, NF = 100) closest to that of *S. confluentus*. The karyotype of the latter species could be derived from that of *S. kronocius* through one additional chromosomal change.

S. malma and *S. alpinus* both possess a long pair of telocentric chromosomes which is distinctly longer than others in the acrocentric series and breaks the size continuity between the bi-armed and uni-armed chromosomes. It is believed that this is a homologous chromosome in the two species, having evolved by means of two chromosomal rearrangements; first, a centromeric fusion of two acrocentrics and second, a terminalization of the centromere. These two changes resulted in both a reduction in chromosome number and a reduction in arm number. The long telocentric marker pair can be recognized in karyotypes published for *Salvelinus malma* from Kamchatka. The shared derived characteristics found in the karyotypes of *Salvelinus malma* and *Salvelinus alpinus* suggest a close relationship. On the basis of our present state of knowledge, the southern forms of the Dolly Varden found on both sides of the north Pacific basin have an identical karyotype and may represent the same taxon once continuous across the Aleutian chain. The Soviets use the name *S. malma Krasheninnikovi Taranetz* for this taxon but the name *S. malma curilus* (Pallas) may have priority. The degree of difference between the karyotypes of the southern Dolly Varden and that determined for the northern Dolly Varden by Viktorovsky (2n = 76-78, NF = 96) suggest that the two taxa may be specifically distinct and that the northern Dolly Varden *S. malma malma* has a sister relationship with *S. alpinus* (2n = 78, NF = 98).

INTRODUCTION

Chromosome studies on North American species of *Salvelinus* have been restricted to the two species *S. namaycush* (Wahl 1960, Davisson, Wright and Atherton 1973) and *S. fontinalis* (Svärdson 1945; Uyeno 1972; Capanna et al. 1973; Davisson, Wright, and Atherton, 1973; Muramoto et al. 1974; Wright 1955).No chromosome information has been published on North American *S. alpinus, S. malma* or *S. confluentus* from the Pacific Northwest. This is in contrast to the many papers available on the karyology of *Salvelinus* from the western Pacific Areas (see Viktorovsky 1978).

The primary purpose for undertaking this work was to provide karyological information on two North American species of *Salvelinus (S. malma* and *S. confluentus*) which have long been confused taxonomically (Cavender 1978). It was anticipated that details of the karyotypes would provide further evidence for discrimination of these two species. In addition, the three other recognized North American species of *Salvelinus (S. namaycush, S. fontinalis*, and *S. alpinus*) were karytoyped and included in this paper for comparison with published karyotypes of forms of *Salvelinus* on the western side of the North Pacific basin.

The intent of these comparisons is to find information in the gross chromosome morphology that might be of use in determining interrelationships within the genus.

MATERIALS AND METHODS

Five specimens of *Salvelinus malma*, nine specimens of *S. confluentus*, three each of *S. namaycush* and *S. fontinalis*, and four *S. alpinus* were used for this study. These individuals are catalogued in the Ohio State University Museum of Zoology (OSUM) collection of fishes.

The populations sampled for chromosome analysis are as follows: *S. malma*, Soleduck River above Soleduck Falls, Olympic National Park, Clallam Co., Washington; *S. confluentus*, Bear Creek, tributary of Middle Fork of Flathead River bordering Glacier National Park, Flathead Co., Montana; *S. namaycush*, hatchery raised individuals from Jordan National Fish Hatchery, Charlevoix Co., Michigan, origin of breeding stock from Marquette Hatchery, Michigan-Lake Superior strain; *S. fontinalis*, Bear Creek tributary of Middle Fork of Flathead River bordering Glacier National Park, Flathead Co., Montana; *S. alpinus*, hatchery raised individuals from Rockwood Hatchery, Rockwood, Manitoba,

Canada; these fish were spawned from anadromous Arctic charr, Nauyuk Lake, NWT, Canada.

The chromosome preparations were made following the thumb squash technique described by McPhail and Jones (1966) with minor modifications for working in the field. Specimens were injected with 0.05 percent Velban (Lilly Co., Ltd.) one to four hours prior to the removal of gill arches, which were then placed in distilled water for one-half hour. Gills were fixed and stained by acetic orcein (60 percent glacial acetic acid). Chromosome preparations of *S. confluentus* and *S. malma* were made in the field. All karyotypes illustrated in this paper were derived from gill epithelial cells. The chromosome numbers given in the text always refer to the diploid complement. Chromosomes were arranged in the karyograms according to decreasing length. Most attention was paid to pairing the bi-armed chromosomes. Members of each pair were aligned in adjacent rows (above and below). This arrangement facilitates comparison of the relative size of chromosomes in a series. Table 1 gives the number of cells of each taxon which were counted and karyotyped.

RESULTS

A comparison of karyotypes obtained for the North American *Salvelinus* studied is presented in Table 2.

Salvelinus fontinalis and *Salvelinus namaycush* had identical karyotypes (Figs. 1 and 2) with 84 chromosomes and an arm number of 100. There were 8 pairs of metacentric-submetacentrics and 68 acrocentric chromosomes. Submetacentrics are present but difficult to identify in each spread. The longest acrocentric is distinctly shorter than the shortest bi-armed element.

Salvelinus confluentus was found to have 78 chromosomes with an arm number of 102. There are 12 pairs of metacentric-submetacentric and 54 acrocentric chromosomes in the karyotype (Fig. 3). The pairs labelled 6 and 10 are distinctly submetacentric. In every complete metaphase spread the number 10 pair could be easily picked out. In the acrocentric series, there is little, if any, difference in length between adjacent pairs. The longest acrocentric chromosome is slightly longer than the shortest metacentric (number 12).

Salvelinus malma has a karyotype (Fig. 4) with 82 chromosomes and an arm number of 98. There are 8 pairs of metacentric-submetacentric and 66 acrocentric chromosomes. Although submetacentrics occur, they are difficult to identify in all spreads. There is one distinctive long acrocentric pair that usually stands out in metaphase spreads where the chromosomes are less contracted. It is considerably longer than the shortest metacentric and matches the length of the fifth or sixth bi-armed pair.

Table 1. Chromosomes numbers of epithelial cells counted.

Species	Number of fish	75	76	77	78	79	80	81	82	83	84	Number of cells karyotyped
S. malma	5			1		2	1	3	12	1		5
S. confluentus	9	1	5	4	16	1	5	1				8
S. alpinus	4	1		1	9							6
S. namaycush	4				2	2	2		1		8	3
S. fontinalis	3						1		2	1	6	3

Table 2. Comparison of karyotypes of North American *Salvelinus* species.

Species	2n	NF	MSM	A
Salvelinus fontinalis	84	100	16	68
Salvelinus namaycush	84	100	16	68
Salvelinus malma	82	98	16	66
Salvelinus alpinus	78	98	20	58
Salvelinus confluentus	78	102	24	54

Salvelinus alpinus has 78 chromosomes with an arm number of 98. There are 10 pairs of metacentric-submetacentric and 58 acrocentric chromosomes in the karyotype (Fig. 5). As in *S. malma*, it is difficult to consistently pick out the submetacentric chromosomes. The longest acrocentric pair exceed the length of the shortest metacentric pair. In most karyotypes prepared, this long acrocentric is distinctly longer than the second member of the acrocentric series.

DISCUSSION

The results achieved for *Salvelinus fontinalis* and *Salvelinus namaycush* are the same as reported by previous workers (op. cit., p. 1). *S. fontinalis* has been karyotyped several times by different persons using specimens orginating from widely separated geographic areas in North America. The lack of variation is remarkable considering the high degree reported for some other *Salvelinus* species such as *S. malma* in Kamchatka. It was important to include karyotypes of *S. fontinalis* and *S. namaycush* in this paper since these two species are interpreted as having (along with *S. leucomaenis*) the primitive karyotype for *Salvelinus*: a chromosome number of 84 and an arm number of 100. The sixteen metacentric-submetacentric chromosomes and the uniform gradation in length of chromosomes from the bi-armed to the uni-armed series are also held to be primitive characteristics of the *Salvelinus* karyotype.

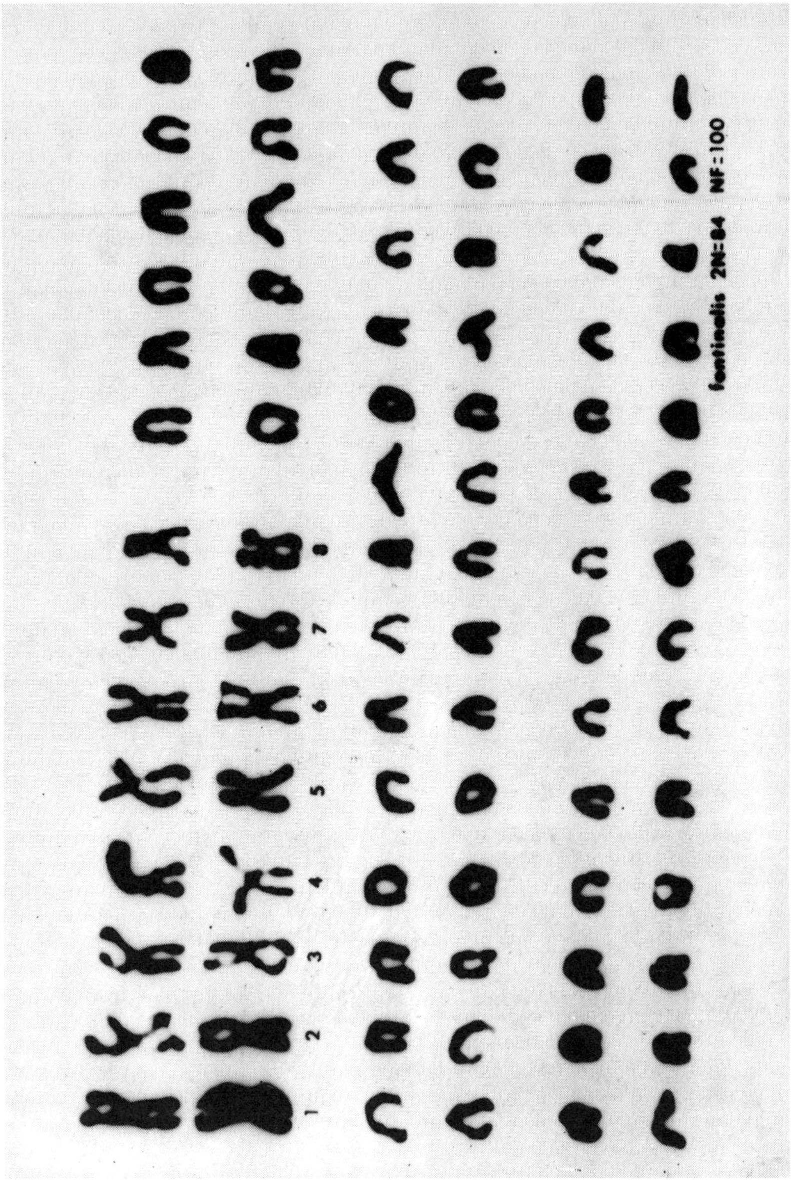

FIG. 1 Karyotype of *Salvelinus fontinalis* from the Flathead River drainage, Montana. Metacentric-submetacentric pairs are numbered (1-9). Other chromosomes are acrocentric.

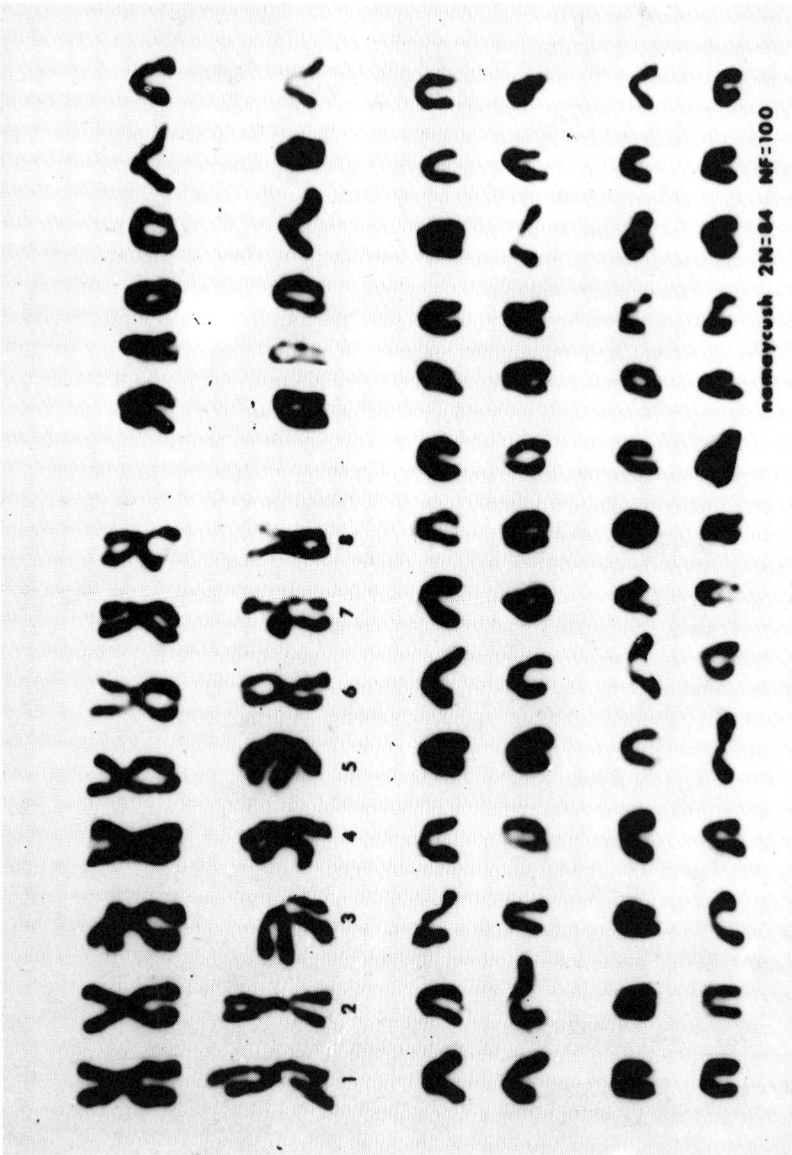

FIG. 2 Karyotype of *Salvelinus namaycush* from hatchery raised individuals originating from Lake Superior stock. Metacentric-submetacentric chromosome pairs are numbered (1-8).

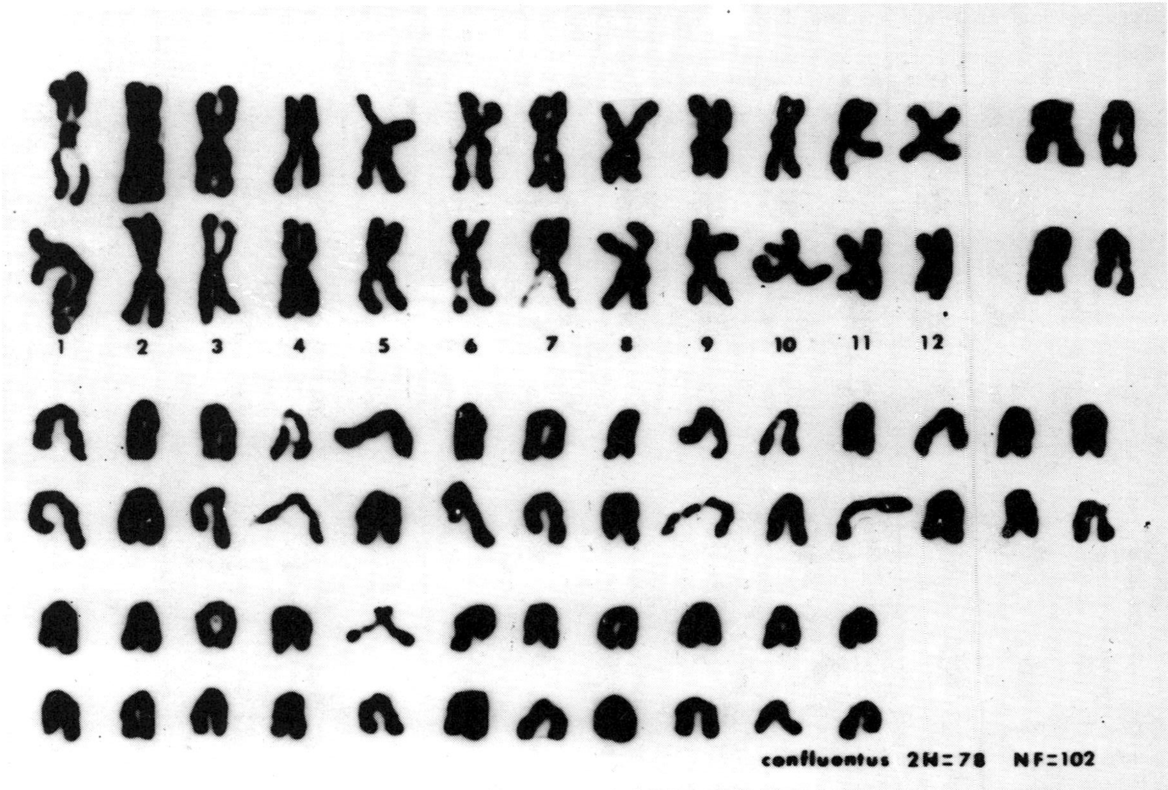

FIG. 3 Karyotype of *Salvelinus confluentus* from the Flathead River drainage, Montana. Metacentric-submetacentric chromosome pairs are numbered (1-12); remaining chromosomes are acrocentric.

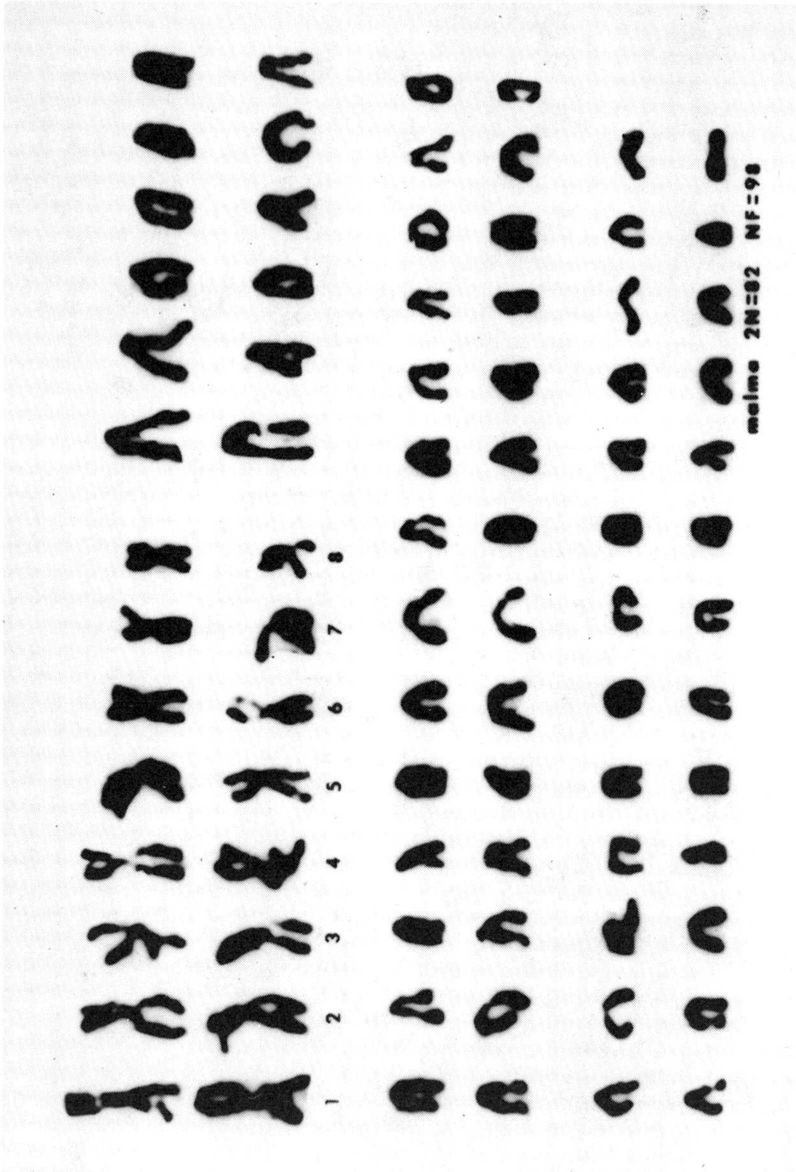

FIG. 4 Karyotype of landlocked *Salvelinus malma* from the Soleduck River, Washington. Metacentric-submetacentric chromosome pairs are numbered (1-10); other chromosomes are acrocentric.

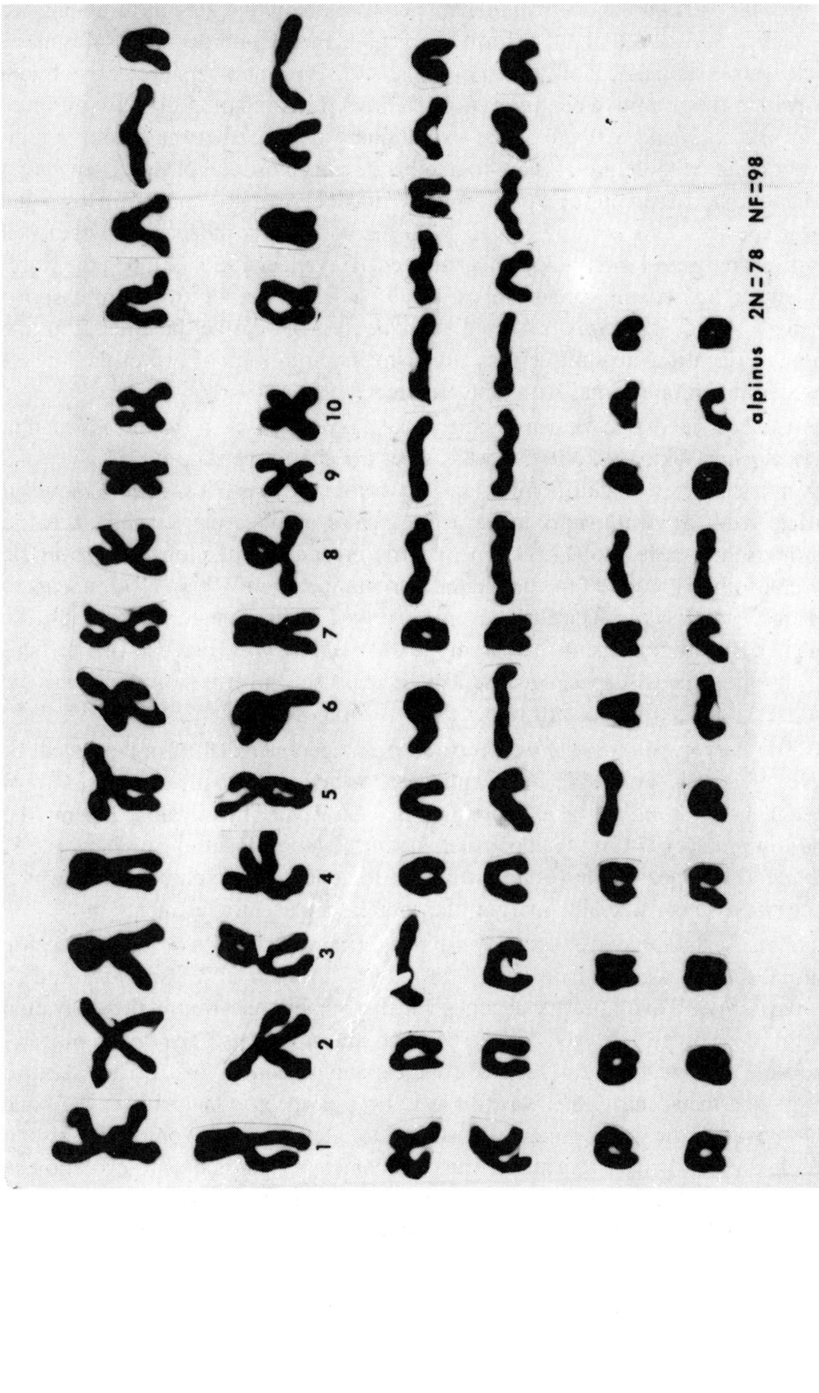

FIG. 5 Karyotype of anadromous *Salvelinus alpinus* from Nauyuk Lake, Northwest Territories, Canada.

As far as is known, *S. confluentus* possesses a unique karyotype among the species of *Salvelinus*: no other form has as high an arm number (102) or as many metacentric-submetacentric chromosomes (24). Both numbers are derived with respect to the primitive condition in *Salvelinus*. In comparing the karyotypes of *S. confluentus* and *S. malma*, the evolutionary distance between them is relatively great as there are at least four separate chromosomal changes separating the two taxa. Based on the chromosomal evidence, the two cannot be considered sister species. This is in agreement with the conclusion independently reached from an analysis of morphological characters (Cavender 1980). Among the North American *Salvelinus* species investigated, a minimum of two chromosomal changes separate *S. alpinus* from *S. confluentus*. To derive the latter from the former, a centromere shift or pericentric inversion in each of two of the longest acrocentric chromosomes would be required.

The best candidate for a sister species of *S. confluentus*, however, is found in *S. kronocius* (Viktorovsky 1975a) which lives in Lake Kronotskoe of the Kamchatka River basin, Kamchatka Peninsula. This form also has 78 chromosomes and differs from *S. confluentus* in having 11 pairs of metacentric-submetacentric chromosomes instead of 12. One pericentric inversion of the longest acrocentric chromosome is required to change the arm number from 100 in *S. kronocius* to 102 in *S. confluentus*. The chromosome involved in this change was probably the number 10 chromosome in the karyotype of *S. confluentus* (Fig. 3). This particular chromosome is suspect because its very short second arm indicates it was the product of a centromere shift rather than a fusion of acrocentrics.

The karyotype of *S. malma* prepared from specimens taken in the Soleduck River, Olympic Peninsula of Washington, is identical to that prepared from specimens of *S. malma* taken from the Shibetsu River of Hokkaido, Japan (Abe and Muramoto 1974) on the opposite side of the North Pacific. The Japanese *S. malma* also appears to have the distinctive long acrocentric chromosome which exceeds, by a considerable margin, the shortest metacentric element.

Because the karyotype of *S. malma* has eight pairs of metacentric-submetacentric chromosomes, which grade in size exactly as they do in *S. fontinalis* and *S. namaycush*, the evolutionary changes which probably occurred in the derivation of this karyotype involved the acrocentric chromosomes. The mechanism of probable change first involved a Robertsonian fusion of two acrocentrics to create one metacentric; this was followed by a pericentric inversion in the same chromosome, the centromere being shifted to a terminal position, giving rise to the distinctive long acrocentric. This mechanism of chromosome evolution in salmonines is not unique since it must have occurred several times in the formation of the long telocentric chromosomes found in the karyotype of *Salmo salar* (Roberts 1970).

For the purpose of cytotaxonomic analysis I have considered the creation of the long telocentric chromosome in *Salvelinus* to be an event that occurred but once in the chromosomal evolution of this genus. In contrast, the creation of additional metacentric-submetacentric chromosomes beyond the original eight, through the mechanism of Robertsonian fusion, must have occurred independently in at least two different phyletic lines; one line represented by *S. malma* and *S. alpinus*, the other represented by *S. confluentus* and *S. kronocius*. A pair of long telocentric chromosomes is apparent in the karyotypes published by Viktorovsky (1978) for the nosed charr, *S. malma schmidti* Viktorovsky, and for *S. malma malma*. Both these charrs are from Kamchatka.

The karyotype achieved in the present study for *S. alpinus* (2n = 78; NF = 98) was unexpected since the Swedish reports (Svärdson 1945; Nygren et al. 1971) of 80 chromosomes for this species have been widely accepted. However, these two publications disagreed on the arm number (96 and 100 respectively for the references cited). In addition, the metaphase spread that was used to illustrate the chromosomes of *S. alpinus* by Nygren et al. (1971) has 78 chromosomes (as correctly reported in the legend to this photograph: p. 265, Fig. 15) but this same spread when karyotyped (p. 264, Fig. 11) was shown to have 80 chromosomes. An explanation for the addition of two chromosomes to this cell was not given even though there are two elements in this karyotype that look like fragments. In order to check this discrepancy more carefully, I photocopied the illustrated metaphase spread in question (Fig. 15 in Nygren et al. 1971) and constructed a karyogram. The resulting karyotype conforms closely to that illustrated here for *S. alpinus* from N.W.T., Canada.

S. alpinus from the Northwest Territories, Canada, has 20 metacentric-submetacentric chromosomes in agreement with Nygren et al. (1971). Depending on the amount of contraction of the chromosomes, the longest telocentric in this species exceeds the shortest metacentric.

Working with *S. alpinus* from Norway, Gjedrem et al. (1977) found a number of cells with 78 chromosomes and a mean of 77.

Comparison with Western Pacific *Salvelinus*

Many of the publications dealing with the karyology of *Salvelinus* from western Pacific land areas have been written by R.M. Viktorovsky. He has summarized his chromosome work in a paper on the *Salvelinus* from Lake Kronotskoe (Viktorovsky 1978). Vasilyev (1975) has also studied the karyology of Kamchatkan *Salvelinus* with conflicting results.

For the southern *Salvelinus malma* from the Kedrova River, Maritime Territory (named *S. malma Krascheninnikovi Taranetz*) Chernenko and Viktorovsky (1971) found 84 chromosomes and an arm number of 96. This differs from the southern form of *Salvelinus malma* from Hokkaido studied by Abe and Muramoto (1974) and Muramoto et al. (1974) which had 82 chromosomes and 98 arms.

For the northern form of Dolly Varden named *S. malma malma* (Walbaum), Viktorovsky (1975b) reported 76 to 78 chromosomes with 96 arms. The sample included freshwater and sea-run fish from Kamchatka. In contrast to Viktorovsky's results, Vasilyev (1975) reported anadromous *Salvelinus* (probably northern Dolly Varden from the Bolshaya and Kamchatka Rivers but called *Salvelinus alpinus* by Vasilyev) with 82 chromosomes and 100 arms. Glubokovskii (1977a) does not consider *S. alpinus* to be one of the *Salvelinus* species that occupies the Kamchatka River basin but Behnke (1980) believes that the resident lake form as found in Lake Nachikinskoye and Lake Dalneye may be *S. alpinus*. Vasilyev (1975) reported the resident lake form from Lake Nachikinskoye with 82 chromosomes and 100 arms. Chernenko (1969) determined that the Lake Dalneye *Salvelinus* had about 76 chromosomes with 96 arms.

For the freshwater resident *Salvelinus* from southern Sakhalin called *S. malma curilus* (Pallas), Viktorovsky (1978) found 84-86 chromosomes and 100 arms. This karyotype is very similar to that of *S. leucomaenis* from Hokkaido published by Abe and Muramoto (1974) and from Sakhalin reported by Viktorovsky (1975b).

The Soviets consider *S. pluvius* a synonym of *S. malma curilus* (Berg 1948) and the Japanese consider *S. pluvius* a synonym of *S. leucomaenis* (Maekawa, pers. comm.); the fish called *S. pluvius* in the rivers of Honshu is regarded by the Japanese workers to be a red-spotted landlocked form of the white-spotted charr, *S. leucomaenis*. Based on morphological evidence, *S. pluvius* is intimately related to *S. leucomaenis* (Cavender 1980).

To date not one Soviet paper has been published specifically dealing with the karyology of an undoubted population of *Salvelinus alpinus*. A good choice would be the *S. alpinus* from the Bering Sea Coast of the Chukchi Peninsula where it occurs sympatrically with *S. malma malma* (Chereshnev 1978).

Confusion in the Soviet literature concerning the karyology of Kamchatkan *Salvelinus* may be partly due to taxonomic problems as yet unresolved. For example, the "longhead charr" from Lake Kronotskoe described as *S. kronocius* by Viktorovsky (1978) has an identical karyotype to the "predatory charr" of Lake Azabache (2n = 78, NF = 100) (Vasilyev 1975). Glubokovskii (1977b) recently described a new species, *Salvelinus albus*, from the Kamchatka River which Viktorovsky (1978) reported as having 78-80 chromosomes with 98 arms. The difficulty some workers have in separating the lake resident *S. alpinus* from the anadromous *S. malma* in the north Pacific land areas may also be part of the problem.

In utilizing the chromosomal evidence reported here as a possible tool for interpreting the evolutionary relationships of North American *Salvelinus*, I have considered two alternative hypotheses. The first is that the main phyletic lines, and thus the branching points on the cladogram (Fig. 6), can be determined from the progressive changes in both chromosome number and arm number. This assumes little or no parallel evolution of these characters which is probably not the case. The second hypothesis considers that the main evolutionary branches occurred before there was any change in the primitive karyotype of 100 chromosome arms. Therefore, ancestors on the separate lines represented by *S. fontinalis, S. namaycush, S. leucomaenis, S. alpinus* and *S. malma* all originally possessed similar karyotypes. With the latter hypothesis no cladistic information is available since identical chromosomal changes have occurred independently in both the *alpinus* and *malma* phyletic lines, whereas in the *fontinalis, namaycush* and *leucomaenis* lines, the gross chromosome morphology has apparently remained stable. Regardless of which view is held, it is apparent that the evolution of new taxa within the *alpinus* and malma complexes involved both change in chromosome number through the mechanism of Robertsonian fusion as well as change in centromere position through the mechanism of pericentric inversion.

In an earlier assessment of relationships within North American *Salvelinus*, based on morphological evidence (Cavender 1980), it was stated that *S. alpinus* and *S. malma* have a sister relationship and *S. confluentus* is sister to these two. The cladogram (Fig. 6B) based on chromosome characters also supports this conclusion.

ACKNOWLEDGEMENTS

I wish to thank John Rupert, Ohio State University, for his help in obtaining specimens of *Salvelinus namaycush*. I am grateful to Lionel Johnson, Freshwater Institute, Department of Fisheries and Oceans, Canada, and Murray Foster at the Rockwood Fish Hatchery (Freshwater Institute) for specimens of *Salvelinus alpinus*. The assistance of Judith Cavender in typing the manuscript is much appreciated.

REFERENCES

Abe, S. and J. Muramoto. 1974. Differential staining of chromosomes of two salmonid species, *Salvelinus leucomaenis* and *Salvelinus malma*. Poc. Jpn. Acad. 50(7):507-511.

Behnke, R.J. 1980. A systematic review of the genus *Salvelinus*. p. 441-480. *In* E.K. Balon [ed.] Charrs: salmonid fishes of the genus *Salvelinus*. Dr. W. Junk Publishers, The Hague.

Berg, L.S. 1948. Freshwater fishes of the USSR and adjacent countries. 4th ed. Zool. Inst. Akad. Nauk SSSR, Vol. 1:466 p. (I.P.S.T. Translation, 1962).

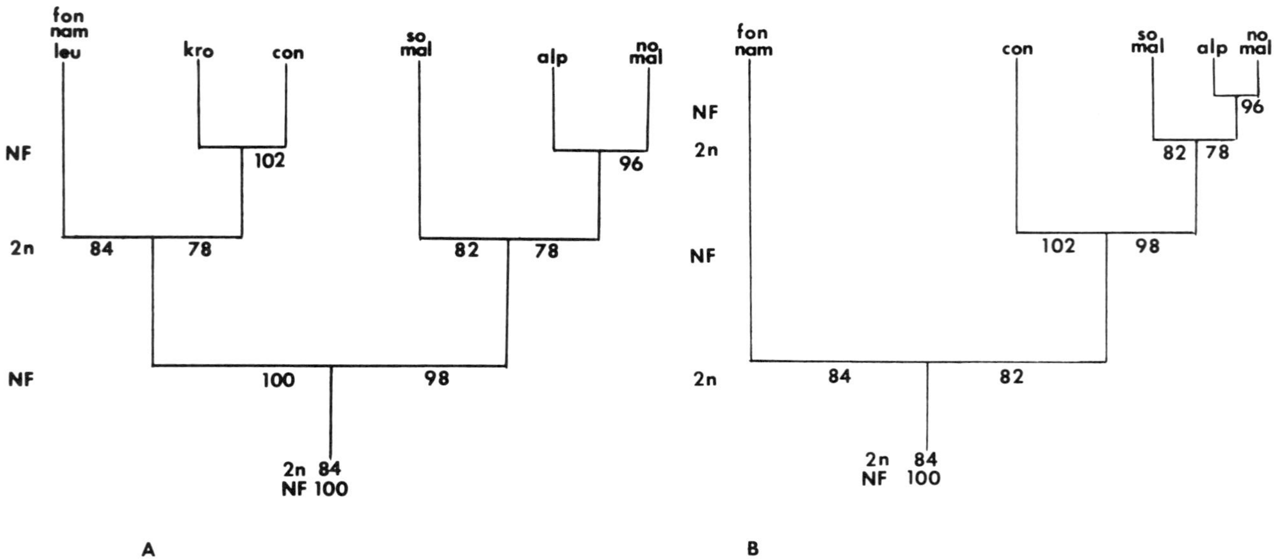

FIG. 6 Two alternative cladograms showing cladistic relationships of the *Salvelinus* species studied. Branching points are based on chromosome number (2n) and arm number (NF).

con = *confluentus*	mal = *malma*	no = northern form
alp = *alpinus*	nam = *namaycush*	leu = *leucomaenis*
fon = *fontinalis*	so = southern form	kro = *kronocius*

Capanna, E., S. Cataudella and T.G. DeFronza. 1973. Some remarks on the karyotype of a intergeneric hybrid, *Salmo trutta* × *Salvelinus fontinalis* (Pisces: Salmoniformes). Genetica 44(2):194-206.

Cavender, T.M. 1978. Taxonomy and distribution of the bull trout, *Salvelinus confluentus* (Suckley), from the American northwest. Calif. Fish Game 64(3): 139-174.

Cavender, T.M. 1980. Systematics of *Salvelinus* from the north Pacific basin, p. 295-322. *In* E.K. Balon [ed.] Charrs: salmonid fishes of the genus *Salvelinus*. Dr. W. Junk Publishers, The Hague.

Chereshnev, I.A. 1978. Systematic position of chars of the genus *Salvelinus* from the Bering Sea Coast of Chukchi Peninsula. Biol. Morya (Vladivost.) 1978(1):36-46.

Chernenko, Ye. U. and R.M. Viktorovsky. 1971. Chromosome sets of the masu salmon, the Siberian char and the southern Dolly Varden char. Nauchn. soobshch. in-ta Biol. Morya (Vladivostok) 1971(2).

Davisson, M.T., J.E. Wright and L.M. Atherton. 1973. Cytogenetic studies of pseudolinkage of LDH loci on the teleost genus *Salvelinus*. Genetics 73:645-658.

Gjedrem, T., A. Eggum and T. Refstie. 1977. Chromosomes of some salmonids and salmonid hybrids. Aquaculture 11(1977):335-348.

Glubokovskii, M.K. 1977a. Taxonomic relationships of chars of the genus *Salvelinus* in the basin of the Kamchatka River. Biol. Morya (Vladivost.) 1977(3):24-35.

Glubokovskii, M.K. 1977b. *Salvelinus albus* sp.m. from the Kamchatka River. Biol. Morya (Vladivost.) 1977(4):49-57.

McPhail, J.D. and R. Jones. 1966. A simple technique for obtaining chromosomes from teleost fishes. J. Fish. Res. Board Canada 23:767-768.

Muramoto, J., J. Azumi and H. Fukuoka. 1974. Karyotypes of 9 species of the Salmonidae. Chrom. Info. Serv. 1974, No. 17:20-23.

Nygren, A., B. Nilsson and M. Jahnke. 1971. Cytological studies in *Salmo trutta* and *Salvelinus alpinus*. Hereditas 67(2):259-268.

Roberts, F.L. 1970. Atlantic salmon (*Salmo salar*) chromosomes and speciation. Trans. Am. Fish. Soc. 99(1):105-111.

Svärdson, G. 1945. Chromosome studies on Salmonidae. Inst. Freshw. Res. Drottningholm Rep. 23:1-151.

Taranets, V. Ya. 1933. Some new freshwater fishes of the Far Eastern region. Dokl. Akad. Nauk. SSSR 2:83-85.

Uyeno, T. 1972. Chromosomes of offspring resulting from crossing coho salmon and brook trout. Jap. J. of Ichthyology 19(3):166-171.

Vasilyev, V.P. 1975. Karyotypes of some forms of Arctic char from Kamchatka. J. of Ichthyol. 15(3):374-386.

Viktorovsky, R.M. 1975a. Karyotypes of the endemic chars from Kronotski Lake. Tsitologiya 17(4):464-466. (In Russian).

Viktorovsky, R.M. 1975b. Karyotypes of the Kundzha (*Salvelinus leucomaenis*) and the malma (*S. malma*) (Salmoniformes, Salmonidae). Zool. Zh. 54(5):787-789.

Viktorovsky, R.M. 1978. Mechanism of speciation in chars from Lake Kronotskoe. Moskva, Izd. Nauk. 106 p. (In Russian).

Wahl, R.W. 1960. Chromosome morphology in lake trout *Salvelinus namaycush*. Copeia 1960: 16-19.

Wright, J.E. 1955. Chromosome numbers in trout. Prog. Fish-Cult. 17:172-176.

Management of the Commercial Fishery for Anadromous Arctic Charr in the Cambridge Bay Area, Northwest Territories, Canada

A.H. KRISTOFFERSON, D.K. McGOWAN and G.W. CARDER

Department of Fisheries and Oceans, Winnipeg, Manitoba, Canada

KRISTOFFERSON, A.H., D.K. MCGOWAN AND G.W. CARDER. 1984. Management of the commercial fishery for anadromous Arctic charr in the Cambridge Bay area, Northwest Territories, Canada, p. 447–461. *In* L. Johnson and B.L. Burns [eds.] Biology of the Arctic charr, Proceedings of the International Symposium on Arctic Charr, Winnipeg, Manitoba, May 1981. Univ. Manitoba Press, Winnipeg.

The largest commercial fishery for anadromous Arctic charr in the Northwest Territories of the Canadian Arctic takes place in the Cambridge Bay area, Victoria Island. Fishing has been carried out since 1960. In 1978 a five year study was initiated to provide the necessary information for its effective management. Results from monitoring, enumeration and tagging suggest some stocks may be heavily exploited as a result of successive harvesting of the same stock at different locations during the open water season. Problems are discussed and future plans outlined.

INTRODUCTION

Commercial fishing for Arctic charr began in Canada along the Labrador coast in the early 1940's (Hunter 1966). In Canada's Northwest Territories commercial fishing for this species first took place in 1947 at Frobisher Bay on Baffin Island (Hunter 1976). It was not until 1960 that such activity developed in the central Arctic at Cambridge Bay on Victoria Island (Barlishen and Webber 1973). Today the Cambridge Bay commercial fishery for anadromous Arctic charr is the largest in the Northwest Territories, producing over 50 tonnes per year.

Seven rivers are fished by this fishery (Fig. 1): the Ekalluk, Paliryuak, Halovik, Lauchlan, Perry, Ellice and Jayco rivers. Fishing takes place in open water and begins after the ice break-up in early July in an effort to fish the downstream migratory runs of charr as they enter the sea. Fishing the upstream run starts about mid August, ending in early September (Kristofferson and Carder 1980). The fishery uses nylon gillnets (140 to 165 mm mesh size, stretched measure), set in estuaries or river mouths. Nets are lifted at least twice daily; the catch is dressed on site (gills and viscera removed) and flown to the freezer plant in Cambridge Bay by small, float-equipped aircraft. There the fish are washed, sorted, fast frozen, then shipped to market in southern Canada and the United States by scheduled air carrier.

Since 1971 Canada's federal Department of Fisheries and Oceans has monitored the catch taken by this commercial fishery (Kristofferson and Carder 1980). A five year program was initiated in 1978 to identify separate charr stocks, measure the size of the migrating component of each and determine the relative contribution of each to the commercial fishery. Although the study is not yet complete, results to date suggest present management strategy may have to be modified to achieve more effective utilization of this resource. This paper discusses the results and interpretation of these data.

MATERIALS AND METHODS

Samples of the commercial catch have been taken from some of the rivers since 1971. These samples were taken randomly, either at the fishing site or at the processing plant in Cambridge Bay. Fork length (± 1.0 mm), round weight and/or dressed weight (± 50 g) were measured and saggital otoliths were removed for age determination.

A field camp was established at the mouth of the Ekalluk River during August 1978, with the intention of erecting a fish weir across the river to enumerate the upstream migration of charr. High water and fast current, due to a late break-up, precluded operating the weir. The commercial gillnet fishery took place at this site from August 19-29, during the upstream run of charr.

Since the weir could not be operated, charr were captured by angling and tagged with Floy spaghetti-type tags anchored between the ptergiophores at the base of the dorsal fin. Over one hundred were tagged by this method between August 25-29. The commercial fishery concluded on August 29, after which charr were taken by gillnet and tagged up to and including September 5; at this time the run appeared to be over. The gillnets were identical to those used in the commercial fishery and were lifted on a continuous basis. All charr tagged were captured in freshwater during what is presumed to be an upstream migration to the overwintering areas. In this manner a total of 638 charr were tagged.

During the summer of 1979 another tagging program was carried out, this time at Starvation Cove (Fig. 4). Charr were captured by nylon gillnets (140 mm mesh size, stretched measure). The program began August 1 and ended August 12, with a total of 567 charr tagged.

During August 1979 a fish weir was successfully operated at the Ekalluk River. The upsteam migration was enumerated and a total of 992 charr from this migration were tagged on their way upstream.

Tag recoveries were recorded from the 1979 and 1980 commercial fisheries following each tagging program. All commercial fishermen were questioned after each season ended and were paid a reward of $3.00 per tag returned. Dates and locations of recaptures were provided by the fishermen.

RESULTS AND DISCUSSION

The fishing locations are shown in Figure 1. Those in close proximity to the Ekalluk River are the Paliryuak, Halovik and Lauchlan rivers in the Wellington Bay area; the Ellice and Perry rivers are almost 200 and 300 km away, respectively, on the mainland; the last major river fished, the Jayco, is at an even greater distance, located high up in Albert Edward Bay.

Age and length compositions taken from samples of the 1979 commercial catch are shown in Figure 2. The age and length compositions of the catch from the Ekalluk, Paliryuak, Halovik and Lauchlan rivers are quite similar to one another. Those from the Perry and Ellice rivers are somewhat different, the fish being younger and smaller. The catch from the Jayco River has a broad age distribution and appears to be made up of large charr in comparison to the other fisheries. Since the gear used by these fisheries are identical, the data suggest that there are at least three distinct stocks of charr utilized by the Cambridge Bay fishery. One appears to occupy Wellington Bay, another the mainland area and the third, Albert Edward Bay. Moore (1975) concluded that two distinct charr populations occupied separate rivers whose mouths were only 40 km apart, based on differences in length frequency characteristics of fish captured near the mouths of the rivers. Although conclusive evidence is not yet available to prove that the Cambridge Bay fishery relies on at least three separate charr stocks, other aspects of the fishery, when considered, provide support for this hypothesis. The two mainland rivers, the Perry and Ellice, were heavily exploited in 1977, the quota on the Ellice being exceeded by over 7000 kg; this appears to be reflected in the age and size distribution. The Jayco River, where the largest and oldest fish were taken, is the most recently established of the fisheries, being first harvested in 1975 (Kristofferson and Carder 1980). To date it appears to be lightly exploited. The similarity of the catch composition from each of the four rivers in the Wellington Bay area suggests they may all be part of a single stock. However, the present management strategy considers that each river has a distinct charr stock and each has its own exploitation quota assigned to it.

Since the commercial fishery began in 1960, changes in management strategy have been necessary, leading to the separate stock approach used at present. The oldest fishery is that based at the Ekalluk River. Figure 3 outlines the history of this fishery over a 20 year period. Since the Ekalluk was the only river fished in the early days an individual quota was assigned to it. This was maintained at

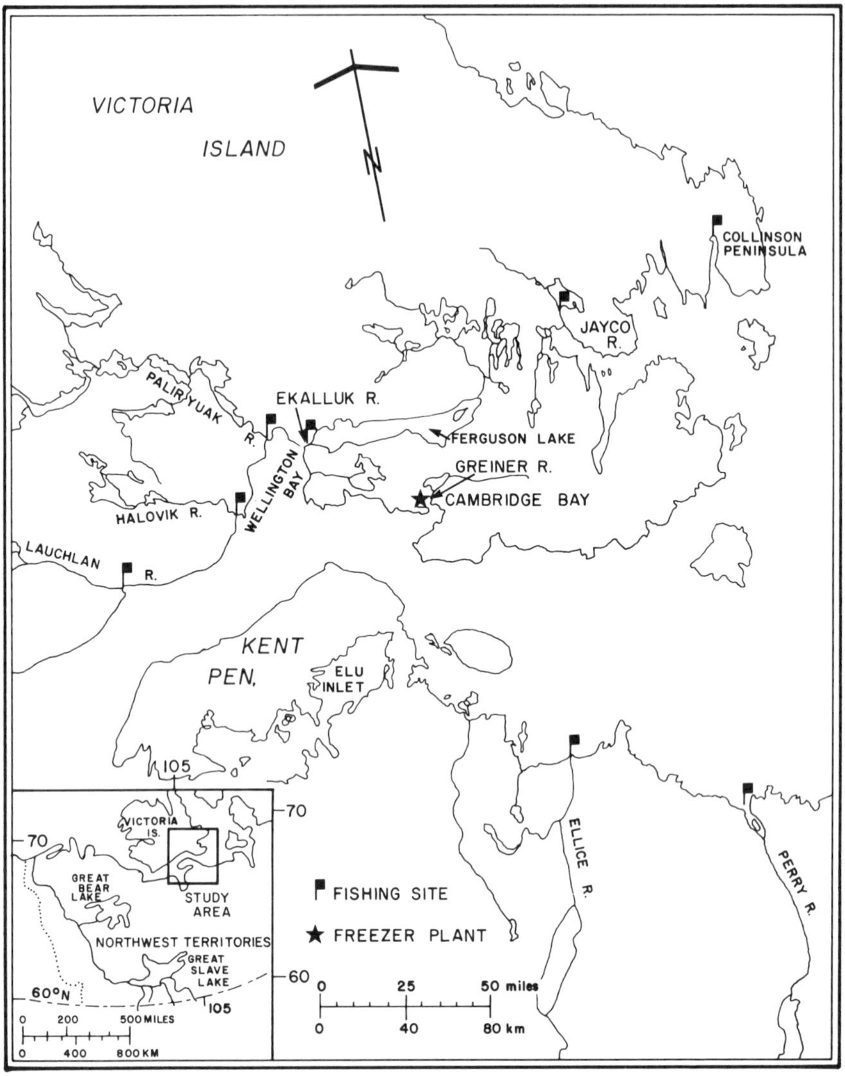

FIG. 1 Map of the Cambridge Bay area showing location of commercial fishing sites for Arctic charr.

about 18,000 kg until 1967. During 1967 the fishermen requested that a single quota be assigned to the Wellington Bay area from Byron Bay to Cape Enterprise (Barlishen and Webber 1973) (Fig. 4) so fishing could take place at the Paliryuak, Halovik and Lauchlan rivers, which this area included. This was granted but

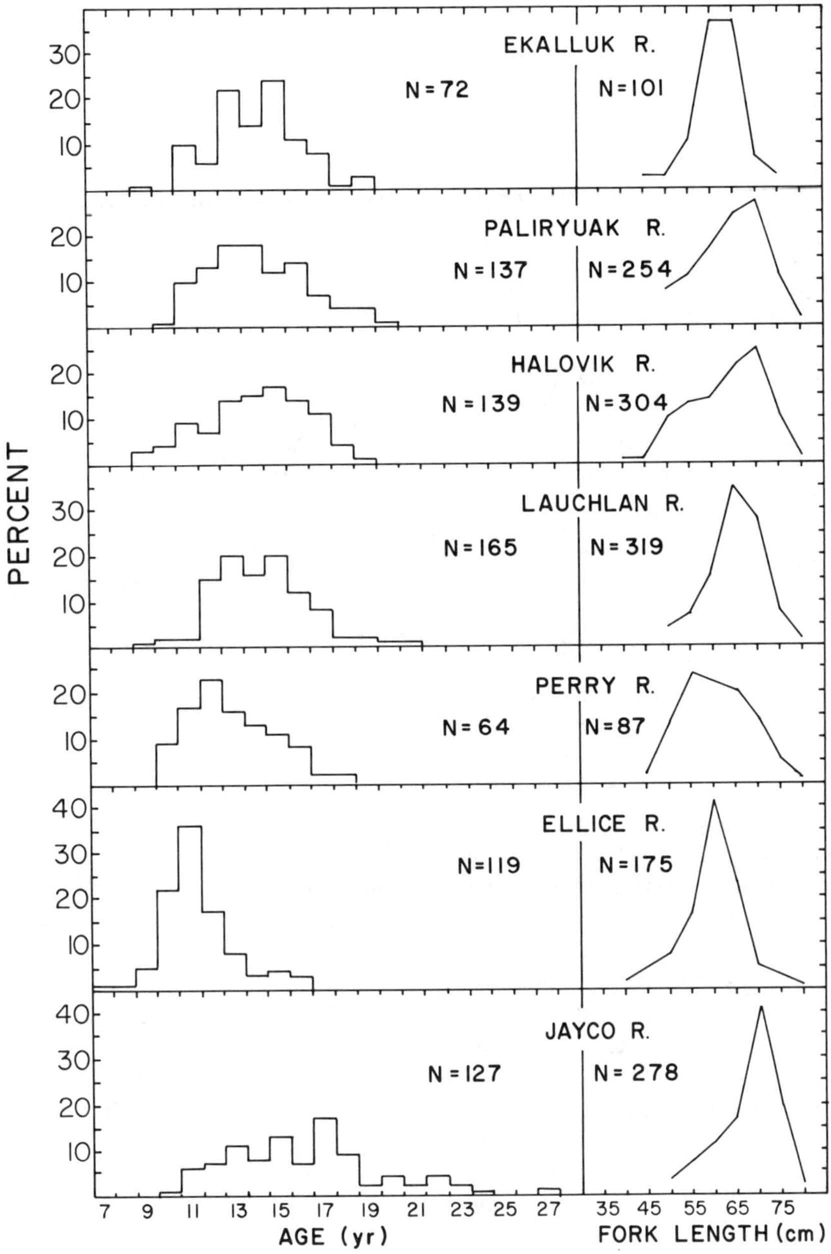

FIG. 2 Age and length composition of samples of Arctic charr taken from the commercial catch at the seven major rivers fished during 1979.

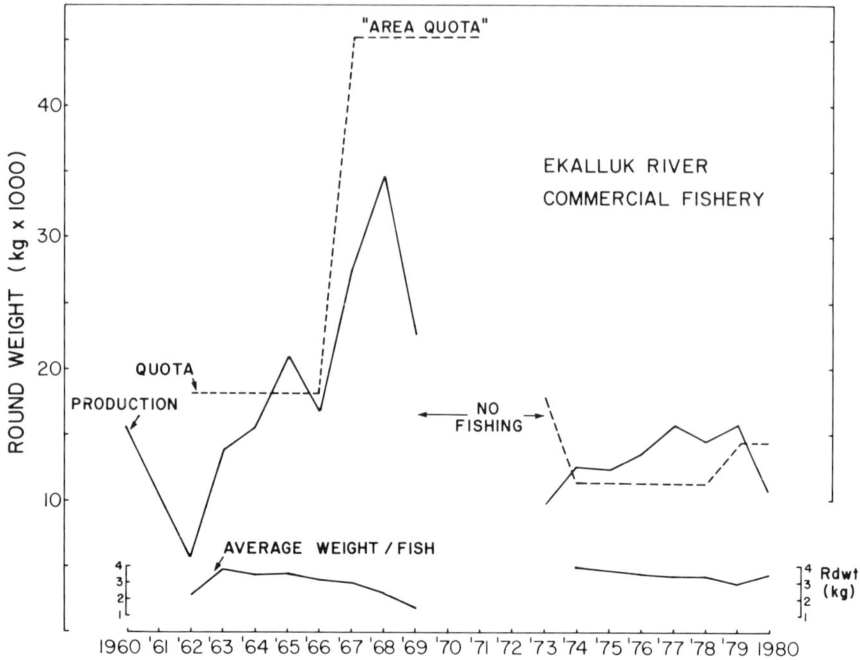

FIG. 3 History of the Ekalluk River commercial charr fishery, 1960-80, showing quotas, annual production and average weight per fish in the catch.

fishing effort in 1967 was confined to the Ekalluk system resulting in a sharp increase in exploitation (Fig. 3). This pattern continued in 1968 when a peak production of over 34,000 kg was realized. Fishing began in 1968 at the Halovik and Paliryuak rivers but production here was less than 9000 kg in total. Between 1963-69 a steady decrease in the average size of fish caught at the Ekalluk was noted (Fig. 3). Finally, in 1969 when production dropped over 11,000 kg from the previous year and the average weight per charr fell from 3.4 kg in 1967 to 1.4 kg in 1969 (Fig. 3, Table 1) the river was closed to commercial fishing. It is very interesting to note that fishing effort was increased at the Halovik River during 1969; the fishery produced a yield of 25,850 kg, comprised of charr with an average weight of 3.2 kg compared with 1.4 kg at the Ekalluk River (Table 1). Both fisheries were carried out at the same time in August and early September. This difference in size of the charr captured by simultaneous fisheries suggests the existence of another, previously unexploited, stock in the Wellington Bay area. During the three years the Ekalluk River was closed to commercial fishing, the fishery was sustained by fishing at the Halovik and Lauchlan rivers and to a

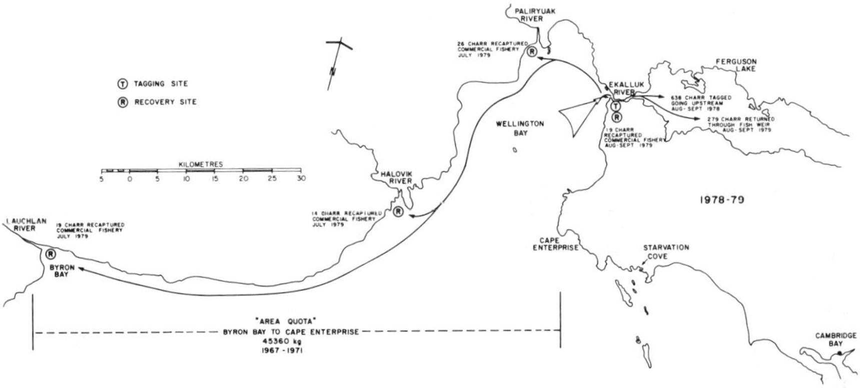

FIG. 4 Tagging of Arctic charr during the upstream migration at Ekalluk River during August-September 1978 and subsequent recoveries in the commercial fisheries during 1979.

lesser extent at the Paliryuak River (Table 1). The area quota was maintained until 1971 when this strategy was abandoned and individual quotas for each river were established, a policy which has remained in force until the present.

Commercial fishing on the Ekalluk River was resumed in 1973, the stock having recovered (Fig. 3). The quota has since been reduced and maintained. Production to the present has remained fairly stable, as has the average fish size. The apparent drop in production in 1980 is attributed to an early upstream migration that the commercial fishery did not take full advantage of.

A new processing plant with a blast freezer went into operation at Cambridge Bay in 1971. Since that time fishing effort has increased steadily. In 1977 a total of nine locations were fished. Due to the limited capacity of the blast freezer fishermen were allowed to harvest some systems during the spring fishery in pursuit of the downstream seaward migration of Arctic charr after break-up, in order to utilize their facilities to the full. This they have continued to do until present on the Paliryuak, Halovik and Lauchlan rivers.

In response to the increase in fishing intensity and as a result of the problem encountered in the past using the area quota strategy, the Department of Fisheries and Oceans initiated this study to improve management strategy. As previously discussed, the first tagging program took place at the Ekalluk River during August 1978, when 638 charr were tagged going upstream. The first returns came in the next year, not from the Ekalluk river fishery but from the spring fisheries at the Paliryuak, Halovik and Lauchlan rivers (Fig. 4). The fall fishery on the upstream migration at the Ekalluk River also captured tagged fish; 279 of the 638 charr tagged going upstream in this river the previous fall were counted through the weir on their way up river again in 1979. A similar pattern was

Table 1. Production of Arctic charr from Wellington Bay, 1967-72; average weight per fish shown in brackets.

RIVER	PRODUCTION - ROUND WEIGHT (KG)					
	1967	1968	1969	1970	1971	1972
EKALLUK	27670(3.0)	34300(2.4)	22680(1.4)	-	-	-
PALIRYUAK	-	6460(1.9)	-	5880(2.4)	-	-
HALOVIK	-	2614(3.4)	25850(3.2)	26200(3.4)	10430(?)	6480(2.5)
LAUCHLAN	-	-	-	2420(2.6)	19050(?)	20990(4.3)
TOTAL	27670	43374	48530	34500	29480	H 6480[1] L 20990[2]
AREA QUOTA	45360	45360	45360	45360	45360	H 9070 L 18140

1 HALOVIK RIVER
2 LAUCHLAN RIVER

observed in 1980 (Fig. 5). A total of 992 charr were tagged going upstream through the weir at the Ekalluk River in 1979. Initial returns were from the spring fisheries in 1980 at the Paliryuak, Halovik and Lauchlan rivers. Recaptures were also made during the fall commercial fishery at the Ekalluk River in 1980.

It is most interesting to examine the proportion of recaptures from each fishery. Catch and tag recapture statistics are summarized in Table 2. Examination of this table shows that the number of marked to unmarked charr taken in each fishery is very similar, even though all tagging was done on charr migrating upstream in the Ekalluk River. This pattern is evident both in 1979 and 1980. Since the harvests vary in size, this is most evident when numbers of recaptures are adjusted. It is perhaps possible as these data suggest, that virtually all the charr taken in the fisheries at the Paliryuak and Halovik rivers, and a lesser extent at the Lauchlan River (the most distant from the Ekalluk), were part of the upstream migration into the Ekalluk River the previous fall.

In response to a request by the fishermen to fish for charr in mid-summer along the coast at Starvation Cove (Fig. 6), the Department of Fisheries and Oceans carried out a tagging program at this location in early August 1979. Since the weir was to be operated at the Ekalluk River, approximately 40 km to the north, an ideal recovery mechanism was available a few weeks after tagging was done. It was suspected that fishing during mid-summer would harvest charr from stocks already exploited by the other fisheries and indeed results suggest that this did happen. A total of 120 of the 567 charr tagged at Starvation Cove were taken either in the commercial fishery at the Ekalluk River, a few weeks later, or

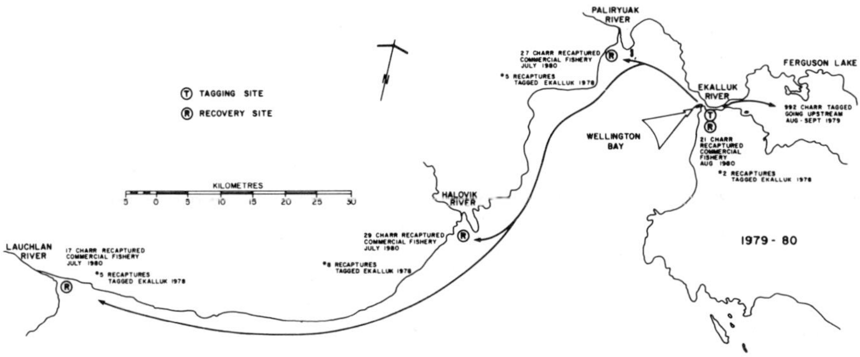

FIG. 5 Tagging of Arctic charr during the upstream migration at Ekalluk River during August-September 1979 and subsequent recoveries in the commercial fisheries during 1980.

counted through the weir on their way upstream (Fig. 6); in addition, a number of recoveries showed up in each of the other three fisheries in the area in 1980.

These tagging results to date indicate that the timing and location of commercial fisheries are important factors in determining whether multiple exploitation of a single stock is taking place or whether separate stocks are being exploited. Johnson (1980) outlines the seasonal movements of Arctic charr, and if such patterns are followed by the charr in Wellington Bay, it would then appear that timing of commercial fisheries is very important, if exploitation of separate stocks is desired.

The downstream, seaward migration of Arctic charr can begin early, often while the surface of the rivers is still frozen (Johnson 1980, Moore 1975, Grainger 1953). Johnson and Campbell (1975) observed that the largest charr moved downstream first as did deMarch et al. (1977); Moore (1975) estimated that the distance travelled from the parent stream was directly proportional to the length and age of the fish. Estimates of the speed at which charr migrate vary from 0.6 to 9.7 km·day^{-1} (Johnson 1980). Seaward migration of charr down the Ekalluk River probably begins at the same time as it does in the Nauyuk Lake system on the Kent Peninsula (Fig. 1). Johnson (1980) reports that the downstream migration of charr starts between June 6-22. The spring commercial fisheries at the Paliryuak, Halovik and Lauchlan rivers started no earlier than July 10 in 1979 and 1980 and usually continued until at least July 23. At minimum this would allow 18 days for charr to be at sea before entering the commercial fishery. Travelling at about 1 km day^{-1}, charr coming downstream in the Ekalluk could easily reach the Paliryuak River, 15 km away, in time to be harvested by the spring fishery. Similarly they would have no trouble reaching the Halovik River approximately 50 km away (shoreline distance). The Lauchlan

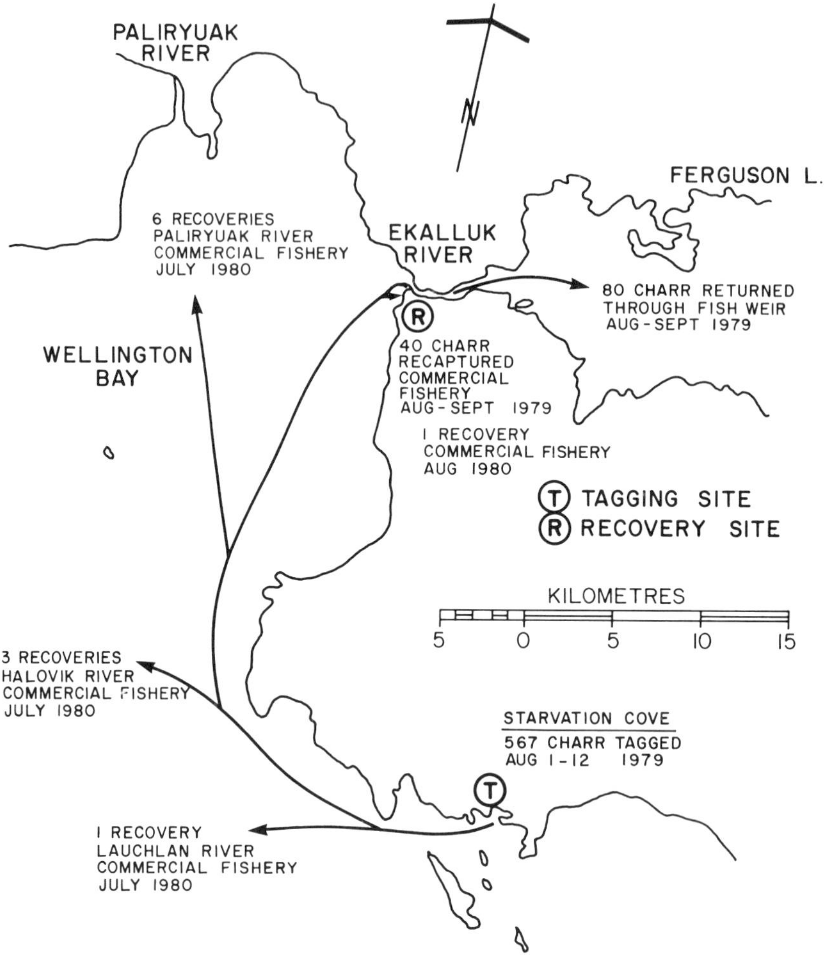

FIG. 6 Tagging of Arctic charr at Starvation Cove, 1979 and subsequent recoveries in 1979 and 1980.

River is over 120 km away (Fig. 4), and in 1980 the proportion of marked to unmarked fish taken there was somewhat less than the others (Table 2). However, given the time at sea and speed of charr, it is still within their reach.

The high degree of homing amongst spawning charr has been documented (Alm 1951, LeCren and Kipling 1963, Frost 1963, Glova and McCart 1974). However, as Johnson (1980) points out, the annual seaward migration of Arctic charr is not a spawning migration, so the degree of homing may not be as intense. Armstrong (1974) suggests that Dolly Varden charr may overwinter in waterbodies other than their natal streams during years when they are not spawning. The possibility of immigration of charr from other systems during non-spawning periods must be considered. Johnson (1980) has observed this in the Nauyuk Lake system. In fact, if distinct stocks of Arctic charr utilize Wellington Bay as the data suggest, there may be considerable mixing of overwintering stocks during years when they are not spawning. Considering this, the tagging results received to date must be viewed with caution. Perhaps the safest interpretation of these tagging results is that the stock of Arctic charr exploited by the Ekalluk River commercial fishery in the fall also contributes significantly to the three spring fisheries at the Paliryuak, Halovik and Lauchlan rivers.

In summary, there is evidence that the Cambridge Bay commercial fishery utilizes several distinct stocks of Arctic charr (Fig. 2). The area quota strategy applied to the Wellington Bay fishery during 1967-71 was not suitable and probably contributed to the overharvest of the Ekalluk River system (Fig. 3). The Ekalluk River charr run, enumerated in 1979, is indeed large (Fig. 7) with over 183,000 charr (549,000 kg) counted going upstream. However, this system showed a sharp decline in 1969 after supporting a mean harvest of 28,000 kg from 1967-69. There is evidence that at least one other distinct stock of Arctic charr was available to the commercial fishery during the three years the Ekalluk River was closed to commercial fishing (Table 1). These fish were taken at the Paliryuak, Halovik and Lauchlan rivers during late August and early September. If fishing is allowed during mid-season, charr can be captured that would otherwise contribute to fisheries carried out at a later date (Fig. 6) so timing and location of fisheries is important to overall management strategy. If the tagging results shown in Figures 4 and 5 and Table 2 are considered at face value it is possible that the harvest on the Ekalluk River stock is greater than the 13,000 kg that it is presently believed to produce, as a result of allowing spring fisheries to take place at the Paliryuak, Halovik and Lauchlan rivers. Presently the Ekalluk River charr stock appears healthy (Figs. 3 and 7). As Johnson points out, there is little information on the capacity of Arctic charr to yield a harvest under exploitation. In order that this stock remain healthy it is necessary to monitor it closely and to determine the extent to which it contributes to the other fisheries. To confirm the existence of another stock(s) in Wellington Bay the upstream fall migration of charr in each of the Paliryuak, Halovik and Lauchlan rivers should be investigated. Enumeration and tagging during these upstream runs, and subsequent returns in the commercial fishery, will help to determine the existence, and relative contribution, of each stock(s) to the commercial fishery.

It is interesting to note that a gillnetting program carried out on the Paliryuak River from August 26 to September 8, 1980 failed to provide evidence of an

Table 2. Comparison of the number of tagged Arctic charr captured by the commercial fisheries at Paliryuak, Halovik, Lauchlan and Ekalluk Rivers. All fish were tagged the previous year going upstream in the Ekalluk River.

	RIVER	PRODUCTION (KG)	AVERAGE WEIGHT PER FISH (KG)	CALCULATED NUMBER OF FISH	NUMBER OF RECAPTURES	ADJUSTED NUMBER OF RECAPTURES[1]	PERCENT POSSIBLY EKALLUK CHARR[2]
1979	EKALLUK	15806	3.0	5269	19	-	-
	PALIRYUAK	11816	3.0	3939	26	34	100
	HALOVIK	7316	2.9	2523	14	29	100
	LAUCHLAN	10845	3.2	3389	19	29	100
1980	EKALLUK	10519	3.5	3005	21	-	-
	PALIRYUAK	7497	3.0	2499	27	32	100
	HALOVIK	7481	3.2	2338	29	37	100
	LAUCHLAN	9151	3.3	2773	17	18	86

1 (NO. FISH CAPTURED IN EKALLUK FISHERY ÷ NO. FISH IN OTHER FISHERY) X NO. RECAPTURES IN OTHER FISHERY.

2 (ADJUSTED NO. RECAPTURES ÷ NO. RECAPTURES AT EKALLUK) X 100.

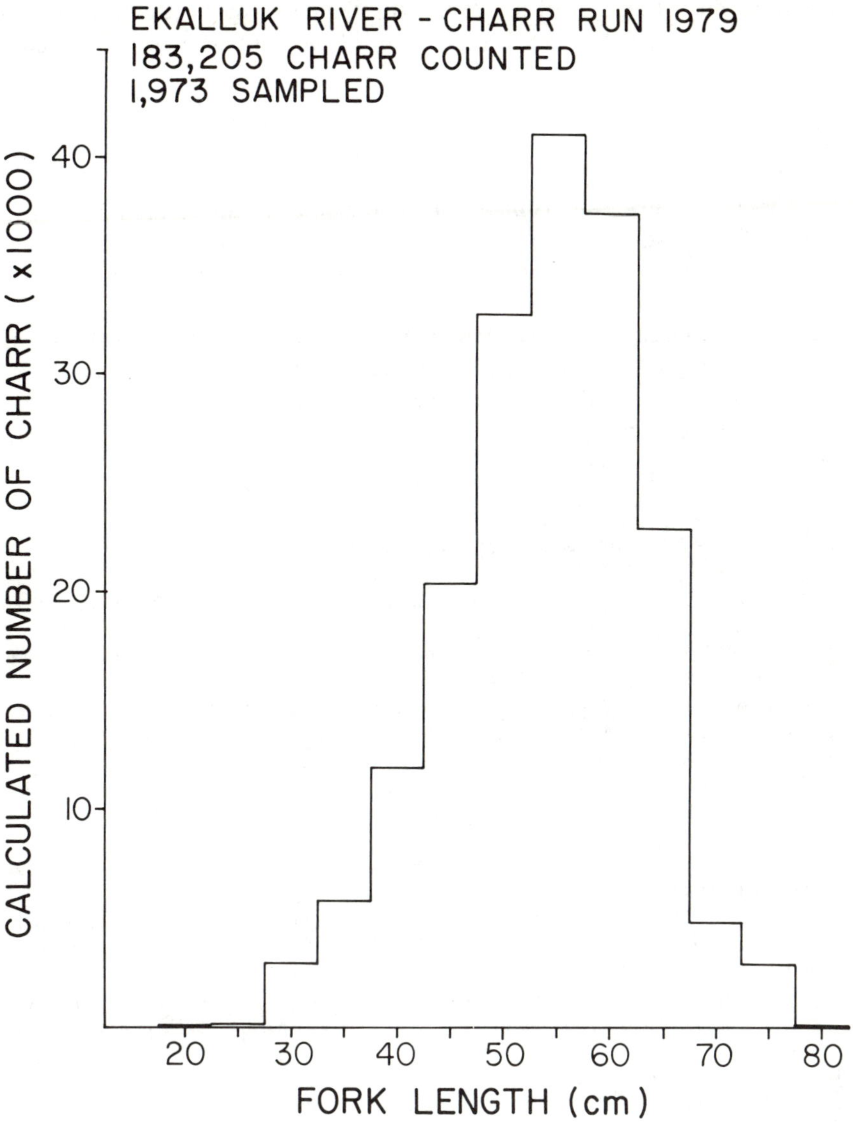

FIG. 7 Size composition of the upstream migration of Arctic charr counted through the weir at Ekalluk River, August-September 1979.

upstream migration of charr in this system. Under present management strategy an annual quota of 9000 kg is assigned to this system. Admittedly there is a possibility that the run was missed or that the run is very late, as local residents claim, so this system is still under investigation. An enumeration and tagging program is planned for the Halovik River in August-September 1981 and a similar program for the Lauchlan River in 1982.

ACKNOWLEDGEMENTS

The authors thank the commercial fishermen and staff of the Ikaluktutiak Eskimo Co-operative in Cambridge Bay for their cooperation during the collection of data. Mr. W. Lyall is thanked for his assistance at the Ekalluk River, as is Dr. T. Dick.

Field assistance was provided by R. Mulders, R. Hagen, C. Nakoshook, H. Jobbins, D. Leroux, B. Sterling, W. Lazarowich, E. Rabesca, M. Oakoak, B. Mackenzie, P. Briggs, D. MacDonnell and G. Pedersen.

Drafting was done by Graphic Services, Department of Fisheries and Oceans, Winnipeg. Typing was done by D. Holland.

REFERENCES

Alm, G. 1951. The tagging of char (*Salmo alpinus*, Linné) in Lake Vattern. Inst. Freshw. Res. Drottningholm Rep. 32:15-31.

Armstrong, R.H. 1974. Migration of anadromous Dolly Varden (*Salvelinus malma)* in southeastern Alaska. J. Fish. Res. Board Can. 31:435-444.

Barlishen, W.J., and T.N. Webber. 1973. A history of the development of commercial fishing in the Cambridge Bay area of the Northwest Territories. Unpub. report for the Federal-Territorial, Task Force Report on Fisheries Development in the Northwest Territories. 37 p.

deMarch, L., B. deMarch and W. Eddy. 1978. Limnological, fisheries and stream zoobenthic studies at Stanwell-Fletcher Lake, a large high Arctic lake. Can. Dep. Ind. North Aff. ESCOM Report A1-04, xiv + 84 p.

Frost, W.E. 1963. The homing of char, *Salvelinus willughbii* (Gunter), in Windermere. Anim. Behav. 11:74-82.

Glova, G. and P. McCart. 1974. Life history of Arctic char (*Salvelinus alpinus)* in the Firth River, Yukon Territory. vii + 50 p. *In* P.J. McCart [ed.] Life histories of anadromous and freshwater fish in the western Arctic. Canadian Arctic Gas Study Ltd., Biological Report Series Vol. 20 (3).

Grainger, E.H. 1953. On the age, growth, migration, reproductive potential and feeding habits of the Arctic char (*Salvelinus alpinus*) of Frobisher Bay, Baffin Island, Canada. J. Fish. Res. Board Can. 10:326-370.

Hunter, J.G. 1966. The Arctic char. Fish. Canada 19(3): 17-19.

Hunter, J.G. 1976. Arctic char and hydroelectric power in the Sylvia Grinnell River. Fish. Res. Board Can. Manuscript report series. 1376:1-21.

Johnson, L. 1980. The Arctic charr, *Salvelinus alpinus*, p. 15-98. *In* E.K. Balon [ed.] Charrs: salmonid fishes of the genus *Salvelinus*. Dr. W. Junk Publishers, The Hague.

Johnson, L. and J.S. Campbell. 1975. The seaward migration of Arctic char (*Salvelinus alpinus*) out of Nauyuk Lake, Kent Peninsula, Northwest Territories. Pap. Presented at the Can. Con. Freshwater Fish Res., Vancouver, B.C. 13 p.

Kristofferson, A.H. and G.W. Carder. 1980. Data from the commercial fishery for Arctic charr, *Salvelinus alpinus* (Linnaeus), in the Cambridge Bay area, Northwest Territories, 1971-78. Can. Data Rep. Fish. Aquat. Sci. 184: v + 25 p.

LeCren, E.D. and C. Kipling. 1963. Some marking experiments on spawning populations of char, *S. willughbii*, in Windermere, England. Int. Comm. Northwest Atl. Fish. Spec. Publ. 4:130-139.

Moore, J.W. 1975. Distribution, movements, and mortality of anadromous Arctic char, *Salvelinus alpinus* L., in the Cumberland Sound area of Baffin Island. J. Fish. Biol. 7:339-348.

The Selection of Arctic Charr *Salvelinus alpinus* L. By Nylon Gillnets

JOHN WILLAS JENSEN
The Royal Norwegian Society of Sciences and Letters,
The Museum, N-7000 Trondheim, Norway

JENSEN, J.W. 1984. The selection of Arctic charr *Salvelinus alpinus* L. by nylon gillnets, p. 463–469. *In* L. Johnson and B.L. Burns [eds.] Biology of the Arctic charr, Proceedings of the International Symposium on Arctic Charr, Winnipeg, Manitoba, May 1981. Univ. Manitoba Press, Winnipeg.

The selection by nylon twine gillnets on 2941 landlocked Arctic charr taken in the Essand-Nesjø reservoir in Norway, was analyzed by the graphical method of Gulland and Harding (1961). The mean condition factor of charr < 280 mm was 1.15 and of larger charr 1.35. By linear regressions the following equations between mesh size m (mm) (knot to knot) and the modal length of the fish *l* (mm) were found:

$l < 280$ mm $m = 0.107\,l + 1.1$

$l > 280$ mm $m = 0.120\,l - 0.2$

Based on these equations a selectivity curve was drawn through two iterations. Examples are given of different gangs of nets and their effectiveness with respect to charr in different size ranges at a condition of 1.15. By choosing adjacent mesh sizes so that both ascending and descending arms of the selectivity curve intersect at 50 percent relative efficiency, a geometrical series was obtained by multiplying by the constant 1.256; the maximum deviations of which are 85 and 114 from the mean taken at 100. By choosing intersects at 60 percent, a corresponding geometrical series with the constant 1.203 is derived, giving maximum deviations of 94 and 105. The commercially available supply of nets may cause problems when building up geometrical series of mesh sizes, and the characteristics of some alternative series are shown. The gillnet selection on brown trout, Arctic charr, and whitefish seems quite comparable; each is more dependent on the condition of the fish than on the species.

INTRODUCTION

The selectivity of gillnets to whitefish, *Coregonus clupeaformis* has been considered by McCombie and Fry (1960) and Regier and Robson (1966) and to brown trout, *Salmo trutta*, by Jensen (1977). Through the years 1972-1980 landlocked Arctic charr were netted yearly in the Essand-Nesjø reservoir in Norway. The reservoir is located at 63°N and 12°E. The 27 km^2 Essand reservoir was impounded to an increased depth of 6 m in 1948. The Nesjø reservoir is a completely new impoundment of 39 km^2 constructed in 1970. At water levels between altitudes 729 and 726 m, the summer situation, the reservoirs have a common surface. During winter drawdown the Essand reservoir is lowered to a level of 726 m and the Nesjø reservoir even lower.

MATERIALS AND METHODS

The charr were caught by routine netting, using gangs of 8 nets of the following mesh sizes (knot to knot): 19.5, 22.5, 26, 29, 31, 39, and 45 mm. Each net was 25 m long and had a depth of 1.5 m. The nets were set in the littoral zone on the reservoir bottom. The analysis is based on 2941 charr. In accordance with Jensen (1977) the graphical method presented by Gulland and Harding (1961) and reviewed by Regier and Robson (1966) was used. The length of the charr was measured as fork length, as defined by Lagler (1978). The smallest mesh, 19.5 mm, caught charr of 170 mm most efficiently. Smaller fish were excluded, as were the few charr larger than 390 mm.

RESULTS

The method of Gulland and Harding (1961) is based on the assumption that the relative efficiency of mesh size (m) to fish length (l) is equal for those combinations of m and l for which $\frac{m}{al + b}$ are equal, where a and b are constant over all mesh sizes and fish lengths. To estimate a and b, the frequency distribution of the catches was plotted for each size of fish, grouped in ten-millimetre intervals. A smooth curve was fitted for each ten-mm-group. The peak of the curve was taken as the most efficient mesh size to that length group, and the figures plotted against each other (Fig. 1).

The maximum length of Essand charr was 290 mm, and the mean condition factor was 1.15. Charr of larger size had visited the Nesjø reservoir, and the mean condition was 1.35. Therefore a break exists in the mesh size-length relationship at 280 mm fish length. By linear regression the following equations were found:

$l > 280$ mm m = 0.107 l + 1.1 (r = 0.992)

$l > 280$ mm m = 0.120 l - 0.2 (r = 0.997)

The best mesh size for each length group was calculated according to these equations. The catch of each length group by each mesh size, expressed as the proportion of the peak catch of that size, was plotted against the fraction best mesh/mesh used. A smooth curve was drawn by eye through the points (Fig. 2). From this curve the relative efficiency of any mesh size on a given fish length can be read. The pooled number of fish within each ten-mm-group was divided by the corresponding pooled efficiency. Thus a better estimate of the relative abundance of each length group was obtained. Based on these factors, a better estimate of the efficiency of each different mesh size for the different length

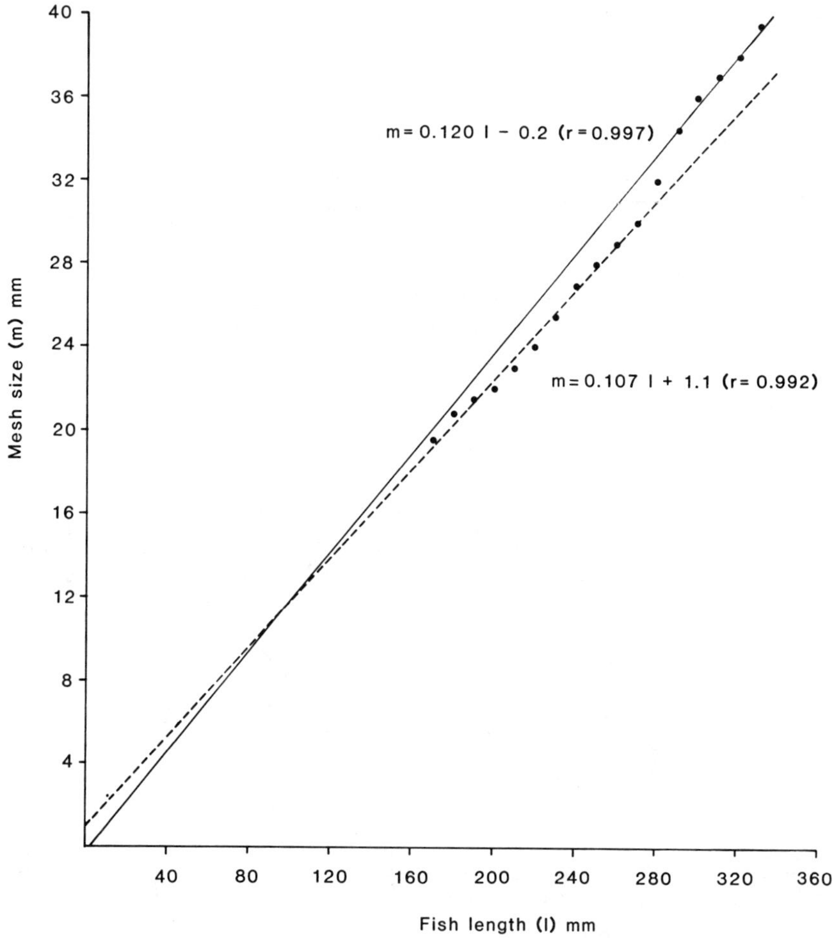

FIG. 1 Relationship between length of charr and the most efficient mesh size.

groups was calculated and plotted in Figure 3. Only small changes were recorded, in either abundance or efficiency; further interactions led to insignificant improvements. The second curve (Fig. 3) was therefore accepted as the best estimate. The relative efficiency values were transformed to percentages, and the fraction best mesh/used mesh with the equivalent *l*/*l*m. lm is the modal length, the fish length caught with highest efficiency by a certain mesh size.

This selectivity curve has a pronounced positive skew. Certain mesh sizes catch charr longer than modal length more effectively than, correspondingly, they catch smaller charr.

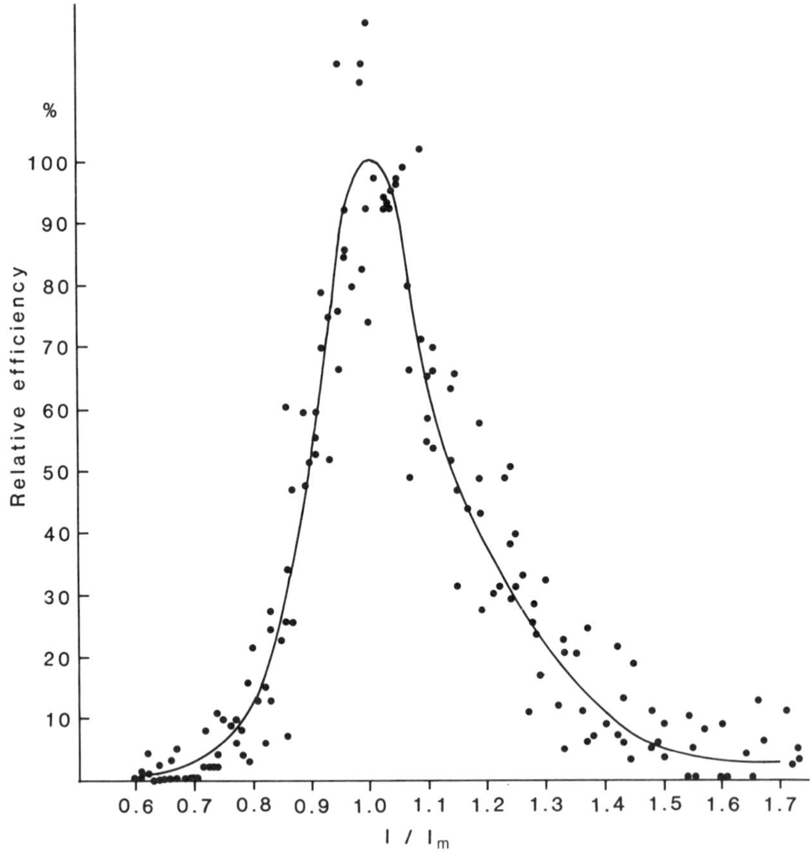

FIG. 2 Relationship between the ratio best mesh/mesh used and the relative efficiency. Triangles are charr > 280 mm, circles charr $\leq$ 280 mm.

DISCUSSION

This selectivity curve for Arctic charr is almost identical with the one Jensen (1977) found for brown trout. The latter, however, indicates a higher efficiency when $l/l_m > 1.2$. The findings are in accord with those achieved for whitefish by McCombie and Fry (1960). The curve for whitefish presented by Regier and Robson (1966), based on their computational method and a skew-normal model, is rather narrow.

It is important to compose gangs of nets of equal efficiency towards all fish lengths of interest. By choosing adjacent mesh sizes so that both ascending and

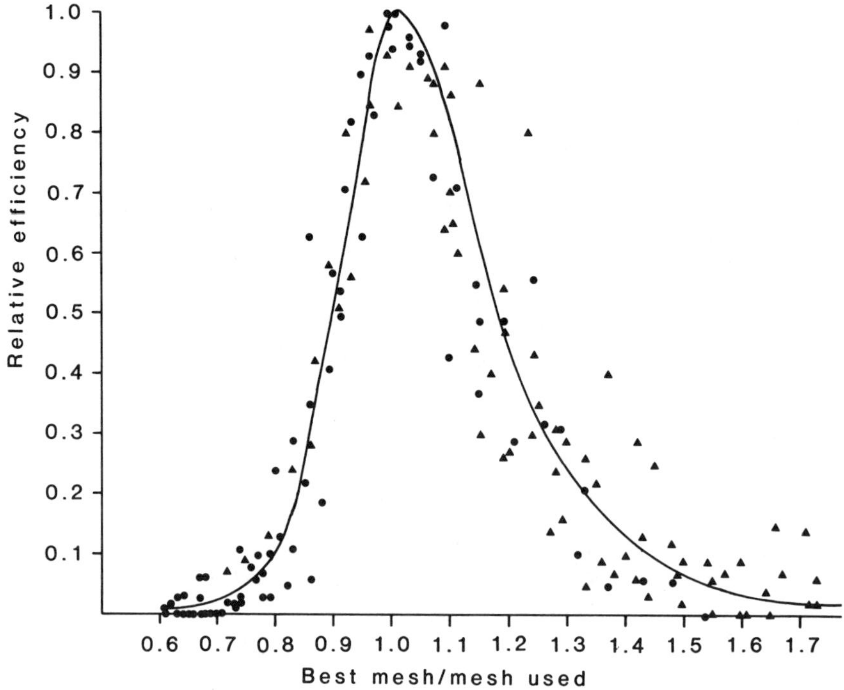

FIG. 3 Final selection curve. The relative efficiency of a net on charr of length l compared to the modal length l_m.

descending arms of the selectivity curves intersect at 50 percent relative efficiency, Regier and Robson derived the following series (stretched meshes in inches): 1.5, 1.8, 2.3, 2.9, 3.7, and 4.7. Its maximum deviations were 88 and 108 with mean efficiency taken at 100 towards whitefish of lengths from 180 to 450 mm.

On the same premise my selectivity curve for Arctic charr gives Series I (Table 1) mesh size in millimetres from knot to knot, for charr of condition 1.15. This series is very close to the geometric series obtained by multiplying by the constant 1.256. It is also identical to the Regier and Robson series mentioned above, but more variable efficiency is implied. Such a series is effective between the modal lengths yielded by the smallest and largest mesh size. The efficiency still undulates between the peaks of the modal lengths of the different meshes and the intersecting points of adjacent meshes. By choosing intersects at 60 percent relative efficiency Series II is derived, making a geometrical series with the constant 1.203. It catches very uniformly between modal lengths, and is highly recommended.

Table 1. Different series of gillnets and their combined relative efficiency towards Arctic charr of condition 1.15.

Series	Mesh size, mm knot to knot	Charr length mm	Max. deviation from mean = 100
I	15.0, 18.8, 23.6, 29.6, 37.3, 47.0	130- 450	85 - 114
II	15.0, 18.0, 21.5, 25.9, 31.3, 37.7, 45.7, 50.3	160- 440	94 - 105
III	16, 19.5, 24, 31, 39, 52	140- 380	82 - 118
IV	2·21, 26, 29, 35, 45, 52	180- 460	87 - 110
V	2·16, 19.5, 22.5, 26, 29, 35, 39, 45, 52	140- 460	88 - 110

The commercially available mesh sizes in Norway make it difficult to build up geometric series. But the six nets of series III catch fairly accurately on lengths from 140 to 380 mm. If the 52 mm mesh is replaced by a 50 mm, this series is equally effective up to 480 mm fish length. To attain smaller deviations, the difference between adjacent meshes has to be reduced. Series IV is the one recommended by the Norwegian Directorate for Wildlife and Freshwater Fish for salmonids from 200 to 450 mm. If smaller fish are of interest, Series V catches charr from 140 mm with the same accuracy.

Fatter fish require larger meshes. Series V catches charr of condition 1.35, of length from 130 to 420 mm with the same accuracy. According to the available information the gillnet selection on brown trout, Arctic charr, and whitefish is quite comparable, but more dependent on the condition of the fishes than on the species.

To check the relative efficiency of a series of nets, follow this procedure. Until further data exists, find the relationship between your mesh sizes and the modal lengths of the fish according to their condition, by interpolating in Figure 1. Based on the modal lengths, the relative efficiency of each mesh size towards the length groups of interest is found in Figure 3. To obtain the relative efficiency of the net series, pool the figures for each length group.

REFERENCES

Gulland, J.A. and D. Harding. 1961. The selection of *Clarias mossambicus* (Peters) by nylon gillnets. J. Cons. 26(2): 215-222.

Jensen, K.W. 1977. On the dynamics and exploitation of the population of brown trout, *Salmo trutta* L. in Lake Øvre Heimdalsvatn, Southern Norway. Inst. Freshw. Drottningholm Rep. 56: 18-69.

Lagler, K.F. 1978. Capture, sampling and examination of fishes, p. 7-47. *In* T. Bagenal [ed.] Methods for assessment of fish production in fresh waters. IBP Handbook No. 3.

McCombie, A.M. and F.E.J. Fry. 1960. Selectivity of gillnets for lake whitefish, *Coregonus clupeaformis*. Trans. Amer. Fish. Soc. 89(2): 176-184.

Regier, H.A. and D.S. Robson. 1966. Selectivity of gillnets, especially to lake whitefish. J. Fish. Res. Board Canada 26(3): 423-454.

The Charrs (*Salvelinus alpinus*) Of Lake Constance, A Lake Undergoing Cultural Eutrophication

JÜRGEN HARTMANN

Institut für Seenforschung und Fischereiwesen, 7994 Langenargen, Federal Republic of Germany

HARTMANN, J. 1984. The charrs (*Salvelinus alpinus*) of Lake Constance, a lake undergoing cultural eutrophication, p. 471–486. *In* L. Johnson and B.L. Burns [eds.] Biology of the Arctic charr, Proceedings of the International Symposium on Arctic Charr, Winnipeg, Manitoba, May 1981. Univ. Manitoba Press, Winnipeg.

The two forms of *Salvelinus alpinus* in Lake Constance, distinguished by morphological features, are also separated ecologically by differences in spawning time, food and vertical and horizontal distribution.

Coincidentally with the eutrophication of the lake the following changes have taken place: increase and then decrease in yield, reduction in number, reduced number of catch sites, intensified growth, improved condition (K), shift in size distribution, less fish as food, more planktonic food, a shift within planktonic food organisms and littoralization of the profundal community. The dace, *Leuciscus leuciscus*, now occupies parts of the former charr niche. Parallel changes have been found in other species of the lake. In deep lakes undergoing eutrophication, the charr yield often follows the common increase-decrease pattern of plankton-eaters. Within a lake, the distribution of charr and its food organisms may indicate local levels of organic pollution.

INTRODUCTION

The Arctic charr is considered to be one of the fish species most sensitive to eutrophication (Grimaldi and Nümann 1972), but apart from yield reduction (Bless 1978) little seems to be known about the effects of eutrophication on this salmonid. Therefore, in this paper, published and unpublished data on the charrs of Lake Constance are reviewed and compared with those of the other fish species of the lake (Hartmann 1979). Additionally, yield data from other lakes of the European Alps are compared with regard to the effects of eutrophication.

DESCRIPTION OF THE LAKE

Lake Constance is situated at the northern edge (47° N 09°E) of the Alps. The Upper Lake consists of two basins (Fig. 1) covering 476 km^2 with a mean depth

Table 1. Fish yield and composition of catch during oligotrophic and mesotrophic period of Lake Constance (Upper Lake).

		1920 - 35	1970 - 79
Arctic charr	%	0.1	0.02
Coregonids	%	75	32
Cyprinids	%	12	27
Perch	%	9	37
total	$kg.ha^{-1}$	8	30

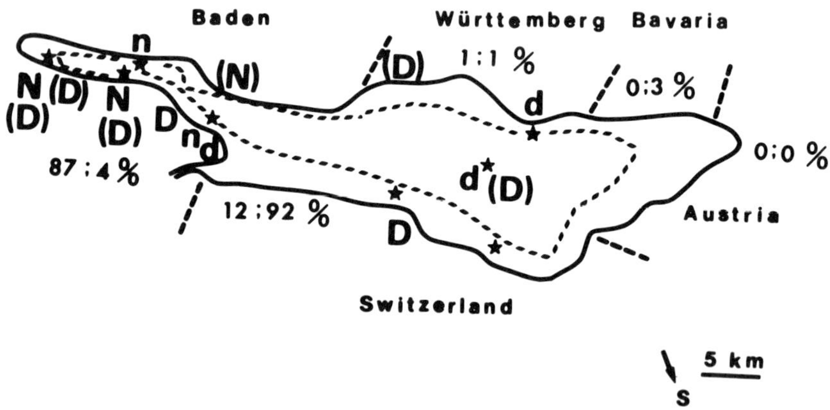

FIG. 1 Distribution of the catches of Arctic charr during the oligotrophic and mesotrophic period of Lake Constance. % = regional share of the total annual yield of charr (1909 - 1938; 1956 - 1979); - - - = borders of regions; 100 m depth line also drawn; () = less than 6 specimens caught recently; D and N = deep living and normal forms caught recently; n and d = these forms caught before eutrophication; * = catch site before eutrophication.

of 100 m. Originally the Upper Lake, Lake Constance in the narrow sense, was oligotrophic; but at least since 1955 a progressive cultural eutrophication has taken place and it is now considered to be mesotrophic. The morphology of the lake and its limnological changes with eutrophication have been described and tabulated earlier (Hartmann and Nümann 1977). The composition of the fish catches during the oligotrophic and the mesotrophic periods of the lake is listed in Table 1. The economic importance of the charr yield has always been small and recently has dropped to zero.

MATERIALS AND METHODS

In addition to data from publications on the charrs of Lake Constance by Nümann (1939) and Dörfel (1974), information has been taken from unpublished material of the 1930's (Nümann) and 1950's (Kriegsmann) and regional catch statistics from Switzerland, Austria, Bavaria, Baden, and Württemberg of 1909-79, as well as data collected in 1980-1982. Whereas Dörfel collected charr in winter in the western basin only, 43 specimens were also collected from the main (eastern) basin. In 1982 and earlier most charr were collected with bottom gillnets. Nümann also used near-bottom nets. Only four specimens were sampled in the upper pelagic zone in 1970-80. Mesh size was mainly 28 mm and 32 mm in the 1980's.

The aging — somewhat difficult with the deep living form — was based on scales in earlier investigations, but on otoliths (Brenner 1980) and/or opercula in the 1980's; back-calculation was made for age at length, down to the outermost completed growth zone, without correction for allometric growth.

Air temperature was measured by the weather bureau at Friedrichshafen.

Throughout this study the charr caught during the oligotrophic period in the eastern basin are compared with those of the deep living form of the mesotrophic period, as the normal form appears to be restricted (more or less) to the western basin. The word "form" is used here in the ecological sense, not as a taxonomic classification; "charr" means always Arctic charr. Yield data of other lakes are taken from a recent paper by Hartmann (1979).

THE CHARRS OF THE LAKE

The existence of different forms of *Salvelinus alpinus* in Lake Constance had been suspected for a long time, but in the 1930's (Nümann 1939) a large collection from the eastern basin was still ascribed to a single form only: the so-called deep living one. In 1972, the deep living form and the normal one were separated (Dörfel 1972; Table 2) using material from the western basin. Behnke (1972), who examined some specimens from Lake Constance, considered the deep living charr as a full species (*S. profundus*). The material of the 1980's reveals no additional criterion of discrimination. Differences in spawning time and food, but not growth and colouration, are confirmed. The few charr recently caught in the upper pelagic zone belong to the deep living form. The normal form becomes more rare each year.

Since 1977 fingerlings of Arctic charr from various European lakes have been planted to prevent extinction of the species by eutrophication (Table 3). The supposition that the endemic stocks are already extinguished is not corroborated by the actual results. For example, exactly the same mean number of gillrakers

Table 2. Selected differences between the two forms of Arctic charr in Lake Constance.

Form	normal	deep living
mean number of gillrakers	28	22
mouth	terminal or inferior	inferior
food in autumn and winter	chironomids > turbellaria	turbellaria > chironomids
flesh	coloured	coloured or pale
months of observed spawning	Nov - Dec	Oct

was found in 1971/72 and 1980. Besides Arctic charr, the brook trout, *S. fontinalis* is also caught occasionally.

CHANGES COINCIDENTAL WITH EUTROPHICATION

Yield

Contrary to the overall yield of fish species of the Upper Lake (Nümann 1972) the regional catches of charr show clear trends. The catch of the Swiss fishermen rose from the 1950's to the 60's and subsequently fell. The same trend on a much lower level can be observed with respect to the landings of Württemberg plus Bavaria, whereas the catch of Baden (western basin) declined from the 1940's onwards (Fig. 2). Thus the trends in the two basins differ. The reduction in recorded yield reflects a real decline in numbers. In the oligotrophic lake, charr were common (but poorly recorded; Nümann 1939); now they are extremely rare. The decline of the catch in the western basin began 20 years before the official start of accelerated eutrophication, but according to Kriegsmann (1935) the cultural eutrophication of the lake became apparent in the 1930's.

Fishermen connect the decline of catches at certain places with an extinction of the mayfly *Ephemera vulgata* and an increase of mud (Dörfel 1972). No correlation has been found between the year-to-year fluctuations of annual yield and those of the annual 0_2-minima.

The overall yield of fish remained stable for a long period although the proportion of charr decreased coincidentally with eutrophication, from 0.1. percent to 0.02 percent as the yield of other species increased.

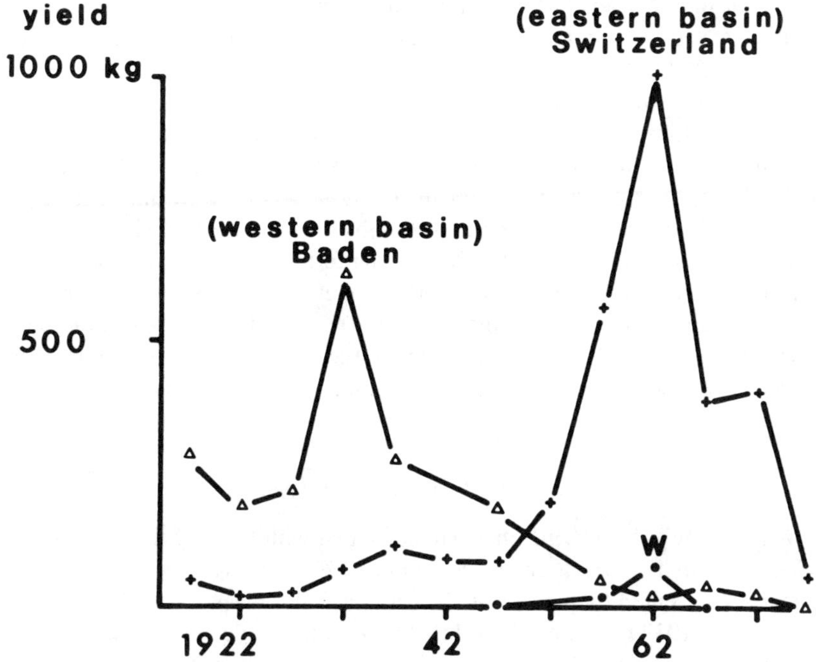

FIG. 2 Local trend of charr yield within Lake Constance. W = Württemberg plus Bavaria (eastern basin). 5 year intervals.

Distribution

After eutrophication the normal form was caught in the western basin only, the deep living one mainly in the eastern part (Fig. 1). There is some evidence that this distribution also existed in the past; for example, many early workers stated that the length of the charr of the eastern basin rarely exceeded 150 or 180 mm, whereas larger specimens were found in or near the western basin (Scheffelt and Schweizer 1926; Moustier 1951). The relative contribution of the western basin to the charr catch declined, while that of the eastern basin increased with eutrophication (Fig. 1).

The places where charr are still caught in numbers are removed from high density human population and protected from prevailing winds. These areas are characterized by high carbonate concentrations in the sediment, low quantities of affluent waters, low values of suspended matter in the affluent water, low densities of tubificids and high densities of turbellaria and chironomids, both of which are important food items of the charr (Hartmann, in press). These observations point to a low level of inorganic (Rossknecht, pers. comm.) and organic pollution at the fishing sites and spawning places of the charr. Conversely a formerly well known site in the northern main basin, having the opposite characteristics, is now deserted.

Besides differences in horizontal distribution there also seem to exist differences in the vertical distribution of the two forms. In autumn and winter the so-called deep living form is caught mainly at depths over 60 m, whereas the normal form is usually caught down to 60 m by the professional fishermen. Also in the 1930's charr ascribed to the deep living form were found in water deeper than 70 m (Nümann 1939). Before and after eutrophication, charr were also found in small numbers in the pelagic zone. Thus apart from a reduction in the number of catch sites there is no apparent change in the distribution, but there is, apparently, an intensification of immigration into the pelagic zone, parallel with that of other species.

Growth, Condition and Age

When today's deep living charr are compared with those of the oligotrophic period, growth appears to have accelerated (Fig. 3). This is confirmed by comparison of length frequencies in the catches. Independent of mesh size, most charr in 1933-1939 measured about 150 mm; now the mode of the "dwarf form" is 250 mm (Fig. 4). The differences in growth of the two forms (Dörfel 1974) are no longer apparent. Thus the terms "dwarf" or "hunger" form no longer fit the deep living charr of Lake Constance.

The weight-length ratio (K) of the fishes also increased. In the length range of 180 to 300 mm (Sept.-Dec.), the charrs caught in the 1980's are 20 g (300 mm fish) to 32 g (180 mm fish) heavier than those caught in the 1930's.

If the age readings are correct (Wagler 1947) the modes of age, in the catch from the main basin, shifted from 2+ - 3+ (78 percent) to 3+ - 5+ (88 percent). This might indicate a reduction of fishing intensity, but an alternative explanation is that, in the more densely populated oligotrophic lake, the growth of most charr ceased at a length of about 150 mm (compare Alm 1946). It must be mentioned here that the year-to-year fluctuations of charr yield correlate ($P<$, 1 percent) with the temperature in May three years earlier Fig. 5). This points to a modal age of three years, which harmonizes with the findings of commercial catches (Dörfel 1974).

Food

In the oligotrophic lake, the larger charr were piscivorous (Nümann 1939) whereas now, in spite of an increase in size, fish are no longer found in charr

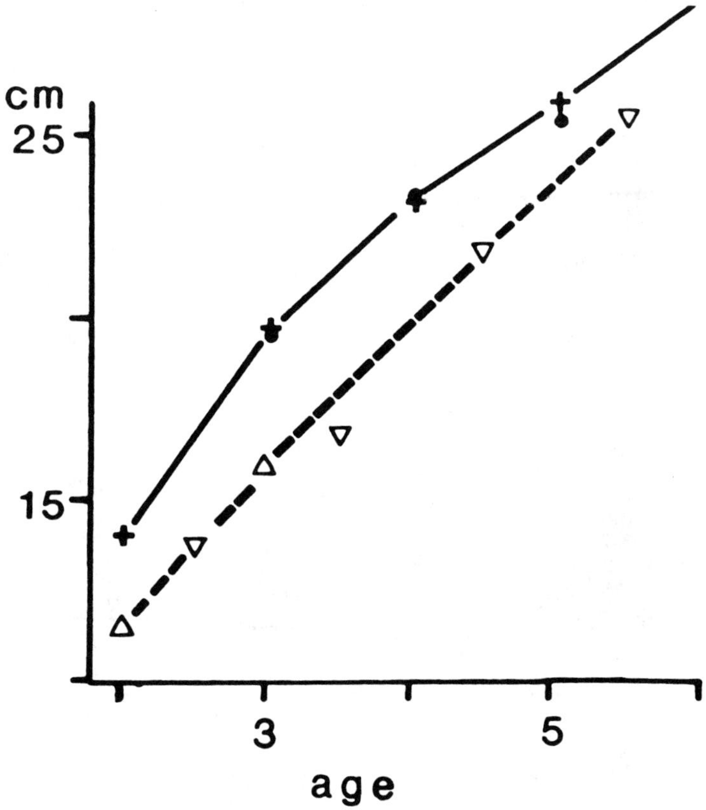

FIG. 3 Growth of Arctic charr during the oligotrophic and mesotrophic period of Lake Constance. —— = deep living form after eutrophication (+ = Dörfel 1974; ● = in 1980). --- = charr before eutrophication (△ = Haakh 1929; ▽ = Nümann 1939).

stomachs. Before eutrophication charr caught in the pelagic zone ate *Heterocope borealis* but in the 1960's this copepod disappeared from the lake, and we now find large cladocerans including *Bythotrephes longimanus*, which had not previously been found in the stomachs (Table 4). A seasonal shift towards planktivorous feeding has occurred in charr as it did with other species. In the 1930's, charr caught within a few meters of the bottom almost exclusively ate benthos (Nümann 1939), whereas now their stomachs contain mainly planktonic cladocera. In the fishermen's opinion charr today stay higher in the water column in summer than in autumn.

The difference in the feeding habits of the two forms, which is confirmed by different flesh colour (Table 2) may be just a consequence of different vertical

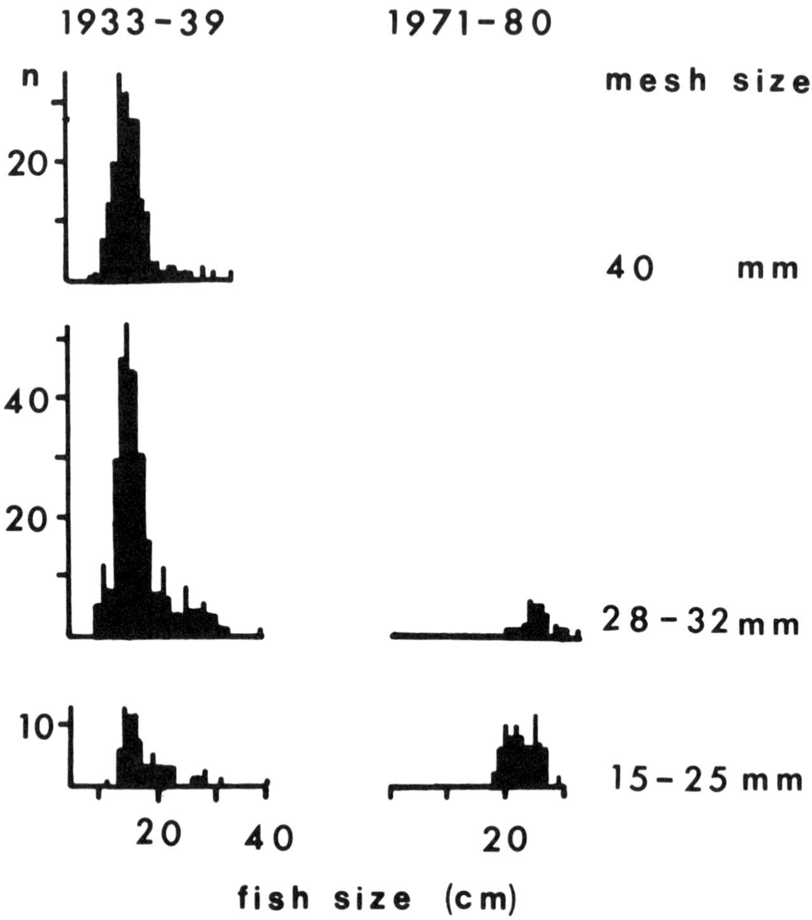

FIG. 4 Length frequencies of Arctic charr caught during the oligotrophic and mesotrophic period of Lake Constance.

distribution. Both chironomids (Reiss 1968) and the normal form of charr were caught 30-60 m deep in the western basin.

Community

At a well known charr fishing site in the northern part of the main basin (Fig. 1) three fish species were found in the profundal zone (> 80 m) in winter before eutrophication: charr, whitefish, *Coregonus pidschian*, and burbot, *Lota lota* (Nümann 1939). Nowadays we get there (and elsewhere) whitefish, *C. lavaretus*,

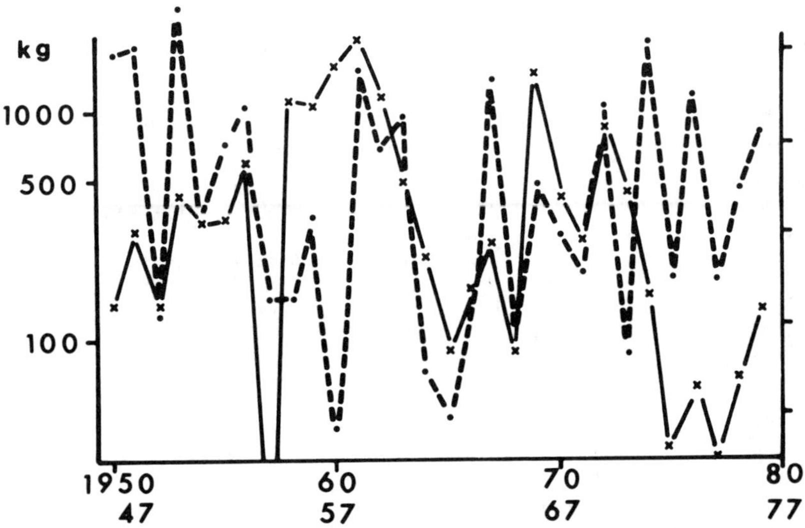

FIG. 5 Annual yield of Arctic charr of Lake Constance and air temperature in May three years before. x = yield; ● = temperature.

perch, *Perca fluviatilis*, roach, *Rutilus rutilus*, dace, *Leuciscus leuciscus* and burbot. Thus, a niche shift, a "littoralization" of the profundal zone has taken place. The molluscs and turbellarians formerly utilized by salmonids are now taken by cyprinids. The dace now appears to occupy parts of the former charr niche (Table 5) in winter.

The yield of charr in the main basin displays the same pattern as that of the other two members of the former profundal community: an increase until a maximum was reached in the 1960's followed by a decrease. This pattern may be compared with the yield trends of the former littoral fishes and the former pelagic fishes (Fig. 6). If the irregularities in the yield of the former pelagic fishes are accepted as fishing effects (Nümann 1973), only the former profundal community deteriorated as the lake eutrophied, although the total number of the fishes in deep water remained unchanged, at least in winter (Hartmann 1980).

DISCUSSION

The two forms of charr in Lake Constance, which can be differentiated by morphological characteristics, are ecologically separated to some extent by differences in distribution, food, and spawning time.

In spite of recent introductions of charr from various other lakes there is no indication yet that the charr in 1980 differ from those of the 1970's (Dörfel 1974).

Table 3. Stocking of Arctic charr in Lake Constance.

Year	n	mm	Probably from
1891-1908	34000		Lake Starnberg, Austria, Bavaria
1977	4000	70 - 90	Scotland
1978	6000	120 - 150	Bavaria and/or Switzerland
1979	10000	100	Bavaria
	5000	60 - 80	Scandinavia
1980	7500	100	Switzerland
	92000		Switzerland
	100		?

But to protect the rest of the endemics (Behnke 1972) the (costly) planting of non-resident forms or species must be stopped immediately, and a closed season should be extended into a closed year. The other populations of the European Alps should be protected from the extensive planting activities that are endangering stocks of the region and even beyond (Table 6). For example, all the populations of German lakes are already considered to be mixed with Swedish imports (Bless 1978). Planting of material from other lakes should be restricted to cases of loss of endemics. So far however, examples of total extinction, as an effect of eutrophication, are surprisingly few in the European Alps.

As shown in Table 7, the biological changes observed in the charr, associated with lake eutrophication, are in line with those described for various other fish species of Lake Constance (Hartmann 1979). According to this table, potential plankton eaters (including charr) can take advantage of eutrophication.

Apart from Lake Constance, little seems to be known about changes in charr biology associated with lake eutrophication. Only some yield data are available: in subalpine European lakes (Fig. 7) undergoing eutrophication, as well as the expected decrease in charr yields, increases also occur, especially in the deeper lakes. In some cases an increase occurred just before the ''official'' start of accelerated eutrophication. Therefore, at least in deep lakes, the yield of charr may often follow the usual increase-decrease pattern. Accordingly, in Tegernsee yield decrease coincided with oligotrophication.

Because yield increase is particularily noticeable in potentially planktivorous species, it may be suspected that the yield trend of an individual charr population depends on the ability of the individual stock to utilize the plankton increased by eutrophication. Beyond Lake Constance we also suspect that differences in biology of the various forms of charr result in differences in sensitivity and vulnerability to stress. In Attersee (Brenner 1980), Lake Constance, Lago Maggiore, Schlierssee and probably also Zürichsee, it is only the deep living form which became dominant or survived. But to date different trends in the two forms, in lakes of the European Alps, have not been determined (Fig. 7); therefore the trend of a particular population is not predictable.

If an international charr watch (Regier and Power 1979) is to be implemented, Lake Constance provides a preliminary reference: a case history of charr in a lake undergoing eutrophication. But also distributional changes within a lake — as in Lake Constance — and changes in charr food organisms, may serve as an indicator of organic pollution or other kinds of stress.

ACKNOWLEDGEMENTS

Information was gathered from many colleagues, especially Dr. T. Brenner, Dr. J. Deufel, Dr. F. Kriegsmann, Dr. O. Moog, L. Probst, Dr. Ch. Ruhlé, Dr. N. Schulz, and Dr. G. Wagner. The difficult collection of the scarce fishes by G. Knöpfler was supported by K. Geiger, A. Göppinger, and H. Traber.

REFERENCES

Alm, G. 1946. Reasons for the occurrence of stunted fish populations. Inst. Freshw. Res. Drottningholm Rep. 25: 1-146.

Behnke, R.J. 1972. The systematics of salmonid fishes of recently glaciated lakes. J. Fish. Res. Board Can. 29 : 639-671.

Bless, R. 1978. Beständsanderungen der Fischfauna in der Bundesrepublik Deutschland. Naturschutz aktuell 2. Kilda, Greven 66 p.

Brenner, T. 1980. The Arctic charr, *Salvelinus alpinus salvelinus*, in the prealpine Attersee, Austria, p. 765-772. *In* E.K. Balon [ed.] Charrs: salmonid fishes of the genus *Salvelinus*. Dr. W. Junk Publishers, The Hague.

Dörfel, H.J. 1972. Zur Problematik der Saiblingpopulationen im Überlinger See. Staatsexamensarb. Freiburg i. Br. TS (partly seen).

Dörfel, H.J. 1974. Untersuchungen zur Problematik der Saiblingpopulationen im Überlinger See (Bodensee). Arch. Hydrobiol. Suppl. 47 : 80-105.

Grimaldi, E. and W. Nümann. 1972. The future of salmonid communities in the European subalpine lakes. J. Fish. Res. Board Can. 29 : 931-936.

Haakh, T. 1929. Studien über Alter und Wachstum der Bodenseefische. Arch. Hydrobiol. 20: 214-295.

Hartmann, J. 1979. Unterschiedliche Adaptionsfähigkeit der Fische an Eutrophierung. Schweiz. Z. Hydrol. 41: 374-382.

Table 4. Food of the deep living form of charr during oligotrophic and mesotrophic period of Lake Constance.

author	year	months	method of analysis	turbellaria	cladocerans[x]	Heterocope	fish
				percent			
Nümann	1935 - 36	Jul - Sep Oct - Mar	occurrence	50 42	19 0	16 8	19 13
Dörfel	1971 - 72	Nov - Jan	numerical	80	0	0	0
	1980 - 82	Apr - Sep Oct - Mar	dominance	25 53	41 2	0 0	0 0

[x] *Daphnia*, *Leptodora*, *Bythotrephes*

Table 5. Changes in the fish community of the profundal region (80 - 100 m) at Langenargen in winter. 1 - 5 = rank of importance in terms of number of fishes and importance as food of fishes respectively.

eutrophication	before			after				
rank	1	2	3	1	2	3	4	5
fish species	*Coregonus pidschian*	*Lota lota*	*Salvelinus alpinus*	*Perca fluviatilis*	*Rutilus rutilus*	*Leuciscus leuciscus*	*Coregonus lavaretus*	*Lota lota*
molluscs	1		4		1	1		
chironomids	2	3	3	1	3			2
cocoons of turbellaria	3		1			2		
copepods			2					
fish		1		2				1
amphipods		2						3
cladocerans				3	2			
tubificids						3		

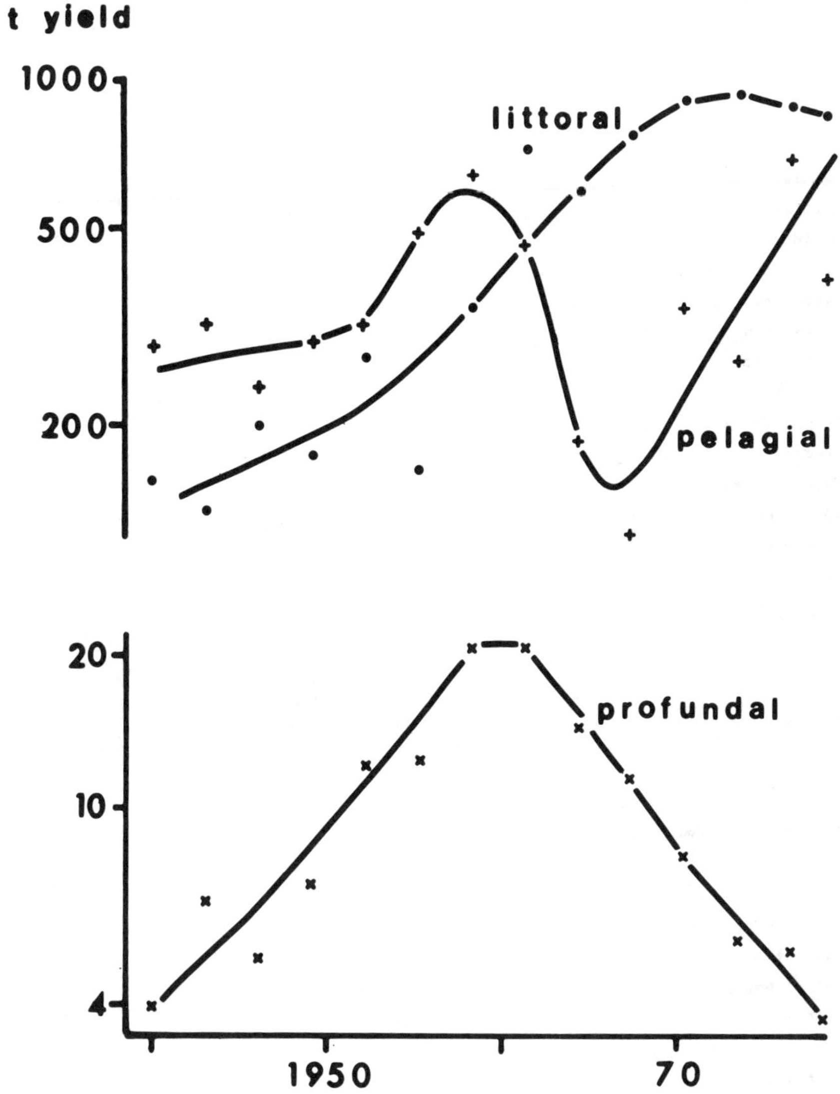

FIG. 6 Trends in yield of fishes clustered according to their belonging to the following communities in oligotrophic Lake Constance: pelagic *(Coregonus lavaretus wartmanni, Salmo trutta)*; littoral *(Perca fluviatilis, Rutilus rutilus, Abramis brama, C. l. macrophthalmus, C. pidschian fera)*; profundal *(S. alpinus, C. p. acronius, Lota lota)*.

Table 6. Stocking of Arctic charr into lakes of the European Alps. Data selected from 22 authors. 0 = stocking before 1910. For example 7 = stocking in the 1970's; + = year of stocking unknown.

from / into	Attersee	Erlauffsee	Puschlsee	Grabenseen	Grundlsee	Hintersee	Irrsee	Jägersee	Lac Leman	Luganer See	Millstättersee	Ossiachersee	Scotland	Seekirchnersee	Starnberger See	Sweden	Traunsee	Sünser See	Walchensee	Walensee	Vierwaldstättersee	Zugersee	Unknown lake (s)
Attersee																				5			
Austria																		+	+			+	
Grundlsee	3																						
Hallstättersee	1	1			1						1	1		1			1						
Hintersee	3				+										8		1						
Yugoslawia				7																5			
Langbathsee					3																		
Mondsee	+																						
Sempacher See																				4			
Sweden															7				7				7
Tappenkarsee								0															
Tegernsee																			0				
Zugersee										0													
and into / from																							
Atter- & Mondsee & Zürser See					+																		
Brienzersee & Sempacher & Thuner See																						0	
Chiemsee															1	+							
Erlaff- & Königssee																							+
Faistenauer Hintersee								0															
Millstätter See					1																		
Lac d'Annecy									+														0
Lago Maggiore										0												0	
Lago Lucendo & Ritom- & Tomsee										+													
Lake Constance													7		0	7			7	7	8		0
Obertrummer See							+																
Schliersee & Sünser See			+																				+
Thunersee																						7	
Walensee										4													
Unknown lake (s)									+	+						7			7				

Table 7. Changes in fishes coincidental with eutrophication of Lake Constance.

Changes	Coregonus lavaretus	Coregonus pidschian	Perca fluviatilis	Rutilus rutilus	Salvelinus alpinus	Alburnus alburnus	Lota lota	Abramis brama	four demersal fish species	invertebrates
Increased growth rates	+	+	+	+	+	+	–	–	–	+
Higher condition factor	+				+					
Less fish eaten	+	·	+	·	+	·	+	·	·	·
Daphnia eaten instead of Heterocope	+	·	+	·	+		·	·	·	·
Less benthos eaten	·	+	+	+	+	·	–	+	–	·
Intensified immigration into the pelagial	·	+	+	+	+	·		+		+
Temporary yield increase	+	+	+	+	+		+	+	–	+

\+ = yes – = no · = insertion absurd

Hartmann, J. 1980. Abfolge der Ernährungsweise der Fische bei Produktionsänderung des Bodensees. Fischwirt 30: 46-48.

Hartmann, J. In press. Charr as indicators of pollution. Fisheries.

Hartmann, J. and H. Löffler. 1978. Saisonale bodennahe Verteilung von Fischen im eutrophierten Bodensee. Arch. Hydrobiol. 83: 69-79.

Hartmann, J. and H. Löffler. 1980. Grössere Wintertiefe von Fischen mit der beschleunigten Eutrophierung des Bodensees. Fischwirt 30: 22-23.

Hartmann, J. and W. Nümann. 1977. Percids of Lake Constance, a lake undergoing eutrophication. J. Fish. Res. Board Can. 34: 1670-1677.

Kriegsmann, F. 1955. Der Wechsel in der Vergesellschaftung der Fischarten des Oberund Untersees und die Veränderungen des See-Reagierens. Arch. Hydrobiol. Suppl. 22: 397-408.

Moustier, J. de. 1951. Apercu general sur le Lac de Constance. 138 p. TS.

Nümann, W. 1939. Untersuchungen uber die Biologie einiger Bodenseefische in der Uferregion und den Randgebieten des freien Sees. Z. Fisch. 37:637-688.

Nümann, W. 1973. Versuch einer Begründung für den Wandel in der qualitativen und quantitativen Zusammensetzung des Fischbestandes im Bodensee während der letzten 60 Jahre und eine Bewertung der Besatzmassnahmen. Schweiz. Z. Hydrol. 35: 206-238.

Reiss, F. 1968. Ökologische und systematische Untersuchungen an Chironomiden (Diptera) des Bodensees. Arch. Hydrobiol. 64: 176-323.

Regier, H.A. and G. Power. 1979. Wild trout and charr watch. Wild Trout Symp. Yellowstone National Park, TS 5 p.

Scheffelt, E. and W. Schweizer. 1926. Fische und Fischerei im Bodensee. 173 p. Enke, Stuttgart.

Wagler, E. 1946. Fische und Fischerei in den bayerischen Voralpenseen. III Salmoniden 4. Der Seesaibling (*Salmo salvelinus* L.). Allg. Fisch. — Ztg. 71 (17): 5-7.

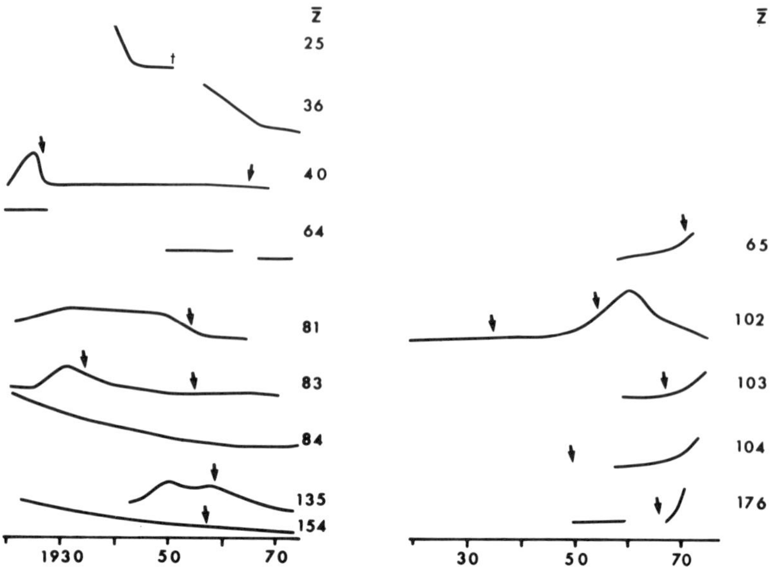

FIG. 7 Trends in Arctic charr yield of European subalpine lakes. $\overline{Z}$ = mean depth; arrow from left to right = start of accelerated eutrophication; arrow from right to left = oligotrophication; † = extinct; (d) = deep living form in this lake; (n) = other forms in this lake; 25 = Schliersee (n); 36 = Mondsee (n); 40 = Tegernsee (d+n); 64 = Neuenburger See (d); 65 = Hallstätter See; 81 = Lac du Bourget; 83 = western basin of Lake Constance (d+n); 84 = Zugersee (n); 102 = eastern basin of Lake Constance (d); 103 = Walensee (d+n); 104 = Vierwaldstättersee (d); 135 = Thunersee (n); 154 = Lac Leman (n); 176 = Brienzer See (n). Charrs also disappeared from the Zuricher See (44 m) and declined in Irrsee (15 m) and Walchensee (82 m).

The Management of Arctic Charr (*Salvelinus alpinus* L.) in Eutrophied Lake Zug[1]

CHRISTIAN RUHLÉ

Fish and Wildlife Service, St. Gallen, Switzerland

RUHLÉ, C. 1984. The management of Arctic charr (*Salvelinus alpinus* L.) in eutrophied Lake Zug, p. 487–492. *In* L. Johnson and B.L. Burns [eds.] Biology of the Arctic charr, Proceedings of the International Symposium on Arctic Charr, Winnipeg, Manitoba, May 1981. Univ. Manitoba Press, Winnipeg.

At the end of the last century Arctic charr dominated the Lake Zug fishery, forming 85 percent of the total harvest. Due to increasing eutrophication over the last 80 years, the annual catch has dropped from 100,000 fish to less than 5,000, while the relative proportion of the catch is now only 2 to 3 percent of its former level.

In the present highly eutrophic state of the lake, the Arctic charr population can be maintained only by heavy stocking. The main catches are made during the spawning season, and in consequence the success of the fishery depends to a great extent on the nature of the spawning grounds. To compensate for the effects of autochthonous sedimentation, spawning grounds have to be revitalized by the dumping of fresh gravel. The old spawning grounds, now located in the poorly oxygenated hypolimnetic waters, have been abandoned and replaced by ones in epilimnetic waters.

[1] This paper is partly based on a thesis presented in 1976 at the Federal Institute of Technology, Zurich, Switzerland (Ruhlé 1977).

INTRODUCTION

In earlier centuries, the high quality of the Arctic charr (*Salvelinus alpinus* L.) of Lake Zug gave this fish an extraordinary economic value, imparting a particular cultural and historical importance to the region around the lake. This justifies support for the remaining small populations of the presently eutrophied lake through the application of usual and unorthodox management techniques.

DESCRIPTION OF THE LAKE

Lake Zug is located in the Swiss pre-alpine region; its shape is similar to that of a kidney; this determines its hydrology, limnology and consequently, its fishery.

A large peninsula pointing out from the western shore divides the lake into two morphologically different basins. Only a small part of the water inflow supplies the upper basin, for which the theoretical renewal time has been determined to be 20.7 years, whereas, for the whole lake, it is only 13 years. All attempts at lake restoration through water purification, therefore, produce slow results in the upper basin; this is the important part of the lake for the charr fishery.

Based on sediment analysis, eutrophication began in the early 20th century (Züllig 1956). At that time also, the first mass developments of phytoplankton populations were recorded (Brutschy 1913). Lake Zug must be considered today as a highly eutrophied lake: the nutrient contents exceed by far, both in the circulated water body and in the water inflow, the limits by which eutrophication is usually defined.

DEVELOPMENT OF THE CHARR POPULATION

Over a long period, the total yield of the fishery of Lake Zug was predominantly charr. As the most sensitive species of the indigenous fish fauna (Grimaldi and Nümann 1972) the Arctic charr was the first species to respond to the well-known increase-decrease pattern heralding the beginning of eutrophication. From an absolute and relative maximum, attained in about 1910, catches have dropped more or less continuously over the last seventy years (Figs. 1 and 2).

Today, charr catches are about 1-2 percent of the total yield although the total yield has increased eightfold in the last forty years because of the much higher yield of cyprinids (*Alburnus alburnus* L., *Rutilus rutilus* L.). At present it is 155 metric tons or 47 kg • ha^{-1} (mean annual yield 1973-1978). The absolute yield of the charr fishery dropped during the last seventy years from approximately 100,000 fish per year to about 5,000 fish per year.

MANAGEMENT CONSIDERATIONS

So far, no extensive information can be given on the influence of eutrophication on the charr population of Lake Zug. Nevertheless, it must be considered as fact that recruitment has been affected by the direct and indirect consequences of this alteration. For this reason, management techniques are primarily directed towards the maintenance of the population, the most important being the stocking of young fish and the periodic renewal of spawning grounds by the deposition of new gravel.

From the end of the last century until the mid-fifties, the lake was stocked with charr fry. As no convincing effect on the population appeared by the end of this period, a change was made to stocking fingerlings in the following years. As

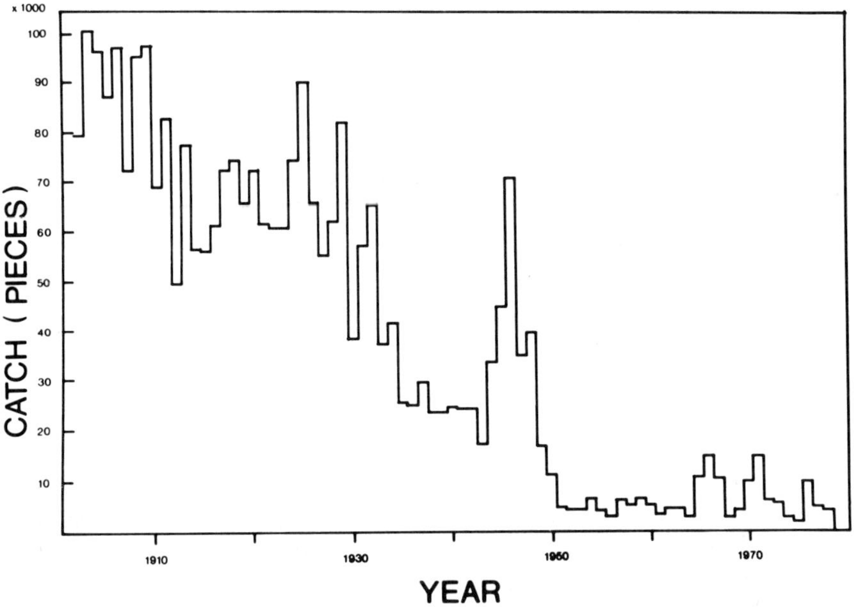

FIG. 1 Yields (number of fish) of the charr fishery 1902 - 1978.

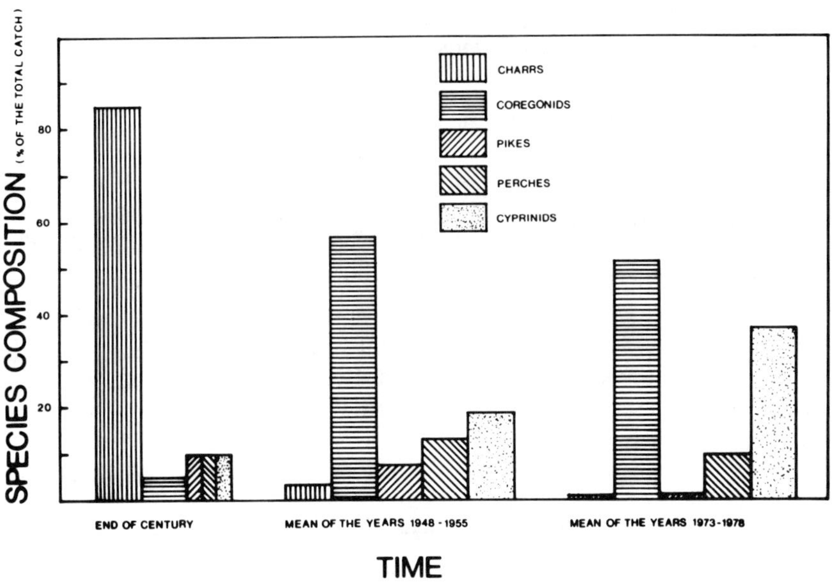

FIG. 2 Species composition of the catches of 1900, 1948 - 1955 and 1973 - 1978.

even better results were obtained by stocking yearlings, attempts were made to enhance production of charr of this age.

From the correlation between stock and catch it can be deduced that practically the total recruitment depends on artificial stocking. This was confirmed by an experiment in which fertilized eggs were placed on the spawning grounds in cages similar to those of Vibert (1959); at the end of the hatching period only a very small number of larvae had attained the 'eyed' stage and practically none had developed further.

The connection between stock and catch allowed an estimation of the catch of fishes stocked at different sizes or ages: for fry (1950s), this is approximately 0.5 percent, for fingerlings (1960s), approximately 2.5 percent, and for yearlings (tagged, 1974), approximately 20 percent.

The production of autochthonous fishes for stocking purposes implies an effective fishery on well-maintained spawning grounds. To keep spawning grounds in the required condition, the fishermen renew them periodically (approx. every third year). From a barge anchored exactly over a spawning ground, they deposit sufficient gravel to cover the old layer altered by the autochthonous sedimentation (sedimentation rate approx. 2 mm • yr^{-1}) (Züllig 1956). On each spawning ground, having a mean surface area of only 200 m^2 (r = 8m), 15 to 20 cubic meters of gravel are deposited. It is of great importance that the gravel be immersed carefully, otherwise the very steep slope of the spawning grounds (max. gradient over 50 percent) would ensure that the gravel to slides off.

Photographic investigation of spawning grounds located in depths of approximately 80 m showed (in combination with catch results) the effectiveness of these gravel depositions. A well defined relationship between the quality of the spawning grounds (expressed in percentage of clean gravel on the total bottom surface) and the number of fishes caught per unit net was found (Fig. 3). This result confirms the investigations made earlier showing that the attractiveness of the spawning grounds to charr depends greatly on their composition (Fabricius and Gustafson 1954). The great difference within the catches on well-maintained spawning grounds (100 percent clean gravel) is, however, remarkable. It is due to this fact that all poor catches are made on deep spawning grounds where it is difficult to place the bottom nets precisely and where the oxygen concentration is near the avoidance-limit (Whitemore et al. 1960). This result indicates that insufficient oxygen in the deeper parts of the lake (Fig. 4) reduces the suitability of the spawning grounds located there, even if their surface is well-maintained by regular gravel addition. For the long term restoration of Lake Zug, it will be necessary to abandon these spawning grounds and to establish new ones above the metalimnic oxygen minimum.

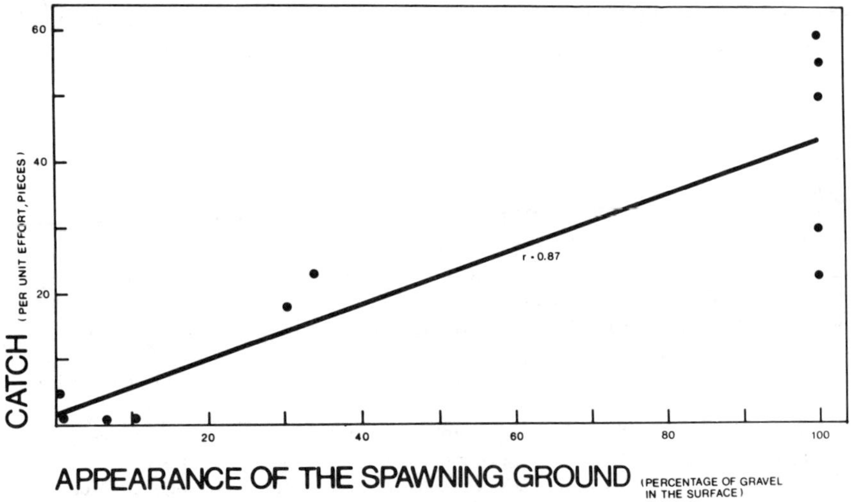

FIG. 3 Relationship between the quality of spawning grounds and their attractiveness to charr.

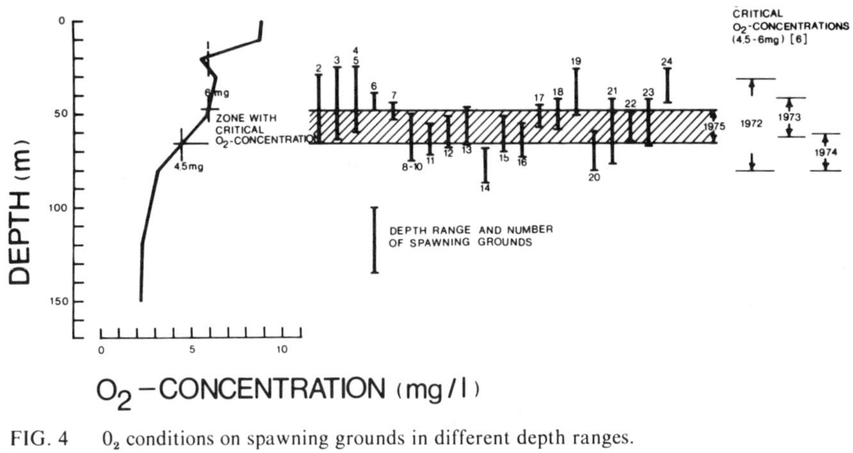

FIG. 4 0_2 conditions on spawning grounds in different depth ranges.

FINAL CONSIDERATIONS

All the measures applied to maintaining (or improving) the stock of Arctic charr in eutrophied Lake Zug are very costly. The expenditures far exceed the profit realized by the fishery. However, the particular cultural and historical importance of this fish in the region around the lake justifies the negative balance. The publicly supported management allows the preservation of this precious fish until such time as a self-supporting population can be restored to the lake.

REFERENCES

Brutschy, A. 1913. Monographische Studien am Zugersee. Arch. Hydrobiol. 8: 43-107.

Fabricius, E. and K.J. Gustafson. 1954. Further aquarium observations on the spawning behaviour of the char, *Salmo alpinus* L. Inst. Freshw. Res. Drottningholm Rep. 35 : 58-104.

Grimaldi, E. and W. Nümann. 1972. The future of salmonid communities in the European subalpine lakes. J. Fish. Res. Board Can. 29 : 931-936.

Ruhlé, Ch. 1977. Biologie und Bewirtschaftung des Seesaiblings (*Salvelinus alpinus* L.) im Zugersee. Schw. Z. Hydrol. 39(1) : 13-45.

Vibert, R. 1959. Possibilités et limites du repeuplement en truites et saumons par boîte d'alevinage. Bull. Off. Int. Cons. Sup. Pêche No. 37, IV.

Whitemore, C.M., C.E. Warren and C. Dudoroff. 1960. Avoidance reactions of salmonid and centrarchid fishes to low oxygen concentrations. Trans. Amer. Fish. Soc. 89: 17-26.

Züllig, H. 1956. Sedimente als Ausdruck des Zustandes eines Gewässers. Schweiz. Z. Hydrol. 18 : 5-144.

Restoration of Charr Populations in Impounded Lakes in Scandinavia By Locally Applied Fertilization

GÖRAN MILBRINK and STAFFAN HOLMGREN

Department of Zoology, University of Uppsala and
Institute of Freshwater Research, Drottningholm, Sweden

MILBRINK, G. AND S. HOLMGREN. 1984. Restoration of charr populations in impounded lakes in Scandinavia by locally applied fertilization, p. 493–508. *In* L. Johnson and B.L. Burns [eds.] Biology of the Arctic charr, Proceedings of the International Symposium on Arctic Charr, Winnipeg, Manitoba, May 1981. Univ. Manitoba Press, Winnipeg.

In order to restore a charr population in particularly poor condition, commercial N- and P-fertilizers were added during four consecutive summer seasons to a small stream leading to a narrow bay of an impounded lake in northern Sweden. The fertilizers promoted the development of an exceedingly rich zooplankton and stream bottom fauna which primarily favoured the charr, both qualitatively and quantitatively, but locally the condition of grayling and brown trout also improved markedly.

The eutrophication gave rise to a most interesting experimental situation; dramatic changes were observed in the intensity of interactive segregation between species — as reflected in the degree of niche-overlap. Some effects of enrichment were retained even during the fifth year of study when no further fertilizers were added, an important consideration in terms of economy. Planktivorous fish were still attracted to the area and the condition of the fish remained at a high level.

A case of seasonal, long-distance feeding-migration of charr was studied in a parallel situation in a series of small lakes, once heavily eutrophied by sewage from a nearby ski resort. Each summer the fish move up a narrow stream from a lake about 5 km downstream to feed on the abundant food. Populations of brown trout and brook trout behave in a similar way. These small lakes are totally devoid of oxygen in the winter.

INTRODUCTION

Controlled nutrient enrichment of oligotrophic waters to improve fish production is an old idea that was revived in the 1970's. The ideal state would be a "balanced eutrophication" where a variety of valuable fish species dominate the fish fauna (cf. Svärdson 1976). In the last decade a trend towards nutrient enrichment of inland waters has developed along parallel lines in Sweden, Norway and Canada. Large-scale as well as small-scale fertilization projects in these countries have provided valuable information in a field which is particularly promising for the improvement of fish production in regulated lakes in Scandinavia (Milbrink and Holmgren 1981a, 1981b; Langeland 1982) and in Canada (Northcote 1972), in inland waters on a small scale (Pyefinch 1960; Fagerström 1966; Filipsson et al. 1968) as well as more extensively (Manzer 1976a; Stockner 1977; Lebrasseur et al. 1978; Robinson and Barraclough 1978).

EFFECTS OF WATER REGULATION IN SCANDINAVIAN LAKES

Since the 1940's Scandinavian waters have been the subject of large-scale hydroelectric development. A majority of the lakes and rivers in northern Sweden are now regulated. In the western part of the province of Jämtland, for instance, where the present studies have been conducted, a greater proportion of the waters than elsewhere in the country are now regulated: at present, about 70 percent of the lake area and an even greater proportion of the running waters.

Due to the considerable water level fluctuation — often 10-20 metres over the year — which is more or less characteristic of impounded lakes, the littoral region becomes increasingly eroded; loose deposits are gradually transported into the deep profundal region where they are trapped. Extreme silting due to erosion may also negatively affect primary productivity in impounded lakes and thereby influence entire food-webs. The bottom fauna, which is normally most abundant in the littoral of the lake, declines rapidly when loose material is carried downwards and the shallow littoral bottoms dry out in the winter, becoming exposed to freezing temperatures.

In the first years following impoundment, however, the productivity may actually increase. The condition of most fish, especially the charr, improves quite significantly in a manner described as the "damming-up effect" (Nilsson 1973). When land is flooded nutrients, organic matter, etc. may dissolve in the water; this enrichment, generally noticeable throughout the entire food chain, is ultimately canalized into fish, but it gradually fades away and, within a few years of damming, a process of "oligotrophication" begins.

When a lake becomes impoverished most fish are forced to feed on whatever bottom fauna is available, i.e. in river outlets, in "pools" connected with these outlets (Grimås 1965; Aass et al. 1972), or on "surface fauna". "Drift", mostly insect larvae transported via streams, is also a possible, although temporary, source of food. The wide range of zooplankton on which charr and pelagic whitefish are normally dependent — when occurring sympatrically with other salmonid fish — is not sufficiently abundant to provide food for large populations.

In order to survive on a lean diet most fish species have to switch to food objects of their second or third choice. The predation pressure exerted on the food items still remaining will be severe and a stage of intensive interactive segregation for food commences (cf. Milbrink and Holmgren 1981b).

There are few biological techniques available for combatting damage to fish caused by water regulation. The fisheries authorities of Sweden have worked actively in the introduction of new fish-food organisms into impounded lakes, such as *Mysis relicta* Lovén, to compensate for a nearly ruined littoral bottom fauna. New fish predators that are better adapted to a pelagic life in impounded lakes, such as lake trout, have likewise been introduced with some success. None of these methods, however, has turned out to be fully successful. By introducing *Mysis*, for instance, no new nourishment has been added to the system.

Compensatory measures of this kind have naturally offered unique possibilities for the study of "interactive segregation" with reference to the sharing of available resources between different fish species such as charr, brown trout, whitefish, etc. and also the utilization of the new fish-food organisms compared with the previously established cladoceran fauna dominated by *Eurycercus lamellatus, Bosmina coregoni* etc. (Northcote 1972; Fürst 1972; Fürst et al. 1978; Lasenby 1979).

DESCRIPTION OF STUDY AREAS

To test whether the addition of fertilizers to a limited area of a large impounded lake could positively affect all levels of the food chain, including the fish, the Institute of Freshwater Research, Drottningholm, together with the County Administrative Board of Jämtland, started a pilot study in Lake Anjan close to the Norwegian border (Fig. 1). The lake has been subjected to long-term regulation since 1940 (area about 26 km^2, amplitude of regulation 9.2 m). That the fish populations were in a very poor state due to impoundment was the main reason for choosing Lake Anjan for these experiments. Since the fish populations are now very sparse and thus difficult to catch, there are no fishing activities in the lake. Other reasons were that the fish fauna, as well as the invertebrate fauna, was comparatively well-known and background information necessary for a study of this kind was available.

Lake Anjan was formerly well-known for its large populations of brown trout and charr. Both species are now scarce and the condition of the fish is below average for the region. Grayling was accidently introduced into the lake in 1952 and spread rapidly in the littoral region and into the inflows. Burbot is a common profundal species and minnows may be found in shallow waters.

By working with the small-scale application of commercial fertilizers and thereby making use of the old practice of "baiting", i.e. attracting fish to the fertilized areas and then keeping them there by offering a super-abundance of food organisms, a number of advantages could be gained, such as an increased

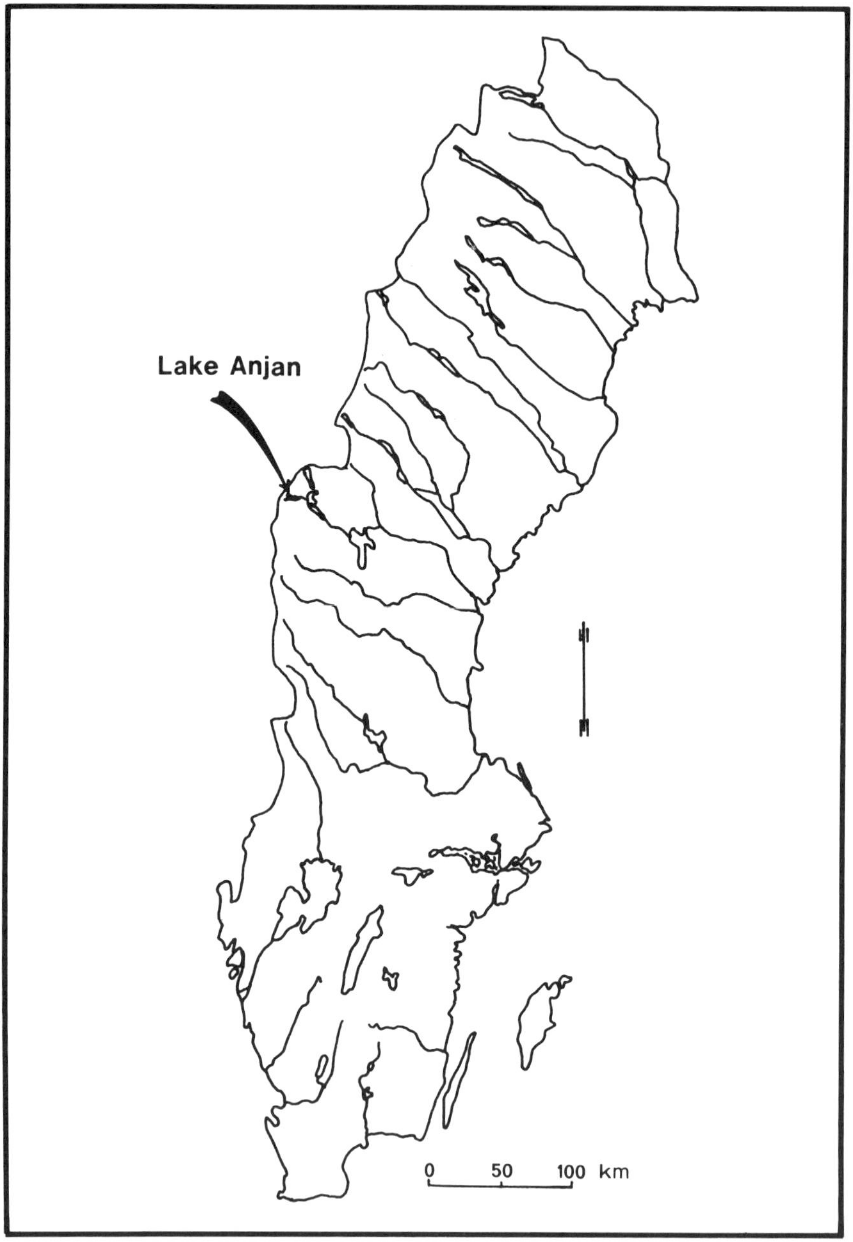

FIG. 1 Study area in the western part of the Province of Jämtland.

catchability of the fish, a better economy than in whole-lake experiments and the improved possibility of fully controlling the effects (Milbrink and Holmgren 1981a).

Since June 1976 phosphates and nitrates have been more or less continuously added to a stream (Fig. 2, point A) entering a narrow bay of Lake Anjan (area about ½ km^2). It was our hope that the morphometry of the bay, with a row of small islands delimiting it from the open lake would, to a considerable extent, prevent the nutrients from diffusing into the main lake. For comparison, an untreated reference bay — with a similar stream and roughly the same characteristics — was selected.

An interesting case of a seasonal, long-distance feeding-migration of fish, i.e. attraction of fish to a source of rich availability of nourishment, was studied by the present authors in a parallel situation in a water system close to Lake Anjan. Preliminary observations have been described in Milbrink and Holmgren (1981a).

Downstream of the vacation resort of Storlien, a series of small lakes (Fig. 3) — Åsvallstjärnarna — have long served for sewage disposal (total area about 6 hectares). At present the sewage from about 5,000 person equivalents$\cdot$year^{-1} is purified in a highly effective purification plant built in 1974. The sediments of

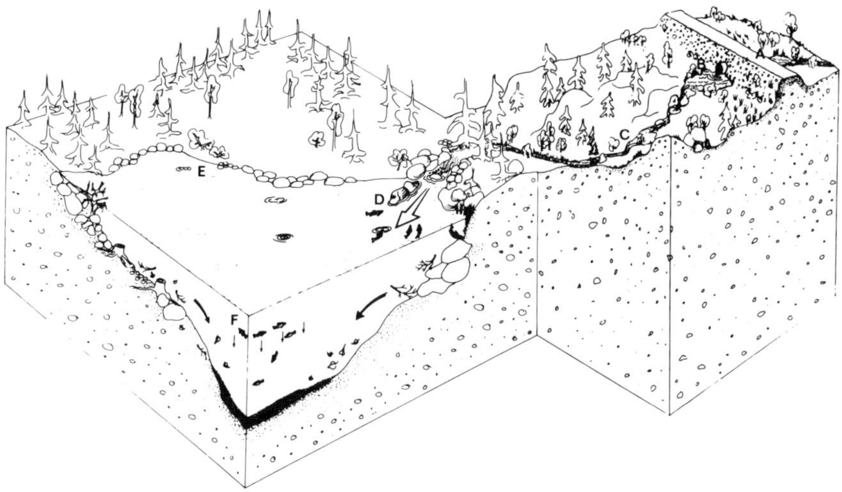

FIG. 2 Section of the fertilized bay of Lake Anjan. Fertilizers were added at point A. Point B is a pool downstream, point C is a natural migration obstacle for fish and point D symbolizes the mouth of the stream where brown trout, in particular, have assembled. In the shallow parts of the bay grayling are frequent (point E) and in the deeper parts, inside a barrier of small islands, charr are most common (from Milbrink and Holmgren 1981 b).

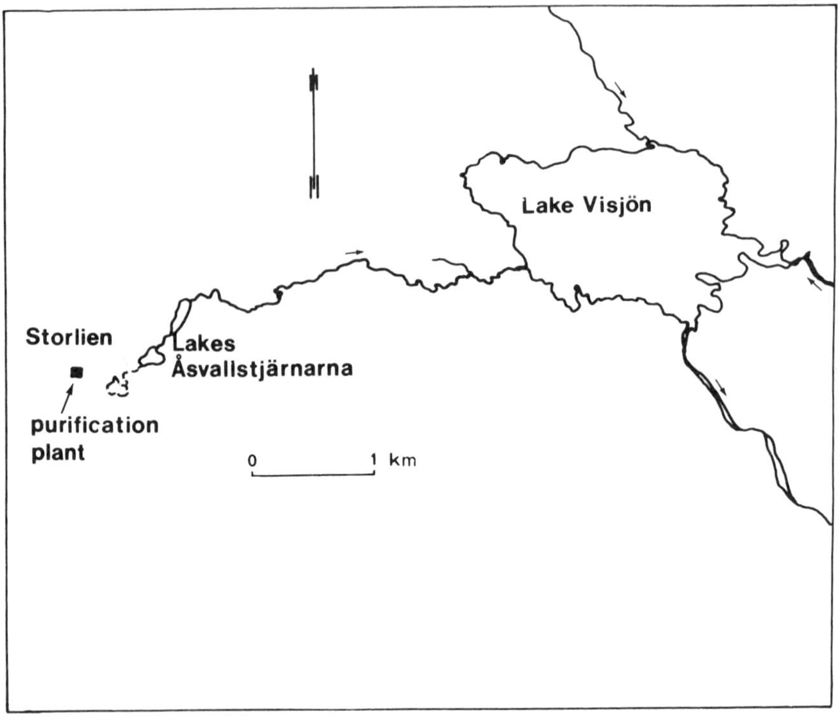

FIG. 3 Lakes Åsvallstjärnarna. These lakes are repopulated each summer by charr, brown trout and brook trout — in all probability migrating from Lake Visjön downstream.

these lakes are still loaded with nutrients and nutrients are continuously released into the water. The lakes drain via a narrow stream, about 5 km long, into Lake Visjön, which is fairly small (2.8 km^2) with a particularly fine population of large charr (condition factor according to Fulton 0.9-1.3, see below). In the summer, Lakes Åsvallstjärnarna hold particularly dense populations of charr, brown trout and brook trout in very good condition, the average coefficient for charr — again according to Fulton's formula — being well above 1.0.

In winter there is almost total depletion of oxygen in these lakes. Testfishing has revealed that the fish leave the lake in late September and return one to two weeks after break-up of the ice in late May. No large fish have ever been caught after September each year. In all probability the fish migrate downstream to Lake Visjön.

The life in Lakes Åsvallstjärnarna is remarkable in several respects. Despite of their location in high-alpine surroundings and the dominance of salmonid fish, these lakes are also characterized by indicators of eutrophy, such as *Nuphar lutea* and "meadows" of *Potamogeton*; they are surrounded by dense stands of *Carex aquatilis* and among bird species otherwise usually associated with eutrophic lowland lakes grey herons and dense populations of waterfowl may be mentioned. The water transparency is generally between 1.5 and 2 metres and the range of values for total-phosphorus is between 50 and 100 $\mu g \cdot l^{-1}$ (back-ground values are around 10 $\mu g \cdot l^{-1}$). The quality of the water may be described as dystrophic-eutrophic.

There are several features common to these two situations, which could both lead to improved sports fishing and a local fishery, at least on a part-time basis.

Unfortunately there is comparatively little information in the literature about seasonal feeding-migration of fish in sub-arctic waters with the exceptions of Curry-Lindahl (1957) and Johnson (1975).

METHODS

From the start of fertilization samples were regularly collected for the analysis of chemical and physical variables, phyto- and zooplankton, littoral bottom fauna and rooted vegetation, sessile algae and bottom fauna of the streams and, lastly, fish (Milbrink and Holmgren 1981a). Among the chemical-physical variables may be mentioned alkalinity, KNMO4, pH, colour, nitrogen and phosphorus.

The fertilizers were added continuously to the stream during the summer season from June 1976 until September 1979. Each year about 500 kg of nitrogen and 50 kg of phosphorus were added, which means a dosage of about 1g $Nm^{-2} \cdot yr^{-1}$ and 100 mg $Pm^{-2} \cdot yr^{-1}$ to the experimental bay — in good agreement with the values recommended by Vollenweider (1968) with due recognition of the fact that a certain proportion of the fertilizers may diffuse out of the bay.

Extensive test-fishing using ordinary survey-nets and pelagic survey-nets (see Milbrink and Holmgren 1981a) was carried out simultaneously in the two selected bays, both before the start of the experiments and then during different stages of enrichment in the years following. Since 1978 parallel fishing has been repeatedly undertaken on comparable reaches of the two bays — a method that should enable fair estimates of the fish populations to be made, despite a fairly uneven bottom topography with old stumps making fishing particularly difficult.

Sampling with a trawling device specially designed for catching fish fry was carried out in 1978 (1 + fish and older) as a complement to the regular nettings.

Fulton's formula for estimating the condition of fish was used (Ricker 1975).

The percentage composition of the food intake of various fish was obtained by calculating the occurrence, on a volumetric basis, of the various food items following the method described by Nilsson (1960).

In the summer of 1976 most interest was devoted to changes in the first levels of the food chain, but in the summers of 1977 and 1978 more time was devoted to the fish. In 1979 — and in 1980 — the entire effort was concentrated on the fish.

RESULTS AND DISCUSSION

The sequence of biological events following nutrient enrichment is generally fairly predictable (cf. Schindler and Fee 1974; Persson et al. 1975; Milbrink and Holmgren 1981a).

After each round of sampling it soon became obvious that the concentrations of phosphorus and nitrogen rapidly decreased in the open water, and accordingly, the nutrients became locked in the primary producers and in the sediments of the stream and the bay itself. Soon after the start of fertilization the phytoplankton increased in a characteristic way; various zooplankton species next responded in "wave after wave". This type of development was repeated during the following years with roughly two maxima for chrysophyseans and cryptophyceans every year; *Chlamydomonas* also increased considerably in 1977 and 1978. No significant increase in blue-green algae, which could possibly depress the degree of utilization of the remainder of the primary production, was observed.

The phytoplankton biomass was always larger in the fertilized bay than in the reference area (2-3 times on an average). This difference is rather small when compared to other fertilized waters where five to twenty-fold increases have been observed (Kalff and Welch 1974; Kuokkel Progress Reports 3-7). The difference is almost certainly due to the masses of zooplankton grazing heavily on the algae. Unfortunately, there are no accurate measures of the actual production rate of the phytoplankton in the enriched bay.

Among the zooplankton, the cladocerans *Bosmina coregoni, Daphnia galeata, Holopedium gibberum* and *Bythotrephes longimanus* all became very abundant; the average numbers of animals were about 5-10 times larger in the enriched bay relative to the reference bay. These species together with the semi-benthic cladoceran *Eurycercus lamellatus*, which is a most characteristic species in the "damming-up phase" after impoundment, are heavily preyed upon by fish. This also means that actual numbers of zooplankters by no means reflects the production of the species.

The rooted vegetation of the enriched bay increased tremendously; this benefited the rich invertebrate fauna as did also the sessile algae of the stream, with which masses of chironomid larvae (orthocladines) were associated. A large proportion of the stream benthos reached the fish in the bay as downstream "drift".

In July 1978 the first indications were received that the effects of enrichment had finally reached the fish populations: shoals of yearlings occupied the shallowest reaches of the fertilized bay and were chasing zooplankton. Most fish were undoubtedly minnows; according to observations and diet analyses, they fed on *Eurycercus, Daphnia, Bythotrephes* and *Bosmina*. Other young fish which could be seen moving in the bottom vegetation were age 1+ grayling. Few, if any, of these fish were observed in the reference bay or in the fertilized bay prior to 1978.

In 1978 catches of fish in the fertilized bay increased dramatically in relation to the reference bay. Catches per fishing effort tended to be 200-300 percent higher in the former bay, a tendency which was even more accentuated in 1979 with average values around 400 percent (range 215-1,400 percent). Trawling for fry confirmed a drastic increase in the number of age 1+ and 2+ grayling in the fertilized bay.

Even if total catches in the fertilized bay were still relatively low compared with natural lakes, Table 1 shows that, from 1978 on, catches were considerably higher in the enriched bay relative to the reference and are now fairly close to average catches in Lake Anjan and other subalpine lakes prior to impoundment. Test-fishing results from September 1976 revealed no differences between the bays; catches per net were low (Table 1).

Twelve consecutive parallel test-fishings, carried out in the two bays in 1978 and 1979, all resulted in significantly higher catches in the fertilized bay. Despite intensive efforts to catch all adult fish in the enriched bay, these efforts seemed to have only a limited effect on the fish populations — a likely effect of a large area of recruitment (cf. Milbrink and Holmgren 1981a).

Even if the first indications of a change within the fish populations were noticeable among grayling, the most significant changes took place within the charr populations. The average condition of charr increased from 0.76 in 1976 to 0.89 in 1978 and 1979. This change was highly significant at a probability of $p = 0.001$. Already in 1977 there was a discernable increase in the condition of charr significant at a probability of $p = 0.10$ to $p = 0.05$. Figure 4 illustrates this gradual rise in the condition of charr in the fertilized bay.

The results of a limited, though comparable, test-fishing in September 1980 confirmed the upward trend in the condition of the fish, in spite of the fact that no further nutrients were added during 1980. Two test-fishings performed in July and in September that year showed that catches were still 2-3 times larger in the fertilized bay than the reference bay.

Length-weight relationships are the traditional way of illustrating the condition of fish and Figure 5 illustrates such relationships between fish caught in the late summer in the two bays from 1976 to 1979, with some data from September

Table 1. Test fishing results from parallel situations in the fertilized bay (left) and reference bay in the late summer of 1976, 1978 and 1979. Catches of charr, grayling, brown trout and burbot are included in the total weights. The number of survey nets used on each occasion is given in parentheses.

	Fertilized Bay				Reference Bay			
	Number of charr	Number of grayling	total weight (kg)	weight/net (kg)	Number of charr	Number of grayling	total weight (kg)	weight/net (kg)
1976								
13 - 19 Sept	8	5	2.7 (12)	.22	5	10	3.5 (12)	.29
1978								
22 - 23 Aug	9	8	3.0 (7)	.43	1	1	1.0 (5)	.20
08 Sept	15	6	4.6 (5)	.92	-	1	.65(2)	.33
1979								
27 - 28 Aug	5	6	2.0 (5)	.40	-	-	.3 (5)	.06
05 - 06 Sept	19	6	5.5 (12)	.46	3	-	1.4 (11)	.13

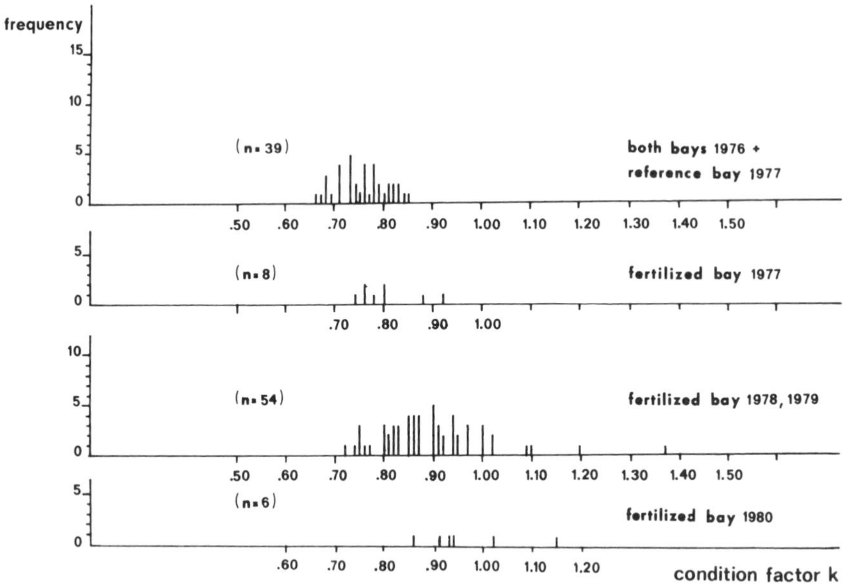

FIG. 4 Changes in the condition of charr (Fulton's formula) from 1976 to 1979. A limited charr material available also from the autumn of 1980 confirms the upward trend (partly from Milbrink and Holmgren 1981 b).

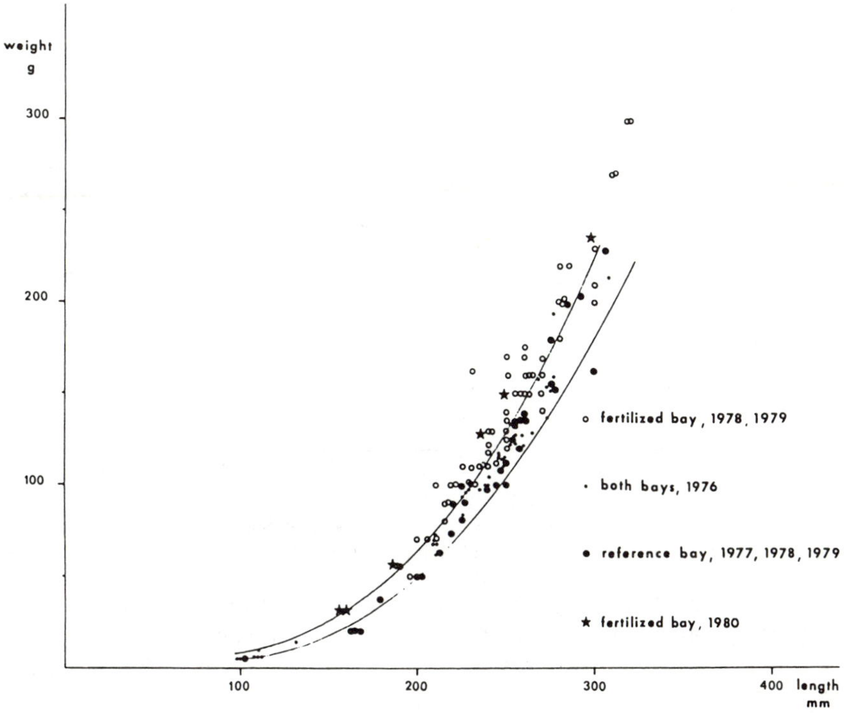

FIG. 5 Length-weight relationships of charr caught in August — September each year. For further details, see text (partly from Milbrink and Holmgren 1981 b).

1980. In the figure the condition of individual charr is plotted. Small dots refer to fish caught in either bay in 1976. Most of these dots lie within the shaded area in the figure. Large dots refer to fish from the reference bay in 1977-1979; they also fall mainly within the same shaded area. Open rings represent charr caught in the fertilized bay during the same periods. It is readily recognized that most charr from the fertilized bay have gained weight. Few circle symbols are found within the shaded area (less than 20 percent). The star symbols, which refer to charr caught in September 1980, clearly fall above the shaded area.

Figure 6 illustrates mean back-calculated lengths of the 1970, 1972 and 1973 year classes of grayling in the bays studied. The mean values from the fertilized bay have been superimposed upon those from the reference bay. There is a readily discernable acceleration in the length increase of fish older than about 4 years. Fish of 6, 7 or 8 years of age apparently enter a new growth period, probably due to the high availability of cladocerans in the shallow reaches of the fertilized bay.

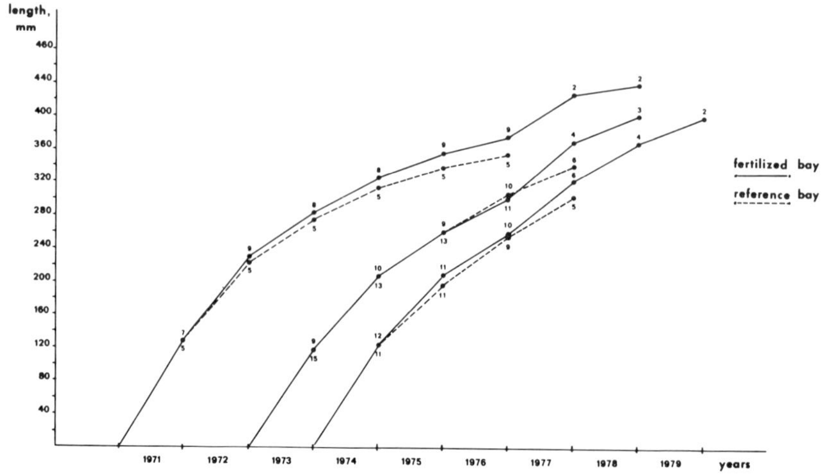

FIG. 6 The growth rate of three different year classes of grayling in the fertilized bay and in the reference bay. Values of n are given above and below each value in the figure, respectively (from Milbrink and Holmgren 1981 b).

Other more subjective indications of improved quality of fish in the fertilized bay were positive changes in the palatability of the charr, changes in the colour of the meat from greyish to orange-red and improvement in the texture of the meat.

Figure 2 gives an idealized view of the positions of the fish species in the fertilized ecosystem, with brown trout close to the mouth of the stream (point C), grayling inhabiting the shallow reaches of the bay (point D) and charr further out below depths of approximately two metres (point E). These positions have changed in the course of the experiments. In 1979 even the charr moved close to the mouth of the stream and all the fish species tended to prefer the same food items, i.e. *E. lamellatus* (see below).

Langeland (1982) reached similar results in a whole-lake fertilization experiment near Trondheim in Norway. Charr responded very quickly and efficiently to a superabundance of zooplankters. Eventually, however, the Norwegian experiment showed a drastic increase in the number of threespine sticklebacks at the expense of the charr; the populations of the larger zooplankters declined while at the same time rotifers, which are not readily eaten by fish, increased in abundance. Threespine sticklebacks likewise "exploded" in a large-scale fertilization project in British Columbia described by Manzer (1976b), possibly effectively competing with young sockeye salmon.

The zooplankton composition of the enriched bay of Lake Anjan changed slightly in the summer of 1980: no further fertilizer was added, and rotifers became more abundant. The fish, however, still had a superabundance of cladocerans to feed on.

The Maucha diagram (Figure 7) demonstrates the food spectra of charr and brown trout, hence the food-niches occupied by the main fish predators in both allopatry and sympatry under different environmental conditions (from Milbrink and Holmgren 1981a). In allopatry both charr and brown trout utilize the bottom fauna to roughly the same extent, as shown by Nilsson in Lakes Skalvattnet and Abelvattnet. The diet of the charr of Lake Anjan, prior to impoundment in August 1938 (Nilsson 1960), was almost identical with the former, while brown trout specialized on "insect larvae and nymphs". After impoundment, charr switched to a diet of euplankton and semi-benthic cladocerans, while trout still fed mainly on "insect larvae and nymphs" plus terrestrial insects (Milbrink and Holmgren 1981a, 1981b). The level of food segregation reached a maximum.

It is interesting to note the role of the grayling under the successive phases of impoundment and nutrient enrichment. Before fertilization, the diet of grayling was very similar to that of brown trout, while after fertilziation the diet moved closer to that of the charr, with a remarkably high percentage of semi-benthic cladocerans (*Eurycercus*). Since the food intake of brown trout after fertilization clearly became modified by an increased proportion of semibenthic cladocerans, the three fish species in 1979 partly occupied the same planktonic food-niche and the level of food segregation reached a minimum owing to the superabundance of zooplankton.

In natural conditions, the level of niche-overlap in subarctic lakes in Scandinavia is often around 30 percent in the late summer (Nilsson 1960). After impoundment, however, the food is generally in short supply and the level of niche-overlap reaches a minimum as the fish species struggle for existence. Considerable overlapping may be shown, as here, in situations where nutrient enrichment has occurred (Milbrink and Holmgren 1981b).

CONCLUSIONS

The present study has verified our main hypothesis that added nutrients pass rapidly from one trophic level to the next and eventually reach zooplankton-feeding fish, among which charr is the most valuable species; but it is also the species most susceptible to water regulation. Since there are very few methods at hand for counteracting the adverse effects of impoundment, "partial fertilization" seems to be an attractive alternative. Under ideal conditions this method is an effective and probably also an economic way of restoring declining fish populations.

It is our purpose to design a method for the enhancement of oligotrophic waters, where increased fish production would make it possible for people in sparsely populated areas to stay and develop a profitable fishery on a part-time

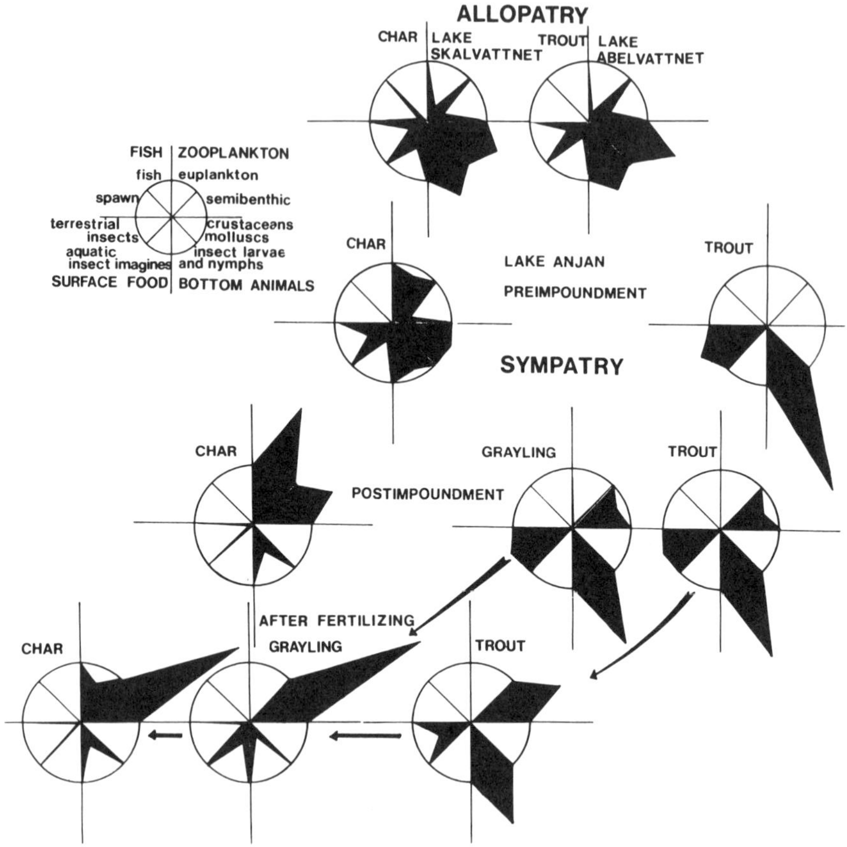

FIG. 7 The food of charr and brown trout in allopatry (Lakes Skalvattnet and Abelvattnet in 1960), and the food segregation between salmonid fish in Lake Anjan before water regulation started (1938) and before (1976) and after (1979) the onset of nutrient enrichment. (Grayling was introduced into Lake Anjan in 1952). All fish considered here were caught in the late summer each year. The radius of each circle represents an average of 25 per cent of the stomach volume (from Milbrink and Holmgren 1981 b).

basis, or for sports fishing organizations to create new and attractive fishing waters through capitalization on the natural tendency of charr and other salmonid fishes to establish feeding migrations. Effects of over-fishing in streams and lakes around recreational centres could also be compensated for by small-scale application of fertilizers.

There is considerable debate as to whether the production of electric energy alone is the most rational way of utilizing the freshwater resources of northern

Scandinavia. Normally, fisheries make up a substantial portion of the livelihood of the residents. By the application of balanced nutrient enrichment to sub-arctic waters two main purposes can be achieved. Firstly, the condition of valuable salmonid fish can be improved and secondly, the buffering capacity against acidification — now rapidly progressing in northern Scandinavia — can be increased. In the light of the information we already possess, indiscriminate installation of advanced sewage treatment — which is generally proclaimed in Sweden — in oligotrophic waters of this kind, seems to be no less than the waste of a natural resource. It is suggested that a modified sewage treatment method be employed that, after necessary sterilization, allows the effluent, high in nutrients, to reach waters that can benefit greatly from it.

ACKNOWLEDGEMENTS

We thank Stiftelsen J.C. Kempes Minne for generous support.

REFERENCES

Aass, P., R. Asplund, and F. Hansson. 1972. Öringens och rödingens näringsval samt betydelsen av en grunddamm i den reglerade sjön Limingen. Information från Sötvattenslaboratoriet, Drottningholm 8: 21 p. (In Swedish with English summary).

Curry-Lindahl, K. 1957. Om rödingen i rinnande vatten. Svensk Fiskeritidskrift 66(5): 4 p. (In Swedish).

Fagerström, Å. 1966. Ett försök att gödsla tjärnar med thomasfosfat. Information fran Sotvattens-laboratoriet, Drottningholm 3 : 26 p. (In Swedish).

Filipsson, O., A. Gönczi, and G. Svärdson. 1968. Ett försök att utfordra fisk i en reglerad sjö. Information från Sötvattenslaboratoriet, Drottningholm 4: 8 p. (In Swedish).

Fürst, M. 1972. Experiments of the transplantation of new fish-food organisms into Swedish impounded lakes. The feeding habits of brown trout and char in Lake Blåsjön. Verh. Int. Ver. Limnol. 18 : 8 p.

Fürst, M., U. Boström, and J. Hammar. 1978. Effekter av nya fisknärings-djur i Blasjon. Information från Sötvattenslaboratoriet, Drottningholm 15 : 94 p. (In Swedish).

Grimås, U. 1965. Inlet impoundments. An attempt to preserve littoral animals in regulated subarctic lakes. Inst. Freshw. Res. Drottningholm Rep. 46: 9 p.

Johnson, L. 1975. The dynamics of Arctic fish populations. Circumpolar Conference on Northern Ecology, Proceedings, Ottawa 1975, III : 10 p.

Kalff, J. and H.E. Welch. 1974. Phytoplankton production in Char Lake, a natural polar lake, and in Meretta Lake, a polluted polar lake, Cornwallis Island, Northwest Territories. J. Fish Res. Board Can. 31: 16 p.

Kuokkel Progress Reports 1-7, 1972-80. (In Swedish with English Summaries).

Langeland, A. 1982. Interactions between zooplankton and fish in a fertilized lake. Holarct. Ecol. 5 : 38 p.

Lasenby, D.C. 1979. A method for determining predation rates of macrozooplankton. Can. J. Zool. 57 : 3 p.

LeBrasseur, R.J., C.D. McAllister, W.E. Barraclough, O.D. Kennedy, J. Manzer, D. Robinson, and K. Stephens. 1978. Enhancement of sockeye salmon (*Oncorhynchus nerka*) by lake fertilization in Great Central Lake: Summary Report. J. Fish Res. Board Can. 35 : 17 p.

Manzer, J.I. 1976a. Preliminary results of studies on the effects of fertilization of an oligotrophic lake on adult sockeye salmon (*Oncorhynchus nerka*) production. Fish. Mar. Serv. Res. Dev. Tech. Rep. 678 : 25 p.

Manzer, J.I. 1976b. Distribution, food and feeding of the threespine stickleback, *Gasterosteus aculeatus*, in Great Central Lake, Vancouver Island, with comments on competition for food with juvenile sockeye salmon, *Oncorhynchus nerka*. Fish. Bull. U.S. 74 : 22 p.

Milbrink, G. and S. Holmgren. 1981a. Addition of artificial fertilizers as a means of reducing negative effects of "oligotrophication" in lakes after impoundment. Inst. Freshw. Res. Drottningholm Rep. 59 : 24 p.

Milbrink, G. and S. Holmgren. 1981b. Fish species interactions in a fertilized reservoir. Inst. Freshw. Res. Drottningholm Rep. 59 : 7 p.

Nilsson, N.-A. 1960. Seasonal fluctuations in the food segregation of trout, charr and whitefish in 14 North-Swedish lakes. Inst. Freshw. Res. Drottningholm Rep. 41: 21 p.

Nilsson, N.-A. 1973. Biological effects of water-power exploitation in Sweden, and means of compensation for damage. Commission Internationale des grands barrages, Madrid 1973: 18 p.

Northcote, T.G. 1972. Some effects of mysid introduction and nutrient enrichment on a large oligotrophic lake and its salmonids. Verh. Internat. Verein. Limnol. 18 : 11 p.

Persson, G., S. Holmgren, M. Jansson, A. Lundgren, B. Nyman, D. Solander, and C. Ånell. 1975. Phosphorus and nitrogen and the regulation of lake ecosystems. Experimental approaches in subarctic Sweden. Circumpolar Conference on Northern Ecology, Ottawa, Canada, 1975 : 13 p.

Pyefinch, K.A. 1960. Trout in Scotland. Her Majesty's Stationary Office, Edinburgh, 70 p.

Ricker, W.E. 1975. Computation and interpretation of biological statistics of fish populations. Bull. Fish. Res. Board Can. 191 : 382 p.

Robinson, D.G. and W.E. Barraclough. 1978. Population estimates of sockeye salmon in an enriched oligotrophic lake. J. Fish Res. Board Can. 35 : 10 p.

Schindler, D.W. and E.J. Fee. 1974. Experimental Lakes Area: whole-lake experiments in eutrophication. J. Fish. Res. Board Can. 31: 17 p.

Stockner, J.G. 1977. Lake fertilization as a means of enchancing sockeye salmon populations: state of art in Pacific NW. Fish Res. Board Tech. Rep. 740.

Svärdson, G. 1976. Interspecific population dominance in fish communities of Scandinavian lakes. Inst. Freshw. Res. Drottningholm Rep. 55 : 27 p.

Vollenweider, R.A. 1968. Scientific fundamentals of the eutrophication of lakes and flowing waters, with particular reference to nitrogen and phosphorus as factors in eutrophication. OECD-report. Water management research, 194 p.

Experiments Towards Improving the Culture of Arctic Charr (*Salvelinus alpinus* L.)

VOLKER STEINER

Institut für Zoologie der Universität, Innsbruck, Austria

STEINER, V. 1984. Experiments towards improving the culture of Arctic charr, (*Salvelinus alpinus* L.), p. 509–521. *In* L. Johnson and B.L. Burns [eds.] Biology of the Arctic charr, Proceedings of the International Symposium on Arctic Charr, Winnipeg, Manitoba, May 1981. Univ. Manitoba Press, Winnipeg.

The deterioration of Arctic charr, *Salvelinus alpinus*, fisheries in Austria has stimulated considerable experimental work aimed at determining the factors involved in the development of satisfactory charr culture. The objective of this programme has been to provide techniques for raising large numbers of fingerlings that would allow the stocking of lakes in order to reverse the decline of the indigenous populations. The most important factors have been found to be: careful selection of parental fish for size and degree of ripeness, creation of optimum conditions for artificial fertilization, the maintenance of strict hatchery sanitation, proper incubation temperatures, and the proper care and feeding of young fish.

Feeding experiments have shown the superiority of natural food for young charr over commercial diets.

The experimental data are used to develop guidelines for the economic and efficient culture of Arctic charr and other salmonid species.

INTRODUCTION

From the beginning of the present century to the middle 1950's many Austrian lakes yielded large quantities of charr. There is extensive documentation on many of the lakes but perhaps the most complete records are those of Grundlsee, a lake with a long history of good production and a very good reputation for its product. Today it has reached the stage where commercial exploitation is no longer profitable.

No serious attempts have been made to maintain or re-establish charr stocks through the development of planting procedures on account of the extreme shortage of suitable material, resulting from the difficulties encountered in the rearing process. The only suitable material is produced by the hatchery of the Bundesinstitut für Gewässerforschung in Scharfling, but the size of the stock produced and the numbers available are inadequate, a state of affairs not confined to Austria (MacCrimmon and Gots 1980). In the hatchery run by the fishermen of Grundlsee, very high mortality, particularly during incubation and early larval stages, has been encountered. Losses in the fry stage were attributed to the use of commercial fish food and other problems were blamed on the parental stock.

As a result of these difficulties an extensive research programme was initiated to examine the factors influencing the early stages of incubation, feeding and growth. The first series of experiments was initiated by the University of Innsbruck, the Grundlsee Hatchery and the Bundesinstitut für Gewässerforschung.

MATERIALS AND METHODS

Various stocks of 'landlocked' Arctic charr from Austrian and Swiss lakes (Grundlsee, Lunzer See, Vierwaldstätter See) were used in these experiments as well as one anadromous stock from Tromsø, Norway. All stocks were autumn spawners, the fish ripening in November. Only fish ripe to the point of 'running' were selected; fertilization was carried out following the 'dry' technique developed by Vrassky (Bagenal and Braun 1971). Incubation was effected in a "Scharflinger" type of incubator, similar to the type employed in the "Californian" system.

Incubation was carried out in total darkness, at a range of temperature between 1.5 and 12° C; at least 1,000 eggs were used in each experimental group. Low temperatures were maintained by recirculation and the use of ice. All trays were cleaned daily and dead eggs removed. Oxygen levels were maintained above 60 percent saturation at all times.

Larvae were measured at hatching and then at intervals of between ten days and one month thereafter. Measurements were performed under anaesthetic.

Feeding was done either by hand or by the use of a "Scharflinger" feeder dispensing a commercial dry food. Feeding regimes were on a daily basis of between 8-12 hours. Other feeds used were either freeze dried plankton, frozen plankton or fresh plankton gathered from the lake and fed immediately. Frozen plankton was fed by an infuser that proved unsatisfactory and this necessitated the development of a completely new unit by the Institut für Fischforschung, Austria. Details on the construction and use of this equipment will be supplied by the author on request. This system reduces the loss of nutriment during thawing, the so-called 'washing-out' effect.

All rations were made up to the equivalent nutritional value of the recommended feeding levels given by the suppliers of the commercial diets (Ewos, Troutvit and Silver Cup (Hokovit)).

The diets used were as follows:

Live plankton: only used for one comparison at Grundlsee hatchery. The plankton consisted mainly of copepods and cladocerans in fluctuating proportions. Individual sizes varied from 0.2 to 0.5 mm. Plankton was caught daily by large nets.

Deep frozen plankton: plankton was caught by large nets in different Austrian lakes (Wallersee, Mondsee, Achensee, Hechtsee) and was frozen immediately, using a special cooling box provided with CO_2 — ice (-70°C). Storage was done at temperatures ranging from -70°C to -25°C prior to feeding.

Freeze dried plankton: deep frozen plankton of the same origin as above was dried in a Heto Sic laboratory freeze dryer down to 5-7 percent water content and stored in glass bottles at freezing temperatures.
Particle sizes x = 0.24 mm (n=50)
68% = 0.17-0.34 mm
80% = 0.15-0.37 mm
protein analysis resulted in a 64 percent raw protein content (analysis by CRINA - Switzerland).

Artificial dry fish food

Commercial Fish Food: products from Ewos, Trouvit, Silver Cup (Hokovit) recommended for trout and salmon.
Particle sizes: x = 0.12 mm (n=50)
68% = 0.06-0.21 mm
80% = 0.05-0.25 mm
protein content as stated by the producers varied from 32-48 (52) percent raw protein.

Other diets: in addition to commercial dry fish food diets special mixtures consisting mainly of curd, blood powder, dried marine crustaceans and vitamin mixtures have been fed to increase feeding activity during the first days of food uptake.

RESULTS

No significant effect on mortality during incubation or early growth could be detected with respect to the different sizes and ages of female parents; however, mortality was high and growth poor when very small or very young males were used (Fig. 1). No significant differences in incubation mortality and hatching mortality could be detected between different egg sizes obtained from small or large females, but there were large differences between large and small eggs selected from within batches; mortality was seven times greater during post-

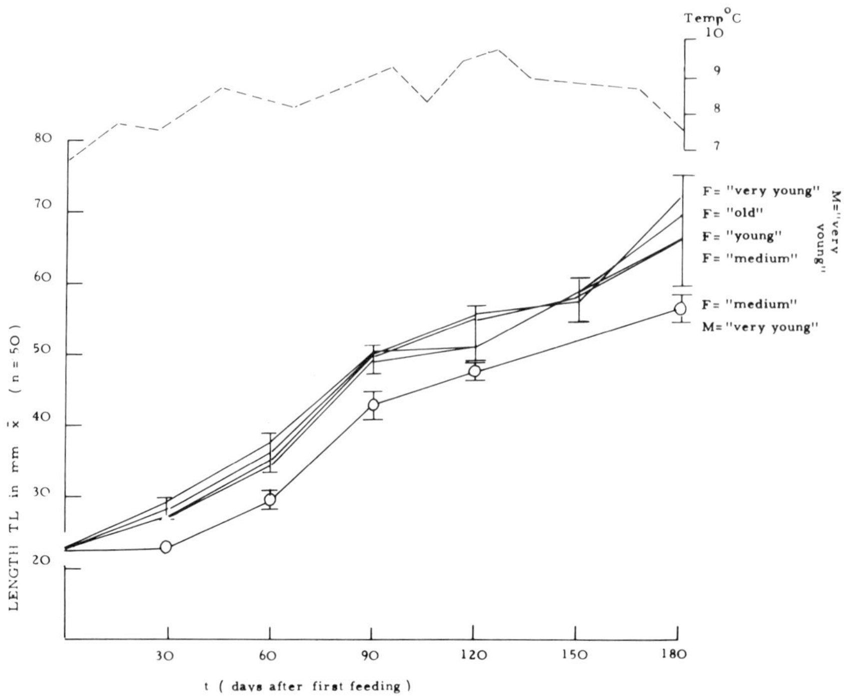

FIG. 1 Early growth of Arctic charr (Grundlsee population). Growth of the progeny from different parental size(age) groups.
M = male, F = female limits are +− 2sm

larval development in fry from small eggs, compared with those from larger eggs (Fig. 2).

The time to hatching increased, as might be expected, with a decrease in incubation temperature (Fig. 3). The important effect of temperature on incubation (Embody and Sadler 1934, Hayes 1942, Gray 1961), specifically discussed for Arctic charr by Swift (1965), was confirmed. Optimum incubation temperature was found to be about 6° C with limits between 3.5 and 8° C (Fig. 4). Mortality increased appreciably above and below these temperatures. This is contrary to the statement of Einsele (1966) that low temperatures are the most satisfactory for the incubation of Arctic charr. Poor environmental conditions such as low oxygen, excessive water flow or rough handling of eggs drastically shortened hatching time and increased mortality at all temperatures.

Feeding was started when the larvae reached the 'swim up' stage; attempting to start feeding before this time did not result in improved growth or reduced mortality.

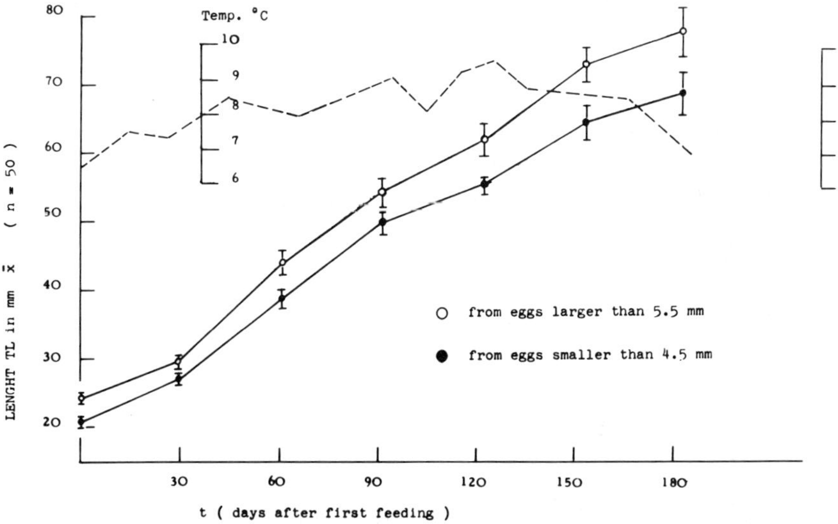

FIG. 2 Early growth of Arctic charr (Grundlsee population). Post-larval growth after selection of maximum-minimum egg sizes.
limits are +− 2sm

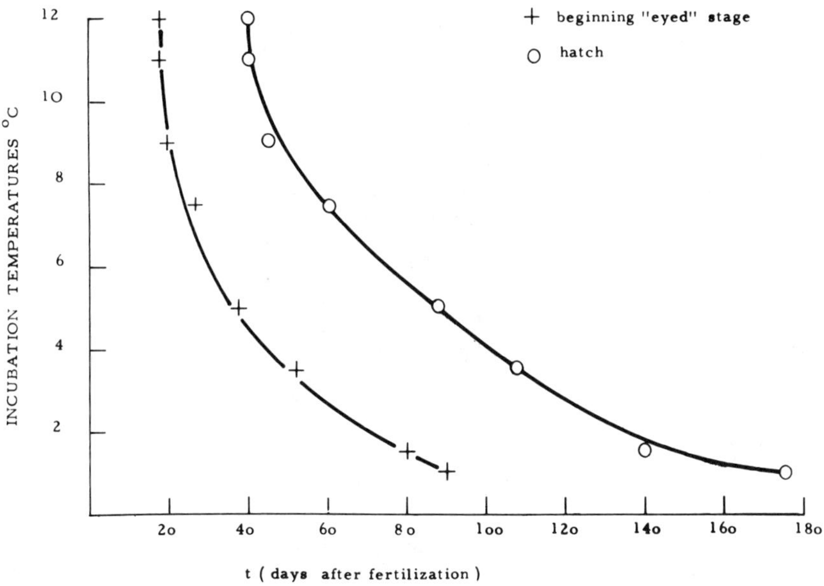

FIG. 3 Egg development time at different temperatures (Grundlsee charr).

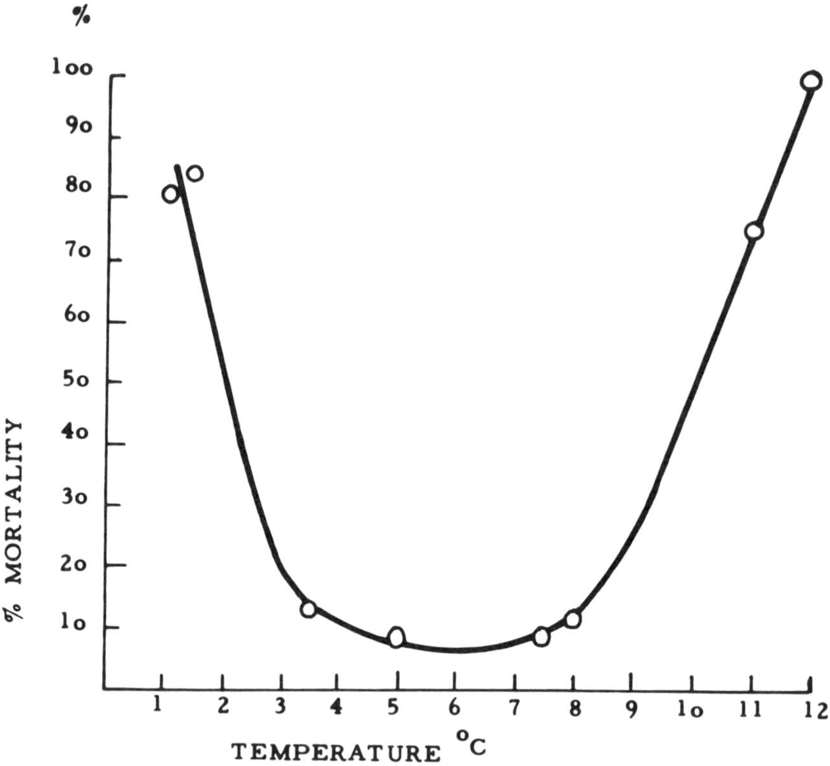

FIG. 4 Mortality during egg development at different temperatures (Grundlsee charr).

In the first series of experiments on feeding, commercial dry food and fresh plankton were compared (Fig. 5); unfortunately difficulties were experienced in the collection of adequate amounts of live plankton at the time feeding commenced. This fact has to modify the results which are presented for the simple reason that no other results on the feeding of Arctic charr on live plankton are available.

In the second series of experiments, freezed dried plankton, and plankton frozen immediately following collection were compared with commercial dry food. Growth on frozen plankton appeared to be greatly superior to that on freeze dried plankton or commercial food (Figs. 6 and 7); mortality on the plankton diet was six times less than on the commercial dry diet, (Table 1) although the actual mortality on the commercial diet was quite low (18.3 percent after 250 days of feeding) when compared with previous results at the Grundlsee hatchery. Frozen plankton clearly gave better survival than freeze dried plankton (less than 1 percent mortality after 50 days). A change of diet from plankton to commercial

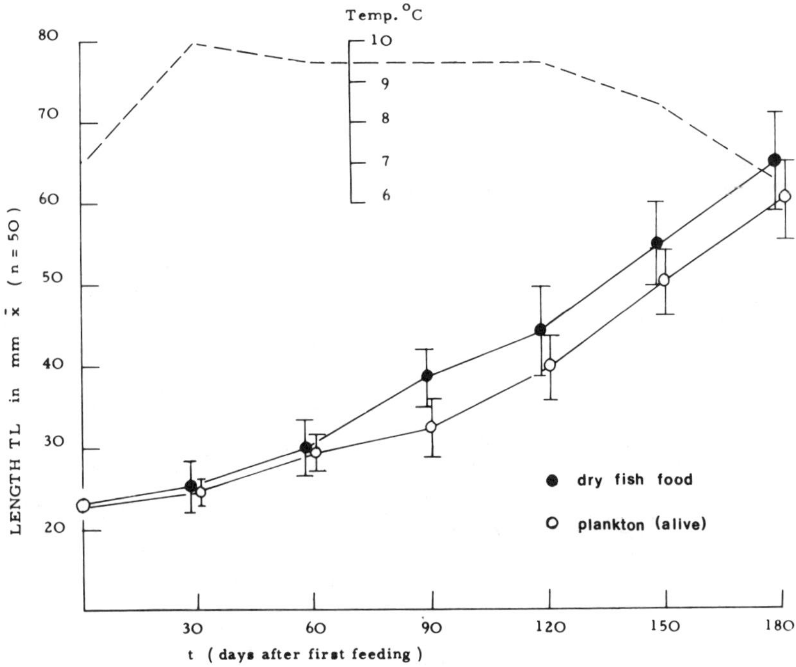

FIG. 5 Early growth of Arctic charr (Grundlsee population) fed natural and artificial diets.

food resulted in some mortality; this was greatly reduced in those fish initially fed frozen plankton rather than freeze dried. Mortality was negligible if the switch in diet was delayed until after the fish had reached a length of 60-70 mm. In order to reach this size on a plankton diet, the fish require large plankton individuals which may be obtained by either selective netting for size or by sieving each plankton haul immediately after collection.

CONCLUSIONS AND RECOMMENDATIONS

As a result of this work and taking into account many previous studies (see Bibliography) it is possible to formulate a series of recommendations for the satisfactory raising of young Arctic charr, suitable for the stocking of Austrian lakes.

Summary of results and recommendations for Arctic charr culture in hatcheries:

1) Parental stock should be selected from medium sized (aged) fish. The use of small (young) males has deleterious results. Elimination of the smallest egg sizes reduces mortality and improves growth and development.

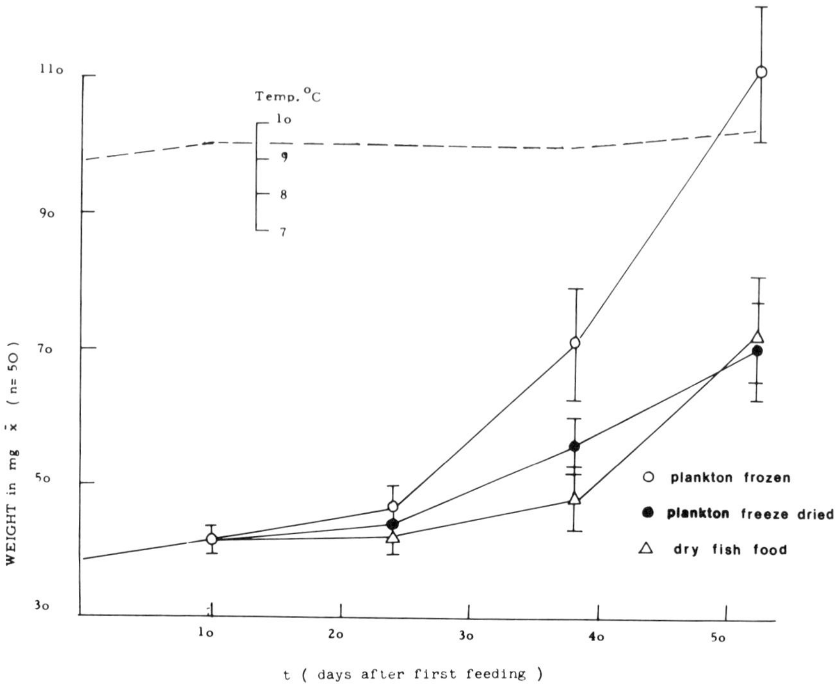

FIG. 6 Early growth of Arctic charr (Tromsö population) fed different diets.

2) At all stages great care must be exercised in maintaining standards of hygiene at a level above that normally used in salmonid culture. The use of self cleaning rearing tanks is strongly recommended for the larval and post-larval states.

3) Incubation temperatures should be between 3.5 and 8.0°C. Larval and post-larval temperatures should be between 10 and 14°C. At later stages in development (above 10-12 cm L_t), Arctic charr show excellent growth and food conversion in colder water (below 10°C). At temperatures above 15°C charr growth is considerably reduced and susceptibility to disease increases.

4) Delay in transferring larvae from incubators to rearing tanks may lead to damage. The use of incubator trays and rearing tanks with rough sides and bottoms should be avoided.

5) Feeding must begin at the moment the larvae leave the bottom and maintain the 'swim up' position. At this stage at least two thirds of the contents of the yolk sac will have been absorbed.

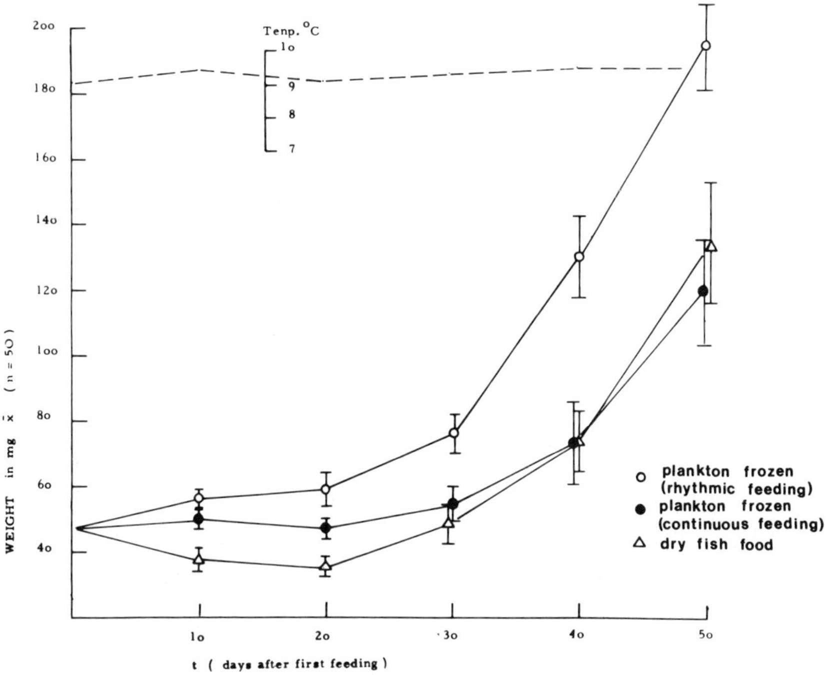

FIG. 7 Early growth of charr (Lunzer Untersee population) under different feeding conditions.

6) The size of particles in the diet must be varied according to the requirements of different sized fish. This is especially important during the first days of external food uptake, and necessitates sieving the commercial dry fish food.
7) Optimum rearing success is attained using a diet of natural crustacean plankton, preferably frozen. Frozen diets necessitate special feeders to avoid loss of nutrients during the thawing process. Care in the collection and proper screening at the time of catching leads to the best results. A diet of frozen copepod plankton (0.2-0.3 mm size class) during the first feeding period (2-3 weeks) provides for optimal growth and low mortality.
8) A diet change from plankton to dry commercial food may cause losses. If this change is done gradually and after the parr have reached a length of about 50 mm, losses will be minimized.

ACKNOWLEDGEMENTS

I thank Professor Dr. H. Janetschek who enabled the "wet" rooms at the Zoology Institute at Innsbruck University to be adapted for this work and

Table 1. Percent mortality during early development of Arctic charr (Grundlsee population) of the progeny from different sized groups of spawners.

Stages	Temp. °C ($\bar{x}$)	M = " medium " F = "very young"	F = " young "	F = " medium "	F = " old "	M = "very young" F = "medium"
egg development (fertilization - hatch)	5.8 (5.5 -7.0)	6.3%	4.4%	8.7%	9.3%	27.5%
yolk sac stage (hatch - first feeding)	6.4 (5.8 -7.5)	1.7%	1.3%	1.9%	1.9%	5.5%
fry stage (first feeding - fingerling(O+) [0)	8.6 (6.5 -10.2)	19.0%	15.0%	14.9%	12.4%	43.1%
total Mortality		27.0%	20.7%	25.5%	23.6%	76.1%

0) raised with artificial food M = Male , F = Female

Professor Dr. R. Pechlaner for his essential contributions over the years of these investigations. I am obliged to the former director of the Bundesinstitut für Gewässerforschung in Scharfling, Dr. E. Bruschek, for hospitality and a lot of advice with respect to my work.

The latest feeding experiments were effected at the Institut für Fischforschung near Innsbruck with help by Dr. W. Wieser.

All experimental work was financed, either partially or fully by the Austrian government — the department of the Forschungsförderungsfonds.

Last, but not least, my sincerest thanks to a number of friends whose help contributed greatly to all experiments.

BIBLIOGRAPHY

Armstrong, R.H. and J.E. Morrow. 1980. The Dolly Varden charr, *Salvelinus malma*, p. 99-141. *In* E.K. Balon [ed.] Charrs: salmonid fishes of the genus *Salvelinus*. Dr. W. Junk Publishers, The Hague.

Bagenal, T.B. and E. Braum. 1971. Eggs and early life history. *In* W.E. Ricker [ed.] Methods for assessment of fish production in fresh waters, IBP Handbook No. 3, second edition.

Bailey, J.E. and D.R. Evans. 1971. The low temperature threshold for pink salmon eggs in relation to a proposed hydroelectric installation. Fishery Bull. Vol. 69, No. 3.

Balon, E.K. 1980. Early ontogeny of the European Arctic charr — altricial form, *Salvelinus (Salvelinus) alpinus alpinus*, p. 607-631. *In* E.K. Balon [ed.] Charrs: salmonid fishes of the genus *Salvelinus*. Dr. W. Junk Publishers, The Hague.

Beamish, F.W.H. 1980. Swimming performance and oxygen consumption of the charrs, p. 765-773. *In* E.K. Balon [ed.] Charrs: salmonid fishes of the genus *Salvelinus*. Dr. W. Junk Publishers, The Hague.

Behnke, R.J. 1980. A systematic review of the genus *Salvelinus*, p. 441-480. *In* E.K. Balon [ed.] Charrs: salmonid fishes of the genus *Salvelinus*. Dr. W. Junk Publishers, The Hague.

Blaxter, J.H.S. 1969. Development: eggs and larvae. Fish physiol. Vol. 3, 177-252, Academic press, New York and London.

Blaxter, J.H.S. and G. Hempel. 1961. Biologische Beobachtung bei der Aufzucht von Heringsbrut. Helgolander wissenschaftliche Meeresunters. 7 : 260-283.

Brenner, T. 1980. The Arctic charr, *Salvelinus alpinus salvelinus*, in the prealpine Attersee, Austria, p. 765-773. *In* E.K. Balon [ed.] Charrs: salmonid fishes of the genus *Salvelinus*. Dr. W. Junk Publishers, The Hague.

Brett, J.R. 1952. Temperature tolerance on young Pacific salmon, genus *Oncorhynchus*. J. Fish. Res. Board Can. 9:265-325.

Brett, J.R. 1971. Growth responses of young sockeye salmon (*Oncorhynchus nerka*) to different diets and planes of nutrition. J. Fish. Res. Board Can. 28: 1635-1643.

Buresch, R. 1925. Studien am Seesaibling mehrerer Alpenseen. Z. Hydrol. 23 : 99-118.

Combs, B.D. 1965. Effect of temperature on the development of salmon eggs. Prog. Fish-Cult. 27: 134-137.

Efimov, V.I. 1962. Survival rate of pink salmon (*Oncorhynchus gorbuscha*) during embryonic development. Abstract in Biol. Abstr. 45 7142, Abstr. No. 89708.

Einsele, W. 1949. Plankton Produktion, Fischernten und Setzlingsaufzucht am Mondsee. Österreichs Fischerei Heft 3: 46-30.

Einsele, W. 1956. Untersuchungen über die Atmungsphysiologie sich entwickelnder Salmonideneier, ihre Anwendung auf die Natur und auf die züchterische Technik. Wiss Suppl. "Österreichs Fischerei" Suppl. Bd. 1 Heft 1 : 1-18.

Einsele, W. 1966. Über die Entwicklung der Eier der wichtigsten Zuchtfische, vor allem über die Dauer der Entwicklung bei verschiedenen Temperaturen. (Kursmauskript aus dem Bundesinstitut für Gewässerforschung in Scharfling — Österreich).

Embody, G.C. and W.D. Sadler. 1934. Relation of temperature to the incubation period of eggs of four species of trout. Trans. Amer. Fish. Soc. 64: 281-192.

Fabricius, E. and K.-J. Gustavson. 1953. Further aquarium observations on the spawning behaviour of the char, *Salmo alpinus* L. Inst. Freshw. Res. Drottningholm Rep. 35: 58-104.

Fry, F.E.J. and M.B. Gibson. 1953. Lethal temperature experiments with speckled trout and lake trout hybrids. J. Hered. 42.

Gray, J. 1928. The growth of fish III. The effect of temperature on the development of eggs of *Salmo fario*. Brit. J. Expt. Biol. 6: 125-130.

Gruber, K. and V. Steiner. 1979. Schutz der Tier — und Pflanzenwelt und natürlicher Lebensräume (Beiträge zur Kampagne des Europarates) — Der Seesaibling - Sonderdruck aus Natur und Land Heft 5/6: 165-170.

Hastings, W.H. 1976. Fish nutrition and fish feed manufacture. FOA Tech. Conf. on Aquaculture 1976 R. 23 : 1-13.

Johnson, L. 1980. The Arctic charr, *Salvelinus alpinus*, p. 15-98. In E.K. Balon [ed.] Charrs: salmonid fishes of the genus *Salvelinus*. Dr. W. Junk Publishers, The Hague.

Kainz, E. and H.P. Gollmann. 1951. Vërsuche zur Anfütterung der Brut von Karpfen (*Cyprinus carpio* L.) und Graskarpfen (*Ctenophyryngodon idella* VAL) mit Trockenfutter. Österreichs Fischerei Heft 4 : 65-73.

Krause, H.R. 1959. Biochemische Untersuchungen uber den postmortalen Abbau von totem Plankton unter aeroben und anaeroben Bedingungen. Arch. f. Hydrobiol. XX IV 3/4: 297-337.

LeBrasseur, R.J. 1969. Growth of juvenile chum salmon (*Oncorhynchus keta*) under different feeding regimes. J. Fish. Res. Board Can. 26: 1631-1645.

McCauley, R.W. 1958. Thermal relations of geographic races of *Salvelinus*. Can. J. Zool. 36: 655-662.

MacCrimmon, H.R. and B.L. Gots. 1980. Fisheries for charrs, p. 797-841. *In* E.K. Balon [ed.] Charrs: salmonid fishes of the genus *Salvelinus*. Dr. W. Junk Publishers, The Hague.

Marr, D.H.A. et al. 1967. Factors affecting the growth of salmon alevins and their survival and growth during the fry stage. International Council for the exploration of the sea C.M. 1967/M : 25 (Separatum).

Mathisen, O.A. and M. Berg. 1968. Growth rates of the charr, *Salvelinus alpinus* (L.) in the Vardnes River, Troms, Northern Norway. Inst. Freshw. Res. Drottningholm Rep. No. 48.

Ricker, W.E. 1971. IBP Handbook No. 3. Methods for assessment of Fish Production. Blackwell Sc. Publ. Oxford and Edinburgh 1971 (second edition).

Price, J.W. 1940. Time-temperature relations in the incubation of the whitefish, *Coregonus clupeaformis* (Mitchell). J. Gen. Physiol. 23 : 449-468.

Ruhlé, Ch. 1977. Biologie und Bewirtschaftung des Seesaiblings (*Salvelinus alpinus* L.) im Zugersee. Swiss J. Hydrology 39 : 12-45.

Steiner, V. and R. Pechlaner. 1975. Mark-recapture experiments with Arctic char, *Salvelinus alpinus* (L.) in Austrian lakes. EIFAC Techn. Rep. 23 (Suppl. 1 Vol. 2) 672-688.

Swift, D.R. 1964. The effect of temperature and oxygen on the growth rate of the Windermere char (*Salvelinus alpinus willughbii*). Comp. Biochem. Physiol. 12: 179-183.

Swift, D.R. 1965. Effect of temperature on mortality and rate of development of the eggs of the Windermere char (*Salvelinus alpinus*). J. Fish. Res. Board Canada 22: 913-917.

Sargent, J.R. et al. 1979. Assimilation of the wax esters of marine zooplankton by herring (*Clupea harengus*) and rainbow trout (*Salmo gairdnerii*). Marine Biology 51: 203-207 (Springer Verlag 1979).

Charr and Man: The Philosophy of Limited Interaction

LIONEL JOHNSON

Department of Fisheries and Oceans, Winnipeg, Manitoba, Canada

JOHNSON, L. 1984. Charr and man: the philosophy of limited interaction, p. 523–532. *In* L. Johnson and B.L. Burns [eds.] Biology of the Arctic charr, Proceedings of the International Symposium on Arctic Charr, Winnipeg, Manitoba, May 1981. Univ. Manitoba Press, Winnipeg.

For thousands of years, charr and man have been associated as they invaded the freshwater and lands exposed during the retreat of the last Pleistocene glaciation. In both cases successful colonization has resulted from an ability to withstand the extreme physiological conditions imposed by low temperature and low energy flow and a capacity to utilize a wide range of food resources. Historical and anthropological studies have shed some light on this association but more studies are needed before the full extent of the partnership can be recognized. Perhaps of even greater importance is the contribution that charr have made, and will continue to make, to the development of human ideas. The study of the genetic and ecological background of the species is in the process of reshaping evolutionary and ecosystem theory. Ultimately the survival of both charr and man depends on the proper development of such theory and its translation into a philosophy of living. Limited interspecific interaction must occupy a prominent position in such a philosophy.

INTRODUCTION

The closing stages of the Pleistocene period witnessed great climatic and faunal changes in both Europe and North America. Sometime between 10,000 and 20,000 years ago, approximately coincident with the glacial maximum, man crossed the Bering land-bridge and appears to have occupied most of the North American continent quite rapidly. Simultaneously, as far as can be established, a large element of the terrestrial megafauna ceased to exist; it has been postulated on good evidence (Martin 1967) that the two events were by no means unconnected. Farther north, retreat of the glaciers exposed vast areas of land and fresh water which immediately became available to those species capable of living in a very harsh environment; such a species is Arctic charr, *Salvelinus alpinus*, and it moved in to occupy the newly available living space.

In view of its capacity for survival close to the limits of existence it may be assumed that Arctic charr closely followed the ice edge as it retreated along the coastlines. The land, removed of its weight of ice, rose relative to the sea level and new fresh waters were formed in basins that previously had been arms of the sea. In this way opportunities were created for the extension of the range of anadromous species. In certain cases further rising of the land caused some lakes to be completely isolated from contact with the ocean, as falls or rapids prevented continued upstream movement; these lakes frequently contained fish species that, perforce, had to relinquish the habit of anadromy. These processes gave rise to anadromous and landlocked stocks living side by side in adjacent waters. At the same time as the dispersal of the charr other species were occupying the new fresh waters: lake trout, *Salvelinus namaycush*; lake whitefish, *Coregonus clupeaformis*; and various species of cisco, *Coregonus* sp.. The distribution of landlocked stocks of fish in the Canadian Arctic strongly indicates that many lakes were initially open to invasion by approximately the same set of species (Johnson 1980); however, in some lakes only Arctic charr remain, in others, lake trout and cisco, and again, some lakes have all the species concerned. There is little doubt that competitive exclusion, or more likely competitive elimination, was responsible for determining the final species complex.

Some considerable time after the flooding of the Bering land-bridge, possibly 3,000-5,000 years before the present, man again crossed from Asia to North America. In this invasion the people were culturally adapted to life on the shores of the Arctic Ocean; they occupied the available sites from what is now Alaska to Greenland. Cultural waves subsequently crossed the inhabited regions recognized sequentially as Thule, Predorset, Dorset and present-day Inuit. With each successive culture there was an improvement in the technology necessary for existence in a harsh environment. Although at first sight the opportunities for technological advance might appear small, owing to limited raw materials, a high degree of ingenuity resulted in significant advances. Unlike the events to the south taking place earlier there is no evidence for the simultaneous extinction of large animals that might have served as a food source during the initial period of expansion; this possibility arose later with the advent of European technology.

THE LEAN ENVIRONMENT

The question of major concern is what qualities or properties do charr and man possess that enable them to exist in a very rigorous environment such as the Arctic and how do they manage to survive in close proximity to each other? What can we learn from this relationship and what are their relationships with the rest of the Arctic ecosystem?

In the Arctic, both charr and man appear as generalists surviving in the harsh conditions by the utilization of all resources available. Each food species may be available only briefly or only capable of being caught opportunistically. Balon (this volume) has shown that Arctic charr is capable of adaptation to circumstances that require a precocial (specialized) or altricial (generalized) mode of existence. All food studies on charr in the Arctic indicate a capacity to take virtually whatever food is available; they may browse on the organisms dwelling in the bottom deposits such as clams (Sphaeridae) or chironomid larvae, feed on mysids, amphipods and isopods on the substratum, or pelagically on plankton or fish (Johnson 1980).

Nilsson (1965, 1967) has shown that when charr are faced with a competitive situation they tend to avoid direct confrontation and seek out alternative food sources. When a resource is abundant, they may share it with brown trout, *Salmo trutta*, but as it declines in abundance over the summer they leave it to the more specialized shoreline fish and move to open waters to feed on plankton. This division of resources, described as interactive segregation, is further discussed by Fraser and Power (this volume) and Heggberget (this volume). In the small Arctic lakes in which only charr now survive it appears that this flexibility has proved to be a most successful feeding strategy; perhaps where other species have survived, eliminating charr, when there is greater opportunity for specialization.

Similarily it seems clear that survival of man in the Arctic is also owing to his generalist capabilities. Although the Inuit were initially dependent on large mammals, both marine and terrestrial (seals, whales, polar bear, musk-ox and caribou), they also utilized seabirds and fish whenever they could be caught. Plants played a less conspicuous role in their diet although the partially digested vegetation in the rumen of the caribou was esteemed a delicacy when caribou had been killed. Except in times of great abundance all parts of the hunted animals were utilized and many items were preserved in a variety of ingenious ways to help tide them over the bleakest days of late winter and early spring.

These generalized food habits of man and charr are closely matched by their lack of physiological specialization. Holeton (1973, 1974) has shown that charr have no observable specialized adaptation to respiration at low temperature as has been postulated for certain Arctic and Antarctic marine species (Scholander et al. 1957; Wolschlag 1964). The respiration rate of charr close to 0°C is quite comparable with that of rainbow trout, *Salmo gairdneri*, as obtained by extrapolation of the curve determined for higher temperatures. Similarly de Vries (1971) considers that anadromous charr have no adaptation that allows them to survive in sea water at temperatures below zero such as are experienced in the Arctic Ocean in winter; such low temperatures are avoided by a return to fresh water in the fall. Comparably Hoar (1976) believes that the genus *Salvelinus* is the anadromous salmonid least adapted to the saltwater life.

The lack of specialization in man is well recognized. In the Arctic as in other inclement climates he has to rely on artefacts, such as clothing and housing, not adaptation, to maintain homeostasis. According to Steffanson (1943) the Inuit developed clothing that allows them to exist in a personal climate that is close to being tropical.

STABILITY

If animals are to survive in a fluctuating environment they must also develop a degree of stability in order to resist the fluctuations. As May (1975) and Johnson (1981) have discussed, an animal develops a degree of robustness in proportion to the harshness of the ecosystem in which it lives. Stability to external perturbations (Hurd et al. 1971) is characterized by resistance to deflection, recovery without oscillation, and thirdly, rapid response time. The first two of these characteristics have been demonstrated on a number of occasions in charr. Kipling (1972, 1984) has shown that severe exploitation of the charr in Windermere, England, in the nineteenth century reduced the mean size to that of fingerlings; but with the imposition of stringent controls the stock recovered its former size and abundance. In the Sylvia Grinnell River, Frobisher Bay, several cycles of depletion and recovery have been observed (Grainger 1953; Hunter 1976); similar recoveries following depletion have been documented for the Tree River (Moshenko et al., 1984) and for the Victoria Island fishery in Wellington Bay (Kristofferson et al., 1984). Recovery following experimental depletion has been closely observed in Keyhole Lake, Victoria Island (Hunter 1970; Johnson 1980 and 1983). In Keyhole Lake, the one case where the rate of recovery has been measured, it has been found to be extremely slow; complete recovery has taken between 10 and 15 years, although the recovery process has taken place without any significant oscillation. The species must therefore be regarded as having two attributes of stability, high biomass and a well-damped recovery but, at the same time, having a slow response time.

Although exhibiting considerable stability with respect to external perturbations (Johnson 1981), Arctic charr appears highly susceptible to internal perturbations that alter the driving variables of the system. Nutrient enrichment and acidification resulting from cultural modifications to the environment seem to eliminate charr before any other north temperate species. In many of the large alpine lakes charr is only being maintained in face of nutrient enrichment and heavy exploitation by ingenious artificial methods (Ruhlé 1977, 1984; Steiner, 1984).

THE CHARR-MAN INTERACTION

The prehistoric relationship between man and charr is a very long and intimate one but we can only guess at the details. The first full description of the traditional Inuit methods of harvest is provided by Jenness (1922). In suitable streams, weirs (saputits or saputs) were constructed out of stones; these weirs contained holding pools into which the charr were led and then speared at leisure. This activity has been very well documented by the National Film Board of Canada (1969). Jenness notes the high degree of effectiveness of the weir and wonders whether a particular stock would be able to survive continued exploitation at the high levels he observed. He was particularly concerned about the exploitation of the smaller size mode in the sea-going stock which are the young migrants in the early years of their seagoing life (Johnson 1981). Dunbar and Hildebrand (1952) also suggested that the high degree of effectiveness of the traditional Inuit methods might lead to overfishing. They believe that it is most likely that such overfishing would lead to the abandonment of a stream when the catch declined below a certain level and that the group would move on to another stream, allowing the first stock to recover. Such movement if carried out on a regular basis would give rise to what might well be regarded as a system of rotational management.

MAN IN HIS ECOSYSTEM

Within the boundary constraints of his niche, man, in his more pristine state, and within a comparatively circumscribed ecosystem, functions as the dominant species or top-predator. He will therefore tend to proceed toward a state of least fluctuation, exerting a damping moment on the rest of the system (Johnson 1981). Below the dominant, the remaining species in the hierarchy come to a state in which their abundances fluctuate around a constant mean; any continuous reduction in the mean value will result in elimination. Overexploitation in these conditions cannot be a feature that can continue indefinitely; any exploited species within a relatively closed ecosystem comes to a steady state or it is eliminated. The process of elimination through competition surely does exist, as can be seen to have occurred in those Arctic lakes in which charr alone occur.

In order to survive, an organism exerts a ''pressure'' on the niche boundaries; those species that survive together can only survive by maintaining this ''pressure'' at the greatest level that they can sustain. The final point of balance is reached when all species are working to their utmost; if the pressure were to be relaxed by one species it would succumb.

The harvest of any resource within a circumscribed system (an autonomous or 'closed' thermodynamic system) can only continue at a certain level; if this level is exceeded the resource declines and is either eliminated or reduced to new level. In effect harvesting is limited and control is exerted through a negative feedback system. If the resource is over-harvested the harvester suffers and is brought back to more appropriate levels by starvation and death. Ultimately the system assumes a steady state of least fluctuation with the dominant species, the head of the ecosystem hierarchy, exhibiting the 'best' integration of energy acquisition and conservation that it can attain. A state of affairs thus results in which only those species that can survive do survive, and the interactions between surviving species achieve equipollence or a state of no change.

To achieve a degree of stability in an ecosystem individual species must function as coherent units, sharing the energy they obtain equally between the individuals of which they are composed. This process of energy equipartitioning has been described under the doctrine of homeokinesis by Soodak and Iberall (1978). Through homeokinesis, the individuals lose some degrees of freedom, thereby enabling them to function as a coherent group and exhibit organization, the capacity to work together.

This process is extremely well exemplified in the Inuit community; energy sharing can readily be discerned in the elaborate protocols and rituals that surround the sharing of the spoils of the hunt within the group, in an equitable manner. Organization can equally well be seen in the working of the group in many hunting, fishing and other communal activities (Balikci 1970). Within their relatively autonomous ecosystem a relatively stable state developed, although frequently perturbed by seasons of unusual severity and hardship that reduced the normal level of resources, resulting in the death of some of those least able to withstand the conditions: the too great a reality of feedback in the autonomous system from which there was no escape.

THE LOSS OF SYSTEM AUTONOMY

The arrival of Europeans effectively opened this closed society to the outside world. In the first instances the Europeans were in search of resources for their ever-expanding economies and this initiated the decline of many of the large animals; bowhead whales and musk-ox were both pushed to the brink of extinction. The universal trend to impose the greatest pressure possible on a resource, combined with the capacity to increase that pressure to a virtually unlimited extent through the rifle and steel harpoon, led to a reduction in the numbers of many animals that brought them close to demise. Only stringent regulations, and in some cases complete prohibition saved some species from extinction. Somewhat less disastrously the introduction of long relatively cheap gillnets made charr stocks vulnerable to heavy exploitation. In previous times only the smaller streams could be heavily exploited because the larger rivers were too deep and swift to have saputits constructed upon them.

It is clear that the unlimited application of modern technology readily results in either the elimination of a species or at least its reduction to relatively insignificant levels. In a relatively closed ecosystem these levels adjust, species are eliminated or not as the case may be, and the system comes to a steady state through the operation of internal feedback processes. With the great increase in communications and transportation man is relieved of the necessity of living within his locally circumscribed niche boundaries. In effect the voyages of the fifteenth and sixteenth century navigators initiated a process in which the whole world began to come together as one ecosystem under the dominant species, man. This process has gathered ever-increasing momentum and will be extremely difficult to arrest until the biological limits of the system are approached. The injection of variable quantities of energy as supplied by fossil fuels gives one reason to believe that there may be considerable overshoot and oscillation before the human population settles down once more to a state of least fluctuation. Oscillation at the limit of existence is likely to be unpleasant as is well demonstrated by the musk-ox: they tend to build up herd strength over a sequence of good years only to be quite drastically reduced in number in exceptionally severe winters (Vibe 1967; Thomas et al. 1981).

CONCLUSION

The only alternative to the fulfilment of these gloomy prognostications is the application of improved theory to ecosystem management. In the first place we need regulation of energy input to an approximately constant level, we need to eliminate further environmental degradation and thirdly we need severe limitation on the application of technology to resource harvesting. Only if some degree of stability can be attained in these areas can we hope to proceed to a state of least fluctuation, energy epipartitioning and the redevelopment of organization. It is only through consciously limiting our interaction with the environment in which we live that we can hope to maintain ecosystems in a semblance of the configurations in which they exist today.

Limited interaction implies ecosystem management to maintain desirable and acceptable levels of departure from the pristine state. These desirable levels can only be arrived at through knowledge, agreements and value judgments. There is no scientific solution, much less is there a numerical solution that can be arrived at independently. To decide on these values necessitates perhaps the greatest moral decision with which the world has ever been faced. Unlimited interaction is war. To paraphrase Clausewitz, war is the furtherance of policy by the unlimited application of technology.

It may appear that this discussion has ranged far and wide from its original theme of charr and man; this is not really the case. It is only in the relatively simple ecosystems of the Arctic that it is possible to discern the complex interactions of cause and effect. The interaction between charr and man re-enacts in microcosm, and in relative stark tones, the dilemma which we face in the harvesting of our natural resources on a world-wide basis. Charr and man clash as two generalized species; there is the possibility of interactive segregation and, within limits, the capacity to at least survive in each other's presence. In the case of more specialized species, the bowhead, the musk-ox or the walrus, there is no avoidance, no segregation; interaction is on an unlimited basis and unless one of the parties gives way there is no possibility that the two species can live sympatrically. It is up to man to move out of the way of these species and segregate himself, for they, in truth, can not. Given the capacity to impose virtually unlimited interactional levels on all species through the development of technological devices, it is in man's own best interest to impose limits on that interaction.

Does philosophy belong in biology? Are such discussions appropriate to learned meetings devoted to the study of animals? I believe that ultimately it all comes back to philosophy; if we do not allow the animals that we study and their relationships with other organisms, including ourselves, to provide us with a valid philosophical approach to the world's problems, we are merely exhibiting biological voyeurism; fun no doubt, but not an adequate contribution to the amelioration of man's rather sorry lot. I firmly believe that the study of Arctic charr, like the capacity of the charr itself to explore the unknown waters left by the receding ice cap, is contributing immensely to the exploration of new intellectual frontiers that, over the next decades, may greatly modify our current biological paradigms.

REFERENCES

Balikci, A. 1970. The Netsulik Eskimo. Natural History Press, Garden City, New York. 264 p.

Balon, E.K. 1984. Life histories of Arctic charrs: an epigenetic explanation of their invading ability and evolution. Paper presented at Int. Symp. on Arctic Charr, Winnipeg, 1981 (This volume).

De Vries, A.L. 1971. Freezing resistance in fishes, p. 157-190. *In* W.S. Hoar and D.J. Randall [ed.] Fish Physiology, Vol. 6. Academic Press, London.

Dunbar, M.J. and H.H. Hildebrand. 1952. Contribution to the fishes of Ungava Bay. J. Fish. Res. Board Can. 9 : 83-128.

Fraser, N. and G. Power. 1984. The interactive segregation of landlocked Arctic charr (*Salvelinus alpinus*) from lake charr (*S. namaycush*) and brook charr (*S. fontinalis*). Paper presented at Int. Symp. on Arctic Charr, Winnipeg, 1981 (This volume).

Grainger, E.H. 1953. On the age, growth, migration, reproductive potential and feeding habits of Arctic char (*Salvelinus alpinus*) of Frobisher Bay, Baffin Island, Canada. J. Fish Res. Board Can. 10(6) : 326-370.

Heggberget, T.G. 1984. Habitat selection and segregation of parr of Arctic charr (*Salvelinus alpinus*), brown trout (*Salmo trutta*) and Atlantic salmon (*Salmo salar*) in two streams in northern Norway. Paper presented at Int. Symp. on Arctic Charr, Winnipeg, 1981 (This volume).

Hoar, W.S. 1976. Smolt transformation, evolution, behaviour and physiology. J. Fish. Res. Board Can. 33 : 1233-1252.

Holeton, G.F. 1973. Respiration of Arctic char, *Salvelinus alpinus*, from a high Arctic lake. J. Fish. Res. Board Can. 30 : 717-723.

Holeton, G.F. 1977. Metabolic cold adaptation of polar fish: fact or artefact? Physiological Zoology 47 : 137-152.

Hunter, J.G. 1970. Production of Arctic char (*Salvelinus alpinus*) in a small Arctic lake / Keyhole Lake, Victoria Island, N.W.T. Fish. Res. Board Can. Tech. Rep. 231: 190 p.

Hunter, J.G. 1976. Arctic char and hydroelectric power in the Sylvia Grinnell River. Fish. Res. Board Can. Manuscript Rep. Ser. 1376.

Hurd, L.E., M.V. Mellinger, L.L. Wolfe and J. McNaughton. 1971. Stability and diversity of three trophic levels in terrestrial successional ecosystems. Science 173 : 1134-1136.

Jenness, D. 1922. Life of the Copper Eskimos, p. 97-109 and 145-157. Rep. Canadian Arctic Expedition 1913-1918, Vol. 12. King's Printer, Ottawa.

Johnson, L. 1980. The Arctic charr, *Salvelinus alpinus*, p. 15-98. *In* E.K. Balon [ed.] Charrs: salmonid fishes of the genus *Salvelinus*. Dr. W. Junk Publishers, The Hague.

Johnson, L. 1981. The thermodynamic origin of ecosystems. Can. J. Fish. Aquat. Sci. 38 : 571-590.

Johnson, L. 1983. Homeostatic characteristics of single species fish stocks in Arctic lakes. Can. J. Fish. Aquat. Sci. 40 : 987-1024.

Kipling, C. 1972. The commercial fisheries of Windermere. Trans. Westmoreland and Cumberland Antiquarian and Archaeological Society. Vol. 72 — New series.

Kipling, C. 1984. Charr fisheries in Windermere, England, during the past four hundred years: organization and management. Paper presented at the Int. Symp. on Arctic Charr, Winnipeg, 1981 (This volume).

Kristofferson, A., D.K. McGowan and G.W. Carder. 1984. Management of the commercial fishery for anadromous Arctic charr in the Cambridge Bay area, Northwest Territories, Canada. Paper presented at Int. Symp. on Arctic Charr, Winnipeg 1981 (This volume).

Martin, P.S. 1967. Prehistoric Overkill, p. 75-120. *In* P.S. Martin and H.E. Wright [eds.] Pleistocene Extinctions. Nat. Acad. Sciences and Nat. Research Council, Yale Univ. Press.

May, R.H. 1975. Stability in ecosystems; some comments, p. 161-168. *In* W.H. van Dobben and R.H. Lowe-McConnell [eds.] Unifying concepts in ecology. Dr. W. Junk Publishers, The Hauge.

Moshenko, R.W., R.F. Peet, L.W. Dahlke and D.H. Dowler, 1984. The Arctic charr sport fishery at Tree River, Northwest Territories, Canada, 1964-78. Paper presented at Int. Symp. on Arctic Charr, Winnipeg, 1981 (This volume).

National Film Board of Canada. 1969. Fishing at the Stone Weir, Parts I and II.

Nilsson, N.-A. 1965. Food segregation between salmonid species in north Sweden. Inst. Freshw. Res. Drottningholm Rep. 46: 58-78.

Nilsson, N.-A. 1967. Interactive segregation between fish species, p. 295-313. *In* S.D. Gerking [ed.] The biological basis of fish production. Blackwells, London.

Ruhlé, Ch.P. 1977. Biologie und Bewirtschaftung des Seesaiblings, in Zugersee. Swiss J. Hydrology 39 : 12-45.

Ruhlé, Ch.P. 1984. Charr management in eutrophic Lake Zug, Switzerland. Paper presented at Int. Symp. on Arctic Charr, Winnipeg, 1981 (This volume).

Scholander, P.R., L. van Dam, J.W. Kanwisher, H.T. Hammel and M.S. Gordon. 1957. Supercooling and osmoregulation in Arctic fish. J. Cellular Comp. Physiol. 49 : 5-24.

Soodak, H. and A. Iberall. 1978. Homeokinetics: a physical science for complex systems. Science 201 : 579-582.

Steffanson, V. 1943. The Friendly Arctic. MacMillan, New York, 812 p.

Steiner, V. 1984. Experiments towards improving the culture of Arctic charr (*Salvelinus alpinus*). Paper presented at Int. Symp. on Arctic Charr, Winnipeg, 1981 (This volume).

Thomas, D.C., F.L. Miller, R.H. Russell and G.R. Parker. 1981. The Bailey Point region and other musk-ox refugia in the Canadian Arctic: A short review. Arctic 34: 34-36.

Vibe, C. 1967. Arctic animals in relation to climate fluctuations. Copenhagen.

Wolschlag, D.E. 1964. Respiratory metabolism and ecological characteristics of some fishes in McMurdo Sound, Antarctica, p. 33-62. *In* M.O. Lee [ed.] Biology of the Antarctic seas. Vol. I, Am. Geophys. Union, Washington, D.C.

Charr Fisheries In Windermere, England, During The Past Four Hundred Years: Organization And Management

CHARLOTTE KIPLING

Freshwater Biological Association, Ambleside, England

KIPLING, C. 1984. Charr fisheries in Windermere, England, during the past four hundred years: organization and management, p. 533–536. *In* L. Johnson and B.L. Burns [eds.] Biology of the Arctic charr, Proceedings of the International Symposium on Arctic Charr, Winnipeg, Manitoba, May 1981. Univ. Manitoba Press, Winnipeg.

Documentary evidence from the 16th century onwards shows that the fisheries in Windermere have been closely controlled for many years. The lake was divided into three areas with a total of twelve fisheries. The fisheries were operated by seines worked from the shore and by long lines. Mesh size, dimensions of nets and the number of boats were regulated. Charr were considered a great delicacy, and were potted and sent to other parts of the country. In the 19th century the old controls became ineffective; over-fishing occurred and the number of sport fishermen increased. A Fishery Board was appointed and a closed season and mesh regulations were enforced, but catches declined drastically and the final commercial netting took place in 1921. Sport fishermen continued to use the plumb line, and today there are about 40 charr anglers. Since 1944 the Freshwater Biological Association has removed each year about 1000 kg of northern pike, *Esox lucius*, and during this period the number of charr, estimated from annual nettings, has increased.

INTRODUCTION

Windermere is a lake in the north of England; it is about 16 km (10 miles) long and less than 1.6 km (1 mile) in width. The main fish species are charr (*Salvelinus alpinus* L.), trout (*Salmo trutta* L.), perch (*Perca fluviatilis* L.), pike (*Esox lucius* L.) and eel (*Anguilla anguilla* L.). These fish, in particular the charr, have always been highly valued, and because of this are mentioned in many legal documents. Thus a list of a lady's possessions, made 700 years ago, includes the fishery of Windermere; it is in Latin and the seal shows King Edward I. In this early period almost all the information about fisheries concerns ownership. Later, for the past 400 years, from the 16th century to the present day, information is available on ownership, organization and management, fishing methods and utilization of the catch.

THE FISHERY

The 16th century was a time of great change in England, and it was then that the fisheries of Windermere were re-organized into a system which continued until the present century.

Windermere was divided into three sections, and each section was called a cubble. The High Cubble extended from the meeting place of the rivers Rothay and Brathay to a line from Ecclerigg Crag to Pinstones Point (Fig. 1). The Middle Cubble extended from this boundary to just south of where the ferry now runs, and the Low Cubble from there to the south end of the lake. It is not known why the lake was thus divided, and there are no topographical or biological reasons for the positions of the boundaries. The earliest mention of the cubbles was in 1570, when the word was already in common use. There was one fishery in the High Cubble, five in the Middle Cubble and six in the Low Cubble. There were thus twelve fisheries in the lake, each confined to a particular cubble. The fisheries could be bought and sold and inherited. Ownership of a fishery conferred the right to net spawning charr at named spawning sites and to catch any species of fish throughout the year within a particular cubble. Some proprietors owned fisheries in more than one cubble, and some of the fisheries were sub-divided. Some proprietors worked the fisheries themselves, while others let them to tenants, but all exercised the right of netting and prevented others from doing so. The fisheries were subject to a rent payable to the Lord of the Fee. Queen Elizabeth I in 1574 received six pounds sterling yearly for "the Fishery and ferrying of the water of Windermere". This amount was never altered, and remained the basic amount from which individual rents were calculated until 1938. Ferrying, that is carriage of persons and goods by boat throughout the lake, was subject to controls until the end of the 18th century. In a sales document of 1697 half a fishery is described as "all charr roades (i.e. spawning places), settings, fishing places, landings, landing places, places for hanging the nets and lines in, with half of the boat". In 1752 a similar list also included stakes, coops and coop stakes. The coops were live boxes, and stakes would be used on shore to dry the nets and in the water to secure the coops.

The main fishing method used by professional fishermen was netting, with what were then called draught nets and are now called seines. These were always set from the shore. At least 10 hauls could be made in a day. The distance from the shore at which the net was set varied from 10 to several hundred metres, and the depth from the surface to 18 m (60 ft). The ropes were made by the fishermen out of cow hair and horse hair, and were extremely light in weight, so that two men could work a net satisfactorily. These nets were used to catch charr, trout, pike and perch. Long lines were also used; these were up to half a mile in length with baited hooks every few yards. The professional fisherman did not fish with rod and line.

Up to the middle of the 19th century the proprietors completely controlled the use of the lake. There were no Fishery Boards or police or similar agencies. The proprietors had a financial interest in maintaining the fish stocks and preventing unauthorized fishing. Their management techniques included limiting the number of boats, controlling the type of gear, regulating mesh size and the dimensions of the nets, and a ban on fishing for a period of years.

Control of boats was strict. Thus in 1697 permission was given to the Braithwaite brothers by John Philipson of Rayrigg and William Robinson of Millerground to keep a boat for three years in the High and Middle Cubbles. The brothers had to agree not to kill any fish nor set any hooks and lines and to use it "most especially for crossing the water to visit and fetch their relations". The penalty for breaking the agreement was twenty shillings.

Gear and mesh were mentioned in an agreement of 1670. The fishermen were permitted to use different nets in summer, with a mesh size of 50 mm (2 in.) or more. These nets, used to fish near the surface, would be bigger horizontally and vertically than those used in deep settings in winter. The mesh would catch charr over 130 mm (5 in.) in length. A hundred years later an agreement in three parts was signed by 16 owners and tenants, including one woman. The first part banned for seven years the capture of case (i.e. autumn spawning) charr on their spawning sites at spawning time. The sites could be netted at other times of year. The second part consisted of a total ban on "case nets". These were clearly different from draught nets, and a description of nets in Ullswater provides a clue. They were gillnets, made with very fine twine, set parallel to the shore, into which charr were driven by boats operating between the net and the shore. The ban was completely effective, as gillnets were not used again in Windermere for 270 years. The third part imposed a minimum mesh size of 80 mm (3 in.) circumference, which would catch charr over 180 mm (7 in.) in length; any new nets were to be inspected and measured before use. Any defaulter was to pay ten pounds sterling, to be divided among the other signatories.

By 1840 the population and tourist trade were increasing, as also was the fishing pressure. Much longer charr nets with smaller mesh were being used, and many more sport fishermen were active. In 1862 someone complained to the local paper saying he had seen in a fisherman's house a basket containing over a thousand small charr "not bigger than your little finger". He finishes the letter "I dare not sign my name, or I should get a stone at my head some night". The fisheries were by then in such a poor way that a committee of proprietors decided to stop all netting from the autumn of 1863. The ban lasted five years. This was the last time that the proprietors acted effectively. There was concern nationally over the state of freshwater fisheries and Boards of Conservators were established, and in 1873 charr were included in the closed season for the first time. A by-law imposed a minimum mesh size of 60 mm (2.5 in.) circumference and limited the dimensions of the nets. The Board rented the fisheries as they became available, acquiring them all by 1915. Catches were poor, and the final netting took place in 1921. Since then, apart from sample nettings by the F.B.A. since 1939, all charr fishing has been by sport fishermen using the plumb line. They number about 40 and they combine pleasure with profit by selling the catch locally. Since 1944 the Freshwater Biological Association has removed about 1000 kg of pike each year, and the number of charr, estimated from annual nettings, has increased.

DISCUSSION

What was the demand for charr? In the 17th and 18th centuries to eat charr was definitely a status symbol. Isaac Walton, in the Complete Angler, stated that charr were a rarity and "of so high esteem with persons of great note". The owner of Rydal Hall used to send charr pies to London for his aristocratic friends. In 1661 he bought four dozen charr for 10 shillings (50p) for this purpose. The fish were cooked in spices, and put in a pastry case for transport by pack-horse. The total package might weigh as much as 150 lbs. In the 18th and 19th centuries pots specially made for the purpose and decorated with pictures of charr, were filled with cooked fish, sealed with butter, and sold to tourists as souvenirs. Windermere charr were considered to be better than those from other lakes, and many pots were sold which contained fish which were neither from Windermere nor charr. A complaint in 1787 mentions Ullswater trout sold as Windermere charr. Many charr were eaten locally, and many sent fresh to other parts of the country, especially after the construction of the railway in 1850. Unfortunately fishermen are very secretive about their catches, and we have no reliable figures of the quantities caught per year; my guess for a good year would be about four tons.

Various reasons can be suggested for the end of the commercial fisheries. Sport fishermen became more numerous and influential, and sport fisheries and tourism were considered of greater importance than net fisheries. Demand for species other than charr declined when cheap sea fish became available, so the pike population (severe predators on charr) was able to increase. Lack of demand for charr however was certainly not among the reasons for the ending of the fisheries.

Historical Development Of The Arctic Charr Fishery In Northern Labrador

LAWRENCE J. LEDREW

Department of Fisheries and Oceans, Research and Resource Services, Newfoundland Region, P.O. Box 5667, St. John's, Newfoundland A1C 5X1

LEDREW, L.J. 1984. Historical development of the Arctic charr fishery in northern Labrador, p. 537–548. *In* L. Johnson and B.L. Burns [eds.] Biology of the Arctic charr, Proceedings of the International Symposium on Arctic Charr, Winnipeg, Manitoba, May 1981. Univ. Manitoba Press, Winnipeg.

Arctic charr, *Salvelinus alpinus*, has been valued as a food source since the Dorset Eskimo occupied the northeastern coast of Labrador. In the period prior to contact with Europeans in 1771, fishing implements employed by the Inuit included spears (leisters), hooks, and stone weirs (saputit). Nets were first introduced to the Inuit by Moravian missionaries in the 18th century, and a gradual transformation from a subsistence fishery to a commercially oriented fishery took place over the following two hundred years. The records of the Moravians show that the first export of Arctic charr from the coast of Labrador was in 1860. Although the cod fishery gained prominence in the late 1860's and became the mainstay of the local economy, good market conditions in the 1930's sparked a renewed interest in the commercial charr fishery. At present the northern Labrador charr fishery is conducted by both Inuit and white settlers, and is the economic base along much of the northern Labrador coast. In 1980, more than 50 percent of the total Canadian commercial production of charr originated in northern Labrador; since 1977 annual landings of charr in Labrador have surpassed 200 metric tonnes.

INTRODUCTION

The commercial Arctic charr (*Salvelinus alpinus*) fishery provides a valuable source of income for the people of northern Labrador (Coady and Best 1976). In excess of 50 percent of the adult population of Nain are employed directly in this fishery, which is the mainstay of the local economy. It is also the oldest fishery of its kind in North America (MacCrimmon and Gots 1980). This paper outlines the historical development of the harvesting of this resource from that of a subsistence nature to the modern commercial fishery of today.

EARLY YEARS

Arctic charr, in all likelihood, has been exploited and utilized in northern Labrador since the arrival of the first Dorset Eskimos, who occupied the region about 2600 to 1000 years ago (Fitzhugh 1977). They are known, from archeological evidence, to have made extensive use of leisters, or fish spears (Fig. 1) which were, in all probability, used for the taking of charr. Taylor (1974) concludes that charr was one of the most important species of fish in the traditional diet of the Inuit: charr were abundant, accessible and predictable in their annual migrations.

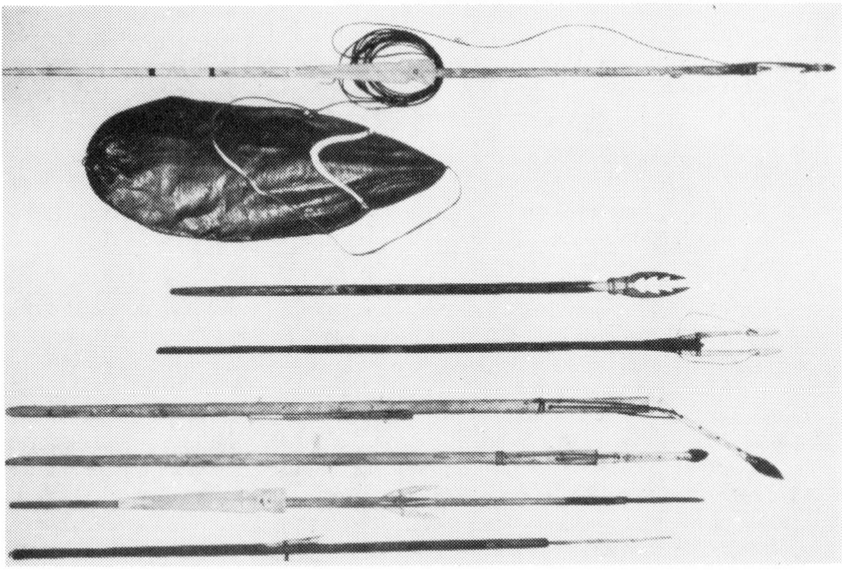

FIG. 1 Leisters, as used by the Dorset Eskimo in northern Labrador (3rd and 4th from the left) (Birket-Smith 1959).

Exploitation by the Inuit was restricted mainly to the freshwater stages of the charr's life cycle, owing to the lack of technology for fishing in the open sea. The methods of fishing varied with the time of year and location. In winter charr were caught in inland lakes and ponds through holes in the ice using a leister and a piece of string with meat attached for bait, the hunter jigging the lure with one hand, and holding the leister ready to strike at a passing fish with the other (Brice-Bennett 1977). In spring, late summer and fall, charr were often speared from overhanging banks at narrow sections of the river. Stone weirs (saputit) were used near the mouths of rivers during the months of August and September (Figs. 2 and 3). In 1773 the Moravian missionary, Jens Haven, described a saputit in Saglek Fiord: "a stone dam was made to hinder the charr from running too far up with the flood [tide], and another dam lower in height downstream over which the charr can get with the flood. At the ebb, charr are kept within the two dams and speared'' (Taylor 1974). The spring and fall fishery was usually carried out by the women and children who dried sufficient quantities for the months ahead (Burch 1960).

FIG. 2 Fishing at the saputit (Birket-Smith 1972).

FIG. 3 Fishing at the saputit (Birket-Smith 1972).

THE MORAVIANS

These traditional methods of exploiting charr remained relatively unchanged until the introduction of nets by the Moravians in 1772. On May 3, 1769, the Moravian Church (also called the Unitas Fratrum or United Brethren) was granted 100,000 acres on the coast of Labrador to set up missions for the purpose of converting the Inuit to Christianity (Hiller 1977). The Moravians established their first mission at Nain in 1771; from that date until the outbreak of the first World War they held undisputed control over northern Labrador's natives, ruling them in two capacities, first, as apostles of Christian civilization and secondly, as unofficial representatives of the Newfoundland government (Jennes 1967). During this period the Moravians held exclusive control of all trading in northern Labrador. Adjacent to each mission, a trading post was established to provide the Inuit with essential goods and to purchase from them items for export such as seal-skins, seal-oil, blubber, cod and cod oil. There was no exchange of money in these transactions as a credit system was used. For more than a century the Moravians covered all the expenses of the Mission in Labrador by trading with the Inuit (Jennes 1967).

In addition to exposing the Inuit to European civilization the Moravians also introduced various items of equipment which helped to improve the life style of these native people; nets for catching seals and fish, introduced in 1772, were among the most important of these (Moravian Diary). This enabled the Inuit to hunt seals for food in the spring, catch enough charr at sea in the summer to serve their needs and still have time to hunt caribou in early fall for clothing. Jorden (1977) states that with the introduction of nets many Inuit families dispersed to the outer islands to fish for charr and salmon in the summer although they continued to use leisters and fish hooks to take charr in the rivers in the fall and spring. What was at one time a group activity became an individual or at least a family practice with the introduction of charr nets (Taylor 1974). This new technology would certainly have increased access to charr stocks in the sea, making possible the June-July resource gathering activity. The Moravians also caught and dried charr in the summer and occasionally purchased dried charr from the Inuit in the fall to trade (give when necessary) back to them in time of need during the ensuing winter. Figure 4 shows the annual resource gathering cycle of the Labrador Inuit during the late 1700's. With the exception of the whale fishery, which had all but disappeared by 1800, this cycle remained unchanged until the early 1860's. Largely due to the decline in the availability of caribou during this period, the Arctic charr fishery for food reached its peak.

In 1857, the first cod fishers arrived on the Labrador coast. During the next 12 years the number of vessels visiting the coast rose to over 500, almost all from Newfoundland (Jennes 1965). This in turn stimulated a greater interest in the cod

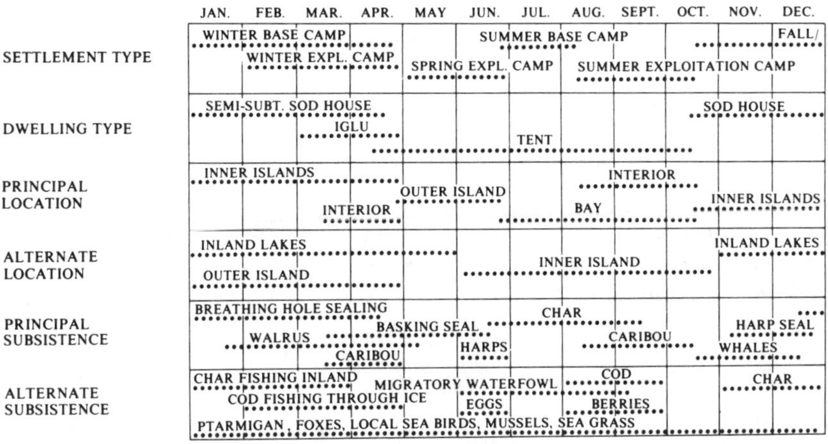

FIG. 4 Annual cycle of the Labrador Inuit, CA 1776 (Cox 1977 in Fitzhugh 1977).

fishery by the Moravians who encouraged the Inuit to spend the summers on the coast and fish for cod. The trend to summer cod fishing began in the late 1860's and by the turn of the century it had become the mainstay of the economy (Brice-Bennett 1977).

In 1860, two barrels each of pickled salmon and charr were sent to England to test market-acceptance. This was the first attempt by the Moravians to export either of these commodities (Moravian Diary). The venture proved successful and was thus the inception of the commercial charr and salmon fishery of northern Labrador, which has continued to the present day. By 1870 markets for pickled charr had also been established in Newfoundland and Boston. The Moravians in 1871 purchased a schooner (the Cordelia) for the collection of fish at the various stations along the coast and for the conveyance of the cargo to Newfoundland. Two vessels now serviced the missions, the Cordelia, and the Harmony, crossing from England in the spring with supplies, and returning in the fall carrying exports.

The Missionaries and store keepers were virtually the only Europeans on the northern Labrador coast in the 18th century. In the first years of the 19th century white settlers began to move into the region, lured by the prospects of rich trapping and trading. The further expansion of these settlers during the next half century slowed down; however, the invasion of cod fishing vessels revived the slow drift northward so that by 1874, 115 settlers were living among the 1,176 Eskimos of northern Labrador (Jennes 1965). Because the Moravians were negatively disposed towards them living within their domain, the majority of settlers lived scattered about the countless bays and inlets of the coast, away from areas traditionally exploited by the Inuit (Kleivan 1966). However, because they wished to live close to good fishing, they settled near prime Arctic charr and Atlantic salmon areas virtually excluding the Inuit from them. By the end of the nineteenth century the settlers supplied a substantial proportion of the exported charr.

In 1888, it was decided by the Moravians that all fish exports to Europe should cease and would, instead, be taken to St. John's, Newfoundland (Moravian Diary). Jennes (1965) gives figures of 579 brls, 787 brls, and 798 brls of charr (1 brl equals 100 kg) shipped by the Moravian mission from Labrador in 1883, 1893 and 1903 respectively (Table 1).

THE TWENTIETH CENTURY

At the turn of the century and up to about 1926, charr exports declined due to the lucrative cod fishery and the decline in export markets during the first World War. In 1926, mounting financial difficulties stemming from low market prices for salted and dried fish, and a large number of outstanding debts, caused the Moravians to decide to lease their trading rights and stores to the Hudson Bay Company for a period of 21 years. The Hudson Bay Company's chief object in trade was the skins of fur bearing animals. Unlike the Moravian mission, they had little interest in promoting the fisheries (Brice-Bennett 1977). In fact they purchased fish only occasionally and effectively forced both Inuit and settlers to spend more time in the country in pursuit of fur (Moores 1980, unpub. rep.). As a result, the commercial charr fishery was reduced to a limited domestic fishery. This period of intensive fur trapping and minimal cod and charr fishing continued until the depression of the 1930's sent fur prices crashing, leaving both Inuit and settlers virtually without income, with seal and trout nets in a state of disrepair (Moravian Diary).

A reduction in the demand for furs, mounting expenses and trade deficits all combined to prompt the Hudson Bay Company to close their northern posts in Labrador in July, 1942; the responsibility for maintaining basic trade and supply services was taken over at that time by the Newfoundland Commission of Government. A new agency, the Northern Labrador Trading Operations (NLTO), under the government's Department of Natural Resources gave immediate priority to the development of a viable resident fishery in northern Labrador: steps were taken to outfit residents with boats and gear, and storage and processing sheds were erected. The cod, seal and Arctic charr fisheries were prosecuted on an increased scale (Coady 1974); but the exploitation of cod varied with its availability and the market price: when either declined, effort towards charr increased.

Table 1. Fish exports from the northern Labrador coast; 1883, 1893, 1903-1907 (Jennes 1965).

Year	Arctic charr Estimated whole wt (kg)	Atlantic salmon Estimated whole wt (kg)	Cod Estimated whole wt (kg)
1883	52,526	6,042	585,467
1893	71,396	816	651,873
1903	72,394	980	878,527
1904			
1905	17,364	277,438	72,571,968
1906	5,769	138,311	54,624,766
1907	13,172	116,865	63,030,318

The impetus given to the fisheries by the Department of Natural Resources (NLTO) did not bring the prosperity anticipated in the early 1940's (Brice-Bennett 1977): during the first 10 years of operation exported charr averaged only 80 metric tonnes per year (Dempson et al. 1980) and cod 1250 metric tonnes (Coady 1974).

In 1951, changes were made, the administration of trade being transferred to the Department of Public Welfare and its agency the Northern Labrador Services Division. At the Labrador Conference, organized by the Newfoundland Government in 1956, it was decided to continue subsidizing the fisheries, to pay higher prices for cod and seal products, and to extend unemployment insurance benefits to cod and charr fishermen. Fishermen were turning to pickling charr during the summer as an alternative to fishing the dwindling size and quantity of the cod that appeared off the coast after the mid-1950's (Brice-Bennett 1977). During this period, and continuing until the late 1960's, the entire commercial catch of charr was pickled. Collection boats operated by the NLTO would leave Nain with barrels, salt and other supplies for distribution along the coast and on the return trip pick up the salt packed fish for delivery to Nain. There it was unpacked and repacked in a heavy brine solution ready for export (Coady 1974).

The first attempt to market fresh charr was in 1960; however, due to poor flying conditions, less than 25 percent of the charr arrived in Montreal in good condition. Subsequent attempts by a charter company in 1961 and 1962 using a vessel with refrigeration capabilities resulted in the export of 14,000 kg and

4,000 kg in the two years respectively. However, the continued breakdown of refrigeration equipment rendered the venture unprofitable and it was therefore discontinued (Coady 1974).

The Northern Labrador Services Division exported smoked charr in 1967 for the first time in response to a promising market. In 1968, 450 kg was distributed to test market-acceptance in Halifax, Montreal, Winnipeg and in Newfoundland. Market acceptance was indeed encouraging and in 1970 a smoking unit was installed in Nain. In the next ten years annual production of smoked charr rose to 6,400 kg.

1970-PRESENT STATUS OF THE FISHERY

Freezing facilities were installed in the Nain fish plant in 1971, the export of fresh frozen charr becoming the main industry in that community. By 1973 the cod fishery had virtually disappeared along the northern coast of Labrador, charr and salmon accounting for 98 percent of the entire commercial fish production (Coady 1974). Figure 5 gives a summary of Arctic charr and cod exports from 1942-1980 (inclusive). The figures for charr represent exports as opposed to actual landings since no white-fleshed charr were marketed from 1948 to 1966. Dempson et al. (1980) state that through this period production in the cod fishery was averaging 1,450 metric tonnes per year. However, with the rapid decline in the availability of cod in 1969 emphasis was subsequently directed towards the charr fishery. Average annual exports increased from 136 metric tonnes during the period 1960-69 to 205 metric tonnes during the past five years.

Prior to 1971 and the installation of freezer facilities in Nain, charr fishing had been carried out along the entire northern Labrador coast; however several factors led to a change in this distribution and fishing effort was concentrated in the immediate Nain region (Coady 1974, Dempson et al. 1980, LeDrew 1980); there was a shift in emphasis from a salted product to a higher priced frozen product; coupled with this were the low market acceptability of white-fleshed and pink-fleshed charr as well as the availability of Atlantic salmon among the outlying islands.

Owing to the intense fishery that developed in the Nain-Okak region, regulations limiting the fishery to residents of northern Labrador were announced in 1973 by the Federal Department of Fisheries and Oceans; gear restrictions were imposed in 1975.

With the Newfoundland Department of Fisheries assuming responsibility for the operation and marketing of the fishery in 1979, more emphasis has been placed on developing markets and the expansion of the present fishery into traditional areas along the northern coast of Labrador. Thus Arctic charr fishing

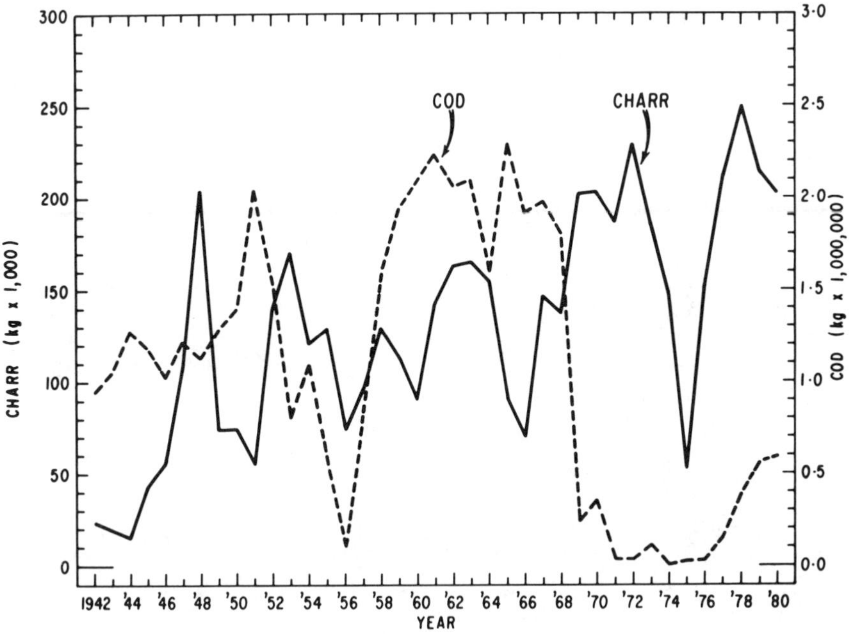

FIG. 5 Summary of Arctic charr and cod exports from the northern Labrador coast; 1942 - 1980 (Coady and Best 1976, and Dempson and LeDrew unpublished data).

has developed from a subsistence fishery to a commercially viable industry, upon which, along with the salmon fishery, the economy of Nain is highly dependent. It is recognized that proper management of the species based upon sound biological information is necessary to ensure the future of this, the oldest commercial Arctic charr fishery in North America.

ACKNOWLEDGEMENTS

The author would like to thank Mr. Brian Dempson for his helpful suggestions and guidance in preparing the text. Thanks are also extended to Vince Taylor, Rex Porter and Mike Chadwick for their useful criticisms concerning the text. Appreciation is extended to Herb Mullett for preparing the figures and to Janice Lannon for typing many drafts of the report.

APPENDIX I Former and Present-day communities in Northern Labrador (Brice-Bennet 1977).

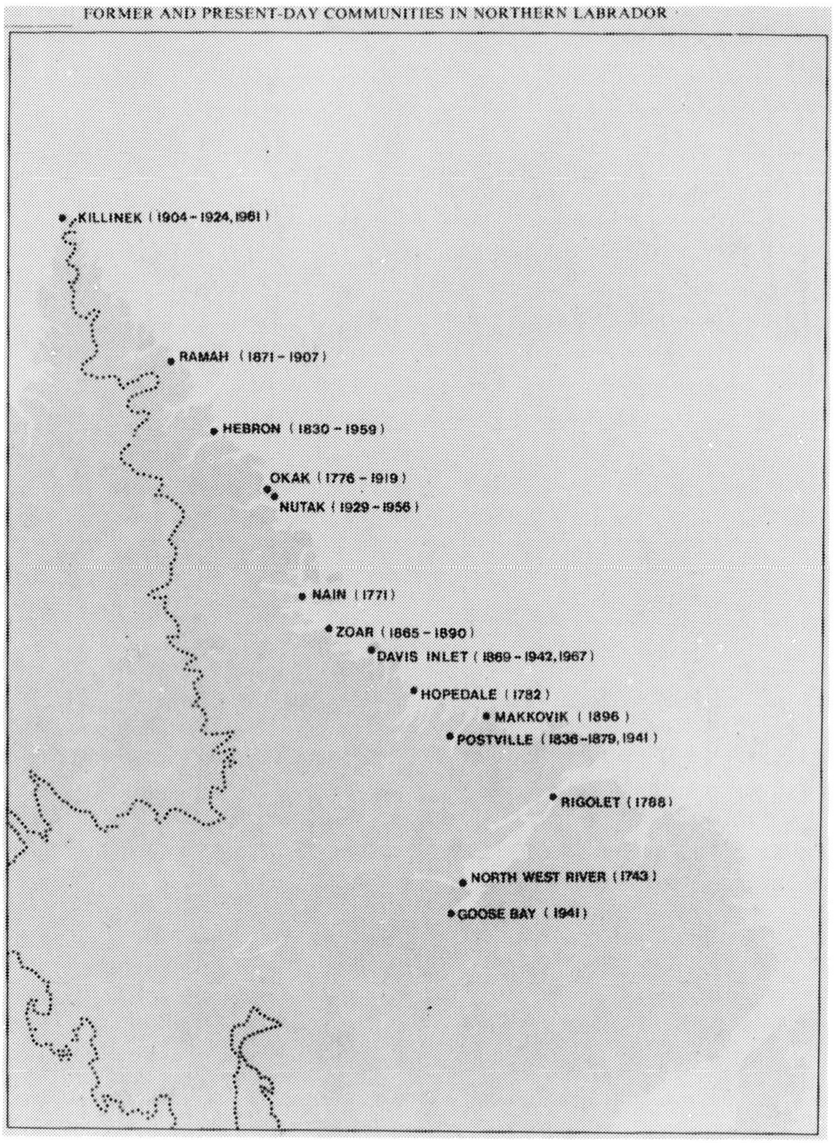

APPENDIX II Chronological order of events relevant to the Labrador charr fishery.

Year	Event
1772	Introduction of charr nets by the Moravians
1860	First export of Arctic charr from Labrador coast to England to test market acceptance
1926	Moravians lease trading rights to the Hudson Bay Company
1942	Trading operation taken over by the Newfoundland government and its agency—Northern Labrador Trading Operations
1951	Administration of trade transferred to Newfoundland Department of Welfare (NLSD)
1970	Smoking unit installed in Nain
1971	Freezing facilities established in Nain fish plant
1973	Fishery limited to residents of northern Labrador
1975	Gear restrictions implemented by Federal Department of Fisheries and Oceans
1979	Newfoundland Department of Fisheries assumed responsibility for the operation and marketing of the fishery

REFERENCES

Andrews, C.W., and E. Lear. 1956. The biology of Arctic char (*Salvelinus alpinus* L.) in northern Labrador. J. Fish Res. Board Can. 13 : 843-860.

Birket-Smith, K. 1959. Eskimos. Methuen Pub. London.

Birket-Smith, K. 1972. Eskimos. Crown Pub. New York, 2nd Ed.

Brice-Bennett, C. 1977. Land use in the Nain and Hopedale regions. *In* Our Footprints are Everywhere: Inuit land use and occupancy in Labrador. Dollco Printing Ltd., Can. p. 97-204.

Burch, Jr. E. 1960. The traditional Labrador Eskimos. Thesis A.B. Degree Princeton Univ. Mass: 250 p.

Coady, L.W. 1974. The Arctic char fishery of northern Labrador. Dept. of Fish. R.D.B. progress Rep. Nfld. Region, No. 104 : 31 p.

Coady, L., and C.W. Best. 1976. Biological and management investigations of the Arctic char fishery at Nain, Labrador. Fish Mar. Serv. Dev. Br. Tech. Rep. 624: 103 p.

Dempson, J.B., L. LeDrew, and T. Nicholls. 1980. Review of biological information on Arctic char (*Salvelinus alpinus*) in Labrador and present status of the commercial fishery. Symposium on Research in the Labrador Coastal and Offshore Region, Mem. Univ. of Nfld. p. 60-74.

Fitzhugh, W. 1977. Indian and Eskimo/Inuit settlement history in Labrador: An archeological view. *In* Our Footprints are Everywhere: Inuit land use and occupancy in Labrador. Dollco Printing Ltd. Can. p. 1-41.

Hiller, J.K. 1977. Moravian land holdings on the Labrador coast: A brief history. *In* Our Footprints are Everywhere: Inuit land use and occupancy in Labrador. Dollco Printing Ltd. Can. p. 83-94.

Jenness, D. 1965. Eskimo Administration. III. Labrador. Arctic Inst. N. Amer. Tech. Rep. No. 16: 94 p.

Jordan, R.H. 1977. Inuit occupation of the central Labrador coast since 1600 A.D. *In* Our Footprints are Everywhere: Inuit land use and occupancy in Labrador. Dollco Printing. Can. p. 43-48.

Kleivan, H. 1966. The Eskimos of northern Labrador: A history of Eskimo-white relations. p. 1771-1955.

LeDrew, L.J. 1980. Research on the Arctic charr fishery of northern Labrador. Can. MS Rep. Fish. Aquat. Sci. 1589: iv + 10 p.

MacCrimmon, H.R. and B.L. Gots. 1980. Fisheries for charrs, p. 797-839. *In* E.K. Balon [ed.] Charrs: salmonid fishes of the genus *Salvelinus*. Dr. W. Junk Publishers, The Hague.

Moravian Records: Handwritten accounts, diaries, etc. of the Moravian dealings on the Labrador coast, 1771-1930. Available, microfilm-Mem. Univ. Nfld.

Taylor, J.G. 1974. Labrador Eskimo settlement of the early contact period. Nat. Mus. Pub. No. 9, 107 p.

Taylor, J. 1977. Traditional land use and occupancy by the Labrador Inuit. *In* Our Footprints are Everywhere: Inuit land use and occupancy in Labrador. Dollco Printing. Can. p. 49-58.

Historical Evidence for the Introduction of Arctic Charr into High-Mountain Lakes of the Alps by Man

ROLAND PECHLANER

Limnology Division, Department of Zoology, University of Innsbruck, Austria

PECHLANER, R. 1984. Historical evidence for the introduction of Arctic charr into high-mountain lakes of the Alps by man, p. 549–557. *In* L. Johnson and B.L. Burns [eds.] Biology of the Arctic charr, Proceedings of the International Symposium on Arctic Charr, Winnipeg, Manitoba, May 1981. Univ. Manitoba Press, Winnipeg.

Many lakes in the Alps, at elevations above the timber-line, have populations of Arctic charr; most, if not all, are believed to have been stocked by man rather than by natural invasion. The impossibility of active invasion by the charr in a specific set of high-mountain lakes in the continental divide of the eastern European Alps is demonstrated. The fishery management practices of late-medieval Tyrolean sovereigns are reviewed as they bear on this problem.

INTRODUCTION

The question of the origin of the Arctic charr populations in high-mountain lakes of the Austrian Tyrol is of considerable interest in a number of spheres. In the first place there is the intrinsic interest in their origin, how in fact did they get there when, today, no direct routes appear to be possible? Secondly they may be able to throw light on the genetic background of charr; three forms of charr (''Schwarzreuter'', ''Normalsaiblinge'' and ''Wildfangsaiblinge'') occur in lakes of the European Alps, at both high and low altitudes, and the question has been raised as to whether they represent genetically different forms or not. This has resulted in a long controversy that is by no means dead as many papers in this symposium attest (Buresch 1925, Haempel 1930, Neresheimer 1937, Schindler 1940, Steinmann 1942, Steinböck 1950, Reisinger 1953). Nordeng (1983) has probably brought this debate to a close by showing conclusively that diadromous, large resident and small resident forms of Arctic charr are morphs rather than genetically differentiated sub-populations of *Salvelinus alpinus* (L.).

The third facet is that isolated charr populations, occurring in a considerable number of relatively small alpine lakes (the terms ''alpine'' and ''high-mountain'' are used here exclusively in relation to the region above timber-line, i.e. the ''alpine horizon'' of botanists), are attractive to scientists interested in mechanisms and results of subspecific differentiation. In this context the length of isolation is critical. Pechlaner (1966, 1969, 1979) postulated that salmonid populations of almost all high-mountain lakes of the Alps must have been transported there by man, and that many of these lakes in the Tyrol were stocked within a few decades of the late 15th and early 16th centuries. This has been doubted recently by Balon and Penczak (1980), who assume that the hydrological and topographical situation during the glacial retreat made high-mountain lakes in the Alps as accessible to natural charr invasion as glacier scoured lakes in the holarctic.

It is therefore imperative to examine the evidence and attempt to reach a conclusion as to the means by which these lakes obtained their original fish populations. The present paper reviews the geomorphological facts that appear to rule out the possible immigration of charr into at least one specific group of alpine lakes; the doubts of Balon and Penczak (1980) concerning the motivation and physical ability of medieval Tyroleans to develop charr fishing in high-mountain lakes calls for an evaluation of the relevant historical documents.

NO ACTIVE INVASION OF CHARR

For a clear answer to the question as to why the natural immigration of charr into high-mountain lakes can be ruled out, it is necessary to focus on a certain set of such lakes. Counter-evidence for the active natural invasion of lakes in the alpine horizon is centered on the following Tyrolean lakes, all with "old" populations of *Salvelinus alpinus*:

a) Vorderer Finstertaler See (2237 m a.s.l.)
b) Hinterer Finstertaler See (2256 m a.s.l.)
c) Mittlerer Plenderlesee (2317 m a.s.l.)
d) Oberer Plenderlesee (2344 m a.s.l.)
e) Berglersee (2466 m a.s.l.)
f) Wannenkarsee (2639 m a.s.l.)
g) Unterer Seekarsee (2655 m a.s.l.)
h) Laubkarsee (2681 m a.s.l.)
i) Geislacher See (2702 m a.s.l.)
k) Schwarzsee ob Sölden (2792 m a.s.l.)

All these lakes lie in the silicious rocks of the central divide of the eastern Alps: lakes (a) to (d) are located in the Stubai Mountains (in the vicinity of the limnological high-mountain station of the University of Innsbruck in Kuehtai), lakes (e) to (k) in the headwater area of the Oetz valley.

The extent of the Würm glaciers, as shown, for example, on the glacial map of the Tyrol by Klebelsberg (1935), leaves no doubt that charrs in this area cannot be relics of any pre- or inter-glacial period. The Würm maximum was achieved about 18,000 years before present, at mean summer temperatures 10°C below the present level (Patzelt 1980). At this time the Inn glacier reached 1800 m thickness at Innsbruck (Heuberger 1974).

In the following millenia there were numerous retreats and advances of the glacier network within the valleys of the Alps, but soon after 13,000 BP the development of woods started in the Inn valley and has not been interrupted since.

The three largest lakes on the present map of the Tyrol — Achensee (929 m a.s.l.), Plansee and Heiterwanger See (both 976 m a.s.l.) — were available for charr colonization at least since 13,000 BP. There is general agreement that Arctic charr migrated into waters of the Alps in the late glacial period (e.g. Thienemann 1950, Reisinger 1953, Johnson 1980). Different opinions about the route of this migration need not be discussed in the present context. In the Alps charrs became strictly lacustrine. For this European lake charr, *Salvelinus alpinus salvelinus* is the most commonly used sub-specific designation (Behnke 1980).

All the alpine charr lakes listed previously did not exist as water bodies during the Egesen stadial (Heuberger 1966, Patzelt pers.comm.), which was the last glacial advance before the retreat to their present limits. Egesen corresponds to early Dryas; its end, which is the end of the late glacial period in both the western and eastern Alps, is characterized by rapid warming, occurring between 10,200 and 10,000 years before present.

According to Patzelt (1980) the Egesen glacial advance was caused by mean summer temperatures being 2° or 3° C colder than at present, the permanent snow-line being 300 to 400 metres lower than now. After Egesen there were at least eight periods of glacial advance, but in 10,000 years both snow- and tree-lines have changed their mean elevation by less than 250 m.

Fluvial erosion has deepened the valleys in the igneous rock of the central Alps by less than 30 m within the period of 10,000 years after the alpine charr lakes had become free of glacial ice (Patzelt pers.comm.); therefore, the present topographical situation of the selected lakes gives an adequate picture of the conditions at the time of glacial retreat.

There may be some question as to the maximum discharge of water from these lakes at different times during the late glacial period as, at present, Schwarzsee ob Sölden and Geislacher See have no surface outlet at all, and the small streams leaving Oberer and Mittlerer Plenderlesee disappear into the ground far above the main valley. Slopes of preglacial origin, too steep for active charr migration separate the lakes in the Kuehtai area (at least by 100 metres), and in the Oetz valley (by several hundred (up to 1000) metres) from any river which might have allowed early invasions of Arctic charr into this region.

EVIDENCE FROM MEDIEVAL DOCUMENTS FOR ARTIFICIAL TRANSPORT OF CHARRS INTO HIGH-MOUNTAIN LAKES

Our knowledge of medieval fishery management in the Tyrol — including stocking of high-mountain lakes with charr benefits from two facts:

1) Both the ownership rights of lakes and the fishing rights in lakes and rivers in medieval times were in the hands of either the sovereigns of the Tyrol, or the bishops of Brixen or Trient in their secular territories. The dukes of Austria, who by 1420 became sovereigns of the Tyrol, chose Innsbruck as their residence. This led to an enforcement of the fishing rights by the administration particularly during the reigns of Erzherzog Sigismund (+ 1496), Kaiser Maximilian I (+ 1519), and Erzherzog Ferdinand II (+ 1595), who claimed the fisheries, demanding delivery of fish for both the pleasure of the court and for helping in its subsistence (Stolz 1936, p. 172/173). Considerable efforts were made to improve fishery management in lakes and rivers and pond fish culture. Owing to this particular concern of the sovereign for fish and fishing the archives of the Tyrolean government are rich sources of relevant information.
2) Several authors have examined the specific historical situation of the fishery in the Tyrol, and have published extensively on their findings. The present review of charr lake management is based mainly on publications of the following authors: Michael Mayr who organized and provided a commentary on a reprinting (1901) of the fishery book of the emperor Maximilian I (written 1504), the manuscript of which is in safe keeping at the Austrian National Library in Vienna; Otto Stolz who published a comprehensive work in 1936 dealing with historical information on the running and standing waters of the Tyrol, including their utilization for fishing with emphasis being placed on legal aspects; Hermann Diem (1964) who evaluated the respective documents as a biologist. His competence and his method of citing all relevant information about fishing waters and fisheries before discussing its bearing on the subject, made his publication especially valuable for the present outline. The book of Franz Niederwolfsgruber (1965) deals mainly with hunting during the reign of Kaiser Maximilian I. A chapter on "The pleasure of fishing" reviews information given by Diem (1964) and earlier authors, but adds further details by evaluating the biographical book "Weisskunig", which was written by the confidential secretary of Kaiser Maximilian I on the basis of information provided by — and partly dictated by — the emperor.

Considerable stocking of high-mountain lakes in the Tyrol, in the late 15th and early 16th centuries, is evident from a report on fishing waters, written around 1500, in the famous fishery book of Kaiser Maximilian I (1504) and from an "instruction" of 1507 to the fisherman Hans Humel. Erzherzog Sigismund is mentioned explicitly as having ordered the stocking of five highmountain lakes; four alpine lakes, specified as being fishless in 1500 (the date one report was written) are subsequently mentioned as containing brown trout or charr in 1504.

A fishery office was organized in the Tyrol perhaps during the reign of Erzherzog Sigismund, but certainly under Kaiser Maximilian I. The first nomination of a fishery director (leading master fisherman, "Oberstfischmeister") is documented for 1502 (Stolz 1936, p. 359). The fishery director had two fields of responsibility: he had to take care of all fishing waters in use by or for the court and he had to ensure the observance of fishery rules by holders of fishing rights throughout the country. Each fishery director was provided by the sovereign with "Fishery Orders", containing specifications of the natural and artificial waters reserved for the court, with indications as to how to manage fishes in other waters; these included closed seasons, legal sizes of fish, mesh sizes of nets, illegal fishing gear, penalties etc., and "Instructions", giving both general and rather detailed directions for the particular fishery director and his staff (Diem 1964, p. 43 ff.).

Instructions for and reports of fishery directors are good sources of information about the fishing waters used by the sovereign and his court, but on the other hand they do not cover water bodies for which the fishing was, at that time, in other hands. For example good fishing sites in the district of Rattenberg, Kufstein and Kitzbühel are not mentioned since this area was a part of Bavaria until 1504; fishing (and hunting) in the inner parts of the Ziller valley is not considered because the respective rights were owned by the bishops of Salzburg. None of the charr lakes listed previously are mentioned, as the Petersberg district to which they all belonged, was enfeoffed from 1407 until 1587 to the lords of Freundsberg, and Kaiser Maximilian I obviously did not wish to curtail their fishing rights (Stolz 1936, p. 228, 360).

At least 14 high-mountain lakes in the Tyrol are mentioned in late medieval documents as salmonid fishing waters for the court. As stressed by Diem (1964) and Niederwolfsgruber (1965), one of the important motives for fishery management in late medieval times was the interest in a combination of hunting and fishing. Kaiser Maximilian I, famous as one of the greatest hunters of his time, both guarded and hunted ibex and chamois with particular enthusiasm, which explains the efforts of his administration to utilize the fishery resources of rather remote high-mountain lakes. The "fishery book" suggests that when the sovereign is hunting chamois in the "Hundstalsee" (2289 m a.s.l.) region, he might enjoy having trout, and the author recommends gillnets for catching them in this lake (Mayr 1901, p. 22; Pechlaner 1979, p. 8). At "Wildsee Waldenpach", a lake above Kauns not yet identified, it is indicated that when hunting there, the sovereign might take one hundred trout from this lake at any time (Mayr 1901, p. 36; Diem 1964, p. 75.). As to the manner of stocking high-mountain lakes we could find only one direct reference: to Lichtsee (2153 m a.s.l.; 1.3 ha), where the author of a fishery report, dated about 1500, simply says that he got several persons "to carry 200 trouts in there . . ." (Diem 1964, p. 73). For people used to carrying heavy loads in the mountains — and the present generation of limnologists has had such load-carrying experience in spite of recently built roads far into the alpine region, cable cars and helicopters — it is not difficult to understand that transporting fish into high-mountain lakes implied no serious problems for the medieval inhabitants of the Tyrol. A few citations from relevant documents are sufficient to close the chain of necessary events: availability, catching, delivery, transport and planting of charr fry, not only demonstrating that the stocking of high-mountain lakes with charr during the reign of Kaiser Maximilian I was a distinct possibility, but also developing the picture as to how this actually occurred.

The three largest charr lakes of the Tyrol, Achensee 6.8 km^2, Plansee 2.9 km^2 and Heiterwanger See 1.8 km^2, (elevations listed previously) were among the waters managed intensively by the administration of the sovereigns of the Tyrol in late-medieval times. Erzherzog Sigismund was not successful in getting back the full ownership of Achensee, which had been donated to the convent of St. Georgenberg in 1120. In a document of 1469 he reconfirmed the ownership of this convent, but at the same time, taking the lake in feudal tenure for himself and his successors, the monastery was entitled to fish there for its own needs (Stolz 1936, p. 188/189).

A similar situation existed for the charr lakes Plansee and Heiterwanger See, where most of the fishing rights were owned by the convent of Stams. The convent authorities could not refuse the request of Erzherzog Sigismund and Kaiser Maximilian I to have these lakes returned, but they insisted on their right to joint use (Stolz 1936, p. 222-224).

Diem (1964), over the course of 25 pages, deals with fishing methods in the time of Kaiser Maximilian I. Seines, gillnets, hook and line and other equipment were used for fishing charr. Proposals to build cages for storing live fish in the lake occur repeatedly in documents of the early 16th century. In 1504 two such containers, one for charr and one for pike, were suggested for Vilsalpsee and Traualpsee. A report of 1515 says that the present abbot of St. Georgenberg was the first to make a cage for keeping charr for the monastery at Achensee alive and fresh. The fishery director orders, in a document of 1527, the building of a special cage for small charr at Plansee (Diem 1964, p. 100/101).

In a letter of 1529, the fishery director Hanns Phadt recommends that the fishermen on Achensee be asked for charr fry, begged for by the bishop of Brixen; if the fish could not be obtained from there, charr fry could be provided by Heiterwanger See, but, in this case ''not before autumn, because earlier they could not be transported thus far due to high temperatures'' (Diem 1964, p. 83).

The ability to transport living charr long distances is attested to repeatedly in the fishery book of 1504; we can learn with respect to Achensee, that charrs for the sovereign were brought alive to Innsbruck from Easter until Whitsun, and then only after St. Bartholomew's Day (August 24th), when it was again colder (Mayr 1901, p. 3); from Heiterwanger See, charr and trout, coregonids and bream could be transported ''in three days to Augsburg or in two days to Innsbruck alive'' (Mayr 1901, p. 27). The successful transportation of coregonids from Achensee (30 km directly NE of Innsbruck) to Brennersee (29 km south of Innsbruck) has been described by Diem (1964, p. 73); such long distance transportation was performed using large water containers pulled by horses, or carried on rafts, in which case water renewal was particularly easy.

In the book ''Weisskunig'' Kaiser Maximilian I boasts of having invented ''the containers, in which charrs and trouts may be kept fresh and with their proper taste, as if they were in their water'' (Niederwolfsgruber 1965, p. 56). In Figure 1 (out of ''Weisskunig'') a man is depicted putting fish into such a container. Among the coloured pictures in the fishery book of 1504 (Mayr 1901) there is one showing such a container on the shore of a high-mountain lake. Similar wooden containers are in use by sports fishermen nowadays for transporting both fish fry and harvested fish of legal size.

In the times of Erzherzog Sigismund and Kaiser Maximilian I it was certainly not true that ''charrs were not worth transporting beyond the range of lakes with easy access'' (Balon and Penczak 1980, p. 792), but these authors are right for later centuries. As shown by Diem (1964 e.g. pp. 24 and 27) very little was done for the good of fishing in high-mountain lakes after Kaiser Maximilan I died, until, in our own day, limnological investigations have shown how management methods can be developed for the improvement of this valuable recreational fishery.

REFERENCES

Balon, E.K. and T. Penczak. 1980. The dwarfed charr of Dösener See, an alpine lake in Austria, p. 773-794. *In* E.K. Balon [ed.] Charrs: salmonid fishes of the genus *Salvelinus*. Dr. W. Junk Publishers, The Hague.

Behnke, R.J. 1980. A systematic review of the genus *Salvelinus*, p. 441-480. *In* E.K. Balon [ed.] Charrs: salmonid fishes of the genus *Salvelinus*. Dr. W. Junk Publishers, The Hague.

Buresch, R. 1925. Studien am Seesaibling mehrerer Alpenseen. Zeitschrift f. Fischerei 23: 99-118.

FIG. 1 Engraving from the book 'Weisskunig' of Kaiser Maximilian I (1450 - 1519), showing fishing and fish transport methods of late medieval times.

Diem, H. 1964. Beiträge zur Fischerei Nordtirols. Veröff. Museum Ferdinandeum 3: 5-132.

Haempel, O. 1930. Fischereibiologie der Alpenseen. Die Binnengewässer 10: 1-259.

Heuberger, H. 1966. Gletschergeschichtliche Untersuchungen in den Zentralalpen zwischen Sellrain und Ötztal. Wissenschaftl. Alpenvereinshefte 20: 1-125.

Heuberger, H. 1974. Alpine quaternary glaciation, p. 319-338. *In* J.D. Ives and R.G. Barry [eds.] Arctic and Alpine Environments. Methuen and Co. Ltd., London.

Johnson, L. 1980. The Arctic charr, *Salvelinus alpinus*, p. 15-98. *In* E.K. Balon [ed.] Charrs: salmonid fishes of the genus *Salvelinus*. Dr. W. Junk Publishers, The Hague.

Klebelsberg, R. 1935. Geologie von Tirol. Borntraeger, Berlin.

Mayr, M. 1901. Das Fischereibuch Kaiser Maximilian I. Wagner'sche Univ. Buch-handlung Innsbruck.

Neresheimer, E. 1937. Die Lachsartigen (Salmonidae) I. Teil. Handb. d. Binnen-fischerei Mitteleuropas, III. 219-370.

Niederwolfsgruber, F. 1966. Kaiser Maximilian I. Jagd- und Fischereibücher. Jagd und Fischerei in den Alpenländern im 16. Jahrhundert. Pinguin-Verlag Innsbruck: 1-75.

Nordeng, H. 1983. Solution to the "charr problem" based on Arctic charr (*Salvelinus alpinus*) in Norway. Can. J. Fish Aquat. Sci. 40 : 1372-1387.

Patzelt, G. 1980. Neue Ergebnisse der Spät-und Postglazialforschung in Tirol. Österr. Geogr. Ges. Zweigver. Ibk., Jahresber. 76/77 : 11-22.

Pechlaner, R. 1966. Salmonideneinsätze in Hochgebirgsseen und-tümpel der Ostalpen. Verh. Internat. Verein. Limnol. 16: 1182-1191.

Pechlaner, R. 1969. Hochgebirgsseen als Lebensraum für Salmoniden. Zool. Anz., Suppl. 32: 750-757.

Pechlaner, R. 1979. High-mountain lakes in Tyrol. "Tirol immer einen Urlaub wert". Innsbruck. 14: 3-24.

Reisinger, E. 1953. Zum Saiblingsproblem. Carinthia II, 63: 74-102.

Schindler, O. 1940. Die Saiblinge des Königssees. Int. Revue ges. Hydrobiol. 39: 600-627.

Steinböck, O. 1950. Probleme der Ernährung und des Wachstums bei Salmoniden. Schweiz. Fischereizeitung, 1950 : 1-7.

Steinmann, P. 1942. Experimentelle Untersuchungen uber die Wüchsigkeit der Seesaiblinge (Rötel). Schweiz. Fischerei-Zeitung 9: 1-8.

Stolz, O. 1936. Geschichtskunde der Gewässer Tirols. Schlern-Schriften 32 : 1-510.

Thienemann, A. 1950. Vörbreitungsgeschichte der Süsswassertierwelt Europas. Versuch einer historischen Tiergeographie d. europäischen Binnengewässer. Die Binnengewässer 18: 1-809.

Migration Of Anadromous Dolly Varden Charr In Southeastern Alaska — A Manager's Nightmare

ROBERT H. ARMSTRONG

Alaska Cooperative Fishery Research Unit,
University of Alaska, Fairbanks, Alaska

ARMSTRONG, R.H. 1984. Migration of anadromous Dolly Varden charr in southeastern Alaska — a manager's nightmare, p. 559–570. *In* L. Johnson and B.L. Burns [eds.] Biology of the Arctic charr, Proceedings of the International Symposium on Arctic Charr, Winnipeg, Manitoba, May 1981. Univ. Manitoba Press, Winnipeg.

Anadromous Dolly Varden charr *Salvelinus malma* of southeastern Alaska have complex and varying migratory patterns. These patterns differ between those charr originating from a lake system and those originating from a nonlake stream system.

Charr originating from a lake system conduct relatively simple annual migrations to and from the sea. They leave the lake in the spring and return throughout the summer and fall. These migrations consist of smolts, nonspawners (not going to spawn that year) and spawners (will spawn in the fall). The charr spend from a few days to several months feeding at sea before returning to the lake system for the winter. Some of the smolts and nonspawners will enter and leave other nearby freshwater systems.

Charr originating from a nonlake system conduct much more complex migrations than those originating from a lake system. Their migration patterns vary between first year migrants (smolts) and subsequent migrations of the nonspawners and spawners. Smolts leave nonlake streams in the spring and fall; once at sea, these charr enter and leave other nonlake streams until one with a lake is found. After spending the winter in the lake, those charr that are not going to spawn migrate to sea in spring and randomly enter other stream systems; these nonspawners return to sea and migrate to a lake system before winter. Spawners that originated from nonlake streams also migrate to sea in the spring from their lake wintering area. Once at sea, these charr migrate directly to their home stream without entering other streams enroute. After spawning, the survivors return directly to the lake system before winter.

These migration patterns create a manager's nightmare because individual stocks are difficult to recognize. Streams and lakes contain mixed stocks of charr originating from streams over a vast area. Successful management is dependent on recognition of these complex migration patterns.

INTRODUCTION

The migratory habits of anadromous Dolly Varden (*Salvelinus malma*) in southeastern Alaska are complicated and difficult to describe. This poster paper is an attempt, through diagrams, to make these migratory patterns more easily understood. Although the results of studies on the charr's migratory habits have been published (Armstrong 1965, 1970, 1974; Armstrong and Morrow 1980) I felt, because a clear understanding was so important to their management, that a poster paper on the subject was appropriate for this symposium.

MATERIALS AND METHODS

Migrating Dolly Varden have been counted, sampled and marked as they passed upstream and downstream at weirs built across several streams in southeastern Alaska (Figs. 1-3). Throughout this ten year study 328,263 charr have been fin-clipped and measured, approximately 50,000 have been tagged with dart or internal anchor tags and 7,414 have been sampled for age, maturity and stomach content.

RESULTS

The migration patterns in anadromous Dolly Varden vary between smolt, nonspawner and spawner and between those fish originating from nonlake and lake (wintering)[1] systems. The results are presented as theories in diagramatic

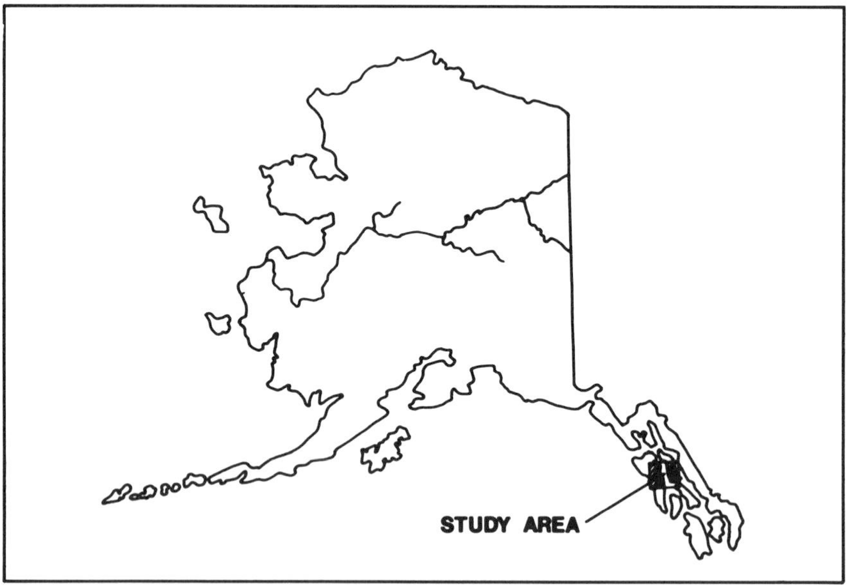

FIG. 1 Study area, Admiralty and Baranof Islands, southeastern Alaska.

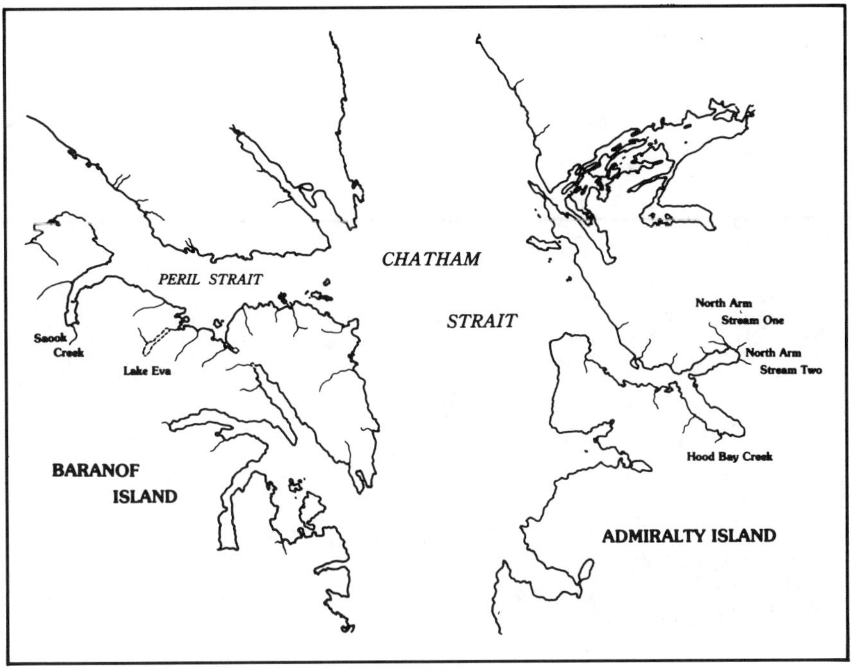

FIG. 2 Locations of the five streams weired for the migration study.

form (Figs. 4-11). The evidence used to develop these theories is briefly presented with each diagram.

MANAGEMENT SUGGESTIONS

1) Recognize the complex nature of the migration patterns in management decisions. Land-use activities that harm one stream or overharvest one stream may affect populations of charr in other systems.
2) Pay particular attention to the protection and preservation of major charr wintering areas.
3) Marking studies should concentrate on smolt, presmolt or spawning charr. For management, identification of individual stocks and the harvest rates imposed on these stocks is extremely important. Marking nonspawning charr or charr in their wintering areas can lead to confusing and erroneous conclusions.

---

[1] Although not common, there are some larger rivers in southeastern Alaska in which Dolly Varden charr overwinter. However, for ease of discussion, the wintering areas will be referred to as lakes.

FIG. 3 Eva Creek — weir operated 3 years — estimated number of immigrant charr per year = 82,940 - 101, 814 — estimated number of spawning females in one year = 1,104 — estimated number of spawning males in one year = 1,029.

Other streams that were weired and the number of Dolly Varden charr captured or estimated were:

Saook Creek — weir operated 1 year — used for sampling purposes only — weir operated 3 days per week — total immigrant charr captured = 3,898.

Hood Bay Creek — weir operated 4 years — estimated number of immigrant charr per year = 5,007 - 7,904 — spawning females per year = 933 - 1,115 — spawning males per year = 334 - 1,165.

North Arm Stream One — weir operated 1 year — estimated number of immigrant charr = 8,285 — spawning females = 533 — spawning males = 512.

North Arm Stream Two — weir operated 1 year — estimated number of immigrant charr = 4,653 — spawning females = 435 — spawning males = 303.

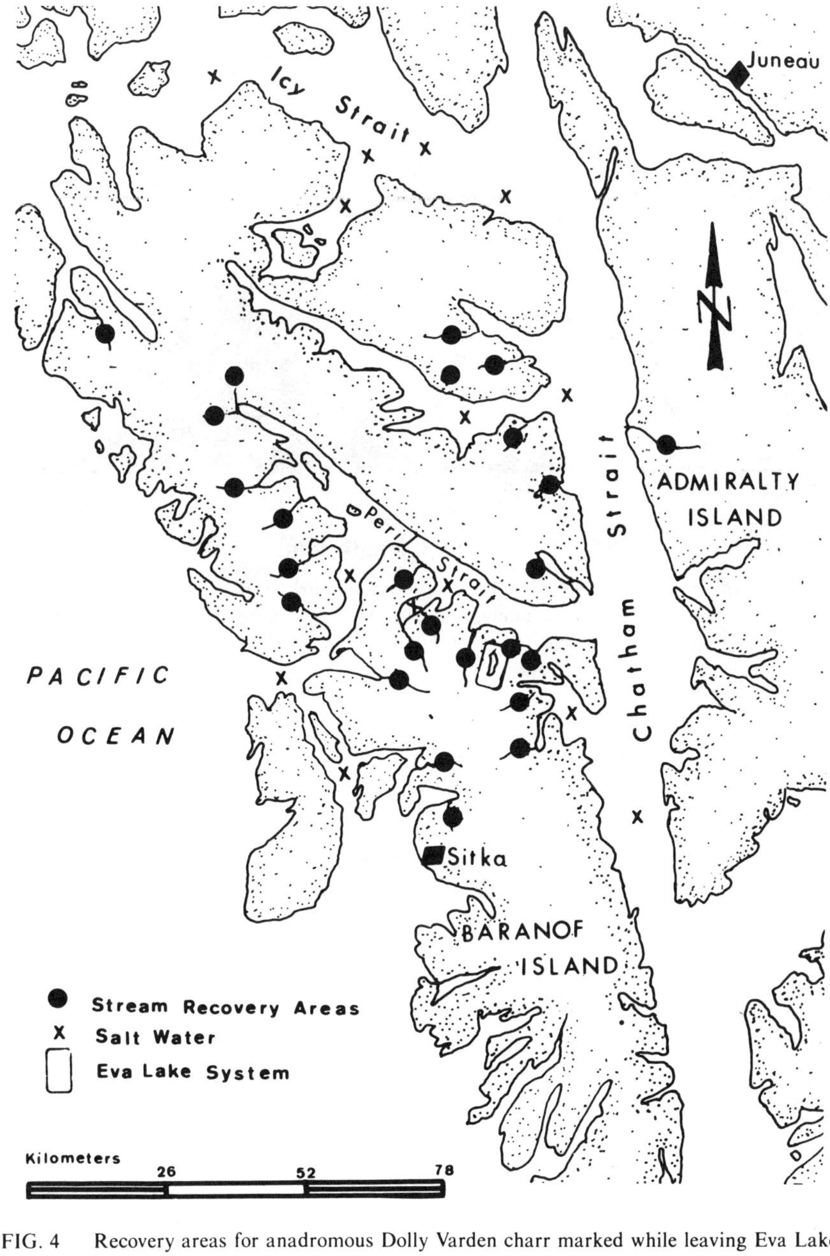

FIG. 4 Recovery areas for anadromous Dolly Varden charr marked while leaving Eva Lake on Baranof Island in southeastern Alaska (from Armstrong 1965).

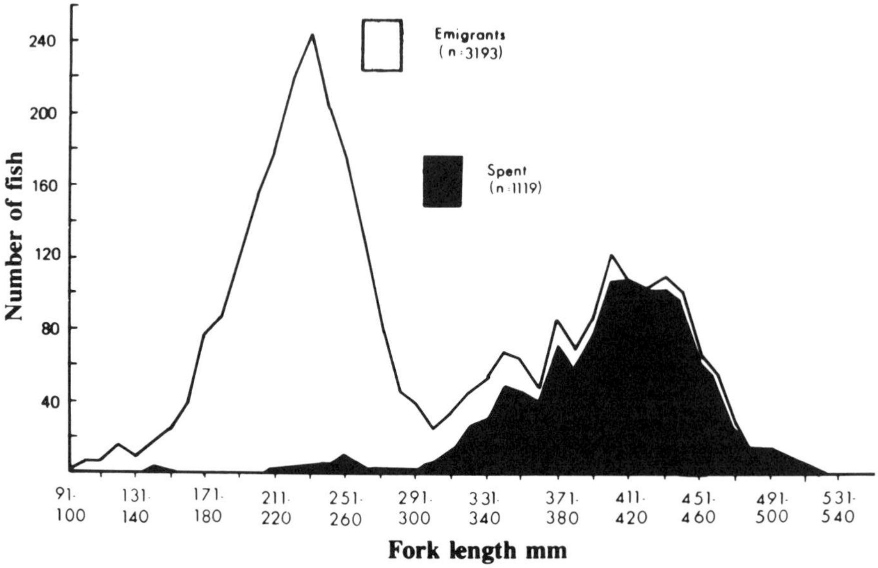

FIG. 5 Length frequency distributions of Dolly Varden charr leaving Hood Bay Creek. It may be possible to recognize immigrant charr spawners belonging to a particular stream by size alone. Similar size relationships were found for charr from other streams in southeastern Alaska (from Armstrong 1974).

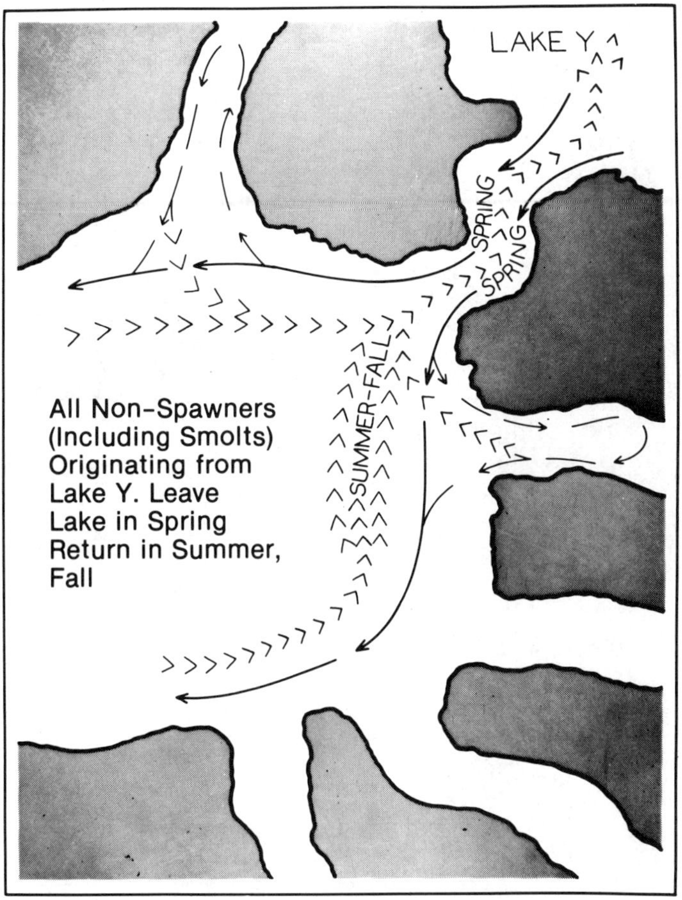

Lake System Origin

FIG. 6 Nonspawner migration theory — migrate to sea in spring, some enter other stream systems — return to lake in summer and fall.

1) Some charr marked as smolt while leaving Eva Lake were recovered in other streams in the same year or in a subsequent year.
2) The small number of recoveries (8) in other systems (4) plus the close proximity (within 10 miles) of these systems to Eva Lake suggest that most nonspawning charr from a lake do not enter other streams or migrate great distances in the ocean.
3) Probable purpose of migration is for feeding.

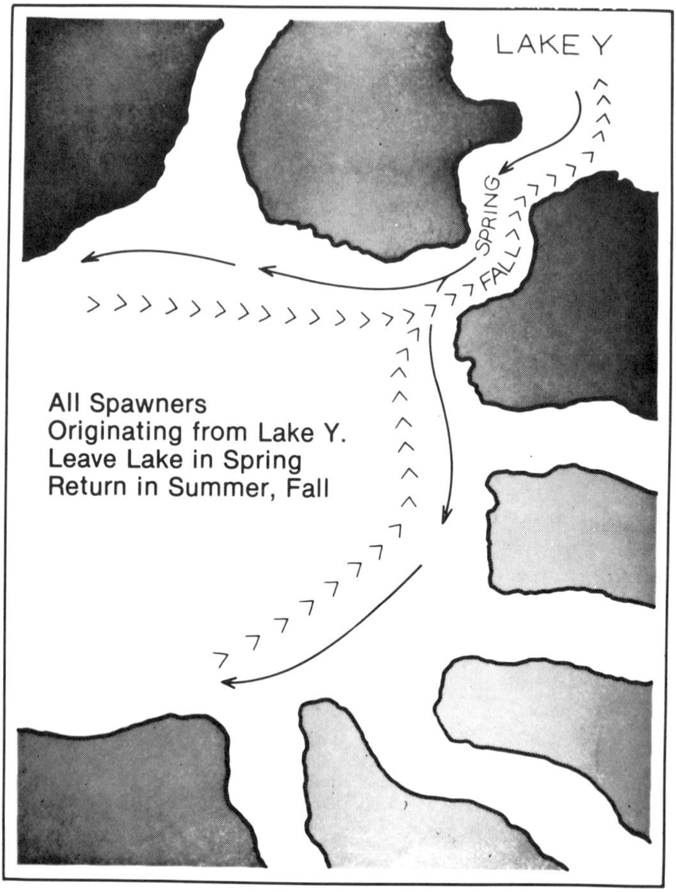

FIG. 7 Spawner migration theory — migrate to sea in spring, do not enter other stream systems — return to lake in summer and fall.

1) No spawners have been found leaving nonlake streams prior to spawning, hence it is unlikely that spawners from lakes are entering these systems.
2) Probable purpose of migration is for feeding.

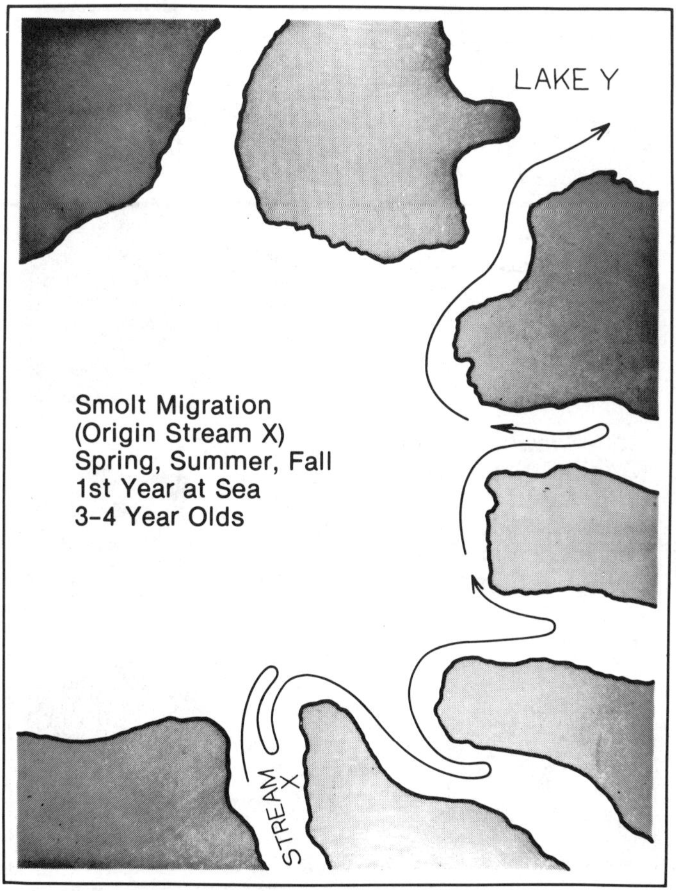

Nonlake System Origin

FIG. 8 Smolt migration theory — smolts move from one freshwater system to another until one with a lake is found.

1) Many (670) smolts marked while leaving Hood Bay Creek were captured entering adjacent study streams.
2) Despite marking all smolts for 3 years, approximately 80 percent of the smolt-sized charr entering Hood Bay Creek were unmarked — hence had originated elsewhere.
3) Most nonspawners leave streams within 10 days of entry.
4) Possible reason for migration — the smolt must search to find a lake in which to overwinter.

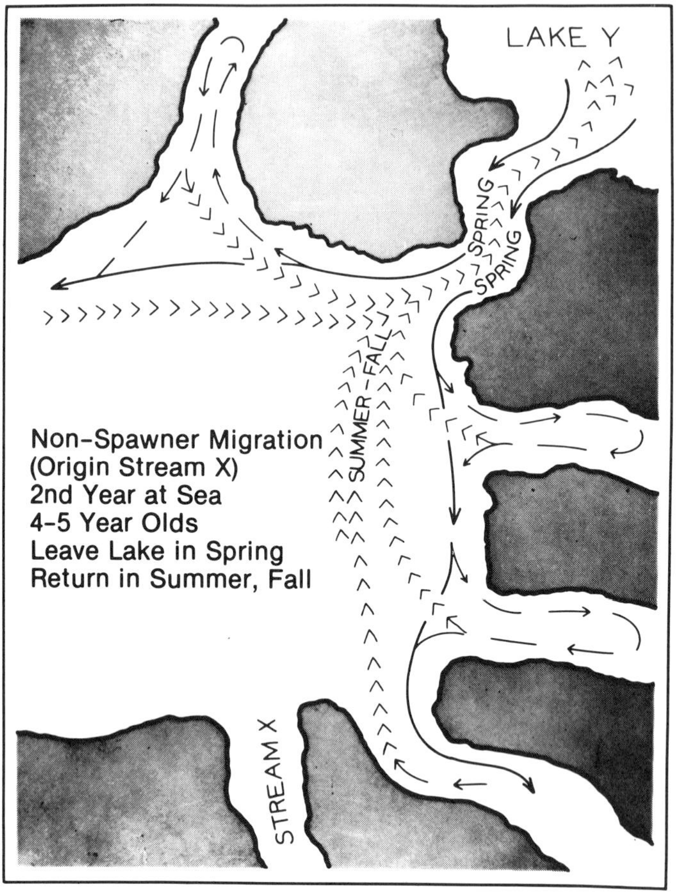

FIG. 9 Nonspawners migration theory — after wintering in lakes, nonspawning charr migrate to sea, enter other stream systems at random and return to the lake system before winter.

1) Nonspawning charr marked while leaving Eva Lake were found in 25 different stream systems up to 226 km from the lake.
2) Of 2,436 Dolly Varden charr marked while leaving Eva Lake and captured while entering Saook Creek, 62 percent returned to Eva Lake before winter (most of these fish were nonspawners).
3) Most nonspawning charr entering Hood Bay Creek had not originated from that system.
4) Probable reason is for feeding — they feed heavily on the eggs of pink and chum salmon.

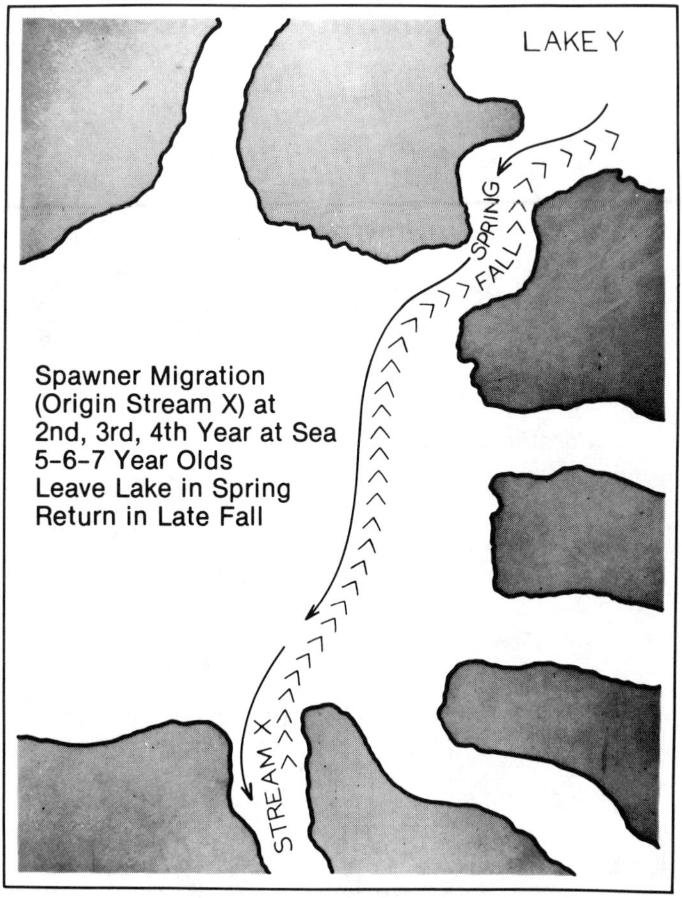

FIG. 10 Spawner migration theory — after wintering in lakes spawners leave in the spring, migrate directly to their home stream, remain until spawning is completed then return directly to the lake system before winter.

1) Homing tendencies strong because:
   a) No straying of charr marked as smolt was observed between adjacent study streams.
   b) All charr spawning in Hood Bay Creek could be accounted as originating from the creek.
   c) Spawning charr displaced from a stream to saltwater returned to the system they were displaced from.
2) No spawning charr left the study streams in Hood Bay before spawning. This indicated a direct movement between the lake and nonlake streams.
3) No spawned-out charr entered any of the nonlake systems studied yet many (2,900) entered the lake system before winter. This indicates a direct migration from the nonlake system to a lake system after spawning.

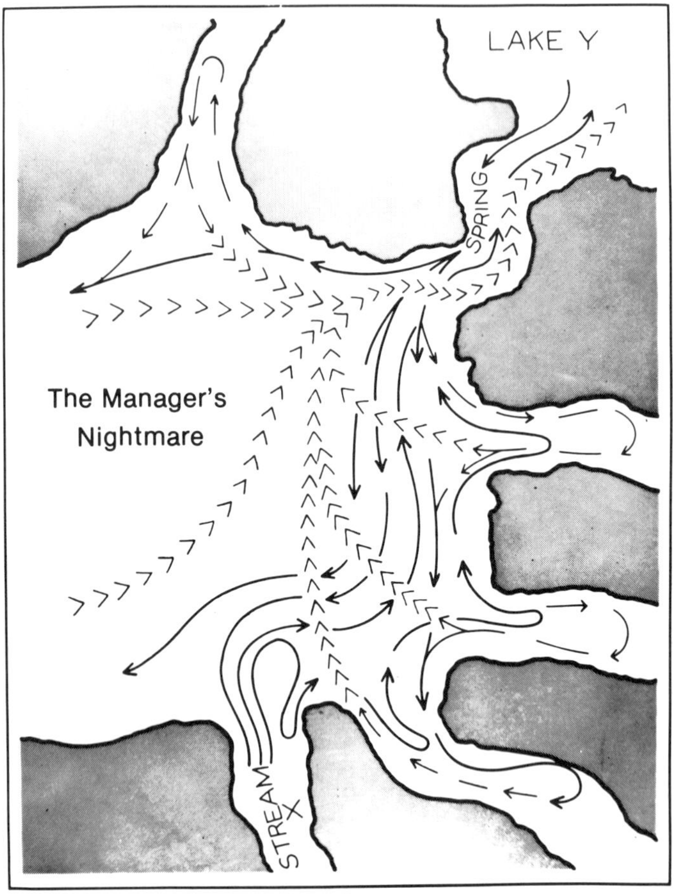

FIG. 11 The managers' nightmare — each stream and lake contains mixed stocks of charr. The same fish may be available for harvest in several nonlake streams, saltwater and a lake system.

REFERENCES

Armstrong, R.H. 1965. Some migratory habits of the anadromous Dolly Varden (*Salvelinus malma*) (Walbaum) in southeastern Alaska. Alaska Dept. of Fish Game Res. Rep. 3 : 36 p.

Armstrong, R.H. 1970. Age, food and migration of Dolly Varden smolts in southeastern Alaska. J. Fish Res. Board Can. 27 : 991-1004.

Armstrong, R.H. 1974. Migration of anadromous Dolly Varden (*Salvelinus malma*) in southeastern Alaska. J. Fish. Res. Board Can. 31 : 435-444.

Armstrong, R.H. and J.E. Morrow. 1980. The Dolly Varden charr, *Salvelinus malma*, p. 99-140. *In* E.K. Balon [ed.] Charrs: salmonid fishes of the genus *Salvelinus*. Dr. W. Junk Publishers, The Hague.

Food Size and Jaw Shape in Arctic Charr, *Salvelinus alpinus*[1]

STEPHEN E. BARBOUR

Department of Forestry and Natural Resources, University of Edinburgh, Edinburgh, Scotland EH9 3JU

BARBOUR, S.E. 1984. Food size and jaw shape in Arctic charr, *Salvelinus alpinus*, p. 571–574. *In* L. Johnson and B.L. Burns [eds.] Biology of the Arctic charr, Proceedings of the International Symposium on Arctic Charr, Winnipeg, Manitoba, May 1981. Univ. Manitoba Press, Winnipeg.

INTRODUCTION

British populations of Arctic charr, *Salvelinus alpinus*, show such a remarkable degree of morphological variation that formerly fifteen separate species were described (Regan 1911), though more recent opinion ascribes them to the single species *Salvelinus alpinus alpinus* (Behnke 1980). It is suggested that the morphometric differences among populations have arisen as specific adaptations to the particular niche of each population.

The two populations compared here are morphometrically distinct and occupy quite different habitats. One is an isolated population from an atypical habitat while the other is from a typical deep, oligotrophic habitat and competes with other salmonids.

[1] This paper is a summary of a visual presentation displayed at the symposium.

STUDY AREAS

Loch Meallt is a small, shallow, meso-eutrophic loch in the Inner Hebrides of Scotland. It is populated by Arctic charr and threespine sticklebacks (*Gasterosteus aculeatus*). The charr do not face competition for food from other fish species; they eat benthic organisms and sticklebacks.

Lock Earn is a large, precipitously deep oligotrophic loch in the central highlands of Scotland. It is populated by Arctic charr, brown trout (*Salmo trutta*)

and introduced rainbow trout (*Salmo gairdneri*). The charr must compete for food with the trout and therefore feed on plankton (*Daphnia* sp.) (Nilsson 1965).

The main limnological features of the two lochs are summarized in Table 1.

METHODS

The charr were captured from both lochs by using experimental gillnets as described by Johnson (1980). Fork length was measured to the nearest millimetre and weight to the nearest gram. The fish were fixed in 10 percent formalin for five to ten days, washed, and then stored in 70 percent ethanol. Subsequently, width of gape was measured using dividers placed on the outside margin of the mandible where it contacted the maxilla, with the mouth approximately half opened. Other morphometric measures were made according to the methods of Hubbs and Lagler (1964).

The food of the charr from Loch Earn was evaluated by direct observation of the stomach contents. The food of the fish from Loch Meallt was studied by Campbell (1976).

RESULTS AND DISCUSSIONS

The charr of Loch Earn adopt the typical characters of adult salmonids as they reach a large size. Sexual dimorphism is marked with some males achieving splendid colouration and a well developed kype at spawning time. The charr of Loch Meallt have a more juvenile appearance with parr marks conspicuous even on large individuals (>300 mm). The head is blunt and the mandible often inferior.

The measurement of mouth gape, expressed as a percentage of head length, differs greatly between the two populations (Table 2).

It has been shown that parr of Atlantic salmon (*Salmo salar*) feed on food particles of a size directly related to their mouth gape (J.E. Thorpe, pers. comm.). In the two populations of Arctic charr in this study, a corollary of this phenomenon seems evident. The charr of Loch Meallt, in the non-competitive niche, feed primarily on benthos and small fish (creatures with body widths between 4-8 mm) and have a larger gape than those of the population of Loch Earn, which, presumably from competition with brown trout (Nilsson 1965) are planktivorous (the plankters having a body width of approximately 1 mm). I suggest that perhaps neotenous individuals, which retain the juvenile blunt headedness, are at a selective advantage in Loch Meallt. Neoteny could have originated if there were no selective pressure for the maintenance of strong sexual dimorphism and other adult salmonid characteristics, characteristics which

Table 1. Summary of limnological characteristics of Loch Meallt and Loch Earn.

	Loch Meallt	Loch Earn
Area	80 ha	1010 ha
Depth (mean)	2 m	42 m
Depth (max.)	3 m	88 m
Altitude	60 m	100 m
pH	7.4	7.2
Conductivity	92.3×10^{-6}	46.5×10^{-6}
Total Phosphorus	$16\ ugl^{-1}$	$5\ ugl^{-1}$

Table 2. Width of gape (as percent of head length) in two populations of Scottish Arctic charr.

Loch	$\bar{x}$	s	range	n	t	p
Meallt	31.5	3.0	26.2 - 37.7	29	9.1	<.001
Earn	24.9	2.8	17.8 - 3[illegible].2	20		

would represent an energy cost during their growth. Thus, the particle size of food would have directed the evolution of gape width in these populations.

ACKNOWLEDGEMENTS

I wish to thank the Elizabeth Sinclair Irvine Fund for a grant to attend the symposium. I am grateful to Drs. D.H. Mills and P. Jones for reviewing and comments, and Miss J. Heath who kindly typed the manuscript.

REFERENCES

Behnke, R.J. 1980. A systematic review of the genus *Salvelinus*, p. 441-480. *In* E.K. Balon [ed.] Charrs: salmonid fishes of the genus *Salvelinus*. Dr. W. Junk Publishers, The Hague.

Campbell, R.N.B. 1976. Examinations of, and comparisons between, the food of two Hebridean loch populations of char (*Salvelinus alpinus* (Linnaeus 1758)) one allopatric, the other sympatric with trout (*Salmo trutta*, Linnaeus 1758). Honours thesis, University of Aberdeen.

Hubbs, C.L. and K.F. Lagler. 1964. Fishes of the Great Lakes Region. University of Michigan Press.

Johnson, L. 1980. The Arctic charr, *Salvelinus alpinus*, p. 15-98. In E.K. Balon [ed.] Charrs: salmonid fishes of the genus *Salvelinus*. Dr. W. Junk Publishers, The Hague.

Nilsson, N.-A. 1965. Food segregation between salmonid species in North Sweden. Inst. Freshw. Res. Drottningholm Rep. 46: 58-73.

Regan, C.T. 1911. The freshwater fishes of the British Isles. Methuen and Co., London.

The Seasonal Movements Of Anadromous Arctic Charr At Nauyuk Lake, Northwest Territories, Canada

ERIC C. GYSELMAN

Department of Fisheries and Oceans, Winnipeg, Manitoba, Canada

GYSELMAN, E.C. 1984. The seasonal movements of anadromous Arctic charr at Nauyuk Lake, Northwest Territories, Canada, p. 575–578. *In* L. Johnson and B.L. Burns [eds.] Biology of the Arctic charr, Proceedings of the International Symposium on Arctic Charr, Winnipeg, Manitoba, May 1981. Univ. Manitoba Press, Winnipeg.

See charts on following three pages.

MIGRATIONS AT SEA

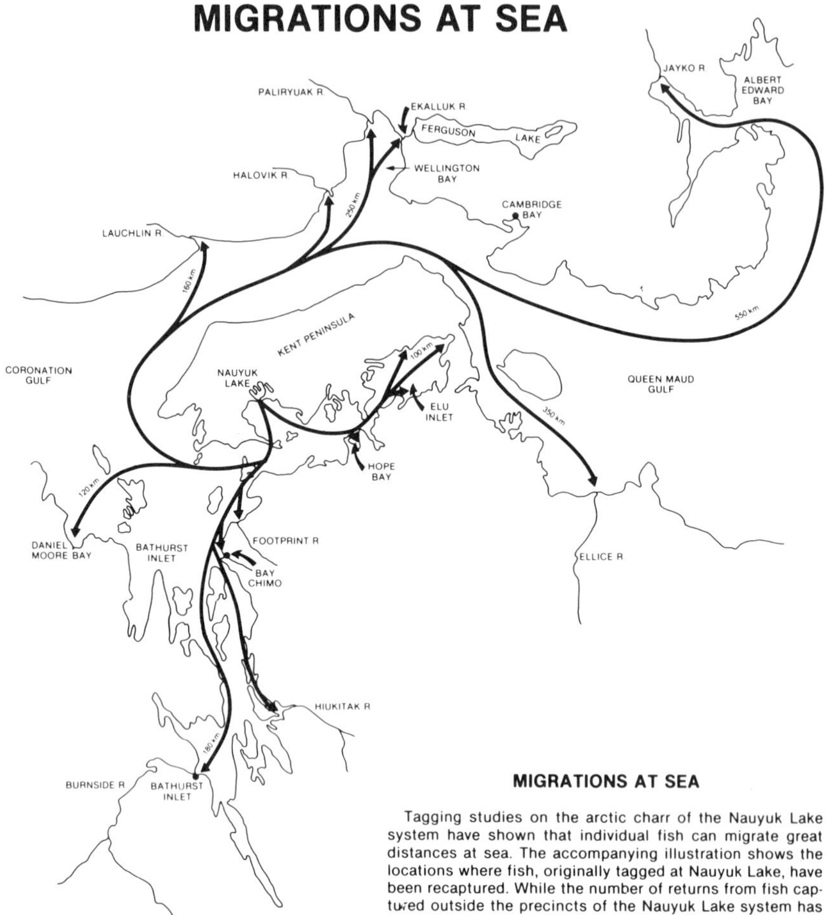

MIGRATIONS AT SEA

Tagging studies on the arctic charr of the Nauyuk Lake system have shown that individual fish can migrate great distances at sea. The accompanying illustration shows the locations where fish, originally tagged at Nauyuk Lake, have been recaptured. While the number of returns from fish captured outside the precincts of the Nauyuk Lake system has not been large (approximately 40) considering that about 4000 fish have been tagged, it must be remembered that fishing pressure on the arctic charr stocks in the area is generally extremely light. The locations indicated, in fact, represent virtually every river system in the area where a significant commercial or domestic fishery takes place. This is important because it demonstrates that there is a considerable interchange among stocks.

The maximum distance that an individual fish can travel during a single summer at sea has been difficult to determine. Fish known to have left Nauyuk Lake in the spring are regularly caught at Bathurst Inlet 180 km to the south in the fall. However, tagged fish caught at the rivers discharging into Wellington Bay system and at the Jayko and Ellice Rivers have always been absent from the Nauyuk Lake system for at least a year. This suggests that distances of 250 km or greater are too far for one summer. It must be remembered, however, that these are straight line distances and that the actual distances travelled would probably have been much greater.

ANADROMOUS MIGRATION

DOWNSTREAM MIGRATION

The seaward migration of arctic charr begins as soon as the outlet from Nauyuk Lake begins to flow in the spring. Typically this occurs during the third week of June but since 1974 it has been as early as June 6 and as late as July 2.

The downstream migration follows a comparatively structure ' sequence. As soon as the ice pan covering the lake melts away from the shore above the outlet, small groups of perhaps 20 to 30 charr begin to congregate along the edge of the ice. Although wary these groups gradually make their way into the river and out to sea. As the run continues more and more fish assemble above the outflow. About 2 weeks after it begins the downstream migration reaches its maximum and begins to taper off until by the end of July it is over.

There is a structured order to the way the fish leave Nauyuk Lake. As the accompanying figure illustrates, the largest fish in the population migrate to sea first. The mean daily length of fish leaving the lake is highest at the beginning of the run, reaches a plateau during the middle period when the majority of the population is migrating and then finally declines toward the end when the smolts, migrating to sea for the first time, are leaving. This sequence has been repeated each year.

UPSTREAM MIGRATION

During the first week of August, arctic charr begin to congregate at the mouth of the Nauyuk River in preparation for the upstream run back into freshwater. As with the downstream run there is a great deal of caution about actually entering the river. Individuals may remain in the estuary for hours or perhaps even days before making the final rush up the short river back into the lake.

The upstream run is not nearly as regimented as is the downstream migration. Both large and small fish return throughout August although perhaps there is a slight tendency for large fish, having left earlier, to return earlier. The number of fish that return each day during the run is quite variable but usually the run peaks about mid-August and can continue well into September.

Of particular interest in the upstream migration is the proportion of fish that, having left Nauyuk Lake in the spring, return in the fall. The probability of returning has proven to be very size dependent. Fish between 40 cm and 60 cm are very faithful to the Nauyuk system returning 85-90% of the time provided they survive the summer fisheries. As the size of the fish increases the probability of their returning to Nauyuk Lake decreases until the fish over 70 cm have less than a 50% chance. Similarly, the rate of return of fish less than 30 cm is very low but this is probably due to an extremely high mortality rate. Overall, about 65% of the downstream run returns in the fall, 30% leaves the Nauyuk Lake system indefinately and about 5% leaves the system for 1 or 2 years.

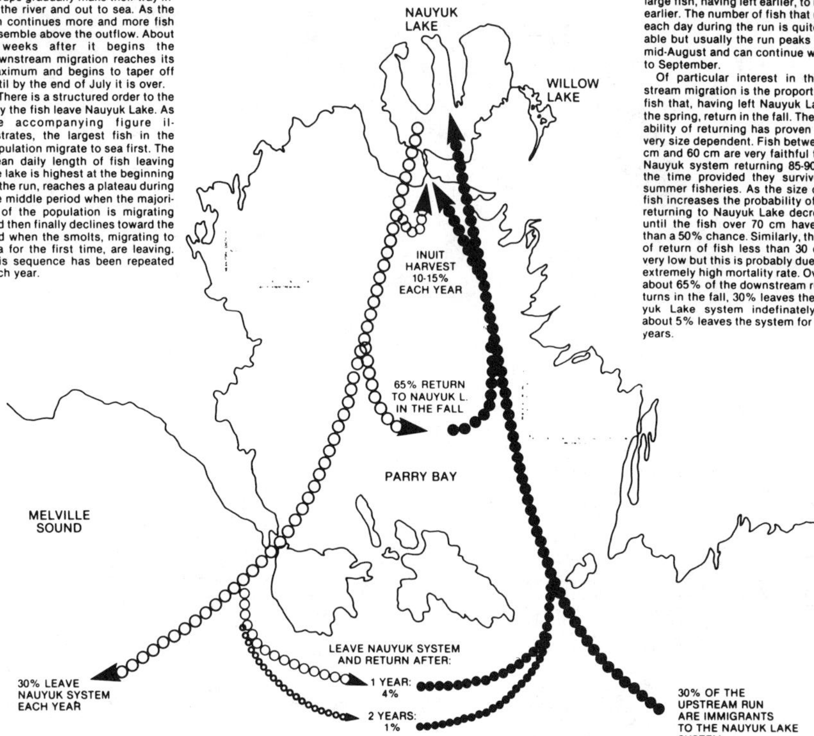

AT SEA

Examination of the stomach contents of anadromous arctic charr from the Nauyuk Lake system has shown that they eat very little while in fresh water. Therefore they must take advantage of the short summer season at sea to accumulate as much energy as possible. Tagging studies have disclosed that many individuals can increase their body weight by 50% in 40 days. In fact some post-spawners have as much as doubled their weight in the same time. There is however a great deal of variability between seasons. For example, in 1978 when the summer was late and short the mean increase was only 12% but in 1975, when it was early and long, the mean increase was 42%. These figures may be compared to an overall average of 30% since 1974.

The length of time spent at sea is highly variable among individual fish. The average number of days spent at sea per year from 1974 to 1978 was 37 but as with weight gain climatic conditions in a particular summer greatly influenced this figure. For example, in 1978, a poor year, the average was 27 days but in 1975, a good year, it was 43 days. Similarly, large fish leave Nauyuk Lake early and may be at sea for 60 or 65 days, whereas smolts, at sea for the first time, may stay for only 2 or 3 days.

At sea, arctic charr from the Nauyuk Lake system feed primarily on capelin *(Mallotus vilosus)* which move into shallow shoreline areas in the summer to spawn. To a lesser extent they also feed on northern sand lance *(Ammodytes dubius)* and occasionally on Greenland cod *(Gadus ogac)*, saffron cod *(Eleginus gracitis)* and the opossum shrimp *(Mysis oculata)*.

SPAWNING MIGRATION

YEAR	NO MIGRATING UPSTREAM	% OF POPULATION
1975	266	2.7%
1976	538	6.7%
1977	228	3.5%
1978	190	4.5%
1979	123	3.5%
1980	235	7.8%

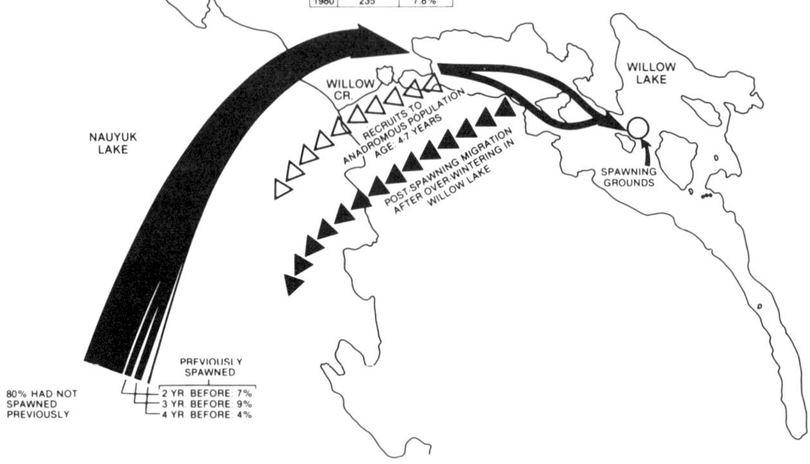

WILLOW CREEK SPAWNING MIGRATIONS

The spawning of anadromous arctic charr in the Nauyuk Lake system has proven to be quite complicated. There has never been any spawning observed in Nauyuk Lake itself but only in a small tributary lake, Willow Lake. Willow Lake is connected to Nauyuk Lake by a narrow braided stream, Willow Creek. Although narrow, Willow Creek is about 1 m deep during the spring melt which allows the movement of pre- and post-spawning anadromous charr into and out of Willow Lake. Because the water level drops quickly once the snow cover has melted fish intent on spawning in the fall must migrate up Willow Creek as soon as it begins to flow in the spring. Similarly, fish that spawned the previous fall are forced to overwinter in Willow Lake and move back down Willow Creek the following spring.

Individual fish in the spawning run tend to be both larger and in better condition than the anadromous population as a whole. The average fork length for pre-spawning males has been 72-77 cm and 63-70 cm for females whereas the mean for the entire population has been 60-63 cm. The mean condition factor (C.F. = $Wx10^5/L^3$) of the pre-spawning fish has been 1.06 compared to 0.92 for the downstream run.

The number of charr that move up Willow Creek in any given spring is highly variable. The accompanying table lists the number and percentage of the population moving upstream each year since 1974. The number of individuals that spawn in a given year is directly related to the conditions that existed during the previous summer. If summer conditions are favourable then more fish are able to accumulate enough energy to spawn the following year. After a poor summer, however, relatively few fish migrate up Willow Creek.

The majority of the post-spawners leave Willow Lake the following spring. However, about 3-4% remain 2 years and about 1% 3 years. It is uncertain whether these fish spawn each year, just the first or only in the last year that they are in Willow Lake.

The hazards involved in running downstream in Willow Creek in the weakened post-spawning condition, followed by a winter on a very low diet, are the major identifiable cause of natural mortality in the adult stock.

SPAWNING AND SPAWNING FREQUENCY

After spending the summer in Willow Lake, the anadromous charr begin to move onto the spawning grounds between late-August and mid-September. By this time they have adopted the morphological characteristics typical of spawning charr in the region. THe males become bright orange-red over their entire body except for white leading edges in the pelvic, pectoral and anal fins; they also develope a pronounced kype. The females may return dark along the back and, of course, do not develop the kype.

Spawning takes place in an area about 100 x 50 m off a low gravel island midway down the lake. The bottom is composed of frost shattered sedimentary gravel 3 cm to 10 cm in diameter interspersed with patches of aquatic vegetation. The water depth ranges from 2-3.5 m.

Spawning of the anadromous fish occurs simultaneously with resident arctic charr that never leave Willow Lake. These fish, although smaller, seem to spawn freely with the anadromous fish. It is speculated that this may be why there are so few male fish in the spawning migration of anadromous charr. The resident population may in fact be acting as a source of males.

The spawning frequency of the anadromous population is highly variable. Because of the spawning sequence involving Willow Lake, it is impossible for individuals to spawn every year. However, spawning every second, third or fourth year is common with every third year being slightly favoured. The key to spawning frequency seems to be the summer conditions in the year prior to that in which the spawning occurs. If the summer after spawning is favourable then many fish will return the following spring to spawn again. If it is not then 2 or 3 summer seasons may be necessary for the fish to accumulate enough energy to begin another spawning cycle.

Anadromous charr in the Nauyuk Lake system first begin to spawn when they are 9 to 11 years of age. They have never been observed to spawn after they are 15 years old although individuals as old as 18 or 20 are not uncommon in the seaward run out of Nauyuk Lake. This suggests that old fish enter a period of sexual senility after the age of 15 years. This being the case, the maximum number of times an individual could spawn would be 4 with 2 or 3 being most likely.

Participants

AASS, Per
Senior Research Biologist
Fiskeforskiningen
P.O. Box 63
N-1432 As-NLH
Norway

ALI, M. Ather
Professeur
University of Montreal
P.O. Box 6128, Station A
Montreal, Quebec
Canada H3C 3J7

ALLARD, Marc
Makivik Corporation
4898 deMaisonneuve West
Westmount, Quebec
Canada H3Z 1M8

ARMSTRONG, Robert H.
Regional Research Supervisor
Alaska Department of Fish & Game
230 South Franklin Street
Juneau, Alaska
United States of America, 99801

AYLES, G. Burton
Freshwater Institute
Department of Fisheries & Oceans
501 University Crescent
Winnipeg, Manitoba
Canada R2T 2N6

BAKER, Randle F.
Graduate Student
Freshwater Institute
Department of Fisheries & Oceans
501 University Crescent
Winnipeg, Manitoba
Canada R3T 2N6

BALON, Eugene K.
Professor
University of Guelph
Guelph, Ontario
Canada N1G 2W1

BARBOUR, Stephen E.
University of Edinburgh
Forestry and Natural Resources
Darwin Building, King's Buildings
Edinburgh, Scotland EH9 3JU

BAUMGARTNER, S.
University of Innsbruck
Institut für Fischforschung
6020 Innsbruck
Peter Mayerstr, Austria

BEHNKE, R.
Department of Fish & Wildlife Biologist
Colorado State University
Fort Collins, Colorado
United States of America 80523

BENDOCK, Terrence N.
Fishery Biologist
Department of Fish & Game
Sport Fish Division
1300 College Road
Fairbanks, Alaska
United States of America 99701

BLACK, Geoff A.
Graduate Student
Department of Zoology
University of Guelph
Guelph, Ontario
Canada N1G 2W1

BODALY, Drew
Research Scientist
Freshwater Institute
Department of Fisheries & Oceans
501 University Crescent
Winnipeg, Manitoba
Canada R3T 2N6

BOND, William
Freshwater Institute
Department of Fisheries & Oceans
501 University Crescent
Winnipeg, Manitoba
Canada R3T 2N6

BRENNER, Tomás B.
Landesanstalt für Fischerei NW,
Albaum
D-5942 Kirchhundem 1
Albaum, GFR
(German Federal Republic)

BURNS, Bonnie
Freshwater Institute
Department of Fisheries & Oceans
501 University Crescent
Winnipeg, Manitoba
Canada R3T 2N6

CAMPBELL, J. Scott
Research Scientist
Freshwater Institute
501 University Crescent
Winnipeg, Manitoba
Canada R3T 2N6

CARDER, Gary
Freshwater Institute
Department of Fisheries & Oceans
501 University Crescent
Winnipeg, Manitoba
Canada R3T 2N6

CAVENDER, Ted M.
Associate Professor of Zoology
Curator of Fishes Museum of Zoology
Ohio State University
Museum of Zoology
1813 North High Street
Columbus, Ohio
United States of America 43210

CLARKE, Redmond
Freshwater Institute
Department of Fisheries & Oceans
501 University Crescent
Winnipeg, Manitoba
Canada R3T 2N6

CLAYTON, J.
Freshwater Institute
Department of Fisheries & Oceans
501 University Crescent
Winnipeg, Manitoba
Canada R3T 2N6

CURTIS, Mark A.
Assistant Professor
Institute of Parasitology
MacDonald Campus of McGill University
Ste-Anne-de-Bellevue, Quebec
Canada H9X 1C0

DAHLKE, Lothar
Freshwater Institute
Department of Fisheries & Oceans
501 University Crescent
Winnipeg, Manitoba
Canada R3T 2N6

DeMARCH, Larry
Freshwater Institute
Department of Fisheries & Oceans
501 University Crescent
Winnipeg, Manitoba
Canada R3T 2N6

DEMPSON, J. Brian
Department of Fisheries & Oceans
P.O. Box 5667
St. John's, Newfoundland
Canada A1C 5X1

DICK, Terry
Department of Zoology
University of Manitoba
Winnipeg, Manitoba
Canada

DONALD, David B.
Canadian Wildlife Service
1000, 9942 - 108th Street
Edmonton, Alberta
Canada T5K 0J6

DUTIL, Jean-Denis
Research Scientist
Department of Fisheries & Oceans
C.P. 15500
901 Cap Diamant
Quebec, Canada G1K 7Y7

FALLIS, Bruce
Freshwater Institute
Department of Fisheries & Oceans
501 University Crescent
Winnipeg, Manitoba
Canada R3T 2N6

FLANNAGAN, John
Freshwater Institute
Department of Fisheries & Oceans
501 University Crescent
Winnipeg, Manitoba
Canada R3T 2N6

FRASER, Noel C.
Masters Degree Candidate
University of Waterloo
Biology Department
Waterloo, Ontario
Canada N2L 3G1

FRANZIN, William
Freshwater Institute
Department of Fisheries & Oceans
501 University Crescent
Winnipeg, Manitoba
Canada R3T 2N6

FÜRST, Magnus
Institute of Freshwater Research
S-170 11 Drottningholm
Sweden

GILLIS, Dave
Makivik Corporation
4898 deMaisonneuve Ave. West
Westmount, Quebec
Canada H3Z 1M8

GROTNES, Per-Egil
Forsteamanuensis
University of Tromsø
P.O. Box 790
N-9001 Tromsø Norway

GYDEMO, Rolf
Institute of Freshwater Research
S-170 11 Drottningholm
Sweden

GYSELMAN, Eric
Freshwater Institute
Department of Fisheries & Oceans
501 University Crescent
Winnipeg, Manitoba
Canada R3T 2N6

HAMMAR, Johan
Research Scientist
Institute of Freshwater Research
S-170 11 Drottningholm
Sweden

HARTMANN, Jürgen
Institut f. Seenforschung und
Fischereiwesen
D-7994 Langenargen
GFR (German Federal Republic)

HEGGBERGET, Tor G.
Research Biologist
Directorate for Wildlife & Freshwater
Fish
Elgeseter gt. 10
7000 Trondheim
Norway

HINDAR, Kjetil
Zoological Institute
University of Oslo
P.O. Box 1050 Blindern
Oslo 3
Norway

IREDALE, Dave
Freshwater Institute
Department of Fisheries & Oceans
501 University Crescent
Winnipeg, Manitoba
Canada R3T 2N6

JENSEN, John W.
Amanusensis
The Royal Norwegian Society of Sciences
and Letters
Erl. Skakkesgt. 47B
N-7000 Trondheim
Norway

JOHNSON, Lionel
Freshwater Institute
Department of Fisheries & Oceans
501 University Crescent
Winnipeg, Manitoba
Canada R3T 2N6

JONSSON, Bror
Zoological Institute
University of Oslo
P.O. Box 1050 Blindern
Oslo 3, Norway

KAWANABE, Hiroya
Professor
Department of Zoology
Faculty of Science
Kyoto University
Kyoto, 606
Japan

KIPLING, Charlotte
Freshwater Biological Association
Ferry House
Far Sawrey
Ambleside LA22 OLP
United Kingdom

KLEMETSEN, Anders
Forsteamanuensis
University of Tromsø
P.O. Box 790
N-9001 Tromsø
Norway

KOMIYAMA, Eishige K.
Faculty of Agriculture
Department of Applied Zoology
Hokkaido University
N9 W9, Kita-ku
Sapporo, 060 Japan

KRISTJÁNSSON, Jon
Institute of Freshwater Fisheries
P.O Box 754
Reykjavik, Iceland

KRISTOFFERSON, Al
Freshwater Institute
Department of Fisheries & Oceans
501 University Crescent
Winnipeg, Manitoba
Canada R3T 2N6

LeDREW, Lawrence J.
Department of Fisheries & Oceans
P.O. Box 5667
St. John's, Newfoundland
Canada A1C 5X1

LEGGETT, Jack
Regional Fisheries Biologist
Fish & Wildlife
Ministry of Environment
540 Borland Street
Williams Lake, British Columbia
Canada V2G 1R8

LEONHARD, Sharon
Freshwater Institute
Department of Fisheries & Oceans
501 University Crescent
Winnipeg, Manitoba
Canada R3T 2N6

MacCALLUM, Wayne R.
Lake Superior Fisheries Assessment Unit
Ontario Ministry of Natural Resources
435 James Street South
P.O. Box 5000
Thunder Bay, Ontario
Canada

MAEKAWA, Koji M.
School of Dentistry
Hokkaido University
North 13, West 7, Kita-ku
Sapporo, 060 Japan

MAITLAND, Peter S.
Institute of Terrestrial Ecology
78 Craighall Road
Edinburgh
Scotland

MANSFIELD, Arthur W.
Director
Arctic Biological Station
Department of Fisheries & Oceans
555 St. Pierre Boulevard
Ste. Anne-de-Bellevue, Quebec
Canada H9X 3R4

McGOWAN, Dale
Freshwater Institute
Department of Fisheries & Oceans
501 University Crescent
Winnipeg, Manitoba
Canada R3T 2N6

McKINNON, Greg
Freshwater Institute
Department of Fisheries & Oceans
501 University Crescent
Winnipeg, Manitoba
Canada R3T 2N6

MILBRINK, Göran
Uppsala University
Institute of Zoology
P.O. Box 561
751 22 Uppsala
Sweden

MORK, Jarle A.
The Norwegian Fishery Research Council
Trondhjeim Biological Station
7000 Trondheim
Norway

MOSHENKO, Robert
Freshwater Institute
Department of Fisheries & Oceans
501 University Crescent
Winnipeg, Manitoba
Canada R3T 2N6

NORDENG, Hans
Zoological Institute
University of Oslo
P.O. Box 1050 Blindern
Oslo 3
Norway

NYMAN, Lennart
ISACF/Institute of Freshwater Research
S-170 11 Drottningholm
Sweden

PECHLANER, Roland
Professor Institute für Zoologie,
Abt. Limnologie
University-STR. 4, A-6020 Innsbruck
Austria

PEET, Roger
Freshwater Institute
Department of Fisheries & Oceans
501 University Crescent
Winnipeg, Manitoba
Canada R3T 2N6

PORTER, Rex
Department of Fisheries & Oceans
P.O. Box 5667
St. John's, Newfoundland
Canada A1C 5X1

POWER, Geoffrey
Department of Biology
University of Waterloo
Waterloo, Ontario
Canada N2L 3G1

READ, Carol
Freshwater Institute
Department of Fisheries & Oceans
501 University Crescent
Winnipeg, Manitoba
Canada R3T 2N6

RUHLÉ, Christian
Fish & Wildlife Service St. Gallen
Regierungsgebaude
CH-9001 St. Gallen
Switzerland

SANEYOSHI, Mineo S.
Associate Professor
Faculty of Pharmaceutical Sciences
Hokkaido University
Sappora 060, Japan

SCHLICK, Rudolf O.
Fisheries Biologist
Department of Natural Resource, Manitoba
Provincial Office Building
P.O. Box 28, 59 Elizabeth Drive
Thompson, Manitoba
Canada R8N 1X4

SIKSTROM, Calvin
ESSO Resources Calgary
600 - 6th Avenue South West
Calgary, Alberta
Canada T2P 0S1

SMITH, Geoffrey
Hatfield Consultants Ltd.
201 - 1571 Bellevue Avenue
West Vancouver, British Columbia
Canada V7V 3R6

STABELL, Ole B.
Syntex Research, Analytical Research
Institute of Zoophysiology
3401 Hillview Avenue
Palo Alto, California
United States of America 94304

STEINER, Volker
University of Innsbruck
Institut für Fischforschung
6020 Innsbruck
Peter Mayerstr
Austria

WALLACE, Rob
Department of Northern Saskatchewan
P.O. Box 5000
LaRonge, Saskatchewan
Canada S0J 1L0

WHETHERLY, Dyana E.
Freshwater Biological Association
Ferry House
Far Sawrey, Ambleside
Cumbria United Kingdom

WONG, Brian W.H.
Biological Advisor
Department of Fisheries & Oceans
P.O. Box 2310
Yellowknife, Northwest Territories
Canada X1A 2P7

WRIGHT, Dennis
Freshwater Institute
Department of Fisheries & Oceans
501 University Crescent
Winnipeg, Manitoba
Canada R3T 2N6